T0137463

Lecture Notes in Computer Science 14609

Founding Editors

Gerhard Goos
Juris Hartmanis

Editorial Board Members

The series Lecture Notes in Computer Science (LNCS), including its subseries Lecture Notes in Artificial Intelligence (LNAI) and Lecture Notes in Bioinformatics (LNBI), has established itself as a medium for the publication of new developments in computer science and information technology research, teaching, and education.

LNCS enjoys close cooperation with the computer science R & D community, the series counts many renowned academics among its volume editors and paper authors, and collaborates with prestigious societies. Its mission is to serve this international community by providing an invaluable service, mainly focused on the publication of conference and workshop proceedings and postproceedings. LNCS commenced publication in 1973.

Nazli Goharian · Nicola Tonellotto · Yulan He ·
Aldo Lipani · Graham McDonald ·
Craig Macdonald · Iadh Ounis
Editors

Advances in Information Retrieval

46th European Conference on Information Retrieval, ECIR 2024
Glasgow, UK, March 24–28, 2024
Proceedings, Part II

 Springer

Editors
Nazli Goharian
Georgetown University
Washington, WA, USA

Yulan He 🆔
King's College London
London, UK

Graham McDonald 🆔
University of Glasgow
Glasgow, UK

Iadh Ounis 🆔
University of Glasgow
Glasgow, UK

Nicola Tonellotto 🆔
University of Pisa
Pisa, Italy

Aldo Lipani 🆔
University College London
London, UK

Craig Macdonald 🆔
University of Glasgow
Glasgow, UK

ISSN 0302-9743   ISSN 1611-3349 (electronic)
Lecture Notes in Computer Science
ISBN 978-3-031-56059-0   ISBN 978-3-031-56060-6 (eBook)
https://doi.org/10.1007/978-3-031-56060-6

This Springer imprint is published by the registered company Springer Nature Switzerland AG
The registered company address is: Gewerbestrasse 11, 6330 Cham, Switzerland

Paper in this product is recyclable.

Preface

The 46th European Conference on Information Retrieval (ECIR 2024) was held in Glasgow, Scotland, UK, during March 24–28, 2024, and brought together hundreds of researchers from the UK, Europe and abroad. The conference was organised by the University of Glasgow, in cooperation with the British Computer Society's Information Retrieval Specialist Group (BCS IRSG) and with assistance from the Glasgow Convention Bureau.

These proceedings contain the papers related to the presentations, workshops, tutorials, doctoral consortium and other satellite tracks that took place during the conference. This year's ECIR program boasted a variety of novel work from contributors from all around the world. In addition, we introduced a number of novelties in this year's ECIR. First, ECIR 2024 included for the first time a new "Findings" track, which was offered to some full papers that were deemed to be solid, but which could not make the main conference track. Second, ECIR 2024 ran a new special IR4Good track that presented high-quality, high-impact, original IR-related research on societal issues (such as algorithmic bias and fairness, privacy, and transparency) at the interdisciplinary level (e.g., philosophy, law, sociology, civil society), which go beyond the purely technical perspective. Third, ECIR 2024 featured a new innovation called the "Collab-a-thon", intended to provide an opportunity for participants to foster new collaborations that could lead to exciting new research, and forge lasting relationships with like-minded researchers. Finally, ECIR 2024 introduced a new award to encourage and recognise researchers who have made significant contributions in using theory to develop the information retrieval field. The award was named after Professor Cornelis "Keith" van Rijsbergen (University of Glasgow), a pioneer in modern information retrieval, and a strong advocate of the development of models and theories in information retrieval.

The ECIR 2024 program featured a total of 578 papers from authors in 61 countries in its various tracks. The final program included 57 full papers (23% acceptance rate), an additional 18 finding papers, 36 short papers (24% acceptance rate), 26 IR4Good papers (41%), 18 demonstration papers (56% acceptance rate), 9 reproducibility papers (39% acceptance rate), 8 doctoral consortium papers (57% acceptance rate), and 15 invited CLEF papers. All submissions were peer-reviewed by at least three international Program Committee members to ensure that only submissions of the highest relevance and quality were included in the final ECIR 2024 program. The acceptance decisions were further informed by discussions among the reviewers for each submitted paper, led by a Senior Program Committee member. Each track had a final PC meeting where final recommendations were discussed and made, trying to reach a fair and equal outcome for all submissions.

The accepted papers cover the state-of-the-art in information retrieval and recommender systems: user aspects, system and foundational aspects, artificial intelligence & machine learning, applications, evaluation, new social and technical challenges, and other topics of direct or indirect relevance to search and recommendation. As in previous years, the ECIR 2024 program contained a high proportion of papers with students as first authors, as well as papers from a variety of universities, research institutes, and commercial organisations.

In addition to the papers, the program also included 4 keynotes, 7 tutorials, 10 workshops, a doctoral consortium, an IR4Good event, a Collab-a-thon and an industry day. Keynote talks were given by Charles L. A. Clarke (University of Waterloo), Josiane Mothe (Université de Toulouse), Carlos Castillo (Universitat Pompeu Fabra), and this year's Keith van Rijsbergen Award winner, Maarten de Rijke (University of Amsterdam). The tutorials covered a range of topics including explainable recommender systems, sequential recommendation, social good applications, quantum for IR, generative IR, query performance prediction and PhD advice. The workshops brought together participants to discuss narrative extraction (Text2Story), knowledge-enhanced retrieval (KEIR), online misinformation (ROMCIR), understudied users (IR4U2), graph-based IR (IRonGraphs), open web search (WOWS), technology-assisted review (ALTARS), geographic information extraction (GeoExT), bibliometrics (BIR) and search futures (SearchFutures).

The success of ECIR 2024 would not have been possible without all the help from the strong team of volunteers and reviewers. We wish to thank all the reviewers and meta-reviewers who helped to ensure the high quality of the program. We also wish to thank: the reproducibility track chairs Claudia Hauff and Hamed Zamani, the IR4Good track chairs Ludovico Boratto and Mirko Marras, the demo track chairs Giorgio Maria Di Nunzio and Chiara Renso, the industry day chairs Olivier Jeunen and Isabelle Moulinier, the doctoral consortium chairs Yashar Moshfeghi and Gabriella Pasi, the CLEF Labs chair Jake Lever, the workshop chairs Elisabeth Lex, Maria Maistro and Martin Potthast, the tutorial chairs Mohammad Aliannejadi and Johanne R. Trippas, the Collab-a-thon chair Sean MacAvaney, the best paper awards committee chair Raffaele Perego, the sponsorship chairs Dyaa Albakour and Eugene Kharitonov, the proceeding chairs Debasis Ganguly and Richard McCreadie, and the local organisation chairs Zaiqiao Meng and Hitarth Narvala. We would also like to thank all the student volunteers who worked hard to ensure an excellent and memorable experience for participants and attendees. ECIR 2024 was sponsored by a range of learned societies, research institutes and companies. We thank them all for their support. Finally, we wish to thank all of the authors and contributors to the conference.

March 2024

Nazli Goharian
Nicola Tonellotto
Yulan He
Aldo Lipani
Graham McDonald
Craig Macdonald
Iadh Ounis

Organization

General Chairs

Craig Macdonald   University of Glasgow, UK
Graham McDonald   University of Glasgow, UK
Iadh Ounis   University of Glasgow, UK

Program Chairs – Full Papers

Nazli Goharian   Georgetown University, USA
Nicola Tonellotto   University of Pisa, Italy

Program Chairs – Short Papers

Yulan He   King's College London, UK
Aldo Lipani   University College London, UK

Reproducibility Track Chairs

Claudia Hauff   Spotify & TU Delft, Netherlands
Hamed Zamani   University of Massachusetts Amherst, USA

IR4Good Chairs

Ludovico Boratto   University of Cagliari, Italy
Mirko Marras   University of Cagliari, Italy

Demo Chairs

Giorgio Maria Di Nunzio Università degli Studi di Padova, Italy
Chiara Renso   ISTI - CNR, Italy

Industry Day Chairs

Olivier Jeunen   ShareChat, UK
Isabelle Moulinier   Thomson Reuters, USA

Doctoral Consortium Chairs

Yashar Moshfeghi   University of Strathclyde, UK
Gabriella Pasi   Università degli Studi di Milano Bicocca, Italy

CLEF Labs Chair

Jake Lever   University of Glasgow, UK

Workshop Chairs

Elisabeth Lex   Graz University of Technology, Austria
Maria Maistro   University of Copenhagen, Denmark
Martin Potthast   Leipzig University, Germany

Tutorial Chairs

Mohammad Aliannejadi   University of Amsterdam, Netherlands
Johanne R. Trippas   RMIT University, Australia

Collab-a-thon Chair

Sean MacAvaney   University of Glasgow, UK

Best Paper Awards Committee Chair

Raffaele Perego   ISTI-CNR, Italy

Sponsorship Chairs

Dyaa Albakour   Signal AI, UK
Eugene Kharitonov   Google, France

Proceeding Chairs

Debasis Ganguly   University of Glasgow, UK
Richard McCreadie   University of Glasgow, UK

Local Organisation Chairs

Zaiqiao Meng   University of Glasgow, UK
Hitarth Narvala   University of Glasgow, UK

Senior Program Committee

Mohammad Aliannejadi   University of Amsterdam, Netherlands
Omar Alonso   Amazon, USA
Giambattista Amati   Fondazione Ugo Bordoni, Italy
Ioannis Arapakis   Telefonica Research, Spain
Jaime Arguello   The University of North Carolina at Chapel Hill,
                           USA
Javed Aslam   Northeastern University, USA
Krisztian Balog   University of Stavanger & Google Research,
                           Norway
Patrice Bellot   Aix-Marseille Université CNRS (LSIS), France
Michael Bendersky   Google, USA
Mohand Boughanem   IRIT University Paul Sabatier Toulouse, France
Jamie Callan   Carnegie Mellon University, USA
Charles Clarke   University of Waterloo, Canada
Fabio Crestani   Università della Svizzera italiana (USI),
                           Switzerland
Bruce Croft   University of Massachusetts Amherst, USA
Maarten de Rijke   University of Amsterdam, Netherlands
Arjen de Vries   Radboud University, Netherlands
Tommaso Di Noia   Politecnico di Bari, Italy
Carsten Eickhoff   University of Tübingen, Germany
Tamer Elsayed   Qatar University, Qatar

Liana Ermakova	HCTI/Université de Bretagne Occidentale, France
Hui Fang	University of Delaware, USA
Nicola Ferro	University of Padova, Italy
Norbert Fuhr	University of Duisburg-Essen, Germany
Debasis Ganguly	University of Glasgow, UK
Lorraine Goeuriot	Université Grenoble Alpes (CNRS), France
Marcos Goncalves	Federal University of Minas Gerais, Brazil
Julio Gonzalo	UNED, Spain
Jiafeng Guo	Institute of Computing Technology, China
Matthias Hagen	Friedrich-Schiller-Universität, Germany
Allan Hanbury	TU Wien, Austria
Donna Harman	NIST, USA
Claudia Hauff	Spotify, Netherlands
Jiyin He	Signal AI, UK
Ben He	University of Chinese Academy of Sciences, China
Dietmar Jannach	University of Klagenfurt, Germany
Adam Jatowt	University of Innsbruck, Austria
Gareth Jones	Dublin City University, Ireland
Joemon Jose	University of Glasgow, UK
Jaap Kamps	University of Amsterdam, Netherlands
Jussi Karlgren	SiloGen, Finland
Udo Kruschwitz	University of Regensburg, Germany
Jochen Leidner	Coburg University of Applied Sciences, Germany
Yiqun Liu	Tsinghua University, China
Sean MacAvaney	University of Glasgow, UK
Craig Macdonald	University of Glasgow, UK
Joao Magalhaes	Universidade NOVA de Lisboa, Portugal
Giorgio Maria Di Nunzio	University of Padua, Italy
Philipp Mayr	GESIS, Germany
Donald Metzler	Google, USA
Alistair Moffat	The University of Melbourne, Australia
Yashar Moshfeghi	University of Strathclyde, UK
Henning Müller	HES-SO, Switzerland
Julián Urbano	Delft University of Technology, Netherlands
Marc Najork	Google, USA
Jian-Yun Nie	Université de Montreal, Canada
Harrie Oosterhuis	Radboud University, Netherlands
Iadh Ounis	University of Glasgow, UK
Javier Parapar	University of A Coruña, Spain
Gabriella Pasi	University of Milano Bicocca, Italy
Raffaele Perego	ISTI-CNR, Italy

Benjamin Piwowarski   CNRS/ISIR/Sorbonne Université, France
Paolo Rosso   Universitat Politècnica de València, Spain
Mark Sanderson   RMIT University, Australia
Philipp Schaer   TH Köln (University of Applied Sciences),
                                 Germany
Ralf Schenkel   Trier University, Germany
Christin Seifert   University of Marburg, Germany
Gianmaria Silvello   University of Padua, Italy
Fabrizio Silvestri   University of Rome, Italy
Mark Smucker   University of Waterloo, Canada
Laure Soulier   Sorbonne Université-ISIR, France
Torsten Suel   New York University, USA
Hussein Suleman   University of Cape Town, South Africa
Paul Thomas   Microsoft, USA
Theodora Tsikrika   Information Technologies Institute/CERTH,
                                 Greece
Suzan Verberne   LIACS/Leiden University, Netherlands
Marcel Worring   University of Amsterdam, Netherlands
Andrew Yates   University of Amsterdam, Netherlands
Shuo Zhang   Bloomberg, UK
Min Zhang   Tsinghua University, China
Guido Zuccon   The University of Queensland, Australia

Program Committee

Amin Abolghasemi   Leiden University, Netherlands
Sharon Adar   Amazon, USA
Shilpi Agrawal   Linkedin, USA
Mohammad Aliannejadi   University of Amsterdam, Netherlands
Satya Almasian   Heidelberg University, Germany
Giuseppe Amato   ISTI-CNR, Italy
Linda Andersson   Artificial Researcher IT GmbH TU Wien, Austria
Negar Arabzadeh   University of Waterloo, Canada
Marcelo Armentano   ISISTAN (CONICET - UNCPBA), Argentina
Arian Askari   Leiden University, Netherlands
Maurizio Atzori   University of Cagliari, Italy
Sandeep Avula   Amazon, USA
Hosein Azarbonyad   Elsevier, Netherlands
Leif Azzopardi   University of Strathclyde, UK
Andrea Bacciu   Sapienza University of Rome, Italy
Mossaab Bagdouri   Walmart Global Tech, USA

Ebrahim Bagheri   Ryerson University, Canada
Seyedeh Baharan Khatami   University of California San Diego, USA
Giacomo Balloccu   Università degli Studi di Cagliari, Italy
Alberto Barrón-Cedeño   Università di Bologna, Italy
Alvaro Barreiro   University of A Coruña, Spain
Roberto Basili   University of Roma Tor Vergata, Italy
Elias Bassani   Independent Researcher, Italy
Christine Bauer   Paris Lodron University Salzburg, Austria
Alejandro Bellogin   Universidad Autonoma de Madrid, Spain
Alessandro Benedetti   Sease, UK
Klaus Berberich   Saarbruecken University of Applied Sciences,
                                    Germany
Arjun Bhalla   Bloomberg LP, USA
Sumit Bhatia   Adobe Inc., India
Veronika Bogina   Tel Aviv University, Israel
Alexander Bondarenko   Friedrich-Schiller-Universität Jena, Germany
Ludovico Boratto   University of Cagliari, Italy
Gloria Bordogna   National Research Council of Italy - CNR, France
Emanuela Boros   EPFL, Switzerland
Leonid Boytsov   Amazon, USA
Martin Braschler   Zurich University of Applied Sciences,
                                    Switzerland
Pavel Braslavski   Ural Federal University, Russia
Timo Breuer   TH Köln (University of Applied Sciences),
                                    Germany
Sebastian Bruch   Pinecone, USA
Arthur Câmara   Zeta Alpha Vector, Netherlands
Fidel Cacheda   Universidade da Coruña, Spain
Sylvie Calabretto   LIRIS, France
Cesare Campagnano   Sapienza University of Rome, Italy
Ricardo Campos   University of Beira Interior, Portugal
Iván Cantador   Universidad Autónoma de Madrid, Spain
Alberto Carlo Maria Mancino   Politecnico di Bari, Italy
Matteo Catena   Siren, Italy
Abhijnan Chakraborty   IIT Delhi, India
Khushhall Chandra Mahajan   Meta Inc., USA
Shubham Chatterjee   The University of Edinburgh, UK
Despoina Chatzakou   Information Technologies Institute, Greece
Catherine Chavula   University of Texas at Austin, USA
Catherine Chen   Brown University, USA
Jean-Pierre Chevallet   Grenoble Alpes University, France
Adrian-Gabriel Chifu   Aix Marseille University, France

Evgenia Christoforou	CYENS Centre of Excellence, Cyprus
Abu Nowshed Chy	University of Chittagong, Bangladesh
Charles Clarke	University of Waterloo, Canada
Stephane Clinchant	Naver Labs Europe, France
Fabio Crestani	Università della Svizzera Italiana (USI), Switzerland
Shane Culpepper	The University of Queensland, Australia
Hervé Déjean	Naver Labs Europe, France
Célia da Costa Pereira	Université Côte d'Azur, France
Maarten de Rijke	University of Amsterdam, Netherlands
Arjen De Vries	Radboud University, Netherlands
Amra Deli	University of Sarajevo, Bosnia and Herzegovina
Gianluca Demartini	The University of Queensland, Australia
Danilo Dess	Leibniz Institute for the Social Sciences, Germany
Emanuele Di Buccio	University of Padua, Italy
Gaël Dias	Normandie University, France
Vlastislav Dohnal	Masaryk University, Czechia
Gregor Donabauer	University of Regensburg, Germany
Zhicheng Dou	Renmin University of China, China
Carsten Eickhoff	University of Tübingen, Germany
Michael Ekstrand	Drexel University, USA
Dima El Zein	Université Côte d'Azur, France
David Elsweiler	University of Regensburg, Germany
Ralph Ewerth	Leibniz Universität Hannover, Germany
Michael Färber	Karlsruhe Institute of Technology, Germany
Guglielmo Faggioli	University of Padova, Italy
Fabrizio Falchi	ISTI-CNR, Italy
Zhen Fan	Carnegie Mellon University, USA
Anjie Fang	Amazon.com, USA
Hossein Fani	University of Windsor, UK
Henry Field	Endicott College, USA
Yue Feng	UCL, UK
Marcos Fernández Pichel	Universidade de Santiago de Compostela, Spain
Antonio Ferrara	Polytechnic University of Bari, Italy
Komal Florio	Università di Torino - Dipartimento di Informatica, Italy
Thibault Formal	Naver Labs Europe, France
Eduard Fosch Villaronga	Leiden University, Netherlands
Maik Fröbe	Friedrich-Schiller-Universität Jena, Germany
Giacomo Frisoni	University of Bologna, Italy
Xiao Fu	University College London, UK
Norbert Fuhr	University of Duisburg-Essen, Germany

Petra Galuščáková	University of Stavanger, Norway
Debasis Ganguly	University of Glasgow, UK
Eric Gaussier	LIG-UGA, France
Xuri Ge	University of Glasgow, UK
Thomas Gerald	Université Paris Saclay CNRS SATT LISN, France
Kripabandhu Ghosh	ISSER, India
Satanu Ghosh	University of New Hampshire, USA
Daniela Godoy	ISISTAN (CONICET - UNCPBA), Argentina
Carlos-Emiliano González-Gallardo	L3i, France
Michael Granitzer	University of Passau, Germany
Nina Grgic-Hlaca	Max Planck Institute for Software Systems, Germany
Adrien Guille	Université de Lyon, France
Chun Guo	Pandora Media LLC, USA
Shashank Gupta	University of Amsterdam, Netherlands
Matthias Hagen	Friedrich-Schiller-Universität Jena, Germany
Fatima Haouari	Qatar University, Qatar
Maram Hasanain	Qatar University, Qatar
Claudia Hauff	Spotify, Netherlands
Naieme Hazrati	Free University of Bozen-Bolzano, Italy
Daniel Hienert	Leibniz Institute for the Social Sciences, Germany
Frank Hopfgartner	Universität Koblenz, Germany
Gilles Hubert	IRIT, France
Oana Inel	University of Zurich, Switzerland
Bogdan Ionescu	Politehnica University of Bucharest, Romania
Thomas Jaenich	University of Glasgow, UK
Shoaib Jameel	University of Southampton, UK
Faizan Javed	Kaiser Permanente, USA
Olivier Jeunen	ShareChat, UK
Alipio Jorge	University of Porto, Portugal
Toshihiro Kamishima	AIST, Japan
Noriko Kando	National Institute of Informatics, Japan
Sarvnaz Karimi	CSIRO, Australia
Pranav Kasela	University of Milano-Bicocca, Italy
Sumanta Kashyapi	University of New Hampshire, USA
Christin Katharina Kreutz	Cologne University of Applied Sciences, Germany
Abhishek Kaushik	Dublin City University, Ireland
Mesut Kaya	Aalborg University Copenhagen, Denmark
Diane Kelly	University of Tennessee, USA

Jae Keol Choi	Seoul National University, South Korea
Roman Kern	Graz University of Technology, Austria
Pooya Khandel	University of Amsterdam, Netherlands
Johannes Kiesel	Bauhaus-Universität, Germany
Styliani Kleanthous	CYENS CoE & Open University of Cyprus, Cyprus
Anastasiia Klimashevskaia	University of Bergen, Italy
Ivica Kostric	University of Stavanger, Norway
Dominik Kowald	Know-Center & Graz University of Technology, Austria
Hermann Kroll	Technische Universität Braunschweig, Germany
Udo Kruschwitz	University of Regensburg, Germany
Hrishikesh Kulkarni	Georgetown University, USA
Wojciech Kusa	TU Wien, Austria
Mucahid Kutlu	TOBB University of Economics and Technology, Turkey
Saar Kuzi	Amazon, USA
Jochen L. Leidner	Coburg University of Applied Sciences, Germany
Kushal Lakhotia	Outreach, USA
Carlos Lassance	Naver Labs Europe, France
Aonghus Lawlor	University College Dublin, Ireland
Dawn Lawrie	Johns Hopkins University, USA
Chia-Jung Lee	Amazon, USA
Jurek Leonhardt	TU Delft, Germany
Monica Lestari Paramita	University of Sheffield, UK
Hang Li	The University of Queensland, Australia
Ming Li	University of Amsterdam, Netherlands
Qiuchi Li	University of Padua, Italy
Wei Li	University of Roehampton, UK
Minghan Li	University of Waterloo, Canada
Shangsong Liang	MBZUAI, UAE
Nut Limsopatham	Amazon, USA
Marina Litvak	Shamoon College of Engineering, Israel
Siwei Liu	MBZUAI, UAE
Haiming Liu	University of Southampton, UK
Yiqun Liu	Tsinghua University, China
Bulou Liu	Tsinghua University, China
Andreas Lommatzsch	TU Berlin, Germany
David Losada	University of Santiago de Compostela, Spain
Jesus Lovon-Melgarejo	Université Paul Sabatier IRIT, France
Alipio M. Jorge	University of Porto, Portugal
Weizhi Ma	Tsinghua University, China

Georgios Peikos   University of Milano-Bicocca, Italy
Gustavo Penha   Spotify Research, Netherlands
Marinella Petrocchi   IIT-CNR, Italy
Aleksandr Petrov   University of Glasgow, UK
Milo Phillips-Brown   University of Edinburgh, UK
Karen Pinel-Sauvagnat   IRIT, France
Florina Piroi   Vienna University of Technology, Austria
Alessandro Piscopo   BBC, UK
Marco Polignano   Università degli Studi di Bari Aldo Moro, Italy
Claudio Pomo   Polytechnic University of Bari, Italy
Lorenzo Porcaro   Joint Research Centre European Commission,
                                 Italy
Amey Porobo Dharwadker   Meta, USA
Martin Potthast   Leipzig University, Germany
Erasmo Purificato   Otto von Guericke University Magdeburg,
                                 Germany
Xin Qian   University of Maryland, USA
Yifan Qiao   University of California, USA
Georges Quénot   Laboratoire d'Informatique de Grenoble CNRS,
                                 Germany
Alessandro Raganato   University of Milano-Bicocca, Italy
Fiana Raiber   Yahoo Research, Israel
Amifa Raj   Boise State University, USA
Thilina Rajapakse   University of Amsterdam, Netherlands
Jerome Ramos   University College London, UK
David Rau   University of Amsterdam, Netherlands
Gábor Recski   TU Wien, Austria
Navid Rekabsaz   Johannes Kepler University Linz, Austria
Zhaochun Ren   Leiden University, Netherlands
Yongli Ren   RMIT University, Australia
Weilong Ren   Shenzhen Institute of Computing Sciences, China
Chiara Renso   ISTI-CNR, Italy
Kevin Roitero   University of Udine, Italy
Tanya Roosta   Amazon, USA
Cosimo Rulli   University of Pisa, Italy
Valeria Ruscio   Sapienza University of Rome, Italy
Yuta Saito   Cornell University, USA
Tetsuya Sakai   Waseda University, Japan
Shadi Saleh   Microsoft, USA
Eric Sanjuan   Avignon Université, France
Javier Sanz-Cruzado   University of Glasgow, UK
Fabio Saracco   Centro Ricerche Enrico Fermi, Italy

Giovanni Trappolini	Sapienza University, Italy
Jan Trienes	University of Duisburg-Essen, Germany
Andrew Trotman	University of Otago, New Zealand
Chun-Hua Tsai	University of Omaha, USA
Radu Tudor Ionescu	University of Bucharest, Romania
Yannis Tzitzikas	University of Crete and FORTH-ICS, Greece
Venktesh V	TU Delft, Germany
Alberto Veneri	Ca' Foscari University of Venice, Italy
Manisha Verma	Amazon, USA
Federica Vezzani	University of Padua, Italy
João Vinagre	Joint Research Centre - European Commission, Italy
Vishwa Vinay	Adobe Research, India
Marco Viviani	Università degli Studi di Milano-Bicocca, Italy
Sanne Vrijenhoek	Universiteit van Amsterdam, Netherlands
Vito Walter Anelli	Politecnico di Bari, Italy
Jiexin Wang	South China University of Technology, China
Zhihong Wang	Tsinghua University, China
Xi Wang	University College London, UK
Xiao Wang	University of Glasgow, UK
Yaxiong Wu	University of Glasgow, UK
Eugene Yang	Johns Hopkins University, USA
Hao-Ren Yao	National Institutes of Health, USA
Andrew Yates	University of Amsterdam, Netherlands
Fanghua Ye	University College London, UK
Zixuan Yi	University of Glasgow, UK
Elad Yom-Tov	Microsoft, USA
Eva Zangerle	University of Innsbruck, Austria
Markus Zanker	University of Klagenfurt, Germany
Fattane Zarrinkalam	University of Guelph, Canada
Rongting Zhang	Amazon, USA
Xinyu Zhang	University of Waterloo, USA
Yang Zhang	Kyoto University, Japan
Min Zhang	Tsinghua University, China
Tianyu Zhu	Beihang University, China
Jiongli Zhu	University of California San Diego, USA
Shengyao Zhuang	The University of Queensland, Australia
Md Zia Ullah	Edinburgh Napier University, UK
Steven Zimmerman	University of Essex, UK
Lixin Zou	Wuhan University, China
Guido Zuccon	The University of Queensland, Australia

Additional Reviewers

Pablo Castells
Ophir Frieder
Claudia Hauff
Yulan He
Craig Macdonald
Graham McDonald

Iadh Ounis
Maria Soledad Pera
Fabrizio Silvestri
Nicola Tonellotto
Min Zhang

Contents – Part II

Full Papers

Full Papers

Efficient Multi-vector Dense Retrieval with Bit Vectors

Franco Maria Nardini[1] , Cosimo Rulli[1(✉)] , and Rossano Venturini[2]

[1] ISTI-CNR, Pisa, Italy
{francomaria.nardini,cosimo.rulli}@isti.cnr.it
[2] University of Pisa, Pisa, Italy
rossano.venturini@unipi.it

Abstract. Dense retrieval techniques employ pre-trained large language models to build a high-dimensional representation of queries and passages. These representations compute the relevance of a passage w.r.t. to a query using efficient similarity measures. In this line, multi-vector representations show improved effectiveness at the expense of a one-order-of-magnitude increase in memory footprint and query latency by encoding queries and documents on a per-token level. Recently, PLAID has tackled these problems by introducing a centroid-based term representation to reduce the memory impact of multi-vector systems. By exploiting a centroid interaction mechanism, PLAID filters out non-relevant documents, thus reducing the cost of the successive ranking stages. This paper proposes "Efficient Multi-Vector dense retrieval with Bit vectors" (EMVB), a novel framework for efficient query processing in multi-vector dense retrieval. First, EMVB employs a highly efficient pre-filtering step of passages using optimized bit vectors. Second, the computation of the centroid interaction happens column-wise, exploiting SIMD instructions, thus reducing its latency. Third, EMVB leverages Product Quantization (PQ) to reduce the memory footprint of storing vector representations while jointly allowing for fast late interaction. Fourth, we introduce a per-document term filtering method that further improves the efficiency of the last step. Experiments on MS MARCO and LoTTE show that EMVB is up to 2.8× faster while reducing the memory footprint by 1.8× with no loss in retrieval accuracy compared to PLAID.

Keywords: Dense Retrieval · Multi-vector · Efficiency · Bit Vectors

1 Introduction

The introduction of pre-trained large language models (LLM) has remarkably improved the effectiveness of information retrieval systems [2,8,13,25], thanks to the well-known ability of LLMs to model semantic and context [1,3,12]. In dense retrieval, LLMs have been successfully exploited to learn high-dimensional dense representations of passages and queries. These learned representations allow answering the user query through fast similarity operations, i.e., inner product or L2 distance. In this line, multi-vector techniques [14,20] employ an LLM

N. Goharian et al. (Eds.): ECIR 2024, LNCS 14609, pp. 3–17, 2024.
https://doi.org/10.1007/978-3-031-56060-6_1

to build a dense representation for each token of a passage. These approaches offer superior effectiveness compared to single-vector techniques [24, 26] or sparse retrieval techniques [5]. In this context, the similarity between the query and the passage is measured by using the *late interaction* mechanism [14, 20], which works by computing the sum of the maximum similarities between each term of the query and each term of a candidate passage. The improved effectiveness of multi-vector retrieval system comes at the price of its increased computational burden. First, producing a vector for each token causes the number of embeddings to be orders of magnitude larger than in a single-vector representation. Moreover, due to the large number of embeddings, identifying the candidate documents[1] is time-consuming. In addition, the late interaction step requires computing the maximum similarity operator between all the candidate embeddings and the query, which is also time-consuming.

Early multi-vector retrieval systems, i.e., ColBERT [14], exploit an inverted index to store the embeddings and retrieve the candidate passages. Then, the representations of passages are retrieved and employed to compute the max-similarity score with the query. Despite being quite efficient, this approach requires maintaining the full-precision representation of each document term in memory. On MS MARCO [17], a widely adopted benchmark dataset for passage retrieval, the entire collection of embeddings used by ColBERT requires more than 140 GB [14] to be stored. ColBERTv2 [20] introduces a centroid-based compression technique to store the passage embeddings efficiently. Each embedding is stored by saving the *id* of the closest centroid and then compressing the residual (i.e., the element-wise difference) by using 1 or 2 bits per component. ColBERTv2 saves up to 10× space compared to ColBERT while being significantly more inefficient on modern CPUs, requiring up to 3 s to perform query processing on CPU [19]. The reduction of query processing time is achieved by Santhanam *et al.* with PLAID [19]. PLAID takes advantage of the embedding compressor of ColBERTv2 and also uses the centroid-based representation to discard non-relevant passages (*centroid interaction* [19]), thus performing the late interaction exclusively on a carefully selected batch of passages. PLAID allows for massive speedup compared to ColBERTv2, but its average query latency can be up to 400 ms. on CPU with single-thread execution [19].

This paper presents EMVB, a novel framework for efficient query processing with multi-vector dense retrieval. First, we identify the most time-consuming steps of PLAID. These steps are i) extracting the top-*nprobe* closest centroids for each query term during the candidate passage selection, ii) computing the centroid interaction mechanism, and iii) decompression of the quantized residuals. Our method tackles the first and the second steps by introducing a highly efficient passage filtering approach based on optimized bit vectors. Our filter identifies a small set of crucial centroid scores, thus tearing down the cost of top-*nprobe* extraction. At the same time, it reduces the amount of passages for which we have to compute the centroid interaction. Moreover, we introduce a highly efficient column-wise reduction exploiting SIMD instructions to speed up

---

[1] the terms "document" and "passage" are used interchangeably in this paper.

this step. Finally, we improve the efficiency of the late interaction by introducing Product Quantization (PQ) [9]. PQ allows to obtain in pair or superior performance compared to the bitwise compressor of PLAID while being up to 3× faster. Finally, to further improve the efficiency of the last step of our pipeline, we introduce a dynamic passage-term-selection criterion for late interaction, thus reducing the cost of this step up to 30%.

We experimentally evaluate EMVB against PLAID on two datasets: MS MARCO passage [17] (for in-domain evaluation) and LoTTE [20] (for out-of-domain evaluation). Results on MS MARCO show that EMVB is up to 2.8× faster while reducing the memory footprint by 1.8× with no loss in retrieval accuracy compared to PLAID. On the out-of-domain evaluation, EMVB delivers up to 2.9× speedup compared to PLAID, with a minimal loss in retrieval quality.

The rest of this paper is organized as follows. In Sect. 2, we discuss the related work. In Sect. 3 we describe PLAID [19], the current state-of-the-art in multi-vector dense retrieval. We introduce EMVB in Sect. 4 and we experimentally evaluate it against PLAID in Sect. 5. Finally, Sect. 6 concludes our work.

2   Related Work

Dense retrieval encoders can be broadly classified into single-vector and multi-vector techniques. Single-vector encoders allow the encoding of an entire passage in a single dense vector [11]. In this line, ANCE [24] and STAR/ADORE [25] employ hard negatives to improve the training of dense retrievers by teaching them to distinguish between lexically-similar positive and negative passages. Multi-vector encoders have been introduced with ColBERT. The limitations of ColBERT and the efforts done to overcome them (ColBERTv2, PLAID) have been discussed in Sect. 1. COIL [6] rediscover the lessons of classical retrieval systems (e.g., BM25) by limiting the token interactions to lexical matching between queries and documents. CITADEL [16] is a recently proposed approach that introduces conditional token interaction by using dynamic lexical routing. Conditional token interaction means that the relevance of the query of a specific passage is estimated by only looking at some of their tokens. These tokens are selected by the so-called lexical routing, where a module of the ranking architecture is trained to determine which of the keys, i.e., words in the vocabulary, are activated by a query/passage. CITADEL significantly reduces the execution time on GPU, but turns out to be 2× slower than PLAID con CPU, at the same retrieval quality. Multi-vector dense retrieval is also exploited in conjunction with pseudo-relevance feedback both in ColBERT-PRF [23] and in CWPRF [22], showing that their combination boosts the effectiveness of the model.

Our Contribution: This work advances the state of the art of multi-vector dense retrieval by introducing EMVB, a novel framework that allows to speed up the retrieval performance of the PLAID pipeline significantly. To the best of our knowledge, this work is the first in the literature that proposes a highly efficient document filtering approach based on optimized bit vectors, a column-wise SIMD reduction to retrieve candidate passages and a late interaction mechanism that combines product quantization with a per-document term filtering.

3   Multi-vector Dense Retrieval

Consider a passage corpus $\mathcal{P}$ with n_P passages. In a multi-vector dense retrieval scenario, an LLM encodes each token in $\mathcal{P}$ into a dense d-dimensional vector T_j. For each passage P, a dense representation $P = \{T_j\}$, with $j = 0, \ldots, n_t$, is produced, where n_t is the number of tokens in the passage P. Employing a token-level dense representation allows for boosting the effectiveness of the retrieval systems [14,19,20]. On the other hand, it produces significantly large collections of d-dimensional vectors posing challenges to the applicability of such systems in real-world search scenarios both in terms of space (memory requirements) and time (latency of the query processor). To tackle the problem of memory requirements, ColBERTv2 [20] and successively PLAID [19] exploit a centroid-based vector compression technique. First, the K-means algorithm is employed to devise a clustering of the d-dimensional space by identifying the set of k centroids $\mathcal{C} = \{C_i\}_{i=1}^{n_c}$. Then, for each vector x, the residual r between x and its closest centroid $\bar{C}$ is computed so that $r = x - \bar{C}$. The residual r is compressed into $\tilde{r}$ using a b-bit encoder that represents each dimension of r using b bits, with $b \in \{1, 2\}$. The memory impact of storing a d-dimensional vector is given by $\lceil * \rceil \log_2 |\mathcal{C}|$ bits for the centroid index and $d \times b$ bits for the compressed residual. This approach requires a time-expensive decompression phase to restore the approximate full-precision vector representation given the centroid id and the residual coding. For this reason, PLAID aims at decompressing as few candidate documents as possible. This is achieved by introducing a high-quality filtering step based on the centroid-approximated embedding representation, named *centroid interaction* [19]. In detail, the PLAID retrieval engine is composed of four different phases [19]. The first one regards the *retrieval* of the candidate passages. A list of candidate passages is built for each centroid. A passage belongs to a centroid C_i candidate list if one or more tokens have C_i as its closest centroid. For each query term q_i, with $i = 1, \ldots, n_q$, the top-*nprobe* closest centroids are computed, according to the *dot product* similarity measure. The set of unique documents associated with the top-*nprobe* centroids then moves to a second phase that acts as a *filtering* phase. In this phase, a token embedding T_j with $j = 1, \ldots, n_p$ is approximated using its closest centroid $\bar{C}^{T_j}$. Hence, its distance with the i-th query term q_i is approximated with

$$q_i \cdot T_j \simeq q_i \cdot \bar{C}^{T_j} = \tilde{T}_{i,j}. \tag{1}$$

Consider a candidate passage P composed of n_p tokens. The approximated score of P consists in computing the dot product $q_i \cdot \bar{C}^{T_j}$ for all the query terms q_i and all the closest centroids of each token belonging to the passage, i.e.,

$$\bar{S}_{q,P} = \sum_{i=1}^{n_q} \max_{j=1 \ldots n_t} q_i \cdot \bar{C}^{T_j} \tag{2}$$

The third phase, named *decompression*, aims at reconstructing the full-precision representation of P by combining the centroids and the residuals. This is done on the top-*ndocs* passages selected according to the *filtering* phase [19]. In the fourth phase, PLAID recomputes the final score of each passage with respect to the query q using the decompressed—full-precision—representation according to *late interaction* mechanism (Eq. 3). Passages are then ranked according to their similarity score and the top-k passages are selected.

$$S_{q,P} = \sum_{i=1}^{n_q} \max_{j=1...n_t} q_i \cdot T_j. \tag{3}$$

PLAID Execution Time. We provide a breakdown of PLAID execution time across its different phases, namely *retrieval, filtering, decompression,* and *late interaction*. This experiment is conducted using the experimental settings detailed in Sect. 5. We report the execution time for different values of k, i.e., the number of retrieved passages.

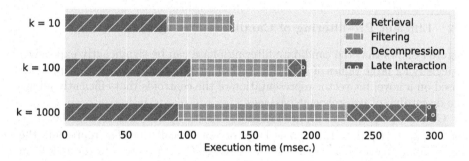

Fig. 1. Breakdown of the PLAID average query latency (in milliseconds) on CPU across its four phases.

4 EMVB

We now present EMVB, our novel framework for efficient multi-vector dense retrieval. First, EMVB introduces a highly efficient pre-filtering phase that exploits optimized bit vectors. Second, we improve the efficiency of the centroid interaction step (Eq. 1) by introducing column-wise max reduction with SIMD instructions. Third, EMVB leverages Product Quantization (PQ) to reduce the memory footprint of storing the vector representations while jointly allowing for a fast late interaction phase. Fourth, PQ is applied in conjunction with a novel per-passage term filtering approach that allows for further improving the efficiency of the late interaction. In the following subsections, we detail these four contributions behind EMVB.

4.1   Retrieval of Candidate Passages

Figure 1 shows that a consistent part of the computation required by PLAID is spent on the retrieval phase. We further break down these steps to evidence its most time-consuming part. The retrieval consists of i) computing the distances between the incoming query and the set of centroids, ii) extracting the top-*nprobe* closest centroids for each query term. The former step is efficiently carried out by leveraging high-performance matrix multiplication tools (e.g., Intel MKL [18,21]). In the latter step, PLAID extracts the top-*nprobe* centroids using the numpy topk function, which implements the *quickselect* algorithm. Selecting the top-*nprobe* within the $|C| = 2^{18}$ centroids for each of the n_q query terms costs up to 3× the matrix multiplication done in the first step. In Sect. 4.2, we show that our pre-filtering inherently speeds up the top-*nprobe* selection by tearing down the number of evaluated elements. In practice, we show how to efficiently filter out those centroids whose score is below a certain threshold and then execute quickselect exclusively on the surviving ones. As a consequence, in EMVB the cost of the top-*nprobe* extraction becomes negligible, being two orders of magnitude faster than the top-*nprobe* extraction on the full set of centroids.

4.2   Efficient Pre-filtering of Candidate Passages

Figure 1 shows that the candidate filtering phase can be significantly expensive, especially for large values of k. In this section, we propose a pre-filtering approach based on a novel bit vector representation of the centroids that efficiently allows the discarding of non-relevant passages.

Given a passage P, our pre-filtering consists in determining whether $\tilde{T}_{i,j}$, for $i = 1, \ldots, n_q$, $j = 1, \ldots, n_t$ is large or not. Recall that $\tilde{T}_{i,j}$ represents the approximate score of the j-th token of passage P with respect to the i-th term of the query q_i, as defined in Eq. 1. This can be obtained by checking whether $\bar{C}_j^T$—the centroid associated with T_j—belongs to the set of the *closest centroids* of q_i. We introduce $\texttt{close}_i^{th}$, the set of centroids whose scores are greater than a certain threshold th with respect to a query term q_i. Given a passage P, we define the list of centroids ids I_P, where I_P^j is the centroid id of $\bar{C}^{T_j}$. The similarity of a passage with respect to a query can be estimated with our novel filtering function $F(P, q) \in [0, n_q]$ with the following equation:

$$F(P,q) = \sum_{i=1}^{n_q} \mathbf{1}(\exists\, j \text{ s.t. } I_P^j \in \texttt{close}_i^{th}). \tag{4}$$

For a passage P, this counts how many query terms have at least one similar passage term in P, where "similar" describes the belonging of T_j to $\texttt{close}_i^{th}$.

In Fig. 2(left), we compare the performance of our novel pre-filter working on top of the centroid interaction mechanism (orange, blue, green lines) against the performance of the centroid interaction mechanism on the entire set of candidate documents (red dashed line) on the MS MARCO dataset. The plot shows that

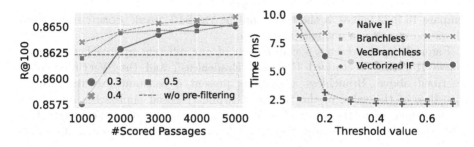

Fig. 2. R@100 with various values of the threshold (**left**). Comparison of different algorithms to construct $\texttt{close}_i^{th}$, for different values of th (**right**). (Color figure online)

our pre-filtering allows to efficiently discard non-relevant passages without harming the recall of the successive centroid interaction phase. For example, we can narrow the candidate passage set to just 1000 elements using $th = 0.4$ without any loss in R@100. In the remainder of this section, we show how to implement this pre-filter efficiently.

Building the Bit Vectors. Given th, the problem of computing $\texttt{close}_i^{th}$ is conceptually simple. Yet, an efficient implementation carefully considering modern CPUs' features is crucial for fast computation of Eq. 4.

Let $CS = q \cdot C^T$, with $CS \in [-1, 1]^{n_q \times |C|}$ be the score matrix between the query q and the set of centroids C (both matrices are L_2 normalized), where n_q is the number of query tokens, and $|C|$ is the number of centroids. In the naïve *if*-based solution, we scan the i-th row of CS and select those j s.t. $CS_{i,j} > th$. It is possible to speed up this approach by taking advantage of SIMD instructions. In particular, the $\_\texttt{mm512\_cmp\_epi32\_mask}$ allows one to compare 16 fp32 values at a time and store the comparison result in a *mask* variable. If *mask* $==$ 0, we can skip to the successive 16 values because the comparison has failed for all the current js. Otherwise, we extract those indexes $J = \{j \in [0, 15] \mid mask_j = 1\}$.

The efficiency of such *if*-based algorithms mainly depends on the *branch misprediction* ratio. Modern CPUs speculate on the outcome of the *if* before the condition itself is computed by recognizing patterns in the execution flow of the algorithm. When the wrong branch is predicted, a *control hazard* happens, and the pipeline is flushed with a delay of 15–20 clock cycles, i.e., about 10 ns. We tackle the inefficiency of branch misprediction by proposing a *branchless* algorithm. The branchless algorithm employs a pointer p addressing a pre-allocated buffer. While scanning $CS_{i,j}$, it writes j in the position indicated by p. Then, it sums to p the result of the comparison: 1 if $CS_{i,j} > th$, 0 otherwise. At the successive iteration, if the result of the comparison was 0, $j+1$ will override j. Otherwise, it will be written in the successive memory location, and j will be saved in the buffer. The branchless selection does not present any *if* instruction and consequently does not contains any branch in its execution flow. The branchless algorithm can be implemented more efficiently by leveraging SIMD instructions. In particular, the above-mentioned $\_\texttt{mm512\_cmp\_epi32\_mask}$ instruction allows to compare 16 fp32 values at the time, and the _mm512_mask_compressstore allows to extract J in a single instruction.

Figure 2(right) presents a comparison of our different approaches, namely "Naïve IF", the "Vectorized IF", the "Branchless", and the "VecBranchless" described above. Branchless algorithms present a constant execution time, regardless of the value of the threshold, while if-based approaches offer better performances as the value of th increases. With $th \geq 0.3$, "Vectorized IF" is the most efficient approach, with a speedup up to 3× compared to its naïve counterpart.

Fast Set Membership. Once $\texttt{close}_i^{th}$ is computed, we have to efficiently compute Eq. 4. Here, given I_P as a list of integers, we have to test if at least one of its members I_P^j belongs to $\texttt{close}_i^{th}$, with $i = 1, \ldots, n_q$. This can be efficiently done using *bit vectors* for representing $\texttt{close}_i^{th}$. A bit vector maps a set of integers up to N into an array of N bits, where the e-th bit is set to one if and only if the integer e belongs to the set. Adding and searching any integer e can be performed in constant time with bit manipulation operators. Moreover, bit vectors require N bits to be stored. In our case, since we have $|C| = 2^{18}$, a bit vector only requires $32K$ bytes to be stored.

Since we search through all the n_q bit vectors at a time, we can further exploit the bit vector representation by stacking the bit vectors vertically (Fig. 3). This allows to search a centroid index through all the $\texttt{close}_i^{th}$ at a time. The bits corresponding to the same centroid for different query terms are consecutive and fit a 32-bit word. This way, we can simultaneously test the membership for all the queries in constant time with a single bitwise operation. In detail, our algorithm works by initializing a mask m of $n_q = 32$ bits at zeros (Step 1, Fig. 3). Then, for each term in the candidate documents, it performs a bitwise xor between the mask and the 32-bit word representing the membership to all the query terms (Step 2, Fig. 3). Hence, Eq. 4 can be obtained by counting the number of 1s in m at the end of the execution with the popcnt operation featured by modern CPUs (Step 3, Fig. 3).

Figure 4(up) shows that our "Vectorized" set membership implementation delivers a speedup ranging from 10× to 16× a "Baseline" relying on a naïve usage of bit vectors. In particular, our bit vector-based pre-filtering can be up to 30× faster than the centroid-interaction proposed in PLAID [19], cf. Fig. 4(down).

4.3   Fast Filtering of Candidate Passages

Our pre-filtering approach allows us to efficiently filter out non-relevant passages and is employed upstream of PLAID's centroid interaction (Eq. 2). We now show how to improve the efficiency of the centroid interaction itself.

Consider a passage P and its associated centroid scores matrix $\tilde{P} = q_i \cdot \bar{C}^{T_j}$. Explicitly building this matrix allows to reuse it in the scoring phase, in place of the costly decompression step (Sect. 4.4). To build $\tilde{P}$, we transpose CS into CS^T of size $|C| \times n_q$. The i-th row of CS^T allows access to all the n_q query terms scores for the i-th centroids. Given the ids of the closest centroids for each passage term

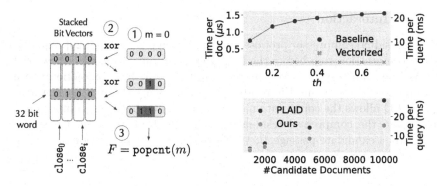

Fig. 3. Vectorized Fast Set Membership algorithm based on bit vectors.

Fig. 4. Vectorized vs naïve Fast Set Membership (**up**). Ours vs PLAID filtering (**down**).

(defined as I_P in Sect. 4.2) we retrieve the scores for each centroid id. We build $\tilde{P}^T$—$\tilde{P}$ transposed—to allow the CPU to read and write contiguous memory locations. This grants more than 2× speedup compared to processing $\tilde{P}$. We now have $\tilde{P}^T$ of shape $n_t \times n_q$. We need to max-reduce along the columns and then sum the obtained values to implement Eq. 2. This is done by iterating on the $\tilde{P}^T$ rows and packing them into AVX512 registers. Given that $n_q = 32$, each AVX512 register can contain $512/32 = 16$ floating point values, so we need 2 registers for each row. We pack the first row into max_l and max_h. All the successive rows are packed into $current\_l$ and $current\_h$. At each iteration, we compare max_l with $current\_l$ and max_h with $current\_h$ using the _mm512_cmp_ps_mask AVX512 instruction described before. The output mask m is used to update the max_l and max_h by employing the _mm512_mask_blend_ps instruction. The _mm512_cmp_ps_mask has throughput 2 on IceLake Xeon CPUs, so each row of $\tilde{P}$ is compared with max_l and max_h in the same clock cycle, on two different ports. The same holds for the _mm512_mask_blend_ps instruction, entailing that the max-reduce operation happens in 2 clock cycles without considering the memory loading. Finally, max_l and max_h are summed together, and the function _mm512_reduce_add_ps is used to ultimate the computation.

We implement PLAID's centroid interaction in C++ and we compare its filtering time against our SIMD-based solution. The results of the comparison are reported for different values of candidate documents in Fig. 4(down). Thanks to the proficient read-write pattern and the highly efficient column-wise max-reduction, our method can be up to 1.8× faster than the filtering proposed in PLAID.

4.4   Late Interaction

The b-bit residual compressor proposed in previous approaches [19,20] requires a costly decompression step before the late interaction phase. Figure 1 shows that in PLAID decompressing the vectors costs up to 5× the late interaction phase.

We propose compressing the residual r by employing Product Quantization (PQ) [9]. PQ allows the computation of the dot product between an input query vector q and the compressed residual r_{pq} without decompression. Consider a query q and a candidate passage P. We decompose the computation of the max similarity operator (Eq. 3) into

$$S_{q,P} = \sum_{i=1}^{n_q} \max_{j=1\ldots n_t} (q_i \cdot \bar{C}^{T_j} + q_i \cdot r^{T_j}) \simeq \sum_{i=1}^{n_q} \max_{j=1\ldots n_t} (q_i \cdot \bar{C}^{T_j} + q_i \cdot r_{pq}^{T_j}), \quad (5)$$

where and $r^{T_j} = T_j - \bar{C}^{T_j}$. On the one hand, this decomposition allows to exploit the pre-computed $\tilde{P}$ matrix. On the other hand, thanks to PQ, it computes the dot product between the query and the residuals without decompression.

We replace PLAID's residual compression with PQ, particularly with JMPQ [4], which optimizes the codes of product quantization during the fine-tuning of the language model for the retrieval task. We tested $m = \{16, 32\}$, where m is the number of sub-spaces used to partition the vectors [9]. We experimentally verify that PQ reduces the latency of the late interaction phase up to 3.6× compared to PLAID b-bit compressor. Moreover, it delivers the same ($m = 16$) or superior performance ($m = 32$) in terms of MRR@10 when leveraging the JMPQ version.

We propose to further improve the efficiency of the scoring phase by hinging on the properties of Eq. 5. We experimentally observe that, in many cases, $q_i \cdot \bar{C}_j^T > q_i \cdot r_{pq}^{T_j}$, meaning that the *max* operator on j, in many cases, is lead by the score between the query term and the centroid, rather than the score between the query term and the residual. We argue that it is possible to compute the scores on the residuals only for a reduced set of document terms $\bar{J}_i$, where i identifies the index of the query term. In particular, $\bar{J}_i = \{j | q_i \cdot \bar{C}_j^T > th_r\}$, where th_r is a second threshold that determines whether the score with the centroid is sufficiently large. With the introduction of this new per-term filter, Eq. 5 now becomes computing the max operator on the set of passages in $\bar{J}_i$, i.e.,

$$S_{q,P} = \sum_{i=1}^{n_q} \max_{j \in \bar{J}_i}(q_i \cdot \bar{C}^{T_j} + q_i \cdot r_{pq}^{T_j}). \quad (6)$$

In practice, we compute the residual scores only for those document terms whose centroid score is large enough. If $\bar{J}_i = \emptyset$, we compute $S_{q,P}$ as in Eq. 5. Figure 5(left) reports the effectiveness of our approach. On the y-axis, we report the percentage of the original effectiveness, computed as the ratio between the MRR@10 computed with Eq. 6 and Eq. 5. Filtering document terms according to Eq. 6 does not harm the retrieval quality, as it delivers substantially the same MRR@10 of Eq. 5. On the right side of Fig. 5, we report the percentage of scored

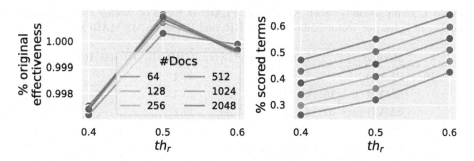

Fig. 5. Performance of our dynamic term-selection filtering for different values of th_r, in terms of percentage of original effectiveness (**left**) and in terms of percentage of original number of scored terms (**right**). The percentage of original effectiveness is computed as the ratio between the MRR@10 computed with Eq. 6 and Eq. 5.

terms compared to the number of document terms computed using Eq. 5. With $th_r = 0.5$, we are able to reduce the number of scored terms of at least 30% (right) without any performance degradation in terms of MRR@10.

5   Experimental Evaluation

Experimental Settings. This section compares our methodology against the state-of-the-art engine for multi-vector dense retrieval, namely PLAID [19]. We conduct experiments on the MS MARCO passages dataset [17] for the in-domain evaluation and on LoTTE [20] for the out-of-domain evaluation. We generate the embeddings for MS MARCO using the ColBERTv2 model. The generated dataset is composed of about 600M d-dimensional vectors, with $d = 128$. Product Quantization is implemented using the FAISS [10] library, and optimized using the JMPQ technique [4] on MS MARCO. The C++ implementation of EMVB will be made publicly available upon publication of this work. We compare EMVB against the original PLAID implementation [19], which also implements its core components in C++. Experiments are conducted on an Intel Xeon Gold 5318Y CPU clocked at 2.10 GHz, equipped with the AVX512 instruction set, with single-thread execution. Code is compiled using GCC 11.3.0 (with –O3 compilation options) on a Linux 5.15.0-72 machine. When running experiments with AVX512 instruction on 512-bit registers, we ensure not to incur in the frequency scaling down event reported for Intel CPUs [15].

Evaluation. Table 1 compares EMVB against PLAID on the MS MARCO dataset, in terms of memory requirements (num. of bytes per embedding), average query latency (in milliseconds), MRR@10, and Recall@100, and 1000.

Results show that EMVB delivers superior performance along both the evaluated trade-offs. With $m = 16$, EMVB almost halves the per-vector memory burden compared to PLAID, while being up to 2.8× faster with almost no performance degradation regarding retrieval effectiveness. By doubling the number

Table 1. Comparison between EMVB and PLAID in terms of average query latency, number of bytes per vector embeddings, MRR, and Recall on MS MARCO.

k	Method	Latency (ms)	Bytes	MRR@10	R@100	R@1000
10	PLAID	131	36	39.4	-	-
	EMVB ($m=16$)	62 (2.1×)	20	39.4	-	-
	EMVB ($m=32$)	61 (2.1×)	36	39.7	-	-
100	PLAID	180	36	39.8	90.6	-
	EMVB ($m=16$)	68 (2.6×)	20	39.5	90.7	-
	EMVB ($m=32$)	80 (2.3×)	36	39.9	90.7	-
1000	PLAID	260	36	39.8	91.3	97.5
	EMVB ($m=16$)	93 (2.8×)	20	39.5	91.4	97.5
	EMVB ($m=32$)	104 (2.5×)	36	39.9	91.4	97.5

of sub-partitions per vector, i.e., $m = 32$, EMVB outperforms the performance of PLAID in terms of MRR and Recall with the same memory footprint with up to 2.5× speed up.

Table 2 compares EMVB and PLAID in the out-of-domain evaluation on the LoTTE dataset. As in PLAID [19], we employ Success@5 and Success@100 as retrieval quality metrics. On this dataset, EMVB offers slightly inferior performance in terms of retrieval quality. Recall that JMPQ [4] cannot be applied in the out-of-domain evaluation due to the lack of training queries. Instead, we employ Optimized Product Quantization (OPQ) [7], which searches for an optimal rotation of the dataset vectors to reduce the quality degradation that comes with PQ. To mitigate the retrieval quality loss, we only experiment PQ with $m = 32$, given that an increased number of partitions offers a better representation of the original vector. On the other hand, EMVB can offer up to 2.9× speedup compared to PLAID. This larger speedup compared to MS MARCO is due to the larger average document lengths in LoTTE. In this context, filtering nonrelevant documents using our bit vector-based approach has a remarkable impact on efficiency. Observe that for the out-of-domain evaluation, our pre-filtering method could be ingested into PLAID. This would allow to maintain the PLAID accuracy together with EMVB efficiency. Combinations of PLAID and EMVB are left for future work.

Table 2. Comparison between EMVB and PLAID in terms of average query latency, number of bytes per vector embeddings, Success@5, and Success@100 on LoTTE.

k	Method	Latency (ms)	Bytes	Success@5	Success@100
10	PLAID	131	36	69.1	-
	EMVB ($m = 32$)	82 (1.6×)	36	69.0	-
100	PLAID	202	36	69.4	89.9
	EMVB ($m = 32$)	129 (1.6×)	36	69.0	89.9
1000	PLAID	411	36	69.6	90.5
	EMVB ($m = 32$)	142 (2.9×)	36	69.0	90.1

6   Conclusion

We presented EMVB, a novel framework for efficient multi-vector dense retrieval. EMVB advances PLAID, the current state-of-the-art approach, by introducing four novel contributions. First, EMVB employs a highly efficient pre-filtering step of passages using optimized bit vectors for speeding up the candidate passage filtering phase. Second, the computation of the centroid interaction is carried out with reduced precision. Third, EMVB leverages Product Quantization to reduce the memory footprint of storing vector representations while jointly allowing for fast late interaction. Fourth, we introduce a per-passage term filter for late interaction, thus reducing the cost of this step of up to 30%. We experimentally evaluate EMVB against PLAID on two publicly available datasets, i.e., MS MARCO and LoTTE. Results show that, in the in-domain evaluation, EMVB is up to 2.8× faster, and it reduces by 1.8× the memory footprint with no loss in retrieval quality compared to PLAID. In the out-of-domain evaluation, EMVB is up to 2.9× faster with little or no retrieval quality degradation.

Acknowledgements. This work was partially supported by the EU - NGEU, by the PNRR - M4C2 - Investimento 1.3, Partenariato Esteso PE00000013 - "FAIR - Future Artificial Intelligence Research" - Spoke 1 "Human-centered AI" funded by the European Commission under the NextGeneration EU program, by the PNRR ECS00000017 Tuscany Health Ecosystem Spoke 6 "Precision medicine & personalized healthcare", by the European Commission under the NextGeneration EU programme, by the Horizon Europe RIA "Extreme Food Risk Analytics" (EFRA), grant agreement n. 101093026, by the "Algorithms, Data Structures and Combinatorics for Machine Learning" (MIUR-PRIN 2017), and by the "Algorithmic Problems and Machine Learning" (MIUR-PRIN 2022).

References

1. Brown, T., et al.: Language models are few-shot learners. In: Advances in Neural Information Processing Systems (NIPS) (2020)
2. Bruch, S., Lucchese, C., Nardini, F.M.: Efficient and Effective Tree-Based and Neural Learning to Rank. Now Foundations and Trends (2023)

3. Dai, D., Sun, Y., Dong, L., Hao, Y., Sui, Z., Wei, F.: Why can GPT learn in-context? Language models secretly perform gradient descent as meta optimizers. arXiv preprint arXiv:2212.10559 (2022)
4. Fang, Y., Zhan, J., Liu, Y., Mao, J., Zhang, M., Ma, S.: Joint optimization of multi-vector representation with product quantization. In: Natural Language Processing and Chinese Computing (2022)
5. Formal, T., Piwowarski, B., Clinchant, S.: SPLADE: sparse lexical and expansion model for first stage ranking. In: Proceedings of the 44th International ACM SIGIR Conference on Research and Development in Information Retrieval (2021)
6. Gao, L., Dai, Z., Callan, J.: COIL: revisit exact lexical match in information retrieval with contextualized inverted list. In: Conference of the North American Chapter of the Association for Computational Linguistics: Human Language Technologies (2021)
7. Ge, T., He, K., Ke, Q., Sun, J.: Optimized product quantization. IEEE Trans. Pattern Anal. Mach. Intel. **36**, 744–755 (2013)
8. Guu, K., Lee, K., Tung, Z., Pasupat, P., Chang, M.: Retrieval augmented language model pre-training. In: Proceedings of the International Conference on Machine Learning (ICML) (2020)
9. Jegou, H., Douze, M., Schmid, C.: Product quantization for nearest neighbor search. IEEE Transa. Pattern Anal. Mach. Intel. **33**, 117–128 (2010)
10. Johnson, J., Douze, M., Jegou, H.: Billion-scale similarity search with GPUs. IEEE Trans. Big Data **7**, 535–547 (2021)
11. Karpukhin, V., et al.: Dense passage retrieval for open-domain question answering. In: Proceedings of the 2020 Conference on Empirical Methods in Natural Language Processing (EMNLP). Association for Computational Linguistics (2020)
12. Devlin, J., Chang, M.-W., Lee, K., Toutanova, K.: BERT: pre-training of deep bidirectional transformers for language understanding. In: Proceedings of NAACL-HLT (2019)
13. Khattab, O., Potts, C., Zaharia, M.: Baleen: robust multi-hop reasoning at scale via condensed retrieval. In: Advances in Neural Information Processing Systems (NIPS) (2021)
14. Khattab, O., Zaharia, M.: ColBERT: efficient and effective passage search via con-textualized late interaction over BERT. In: Proceedings of the 43rd International ACM SIGIR Conference on Research and Development in Information Retrieval, pp. 39–48 (2020)
15. Lemire, D., Downs, T.: AVX-512: when and how to use these new instructions (2023). https://lemire.me/blog/2018/09/07/avx-512-when-and-how-to-use-these-new-instructions/
16. Li, M., et al.: CITADEL: conditional token interaction via dynamic lexical routing for efficient and effective multi-vector retrieval. arXiv e-prints (2022)
17. Nguyen, T., et al.: MS Marco: a human-generated machine reading comprehension dataset (2016)
18. Qian, G., Sural, S., Gu, Y., Pramanik, S.: Similarity between Euclidean and Cosine angle distance for nearest neighbor queries. In: Proceedings of the 2004 ACM Symposium on Applied Computing (2004)
19. Santhanam, K., Khattab, O., Potts, C., Zaharia, M.: PLAID: an efficient engine for late interaction retrieval. In: Proceedings of the 31st ACM International Conference on Information & Knowledge Management (2022)

20. Santhanam, K., Khattab, O., Saad-Falcon, J., Potts, C., Zaharia, M.: ColBERTv2: effective and efficient retrieval via lightweight late interaction. In: Proceedings of the 2022 Conference of the North American Chapter of the Association for Computational Linguistics: Human Language Technologies (2022)
21. Wang, E., et al.: Intel math kernel library. In: High-Performance Computing on the Intel® Xeon Phi^{TM} (2014)
22. Wang, X., MacAvaney, S., Macdonald, C., Ounis, I.: Effective contrastive weighting for dense query expansion. In: Proceedings of the 61st Annual Meeting of the Association for Computational Linguistics (2023)
23. Wang, X., Macdonald, C., Tonellotto, N., Ounis, I.: ColBERT-PRF: semantic pseudo-relevance feedback for dense passage and document retrieval. ACM Trans. Web **17**(1), 1–39 (2023)
24. Xiong, L., et al.: Approximate nearest neighbor negative contrastive learning for dense text retrieval. In: International Conference on Learning Representations (2021)
25. Zhan, J., Mao, J., Liu, Y., Guo, J., Zhang, M., Ma, S.: Optimizing dense retrieval model training with hard negatives. In: Proceedings of the 44th International ACM SIGIR Conference on Research and Development in Information Retrieval (2021)
26. Zhan, J., Mao, J., Liu, Y., Guo, J., Zhang, M., Ma, S.: Learning discrete representations via constrained clustering for effective and efficient dense retrieval. In: Proceedings of the Fifteenth ACM International Conference on Web Search and Data Mining, pp. 1328–1336 (2022)

CrisisKAN: Knowledge-Infused and Explainable Multimodal Attention Network for Crisis Event Classification

Shubham Gupta(✉)[ID], Nandini Saini(✉)[ID], Suman Kundu(✉)[ID],
and Debasis Das(✉)[ID]

Department of Computer and Science Engineering, Indian Institute of Technology
Jodhpur, Jodhpur, India
{gupta.37,saini.9,suman,debasis}@iitj.ac.in

Abstract. Pervasive use of social media has become the emerging source for real-time information (like images, text, or both) to identify various events. Despite the rapid growth of image and text-based event classification, the state-of-the-art (SOTA) models find it challenging to bridge the semantic gap between features of image and text modalities due to inconsistent encoding. Also, the black-box nature of models fails to explain the model's outcomes for building trust in high-stakes situations such as disasters, pandemic. Additionally, the word limit imposed on social media posts can potentially introduce bias towards specific events. To address these issues, we proposed CrisisKAN, a novel **K**nowledge-infused and Explainable Multimodal **A**ttention **N**etwork that entails images and texts in conjunction with external knowledge from Wikipedia to classify crisis events. To enrich the context-specific understanding of textual information, we integrated Wikipedia knowledge using proposed wiki extraction algorithm. Along with this, a guided cross-attention module is implemented to fill the semantic gap in integrating visual and textual data. In order to ensure reliability, we employ a model-specific approach called Gradient-weighted Class Activation Mapping (Grad-CAM) that provides a robust explanation of the predictions of the proposed model. The comprehensive experiments conducted on the CrisisMMD dataset yield in-depth analysis across various crisis-specific tasks and settings. As a result, CrisisKAN outperforms existing SOTA methodologies and provides a novel view in the domain of explainable multimodal event classification. (Code repository: https://github.com/shubhamgpt007/CrisisKAN)

Keywords: Multimodal Network · Explainable · Knowledge Infusion · Crisis Detection

1 Introduction

With the fast-growing popularity of the internet, social media platforms have become vital medium for the early identification and detection of crisis events

N. Goharian et al. (Eds.): ECIR 2024, LNCS 14609, pp. 18–33, 2024.
https://doi.org/10.1007/978-3-031-56060-6_2

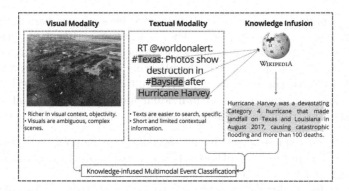

Fig. 1. Illustration of Twitter example using knowledge enhanced multimodal event classification to address challenges in visual-textual modality

such as hurricanes [24], earthquakes [39], and disease outbreaks like COVID-19 [30]. These platforms provide massive number of user generated information in the form of either text, image or both for same event within milliseconds. At the same time, rapidly evolving computer vision [19,35] and natural language processing [10,11] methods are being leveraged to use these data for examining and categorizing crisis events.

Some of the recent works [1,27,33,42] have been proposed for image-text based multimodal event classification using high-level feature fusion strategy. In these models, feature extraction performed using distinct backbone for each modality, and the aggregation method was utilised for final outcome. However, different aggregation methodologies can create a semantic gap between individual modalities due to inconsistent encoding methods. Further, these backbones are based on deep neural network which can be viewed as black-box and not interpretable. Nevertheless, these methods considered knowledge available from the social media post which is limited due to text length constraint imposed by many social media platforms.

In order to address these issues, we present a novel explainable multimodal framework, named as CrisisKAN (**K**nowledge-infused and Explainable Multimodal **A**ttention **N**etwork), to classify crisis events. CrisisKAN consists a series of phases starting with knowledge infusion followed by image-text feature extraction, multimodal classification, and model explanation. The knowledge infusion step leverages external information by integrating Wikipedia knowledge using a proposed wiki extraction algorithm to enhance the knowledge available with the limited text. Figure 1 illustrates an instance of external knowledge extraction through Wikipedia that includes entities information such as 'Hurricane Harvey', 'Texas', and 'Bayside'. Subsequently, feature extraction is applied to both modalities, and their fusion is performed through a guided cross-attention module, that effectively bridges the semantic gap between distinct feature sets. To instill confidence in the model's predictions a model-specific explainable module is also integrated that enables the analysis of feature maps within the black box model. This module also serves as a qualitative performance parameter along-side quantitative (accuracy, F1 etc.). The extensive experiments on crisisMMD [4] dataset demonstrate the superior performance of the CrisisKAN compared to existing state-of-the-art solutions, highlighting its effectiveness in crisis event classification. We also propose a new metric Multi-task Model Strength (MTMS) which provides performance of an individual model across different tasks. In summary, our contribution of this work are four fold:

- First of all, we exploit knowledge infusion in the multimodal crisis event clas-sification. The proposed implementation effectively overcoming limitations of short text and event biases.
- We integrated Guided Cross Attention mechanism to fill the semantic gap among the modalities while aggregating various large pre-trained unimodal classification model.
- Exaplainablity (XAI) of the outcome is incorporated in the model. Proposed XAI layer in the model not only excels on diverse image-text multimodal classification benchmarks but also ensures transparency and interpretability in the outcomes.
- We propose a new metric Multi-task Model Strength (MTMS) that provide model performance in more generalization manner across various tasks.

This paper's overall structure is as follows: In Sect. 2, we add the study about various classification models and explainability. The proposed methodology is detailed in Sect. 3, while Sect. 4 describes the experimental setup and discusses the results. Finally, Sect. 5 provides a summary of the proposed work and iden-tifies directions for future research.

2   Related Work

The goal of multimodal learning is focused on integrating different modalities information into a single representation to enrich the representation of data, enabling more robust and accurate predictions. Recent works such as image-text matching [34], multimodal detection [8,23], sarcasm detection in memes [7] and many more take advantage of the diverse information available in both image and text forms. These studies integrate the different modality of features broadly based on three strategies: high-level, intermediate, and low-level fea-ture fusion. In the first strategy, independent deep neural networks are used to generate high-level features for each modality [15,17]. Following that, fusion occurs at the model's final layers using aggregation methods such as summa-tion [27], tensor fusion [47], OR function based fusion [3] etc. Wang et al. [45] proposed an event detection model that combines low and high-level features to capitalise on their respective advantages and Gupta et al. [16] proposed a com-munity based unsupervised event detection model that forms the interaction graph over text modality and apply community detection algorithm to find out micro-level events. On the contrary, intermediate-level feature fusion strategies were optimised that introduce multimodal BERT by focusing on fine-grained token features of image and text modality [26,31]. However, these strategies are still limited in their capacity to semantically align features for each modality. Also, there is loss of fine-grained information due to limitation on length in social media text.

On the other hand, it is evident from the literature that knowledge infusion methods are advantageous to enhance context-specific understanding to enrich the unimodal representation for better performance. Recent research has expanded knowledge by topic wise contextual knowledge [37] and social profile information based knowledge [41] for tweet classification task to overcome the short text limitation. Later, researchers [5,43] expand tweet representation through Wikipedia to perform sentiment classification task. This motivates us to explore a novel direction in the field of multimodal crisis event classification.

Further, accuracy may not always a sufficient parameter to evaluate the models for critical applications where explainability is a necessary component. Recently, the eXplainable Artificial Intelligence (XAI) has become a prominent research area to understand the black box functionality of the deep neural network in various domains [2,36,38]. In a multimodal context, where the model's decision is influenced by the integration of two or more modalities, eXplainable Artificial Intelligence (XAI) plays a crucial role in diagnosing errors. This aids in enhancing model performance, tackling biases, and refining the entire system [20,25]. The explainability of deep learning models provides justification on predicted outcome of model by means of qualitative or quantitative measures, which can be important for decision-making in high-stakes situation such as disaster, pandemic or medical imaging etc. This shows the importance of explainability in multimodal crisis event classification to effectively explain feature vectors which is responsible in the model's outcome.

3   Methodology

The proposed architecture of CrisisKAN is intended to solve the classification problem for crisis events. This section outlines comprehensively the problem definition and sub-module of developed methodology.

3.1   Problem Definition

Given a training dataset $D = \{(I_i, T_i, y_i)\}_{i=1}^{N}$, where image I_i and text T_i represent a crisis event labeled with $y_i \in C$, the objective is to learn a function f such that $f : f(I_i, T_i) \rightarrow y_i$. Here, C is the set of crisis events. The aim is to enhance predictive accuracy by leveraging information from both modalities.

3.2   Model Overview

A brief overview of the proposed model, CrisisKAN is depicted in Fig. 2. The model takes both text and image content as input and provide a probability distribution of labels over different event classes along with an attention mask

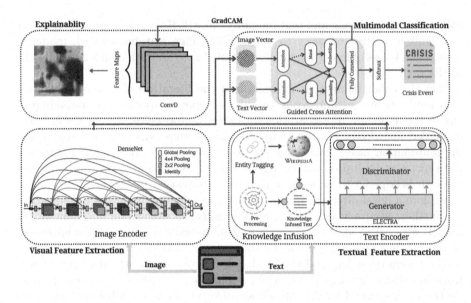

Fig. 2. The overall architecture of CrisisKAN.

for explainability. This framework comprises four key components: Visual feature extraction, Textual feature extraction, Guided Cross Attention Module, and Explainability. Description for each of these components is provided in the following sections.

Visual Feature Extraction. Convolutional Neural Networks (CNNs) is utilized to extract feature maps from images. We use DenseNet [21], as it reduces module sizes while enhancing connections between layers. This addresses parameter redundancy and leads to improved accuracy. Although alternative methods like Ghostnet [18], AlexNet [29] are viable, DenseNet is both efficient and a prevalent option for this purpose. For each input image I, our approach yields the following outcome:

$$\mathcal{Z}_I = \text{DenseNet}(I) \in \mathbb{R}^{c \times h \times w} \tag{1}$$

Here, c, h, and w are the number of channels, height, and width of the feature vector respectively.

Textual Feature Extraction. Given a textual input denoted as T, comprising a set of words $\{w_1, w_2, \cdots, w_n\}$ associated with event-related information, text feature extraction involves two steps. First, it infuses text with external knowledge, followed by feeding this enhanced text into a transformer-based model to generate the final text-based feature vector.

Knowledge Infusion. To address challenges arising from biases inherent in events of similar types and the constraints of limited text, we incorporate external knowledge from Wikipedia using a proposed wiki extraction algorithm. We extract primary entities associated with the event from the given text. Relevant entities from external knowledge are identified using a relatedness score γ_{rel} computed as follows for each word w_i within the text T.

$$\gamma_{rel} = Rel(w_i), \ w_i \in T \tag{2}$$

Function $Rel(\cdot)$ provides relevancy of w_i in Wikipedia. If γ_{rel} is greater than some threshold then it is regarded as pertinent to the specific event and considered as primary entity denoted by e. Implementation details of $Rel(\cdot)$ and thereshold is provided in Sect. 4.1. The assumption here is that text associated with the image contains the entities depicted in the image modality. Subsequent to the extraction of primary entities, we retrieve information about these entities from Wikipedia pages using $Wiki(\cdot)$ function and concatenate the gathered data. This process yields external information denoted as T^{wiki}.

$$T^{wiki} = Concat([e_1^{wiki}, \cdots, e_k^{wiki}]) \tag{3}$$

The detailed algorithm to generate T^{wiki} is presented in the Algorithm 1.

Algorithm 1. Entity2Wiki Knowledge Extraction Algorithm

Input: T, **Input Text**
Output: T^{wiki}, **Wikipedia Text**
1: **for each** $w_i \in T$ **do**
2:      Calculate γ_{rel} on w_i
3:      **if** $\gamma_{rel} > thres$ **then**
4:           $T^{wiki} = T^{wiki} \cdot Wiki(w_i)$
5:      **end if**
6: **end for**
7: Return T^{wiki}

Text Encoder. We utilized the ELECTRA [9] model to encode sequences of tokens in sentences. Although there are many transformer variants such as BERT [12], XLNET [46], etc. are available but ELECTRA is considered more parameter-efficient and faster to train than BERT due to its replaced token detection objective and generator-discriminator setup. Once we collect the domain information in the form of external knowledge T^{wiki} related to the event, we create final text for the input to the final input for ELECTRA by fusing original text T and external knowledge T^{wiki} with a special "[SEP]" tag. Subsequently, these tokens were tokenized and input into the ELECTRA model to derive the final layer embedding h^i in the following manner:

$$S = Concat([T, T^{wiki}]), \quad [h_S^0, h_S^1, \ldots, h_S^{N_s}] = \text{ELECTRA}(S), \mathcal{Z}_S = h_S^0 \quad (4)$$

Here $\mathcal{Z}_S \in \mathbb{R}^{n \times l \times d}$ and n, l, d are batch size, maximum sequence length, dimension of feature vector.

Guided Cross Attention-Based Fusion. We examined the image feature map $\mathcal{Z}_I$ and the text feature map $\mathcal{Z}_S$ derived from the encoding process in both modalities. Notable semantic inconsistencies are identified during training that impact the overall performance of the final model. For example, during tweet classification, a given modality might contain irrelevant or misleading content, potentially leading to unfavorable information exchange. We modified the cross attention module from [1] by incorporating self-attention as a preceding method. This provides guidance for the cross attention module. In other words, our model is meticulously structured to mitigate the influence of one modality on the other, employing guided self-attention as follows:

$$Attn(Q, K, V) = softmax \left(\frac{QK^T}{\sqrt{d}} \right) V \quad (5)$$

Here Q, K, and V are the query, key and value. $\sqrt{d}$ is a normalization factor that helps control the scale of the dot products and ensures more stable gradients during training. Now, new representations of $\mathcal{Z}_I$ and $\mathcal{Z}_S$ are achieved as follows:

$$\overline{\mathcal{Z}_I} = Attn(\mathcal{Z}_I), \quad \overline{\mathcal{Z}_S} = Attn(\mathcal{Z}_S) \quad (6)$$

Now, project new representation of image $\overline{\mathcal{Z}_I}$ and text $\overline{\mathcal{Z}_S}$ into a fixed dimensionality K in following manner:

$$\tilde{\mathcal{Z}}_I = F \left(W_I^T \overline{\mathcal{Z}_I} + b_I \right), \quad \tilde{\mathcal{Z}}_S = F \left(W_S^T \overline{\mathcal{Z}_S} + b_S \right) \quad (7)$$

where F represents an activation function such as ReLU and both $\tilde{\mathcal{Z}}_I$ and $\tilde{\mathcal{Z}}_S$ are of dimension K and K is fixed to 100.

Next to apply cross attention, attention masks $\alpha_{\mathcal{Z}_I}$ and $\alpha_{\mathcal{Z}_S}$ on $\overline{\mathcal{Z}_I}$ and $\overline{\mathcal{Z}_S}$ are calculate as follows:

$$\alpha_{\mathcal{Z}_I} = \sigma \left(W_I'^T \overline{\mathcal{Z}_S} + b_I' \right), \quad \alpha_{\mathcal{Z}_S} = \sigma \left(W_S'^T \overline{\mathcal{Z}_I} + b_S' \right) \quad (8)$$

where σ is the Sigmoid function. $\alpha_{\mathcal{Z}_I}$ for the image is completely dependent on the text embedding $\overline{\mathcal{Z}_S}$, while the attention mask $\alpha_{\mathcal{Z}_S}$ for the text is completely dependent on the image embedding $\overline{\mathcal{Z}_I}$. Once the attention masks $\alpha_{\mathcal{Z}_I}$ and $\alpha_{\mathcal{Z}_S}$ for the image and text are obtained, we can enhance the projected image and text embeddings $\tilde{\mathcal{Z}}_I$ and $\tilde{\mathcal{Z}}_S$ by applying element-wise multiplication with $\alpha_{\mathcal{Z}_I} \cdot \tilde{\mathcal{Z}}_I$ and $\alpha_{\mathcal{Z}_S} \cdot \tilde{\mathcal{Z}}_S$. The final stage of this module involves taking the combined embedding, which represents both the image and text pair, and passing it into fully-connected network and the classification is performed using the conventional softmax cross-entropy loss.

Explainability. The primary motivation for incorporating an explainability module is to analyze errors in predictions and establish a transparent system through visual explanations. We employ explainablity in our CrisisKAN by using a model-specific method, known as Grad-CAM [40]. It is calculated using gradient information flowing into last convolution layer of the model. To employ this, we have added a new convolution layer (ConvD) to identify and visualize importance of each modality feature vector for building decision. In order to obtain, the class-discriminative localization map for a particular crisis event class label y, the method first computes the gradient of the score G^y before the softmax with respect to k number of feature maps A^k. To obtain grad weights (α_k^y), these gradients are global average-pooled using in which (w, h) is the width and height dimensions and Z is normalizing factor. In summary, $\frac{\partial G^y}{\partial A_{ij}^k}$ is gradients value by backpropogation and $\sum_{i=1}^{w} \sum_{j=1}^{h}$ is global average pooling in Eq 9.

$$\alpha_k^y = \frac{1}{Z} \sum_{i=1}^{w} \sum_{j=1}^{h} \frac{\partial G^y}{\partial A_{ij}^k}, \quad I_{\text{Grad-CAM}}^y = ReLU\left(\sum_k \alpha_k^y \cdot A^k\right) \tag{9}$$

Finally, Grad-CAM for image modality $I_{\text{Grad-CAM}}^y$ is obtained by ReLU activation function over weighted sum of feature maps which is shown in Eq. 9. As a result, Grad-CAM provides a class-specific heatmap to visualize important region in the image modality as a visual explanations.

4   Experiments and Results

In order to evaluate the efficacy of CrisisKAN, extensive experiments are conducted using the CrisisMMD dataset [4]. This section reports dataset, experimental settings, results and ablation study.

4.1   Dataset and Settings

CrisisMMD [4] is widely recognised multimodal crisis dataset. This dataset was collected from the social media platform Twitter and contains images and textual information related to seven major crisis events including earthquakes, hurricanes, wildfires, and floods that happened in the year 2017. It is also categorized into three distinct tasks. Task 1 is designed to categorize whether a given image-text pair is informative and non-informative. On the contrary, Task 2 is concerned with categorising the impact of event in five classes, which include infrastructure damage, vehicle damage, rescue efforts, affected individuals (injury, dead, missing, found etc.), and others. Task 3 focuses on severity assessment, categorising as severe, mild, and little/no damage. We evaluate these tasks in two distinct settings similar to SOTA [1] method where Setting A considers only image-text pairs with identical labels, while Setting B incorporates all types of labeled image-text pairs for training, with test data matching that of Setting A. Table 1 shows the distribution of training, validation, and testing data for different tasks and settings.

Table 1. Dataset distribution with different split, tasks and settings

Setting	Task	Train	Val	Test	Total
A	Task 1	9601	1573	1534	12708
	Task 2	2874	477	451	3802
	Task 3	2461	529	530	3520
B	Task 1	13608	1573	1534	16715
	Task 2	8348	477	451	9276

4.2 Experimental Setup and Evaluation Metrics

We trained our model across tasks and settings with the following parameters: (i) base learning rate: 2×10^{-3}, (ii) decay: 10X, (iii) batch size: 64, (iv) optimizer: Adam with $\beta 1 = 0.9$, $\beta 2 = 0.999$, and $\epsilon = 1 \times 10^{-4}$, (v) loss function: categorical cross-entropy, (vi) epochs: 50. We have used $Tagme$ [13] API as $Rel(\cdot)$ function to identify the relatedness of entities in the input text. Also, we have varied $thres$ from 0 to 0.5 and considered 0.1 threshold to label a word as relevant entity. These models are executed on NVIDIA A100 with 40 GB GPU memory. Symbols such as '@', '#' and hyperlinks are removed as part of textual pre-processing. The model's performance is assessed using standard metrics such as classification accuracy, macro F1-score and weighted F1-score. These metrics show the performance of model with respect to each task.

Multi-task Model Strength (MTMS). The metrics defined above are assessed in relation to each task. Therefore, to evaluate the cumulative performance across all tasks, a new performance metric is introduced (Eq. 10) called as Multi-task Model Strength (MTMS) which shows the overall strength of model.

$$\text{MTMS} = \sum_{i=1}^{i=n} \beta_i \times Acc_i \qquad (10)$$

Here $\beta_i = \frac{c_i}{\sum_{i=1}^{i=n} c_i}$, Acc_i is accuracy and c_i is number of classes for task i. The MTMS ranges from 0 to 1, with a value of 0 signifying that accuracy is zero across all tasks. MTMS assigns more importance to tasks with a higher number of classes, as the complexity of classification task increases in correlation with the number of classes they entail.

4.3 Quantitative Results

We evaluate our proposed methodology against several state-of-the-art technaiques including unimodal based and multimodal based methods. We compare our approach with single modality networks DenseNet for image, language models BERT and ELECTRA across the all tasks. The second category includes existing image-text multimodal classification methods [1,6,14,22,26–28,32,44]. Some of the methods [1,32,44] emphasize on multimodal fusion based on global

Table 2. Comparative study on Setting A and B in terms of classification Accuracy (Acc)%, Macro F1-score (M-F1)%, Weighted F1-score (W-F1)% and Multi-task Model Strength (MTMS)%.

Method	Task 1			Task 2			Task 3			MTMS
	Acc	M-F1	W-F1	Acc	M-F1	W-F1	Acc	M-F1	W-F1	
Setting A										
Unimodal DenseNet [21]	81.6	79.1	81.2	83.4	60.5	87.0	62.9	52.3	66.1	76.9
Unimodal BERT [12]	84.9	81.2	83.3	86.1	66.8	87.8	68.2	45.0	61.1	80.5
Unimodal ELECTRA [9]	86.3	82.9	85.5	87.2	67.4	88.2	68.8	49.1	62.4	81.5
Cross-attention [1]	88.4	87.6	88.7	90.0	67.8	90.2	72.9	60.1	69.7	84.5
CentralNet [44]	87.8	85.3	86.1	89.3	64.7	89.8	71.1	57.4	68.7	83.5
GMU [6]	87.2	84.6	85.7	88.7	64.3	89.1	70.6	57.1	68.2	82.9
CBP [14]	87.9	85.6	86.4	90.2	66.1	89.8	65.8	60.4	69.3	82.4
CBGP [27]	88.1	86.7	87.3	84.7	65.1	88.7	67.9	50.7	64.6	80.3
MMBT [26]	86.4	85.3	86.2	88.7	64.9	89.6	70.1	59.2	68.7	82.7
VisualBERT [32]	88.1	86.7	88.6	87.5	64.7	86.1	66.3	56.7	62.1	81.3
PixelBERT [22]	88.7	86.4	87.1	89.1	66.5	88.9	65.2	57.3	63.7	81.8
ViLT [28]	87.6	85.1	88.0	86.7	61.2	87.2	67.6	58.4	65.0	81.2
CrisisKAN(Ours)	**91.7**	**90.3**	**91.2**	**93.6**	**70.3**	**93.4**	**73.1**	**63.1**	**72.2**	**87.1**
Setting B										
Unimodal DenseNet [21]	84.4	82.8	84.6	74.8	60.7	79.9	–	–	–	77.5
Unimodal BERT [12]	83.9	80.4	83.2	82.7	59.2	81.7	–	–	–	83.1
Unimodal ELECTRA [9]	84.6	81.1	83.8	83.4	60.1	83.5	–	–	–	83.8
Cross Attention [1]	85.6	82.3	84.8	89.3	63.4	89.8	–	–	–	88.2
CrisisKAN(ours)	**86.9**	**84.5**	**86.2**	**90.1**	**65.3**	**90.9**	–	–	–	**89.2**

features obtained by each modality backbone, while other studies like [14,27] employ compact bilinear pooling based fusion. The quantitative results for these baselines are presented in Table 2 for Setting A and B. We assess the efficacy of each baseline within the current dataset configuration, as shown in Table 1.

From Table 2, we can observe that all multimodal methods perform better than unimodal models, demonstrating the strength of multimodality learning. Moreover, compared with multimodal baselines, the accuracy of our CrisisKAN is significantly improved by approximately $3\% - 5\%$ in Task 1, $3\% - 7\%$ in Task 2 and $1\% - 8\%$ in Task 3 with Setting A. We also calculate the multi task model strength (MTMS) across three tasks and can interpret that CrisisKAN achieves high MSMT with the score of 87.1%.

In the Table 2, we also investigate our model with addition of noisy data in Setting B where image and text pair are labelled differently for same event. We find that CrisisKAN performs better compare to unimodal models and multimodal baseline both. Hence we can conclude from these results that our model is effectively learning the textual and visual features together.

4.4 Qualitative Results

Along with accuracy, we also present our model's capabilities in terms of explainablity. The Fig. 3 show the comparative study of visual explainations for baseline model [1] and ours model's prediction by highlighting the regions. We show the explainablity results across the three tasks where impact of features increase with the change of color gradient from blue to orange. The red box in the baseline model's results express those regions which is responsible for correct event classification and correctly learned by our proposed model. For task 1, our model's Grad-CAM map for informative class put more focus in the middle region and as by visual verification this result with the original image that regions are important to classify correctly in respective class. Similar way, the map for other tasks shows that effective and contributing features are extracted in the model leveraging guided cross attention. Finally, our visualizations on these examples take a step toward to build more trust on the proposed model.

4.5 Ablation Study

Leveraging Different Image and Text Encoders: To evaluate our model effectiveness, we leverage different image and text encoder such as CrisisKAN+ Ghostnet (**CG**), CrisisKAN+BERT (**CB**), CrisisKAN+ALBERT (**CA**), CrisisKAN+XLNet (**CX**), and CrisisKAN+ELECTRA (**CE**) (ours). This comparative analysis is conducted on Task 1 and Task 2 within Setting A, and the obtained results are depicted in Fig. 4. The graph distinctly illustrates that ELECTRA outperforms the other models, achieving an accuracy of 91.7% and 93.6% for Task 1 and Task 2, respectively. Additionally, it's noteworthy that XLNet and ALBERT yield comparable outcomes, each with an accuracy of 89.9% and 91.2% for Task 1, and 89.1% and 90.7% for Task 2. This study establishes that through the integration of diverse image and text encoders, CrisisKAN significantly surpasses existing multimodal methodologies with significant margin.

Comparison Among CrisisKAN Variants: Furthermore, we conduct a comparative analysis of the components within CrisisKAN (**MS4**) in relation to two specific aspects: the integration of knowledge and the utilization of guided attention mechanism. This study involves three distinct model settings: 1) without external knowledge and without guided attention (**MS1**), 2) without Wikipedia knowledge (**MS2**), and 3) without guided attention (**MS3**). The corresponding outcomes are presented in Fig. 5 which provide the key insight that the inclusion of guided attention leads to an approximate 1% increase over the performance of (**MS1**). Additionally, the introduction of external knowledge results in improvements of around 2% for both Task 1 and Task 2. Ultimately, when both guided attention and external knowledge are combined, ours (**MS4**) demonstrates a notable enhancement of 3–4% over (**MS1**), and 2–3% over both (**MS2**) and (**MS3**), for both tasks.

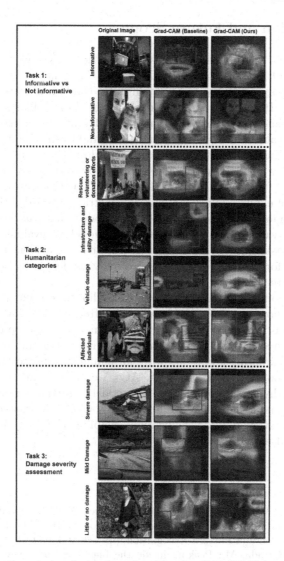

Fig. 3. Comparative study of visual explanation for CrisisKAN (ours) with baseline model [1] across various tasks in Setting A.

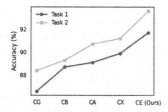

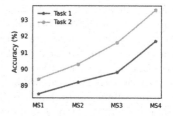

Fig. 4. Comparison on different image and text encoders.

Fig. 5. Comparison on different CrisisKAN Model Settings (MS).

5   Conclusion and Future Scope

In our research, we introduce CrisisKAN, a novel Knowledge-infused and Explainable Multimodal Attention Network. This network is specifically designed to classify crisis events by intelligently combining features from both images and text. In order to address the challenge of event biases, we have integrated external knowledge sourced from Wikipedia into our model. To facilitate better communication among diverse modalities and to sift through irrelevant or potentially misleading data, we propose a guided cross attention module. This module effectively narrows down the semantic gap between modalities, enabling us to selectively fuse only the valuable information. Our research also delves into the realm of explainability. We have taken strides towards bridging the gap between the enigmatic "black box" models to models that can be explained and interpreted. To evaluate the effectiveness of our model, we evaluate our model on different crisis tasks and settings. Form the results, it is evident that our model not only outperforms many image and text only models but also other SOTA multimodal approaches.

While our proposed CrisisKAN model has a limitation due to the availability of external knowledge from Wikipedia only after an event has occurred, this drawback can be mitigated by extending the model to incorporate real-time information from social media during the event itself. For the future direction, we are actively exploring additional ways for enhancing the performance of our image and external knowledge module. Also, we are committed to seamlessly integrating lengthy textual inputs into our model to further strengthen it capabilities. Furthermore, the explainability module will be expanded through the integration of text modality visualization, enabling the analysis of effects or errors associated with each modality.

References

1. Abavisani, M., Wu, L., Hu, S., Tetreault, J., Jaimes, A.: Multimodal categorization of crisis events in social media. In: Proceedings of the IEEE/CVF Conference on Computer Vision and Pattern Recognition, pp. 14679–14689 (2020)
2. Adadi, A., Berrada, M.: Peeking inside the black-box: a survey on explainable artificial intelligence (XAI). IEEE Access **6**, 52138–52160 (2018)
3. Agarwal, M., Leekha, M., Sawhney, R., Ratn Shah, R., Kumar Yadav, R., Kumar Vishwakarma, D.: MEMIS: multimodal emergency management information system. In: Jose, J.M., et al. (eds.) ECIR 2020. LNCS, vol. 12035, pp. 479–494. Springer, Cham (2020). https://doi.org/10.1007/978-3-030-45439-5_32
4. Alam, F., Ofli, F., Imran, M.: CrisisMMD: multimodal twitter datasets from natural disasters. In: Proceedings of the International AAAI Conference on Web and Social Media. vol. 12 (2018)
5. Anonymous: EA2n: Evidence-based AMR attention network for fake news detection. In: Submitted to The Twelfth International Conference on Learning Representations (2023). https://openreview.net/forum?id=5rrYpa2vts, under review
6. Arevalo, J., Solorio, T., Montes-y Gómez, M., González, F.A.: Gated multimodal units for information fusion (2017). arXiv preprint arXiv:1702.01992

7. Bandyopadhyay, D., Kumari, G., Ekbal, A., Pal, S., Chatterjee, A., BN, V.: A knowledge infusion based multitasking system for sarcasm detection in meme. In: Kamps, J., et al. Advances in Information Retrieval. ECIR 2023. LNCS, vol. 13980. Springer, Cham (2023). https://doi.org/10.1007/978-3-031-28244-7_7
8. Chu, S.Y., Lee, M.S.: MT-DETR: robust end-to-end multimodal detection with confidence fusion. In: 2023 IEEE/CVF Winter Conference on Applications of Computer Vision (WACV), pp. 5241–5250 (2023). https://doi.org/10.1109/WACV56688.2023.00522
9. Clark, K., Luong, M., Le, Q.V., Manning, C.D.: ELECTRA: pre-training text encoders as discriminators rather than generators. In: ICLR. OpenReview.net (2020)
10. Dai, Z., Yang, Z., Yang, Y., Carbonell, J., Le, Q.V., Salakhutdinov, R.: Transformer-XL: Attentive language models beyond a fixed-length context (2019). arXiv preprint arXiv:1901.02860
11. Devlin, J., Chang, M.W., Lee, K., Toutanova, K.: BERT: Pre-training of deep bidirectional transformers for language understanding (2018). arXiv preprint arXiv:1810.04805
12. Devlin, J., Chang, M.W., Lee, K., Toutanova, K.: BERT: Pre-training of deep bidirectional transformers for language understanding. In: NAACL Volume 1 (Long and Short Papers), pp. 4171–4186. ACL, Minneapolis, Minnesota (2019)
13. Ferragina, P., Scaiella, U.: TAGME: on-the-fly annotation of short text fragments (by wikipedia entities). In: ICIKM, pp. 1625–1628. CIKM 2010, ACM, New York, NY, USA (2010)
14. Fukui, A., Park, D.H., Yang, D., Rohrbach, A., Darrell, T., Rohrbach, M.: Multi-modal compact bilinear pooling for visual question answering and visual grounding (2016). arXiv preprint arXiv:1606.01847
15. Gallo, I., Ria, G., Landro, N., La Grassa, R.: Image and text fusion for UPMC food-101 using BERT and CNNs. In: 2020 35th International Conference on Image and Vision Computing New Zealand (IVCNZ), pp. 1–6. IEEE (2020)
16. Gupta, S., Kundu, S.: Interaction graph, topical communities, and efficient local event detection from social streams. Expert Syst. Appl. **232**, 120890 (2023)
17. Gupta, S., Yadav, N., Sainath Reddy, S., Kundu, S.: FakEDAMR: Fake news detection using abstract meaning representation (2023)
18. Han, K., Wang, Y., Tian, Q., Guo, J., Xu, C., Xu, C.: GhostNet: more features from cheap operations. In: 2020 IEEE/CVF Conference on Computer Vision and Pattern Recognition (CVPR), pp. 1577–1586 (2020)
19. He, K., Gkioxari, G., Dollár, P., Girshick, R.: Mask R-CNN. In: Proceedings of the IEEE International Conference on Computer Vision, pp. 2961–2969 (2017)
20. Holzinger, A., Malle, B., Saranti, A., Pfeifer, B.: Towards multi-modal causability with graph neural networks enabling information fusion for explainable AI. Inf. Fusion **71**, 28–37 (2021)
21. Huang, G., Liu, Z., Van Der Maaten, L., Weinberger, K.Q.: Densely connected convolutional networks. In: 2017 IEEE Conference on Computer Vision and Pattern Recognition (CVPR), pp. 2261–2269 (2017)
22. Huang, Z., Zeng, Z., Liu, B., Fu, D., Fu, J.: Pixel-BERT: Aligning image pixels with text by deep multi-modal transformers (2020). arXiv preprint arXiv:2004.00849
23. Hubenthal, M., Kumar, S.: Image-text pre-training for logo recognition. In: 2023 IEEE/CVF Winter Conference on Applications of Computer Vision (WACV), pp. 1145–1154 (2023). https://doi.org/10.1109/WACV56688.2023.00120

24. Hunt, K., Wang, B., Zhuang, J.: Misinformation debunking and cross-platform information sharing through Twitter during Hurricanes Harvey and Irma: a case study on shelters and ID checks. Nat. Hazards **103**(1), 861–883 (2020). https://doi.org/10.1007/s11069-020-04016-6

25. Joshi, G., Walambe, R., Kotecha, K.: A review on explainability in multimodal deep neural nets. IEEE Access **9**, 59800–59821 (2021). https://doi.org/10.1109/ACCESS.2021.3070212

26. Kiela, D., Bhooshan, S., Firooz, H., Perez, E., Testuggine, D.: Supervised multimodal bitransformers for classifying images and text (2019). arXiv preprint arXiv:1909.02950

27. Kiela, D., Grave, E., Joulin, A., Mikolov, T.: Efficient large-scale multi-modal classification. In: Proceedings of the AAAI Conference on Artificial Intelligence. vol. 32 (2018)

28. Kim, W., Son, B., Kim, I.: ViLT: vision-and-language transformer without convolution or region supervision. In: International Conference on Machine Learning, pp. 5583–5594. PMLR (2021)

29. Krizhevsky, A.: One weird trick for parallelizing convolutional neural networks (2014). CoRR abs/1404.5997

30. Kwan, J.S.L., Lim, K.H.: Understanding public sentiments, opinions and topics about COVID-19 using twitter. In: 2020 IEEE/ACM International Conference on Advances in Social Networks Analysis and Mining (ASONAM), pp. 623–626. IEEE (2020)

31. Li, G., Duan, N., Fang, Y., Gong, M., Jiang, D.: Unicoder-VL: a universal encoder for vision and language by cross-modal pre-training. In: Proceedings of the AAAI Conference on Artificial Intelligence. vol. 34, pp. 11336–11344 (2020)

32. Li, L.H., Yatskar, M., Yin, D., Hsieh, C.J., Chang, K.W.: VisualBERT: A simple and performant baseline for vision and language (2019). arXiv preprint arXiv:1908.03557

33. Liang, T., Lin, G., Wan, M., Li, T., Ma, G., Lv, F.: Expanding large pre-trained unimodal models with multimodal information injection for image-text multimodal classification. In: 2022 IEEE/CVF Conference on Computer Vision and Pattern Recognition (CVPR), pp. 15471–15480 (2022). https://doi.org/10.1109/CVPR52688.2022.01505

34. Long, S., Han, S.C., Wan, X., Poon, J.: GraDual: graph-based dual-modal representation for image-text matching. In: 2022 IEEE/CVF Winter Conference on Applications of Computer Vision (WACV), pp. 2463–2472 (2022). https://doi.org/10.1109/WACV51458.2022.00252

35. Mao, X., et al.: Towards robust vision transformer. In: Proceedings of the IEEE/CVF Conference on Computer Vision and Pattern Recognition, pp. 12042–12051 (2022)

36. Moraliyage, H., Sumanasena, V., De Silva, D., Nawaratne, R., Sun, L., Alahakoon, D.: Multimodal classification of onion services for proactive cyber threat intelligence using explainable deep learning. IEEE Access **10**, 56044–56056 (2022)

37. Nazura, J., Muralidhara, B.L.: Semantic classification of tweets: a contextual knowledge based approach for tweet classification. In: 2017 8th International Conference on Information, Intelligence, Systems & Applications (IISA), pp. 1–6 (2017). https://doi.org/10.1109/IISA.2017.8316358

38. Petsiuk, V., et al.: Black-box explanation of object detectors via saliency maps. In: Proceedings of the IEEE/CVF Conference on Computer Vision and Pattern Recognition, pp. 11443–11452 (2021)

39. Sakaki, T., Okazaki, M., Matsuo, Y.: Earthquake shakes twitter users: real-time event detection by social sensors. In: Proceedings of the 19th International Conference on World Wide Web, pp. 851–860 (2010)
40. Selvaraju, R.R., Cogswell, M., Das, A., Vedantam, R., Parikh, D., Batra, D.: Grad-CAM: visual explanations from deep networks via gradient-based localization. In: 2017 IEEE International Conference on Computer Vision (ICCV), pp. 618–626 (2017). https://doi.org/10.1109/ICCV.2017.74
41. Shu, K., Zhou, X., Wang, S., Zafarani, R., Liu, H.: The role of user profiles for fake news detection. In: IEEE/ACM ASONAM, pp. 436–439. ASONAM 2019, Association for Computing Machinery, New York, NY, USA (2020)
42. Singh, A., et al.: FLAVA: a foundational language and vision alignment model. In: Proceedings of the IEEE/CVF Conference on Computer Vision and Pattern Recognition, pp. 15638–15650 (2022)
43. Tahayna, B., Ayyasamy, R., Akbar, R.: Context-aware sentiment analysis using tweet expansion method. J. ICT Res. Appl. **16**(2), 138–151 (2022)
44. Vielzeuf, V., Lechervy, A., Pateux, S., Jurie, F.: CentralNet: a multilayer approach for multimodal fusion. In: Proceedings of the European Conference on Computer Vision (ECCV) Workshops (2018)
45. Wang, Y., Xu, X., Yu, W., Xu, R., Cao, Z., Shen, H.T.: Combine early and late fusion together: a hybrid fusion framework for image-text matching. In: 2021 IEEE International Conference on Multimedia and Expo (ICME), pp. 1–6. IEEE (2021)
46. Yang, Z., Dai, Z., Yang, Y., Carbonell, J., Salakhutdinov, R., Le, Q.V.: XLNet: Generalized Autoregressive Pretraining for Language Understanding. Curran Associates Inc., Red Hook, NY, USA (2019)
47. Zadeh, A., Chen, M., Poria, S., Cambria, E., Morency, L.P.: Tensor fusion network for multimodal sentiment analysis (2017). arXiv preprint arXiv:1707.07250

Utilizing Low-Dimensional Molecular Embeddings for Rapid Chemical Similarity Search

Kathryn E. Kirchoff[1], James Wellnitz[2], Joshua E. Hochuli[2], Travis Maxfield[2], Konstantin I. Popov[2], Shawn Gomez[3,4(✉)], and Alexander Tropsha[2(✉)]

[1] Department of Computer Science, UNC Chapel Hill, Chapel Hill, USA
kat@cs.unc.edu
[2] Eshelman School of Pharmacy, UNC Chapel Hill, Chapel Hill, USA
{jwellnitz,joshua_hochuli,tmaxfield,kpopov,alex_tropsha}@unc.edu
[3] Department of Pharmacology, UNC Chapel Hill, Chapel Hill, USA
smgomez@unc.edu
[4] Joint Department of Biomedical Engineering at UNC Chapel Hill and NCSU, Chapel Hill, USA

Abstract. Nearest neighbor-based similarity searching is a common task in chemistry, with notable use cases in drug discovery. Yet, some of the most commonly used approaches for this task still leverage a brute-force approach. In practice this can be computationally costly and overly time-consuming, due in part to the sheer size of modern chemical databases. Previous computational advancements for this task have generally relied on improvements to hardware or dataset-specific tricks that lack generalizability. Approaches that leverage lower-complexity searching algorithms remain relatively underexplored. However, many of these algorithms are approximate solutions and/or struggle with typical high-dimensional chemical embeddings. Here we evaluate whether a combination of low-dimensional chemical embeddings and a k-d tree data structure can achieve fast nearest neighbor queries while maintaining performance on standard chemical similarity search benchmarks. We examine different dimensionality reductions of standard chemical embeddings as well as a learned, structurally-aware embedding—SmallSA—for this task. With this framework, searches on over one billion chemicals execute in less than a second on a single CPU core, five orders of magnitude faster than the brute-force approach. We also demonstrate that SmallSA achieves competitive performance on chemical similarity benchmarks.

Keywords: Cheminformatics · Virtual screening · Drug discovery

1 Introduction

Searching a database for the nearest neighbors to a query is a task that spans many fields, from computer graphics [38,45] to medical diagnostics [18] and

---

K. E. Kirchoff, J. Wellnitz, J. E. Hochuli, T. Maxfield—Authors contributed equally.

© The Author(s), under exclusive license to Springer Nature Switzerland AG 2024
N. Goharian et al. (Eds.): ECIR 2024, LNCS 14609, pp. 34–49, 2024.
https://doi.org/10.1007/978-3-031-56060-6_3

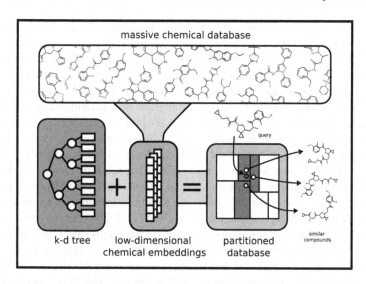

Fig. 1. Overview of the similarity search framework. k-d trees are combined with with low-dimensional chemical embeddings to produce a partitioned chemical space, which can be quickly queried for nearest neighbors.

chemistry [32]. In chemistry, the identification and retrieval of similar compounds plays a vital role in various domains, including drug discovery. Nearest neighbor search algorithms have been widely employed for this purpose, allowing researchers to explore large chemical databases to help discover potential drug candidates. When applied to this task, nearest neighbor searching, or chemical similarity searching, is often called virtual screening, where databases are searched for chemicals similar to ones with known desirable properties, assuming that similar chemicals share similar properties [13]. This approach is widely adopted and has been used successfully to discover potent drug candidates [16]. Currently, most chemical similarity searching methods, like the popular Arthor [2], use a simple, brute-force algorithm to compare all queries to all database chemicals.

However, the sizes of searchable chemical databases have recently undergone dramatic expansion. For example, the latest release of the ZINC database [42] and Enamine's catalog [17] now number close to 40 billion chemicals each. Brute-force approaches to nearest neighbor searching struggle to scale to databases of this size. Often, hundreds of queries need to be searched against the database, resulting in trillions of calculations. As a result, there is a growing need for faster search algorithms that can handle the ever-increasing data volume.

Indeed, there exist algorithms that effect faster runtimes for brute-force searching, one being the k-dimensional tree (k-d tree), which achieves logarithmic searching complexity [10]. To date, k-d trees have been used to achieve algorithmic speed-ups in numerous applications. However, this data structure has yet to be effectively used for the cheminformatics tasks of large-scale chemical similarity search and virtual screening. This lack of use could stem from two issues that arise when utilizing k-d trees for these purposes. First, chemical databases contain tens of billions of entries, and a k-d tree built on such a database of such size cannot exist purely in memory on most systems, thus requiring implementations designed for low-memory usage. Second, k-d trees are only effective at searching low-dimensional data and provide near-zero benefit as the data dimensionality exceeds about 20 [48], a problem since nearly all common chemical representations exceed 20 dimensions.

In fact, typical representations of chemicals have dimensionalities in the thousands [52], which are nowhere near compatible with the k-d tree approach. There is precedent for reducing high-dimensional objects to lower dimensions while preserving relative distance [20], leveraging techniques such as random projection [5] or Principal Components Analysis (PCA) [21]; however, to our knowledge dimensionality reduction techniques have not yet been applied in conjunction with k-d trees to allow for faster chemical similarity searching. Specific to chemistry, chemical representation learning has emerged as a powerful technique to organize chemicals in a meaningful way at lower dimensions [51]. However, previous implementations of these techniques still do not reduce dimensionality low enough, nor are these approaches always trained to organize the embedding space in a meaningful way for the specific task of chemical similarity search.

In this work, we propose an effective framework for chemical similarity searching: combining a meaningful low-dimensional chemical embedding with a custom k-d tree implementation designed to index billion-sized chemical data sets with low memory constraints. Further, we show that utilizing representation learning to generate Small Structurally-Aware embeddings (SmallSA) provides better performance on similarity searching benchmarks compared to dimensional reduction methods applied to existing high-dimensional embeddings.

To our knowledge, this is the first publicly disseminated method that leverages a k-d tree with low-dimensional chemical embeddings for the task of rapid chemical similarity searching with applications to drug discovery.

2   Background

2.1   Nearest Neighbor Searching Algorithms

Nearest neighbor search has been the subject of many algorithmic improvements aimed at lowering complexity. For our purposes, we will distinguish two main classes of such improvements: 1) reducing searchable space and 2) dimensionality reduction techniques.

Reducing Searchable Space

Indexing Methods. Given a search database, indexing methods, like the Ball-Tree [33] and k-dimensional tree (k-d tree) [10], partition it in such a way that individual queries need not be compared to every element in the database. This effectively reduces the search space, leading to at best $O(q \log n)$ complexity, where q is the number of queries and n is the size of the space. However, this best-case speed-up only occurs for sufficiently small data dimension d, [43] collapsing to the brute-force complexity of $O(qn)$ beyond about $d = 20$ [48].

Approximate Nearest Neighbor Search. Indexing methods are exact, but if absolute correctness is not required there are also fast approximate nearest neighbor methods [46]. These methods often involve the use of hashing, as in the case of locality sensitive hashing (LSH) forests [7] and spectral hashing [50], or the use of clustering [14] to group together similar portions of the database. Like the indexing methods described above, these approximate search methods can be described as reducing the effective search database size, n, resulting in sublinear search times. However, the accuracy of these methods is dependent on the quality of the grouping, whether hashing clustering or otherwise [46]. Though fast algorithms exist, effective clustering methods can also have difficulty scaling to massive data sets [29], bringing into question their utility on billion-sized chemical datasets.

Dimensionality Reduction. Dimensionality reduction methods, like PCA [21], t-SNE [44] or UMAP [30], can also be applied to increase search speed by reducing the number of dimensions, d. Accounting for dimensionality, the complexity of brute-force search can be described as $O(qdn)$, and thus a smaller d can increase efficiency. Further, if d is reduced to a small enough dimensionality, it can be applied in tandem with indexing methods, yielding sublinear complexity [6,23,25]. Like the approximate methods described above, accuracy is tied to the ability of the dimensionality reduction to preserve the original spatial relationships between data points, and not all dimensionality reduction methods can scale to billion-sized data sets.

2.2   Chemical Similarity

In cheminformatics, "chemical similarity searching" amounts to searching a database for the nearest n neighbors to a given query molecule. "Similarity" in chemical space is not a well-defined concept, and there is no agreed-upon definition as to what constitutes a pair of similar chemicals [28]. For example, a chemist can claim two chemicals are similar due to similar structure, while another can claim the pair is dissimilar due to different properties. With this in mind, and to establish a baseline for comparison, we choose a broadly applicable definition of chemical similarity: two chemicals are similar if their structures are. In this case, graph edit distance (GED) can be used as the ground truth metric for chemical similarity.

The graph edit distance (GED) [37] between two graphs is the smallest total number of edits to one graph that are needed to make it isomorphic to the second graph. GED is often approximated, as calculating the actual value is

NP-hard [53]. Since a chemical structure can be represented as a graph of atoms (nodes) and bonds (edges), the distance between any two chemicals can be measured with approximate GED [11].

2.3 Virtual Screening

Chemical similarity searching is often used in early stages of drug discovery, in a process called "virtual screening" [26]. This practice is based on the idea that chemicals with similar structures likely share similar biological activity profiles, a concept known as structure-activity relationship (SAR) [13]. Given a single chemical with a desirable biological activity, similarity search can be used to find similar chemicals that are expected to share that activity profile. Generally, virtual screening is carried out with hundreds of query molecules against a database of billions. These large chemical databases are projected to get larger, increasing into the trillions [27].

2.4 Related Works

There are some existing chemical similarity search methods designed to function on large chemical databases, notably Arthor [2,39] and SmallWorld [4], both developed by NextMove Software, and SpaceLight [8]. Arthor uses a brute-force search, but is accelerated through a highly optimized, domain-specific implementation [40] that, while fast, is still linear in complexity. Further, it utilizes a hashing technique that is vulnerable to collisions which can hinder the accuracy. SmallWorld's approach is based on GEDs between chemicals, generating a graph of graphs representing the database, where chemical graphs are connected by an edge if they are 1 GED apart; given a graph in the database, the exact GED to nearby graphs can then be quickly calculated. However, this method requires expensive, high-end hardware to run, and can only be used in cases where the search query is in the database since it relies on an extremely expensive pre-calculation of the database. Spacelight uses an approximate clustering approach that relies on the search database being composed of a smaller number of chemical building blocks combined via known chemical reactions, allowing large parts of the database being searched to be skipped. Not all chemical databases satisfy this constraint and, even among those that do, the required decomposition is often proprietary, limiting the utility. Furthermore, all three methods described here—SmallWorld, Arthor, and Spacelight—are commercial and not open source.

Among open source tools for chemical similarity searching, there are few options. ChemMine [12] is a web tool that implements several similarity search approaches. The most popular, FPSim2 [15], uses the same Baldi bounds [40] approach as Arthor to partition the database and accelerate brute-force search. This partitioning scheme can only generate a maximum number of partitions equal to the dimensionality of the embedding. Additionally the size of the partitions are unequal, with the largest partitions also likely to be those that must be searched for an average query. As a results, this approach is both limited to binary embeddings and suffers from longer run times on large chemical databases. There are also open source libraries for fast approximate neighbor search of high dimensional data, like FAISS (Facebook AI Similarity Search) [19] or Annoy (Approximate Nearest Neighbors Oh Yeah) [1]. However, these implementations struggle on billion-sized databases and require significant modification to accommodate current large chemical databases on minimal hardware. These libraries often rely on some form of dimensionality reduction paired with a partitioning data structure, making their approach similar to the approach implemented in this work.

3 Methods

3.1 Overview of Approach

To efficiently search a large database, we encode chemicals into a low-dimensional embeddings which we then use to construct a k-d tree. Database embeddings are calculated once at tree construction time, while query embeddings are calculated on the fly. Query embeddings are then used to search the k-d tree to find the k nearest neighbors. This process is summarized in Fig. 1.

3.2 SmallSA: Learned Embedding

To generate the learned SmallSA embeddings, we opt to utilize the SALSA framework presented in [24], which is trained to map chemical compounds to an embedding space that respects GED. We modify the SALSA architecture to our specific objective, training a model that compresses the chemical space to dimensionalities below 20, specifically to 16 and 8 dimensions. As a further deviation from the original SALSA, we train our model on a diverse sample of 1.4 million chemical compounds from the Enamine chemical catalog [17]. Further details on our implementation of the SALSA approach can be found in appendix A.

3.3 Other Embeddings

There are several commonly utilized approaches for embedding chemicals. The most common embeddings are Extended-Connectivity Fingerprints (ECFPs) and Extended-Connectivity Count Vectors (ECFCs) [36]. Both embeddings are topological fingerprints indicating presence of molecular substructures; however, ECFPs are *binary* indications of substructures, while ECFCs indicate *counts* of substructures. For both fingerprinting methods, we opted for a radius of two and a resulting 256d vector. Herein, we refer to these versions as "ECFP-256" and "ECFC-256". Furthermore, to calculate embedding distances, we used the Jaccard index (i.e., Tanimoto index in cheminformatics) [41] for ECFP-256 and Euclidean distance for ECFC-256.

Additionally, we obtained low-dimensional versions of either fingerprint through two dimensionality reduction methods: (1) sparse random projections

(SparseRP) and (2) PCA. We computed PCA at both eight and 16 dimensions ("PCA-8" and "PCA-16"), and computed SparseRP at only 16 dimensions ("SparseRP-16"). Other common dimensionality reduction methods, including t-Distributed Stochastic Neighbor Embedding (t-SNE) [44] and Uniform Manifold Approximation and Projection (UMAP) [30], were initially considered but left out due to incompatibility with large datasets.

3.4   *k*-d Tree Implementation

As available open-source k-d tree implementations are not designed to handle the scale of large chemical databases, we implement a custom k-d tree based on a proposal from [9] designed to handle large datasets. Our resulting k-d tree is able to organize *terabytes* of chemical data, but requires only *tens of gigabytes* of memory to search. Implementation details and source code are provided at the GitHub link at the end of the document.

3.5   Large Benchmarking Dataset

To benchmark the performance of our proposed framework for chemical similarity searching on billion-sized chemical databases. A database of 1.3 billion chemicals randomly sampled from the 33.5 billion Enamine REAL Space library [17] was prepared. Additionally, we sampled 100 random compounds from the REAL space to serve as the query set. There is no overlap between the training set used for SmallSA, mentioned previously, and the database or query set.

4   Experiments and Analysis

To assess the effectiveness of our proposed framework for chemical similarity search—utilizing k-d trees with low-dimensional embeddings—we compare performance of low (SmallSA-8, SmallSA-16, PCA-8, PCA-16, SparseRP-16) and high (ECFP-256, ECFC-256) dimensional embeddings on three tasks: 1) GED of retrieved queries to their retrieved hits, 2) performance on the RDKit Virtual Screening benchmark, and 3) computational speed (and hit quality) with respect to computing resources. Results of these experiments are summarized in Table 1.

4.1   Query–Hit Graph Edit Distance

To assess the quality of retrieved compounds, we evaluate the GEDs of the top k retrieved compounds (k nearest neighbors) per query, positing that high quality compounds will be those of low GED from the query. For each method, we query the same set of 100 randomly sampled compounds against the 1.3 billion dataset, and for each query, determine its 100 nearest neighbors. GED was approximated (due to expensive exact computation) using a bipartite graph matching approach [34] with the cost for all edge and node edits set to one. We plot the resulting GEDs, per k neighbors, for each method in Fig. 2.

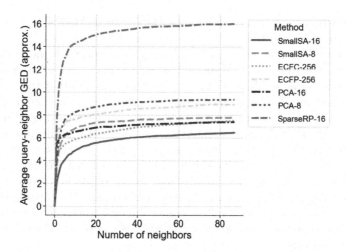

Fig. 2. The average approximate graph edit distance (GED) between query molecule and nearest neighbors, per method, shown as a function of the number of neighbors considered. Lower distances are better. Lines are smoothed using a running average approach for simplicity of analysis.

Results indicate that, for relatively low k, six out of seven embedding techniques are comparably successful at retrieving high quality neighbors. The exception is SparseRP-16, which generally performed worse, obtaining neighbors of higher GED compared to all other methods. Overall, the structurally aware learned embedding, in this case SmallSA-16, performed the best out of all embeddings. Further, SmallSA-8 performed better than PCA-8, and comparably to the best high-dimensional method, ECFC-256.

4.2   RDKit Virtual Screening Benchmark

To assess performance at the virtual screening task, we implement the RDKit Virtual Screening benchmark proposed by Riniker and Landrum [35]. Briefly, this platform works by determining the performance of each similarity searching method in "simulated" virtual screening against 69 proteins (targets) that are targeted by chemicals. Each simulated screening starts with a set of chemicals with known activity against the target ("actives") and a set of chemicals with no activity against the target ("decoys"). A random subset of actives and decoys are used as the database to search against, while the remainder of actives are used as the query.

Each compound in the database is ranked according to its closest distance to any of the actives, and this rank is used as a simple nearest neighbor classification model. The area under the receiver operating curve (AUROC) for this model can be calculated using the known activity class of the compounds in the search database. This is related to how similarity search is used in practice for drug discovery with the key difference being that the true activity class is not known *a priori*, as it is in this benchmark. We plot the resulting AUROCs, per database, for each method in Fig. 3.

The results show that SmallSA-based methods (SmallSA-8 and SmallSA-16) significantly outperform both the high-dimensional and low-dimensional embeddings. Considered alongside the GED results, this indicates that there exist low-dimensional chemical embeddings capable of performing at least equivalently on the virtual screening task compared to the more traditional high-dimensional embeddings, like ECFC-256. It should be noted that these results are not meant to imply that SmallSA-based methods are state-of-the-art among learned representations; indeed, demonstrating this would require a much broader comparison with other learned representations. Instead, these results are meant to simply demonstrate that low-dimensional chemical embeddings are capable of performance on par with some commonplace higher-dimensional embeddings, implying that similarity search with these small embeddings can be useful for downstream drug discovery tasks.

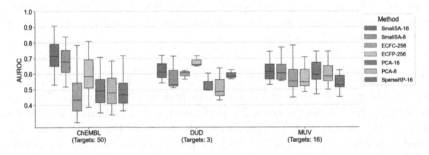

Fig. 3. AUROC achieved by each embedding on the RDKit virtual screening benchmark of 69 query targets, grouped by target database. Each database is indicated on the x-axis. Note that out of the 69 targets, most targets (50) belong to the ChEMBL database.

4.3   Computational Speed

A major goal of our proposed framework is *rapid* determination of nearest neighbors utilizing minimal computing resources. To compare speed performance to other methods, we collected timings for retrieval of the top 100 neighbors for 100 queries on the 1.3 billion database. Queries using low dimensional embeddings were executed using both the k-d tree framework and brute force approach, but only brute force for high-dimensional embeddings.

Table 1. GED, virtual screening, and timing results. Best is **bold**, second best is underlined. Runtime is proportional to the dimensionality, not method of dimensionality reduction, therefore methods with the same dimension have identical query times.

| | | Query accuracy | | Query time (s) | |
Method	D	VS AUC	GED	Brute-force	k-d tree
ECFC-256	256	0.494	<u>6.69</u>	$24,900 \pm 2,200$	N/A
ECFP-256		0.566	8.30	$13,600 \pm 1,000$	
SparseRP-16	16	0.501	15.06	<u>$3,000 \pm 70$</u>	<u>37.5 ± 26</u>
PCA-16		0.524	6.97		
SmallSA-16		**0.692**	**5.74**		
PCA-8	8	0.517	8.78	**$11,400 \pm 98$**	**0.606 ± 0.35**
SmallSA-8		<u>0.662</u>	7.30*		

Nearest neighbor queries were executed on a workstation with three consumer-grade solid states drives in a RAID 5 array and one core of an AMD Threadripper PRO 3955WX. Due to high computational cost, brute force for high-dimensional embeddings were carried out on the UNC Longleaf computing cluster. Timings are reported for single-core performance in Table 1.

Results indicate that combination of low-dimensional embeddings with k-d trees achieves substantially reduced compute time—speed-ups between 10^3 (16-dimensional) to 10^5 (8-dimensional) fold—compared to the combination of high dimensional embeddings with a brute-force approach. Practically, this results in reduced query times, from several hours down to under a minute with no change in hardware. Further, upon visual inspection, we find that this substantial speedup does not coincide with a reduction in perceived hit quality (see example in Fig. 4).

We are unable to compare these timings and benchmarks rigorously to the commercial releases of Arthor, SmallWorld and SpaceLight due to lack of access; however all methods report run-times ranging between several seconds to several minutes, depending on the query [3, 8].

5   Discussion

In this work, we showcase the feasibility of achieving rapid speed ups in similarity search, through a novel application combining low-dimensional embeddings and an efficient, low-memory k-d tree implementation. Further, we show that utilizing a learned, structurally-aware embedding approach (SmallSA) effects a low-dimensional embedding that excels at virtual screening tasks, demonstrating marked improvements over both traditionally high-dimensional fingerprints as well as their dimension-reduced counterparts.

Query

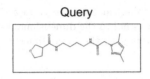

Nearest Neighbors

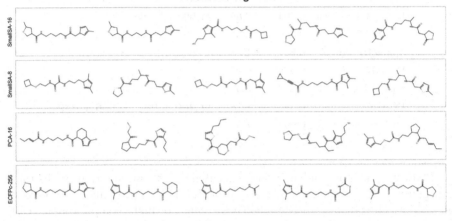

Fig. 4. Example of a query molecule and the hits obtained by select high-performing embeddings.

Alongside our work, others have explored low-dimensional [22, 31] and structure-aware [24, 47] chemical representations for statistical modeling and faster, brute-force similarity searching. There is also extensive literature on methods for sublinear nearest neighbor searching. Ours is the first work, to our knowledge, to show that combining low-dimensional embeddings and k-d trees can achieve sublinear chemical similarity searches on billion-sized datasets while still maintaining sensible results. We do not contend that the combination specifically of SmallSA and k-d trees is the most optimal setup, as we did not investigate other learned, low-dimensional embeddings or other partition-tree based similarity searching algorithms. Instead we point out that leveraging a meaningful, chemically aware, low-dimensional embedding with a searching algorithm that excels with low-dimensional data can provide major speed ups in exact chemical similarity searching.

5.1   Requirement for Exactness

Many of the open-source libraries for similarity searching of large databases leverage approximate searching algorithms. This means they can fail to return the true nearest neighbor. This is a concern for chemists, as often times the goal of a chemical similarity search is to find a small number of compounds most similar to a query for purchase and experimental assessment. Thus, assuming a good structurally-aligned representation of chemicals, a similarity searching tool should be designed to prioritize exactness. Leveraging an exact implementation from available open source methods for high-dimensional similarity searching often results in either brute force searches (e.g. 'IndexFlatL2' in FAISS [19]) or dramatically slower run time. Instead, we chose to implement a k-d tree, which provides exact performance while providing an efficient sublinear search, assuming dimensionality of embeddings is below 20.

5.2 Additional Applications

There are additional cheminformatics tasks that can be accelerated using the framework reported here. The k-d tree can execute range queries to return all chemicals within some bounds in embedding space. Quantitative structure-activity relationship (QSAR) models often use tree-based algorithms (e.g. decision trees) for predicting the biological activity of chemicals. A learned tree-based model built on the same low-dimensional embeddings used in this work can then encode the regions of chemical space where desirable chemicals exist. Thus, range queries on a k-d tree can quickly filter out predicted active compounds from billion-sized chemical databases according to a learned model. Fine-tuning of other statistical models, like deep neural networks, could also make them compatible with this approach if they are built or constricted to partition a space. Such an approach would allow QSAR models to be more efficiently utilized for screening of massive chemical databases.

Data and Code Availability. Source code and dataset information can be found on GitHub.

Acknowledgments. We would like to thank the Research Computing group at the University of North Carolina at Chapel Hill for providing computational resources and support. This work was supported by the National Institutes of Health (Grant R01GM140154). JW is supported by the National Institute of General Medical Sciences of the National Institutes of Health under Award Number T32GM135122. This work was supported by the following grants to SMG from the National Institutes of Health - U24DK116204 (NIDDK), U01CA238475 (NCI), R01CA233811 (NCI), U01CA274298 (NCI). T.M. was supported by the National Institute of General Medical Sciences of the NIH under Award Number T32GM086330. The funders had no role in study design, data collection and analysis, decision to publish, or preparation of the manuscript.

Appendix

A   SALSA Embedding Procedure

In learning molecular representations, chemicals are often initially expressed as alphanumeric sequences, so-called SMILES strings [49] which effectively describe molecular graphs (that is, atoms connected by bonds). SMILES strings enable the use of sequence-to-sequence autoencoders for learning molecular representations. However, autoencoders trained solely on SMILES are insufficient to learn molecular representations that capture the structural (graph-to-graph) similarities between molecules, resulting in disorganized embeddings that potentially hinder downstream applications.

The recently published molecular representation model, SALSA [24], seeks to remedy this limitation of SMILES-based autoencoders by explicitly enforcing structural awareness onto learned representations. This is accomplished by modifying a SMILES-based autoencoder to additionally learn a contrastive objective of mapping structurally similar molecules to similar representations. Here, "structurally similar" is defined as any two molecules having a graph edit distance (GED) of one. This contrastive objective necessitates a dataset comprised of pairs of 1-GED molecules. For their implementation, the SALSA authors chose to use a curated chemical dataset from ChEMBL, and then, from this dataset, generated 1-GED neighbor compounds (or, mutants) for each ChEMBL anchor compound.

The standard operational definition of GED permits a wide array of modifications that may drastically change chemical and biological properties of small molecules. The SALSA methodology operates on a more conservative notion of graph edits aimed at maintaining relative chemical similarity between an anchor compound and mutant compounds. To obtain 1-GED mutants, SALSA defines three node-level (i.e. atom-level) transformations:

1. *Addition*: Append a new node and corresponding edge to an existing node
2. *Substitution*: Change the atom type of an existing node
3. *Deletion*: Remove a singly-attached node and its corresponding edge

Notably, these transformations neither break nor create new cycles in the molecular graphs, and they do not disconnect previously connected graphs. There are additional nuances to the definition of graph edits, and we refer readers to the referenced publication for full details.

We extend the SALSA framework starting with a dataset of approximately 1.3 million chemicals diversity-sampled from the Enamine REAL library of approximately 40 billion chemicals. Additionally, we modified the hyperparameters of the model to output a chemical embedding whose dimensionality would enable the fast indexing search methods described previously. Specifically, we chose embeddings of dimensions 8 and 16.

References

1. ANNOY library. https://github.com/spotify/annoy. Accessed 01 Aug 2017
2. NextMove software | Arthor. https://www.nextmovesoftware.com/arthor.html
3. NextMove software | Arthor. https://www.nextmovesoftware.com/talks/Sayle_EvolutionVsRevolution_ICCS_201805.pdf
4. NextMove software | SmallWorld. https://www.nextmovesoftware.com/smallworld.html
5. Achlioptas, D.: Database-friendly random projections: Johnson-lindenstrauss with binary coins **66**(4), 671–687. https://doi.org/10.1016/S0022-0000(03)00025-4

6. Agrawal, R., Faloutsos, C., Swami, A.: Efficient similarity search in sequence databases, pp. 69–84. https://doi.org/10.1007/3-540-57301-1_5

7. Bawa, M., Condie, T., Ganesan, P.: LSH forest: self-tuning indexes for similarity search, pp. 651–660. https://doi.org/10.1145/1060745.1060840

8. Bellmann, L., Penner, P., Rarey, M.: Topological similarity search in large combinatorial fragment spaces 61(1), 238–251. https://doi.org/10.1021/acs.jcim.0c00850, publisher: American Chemical Society

9. Bentley, J.: Multidimensional binary search trees in database applications. IEEE Trans. Softw. Eng. SE-5(4), 333–340 (1979). https://doi.org/10.1109/TSE.1979.234200

10. Bentley, J.L.: Multidimensional binary search trees used for associative searching. Commun. ACM 18(9), 509–517 (1975). https://doi.org/10.1145/361002.361007

11. Birchall, K., Gillet, V.J., Harper, G., Pickett, S.D.: Training similarity measures for specific activities: Application to reduced graphs 46(2), 577–586. https://doi.org/10.1021/ci050465e, publisher: American Chemical Society

12. Cao, Y., Jiang, T., Girke, T.: Accelerated similarity searching and clustering of large compound sets by geometric embedding and locality sensitive hashing. Bioinformatics 26(7), 953–959 (2010). https://doi.org/10.1093/bioinformatics/btq067

13. Crum-Brown, A., Fraser, T.R.: The connection of chemical constitution and physiological action. Trans. R. Soc. Edinb. 25(1968–1969), 257 (1865)

14. Deng, Z., Zhu, X., Cheng, D., Zong, M., Zhang, S.: Efficient kNN classification algorithm for big data 195, 143–148. https://doi.org/10.1016/j.neucom.2015.08.112,https://www.sciencedirect.com/science/article/pii/S0925231216001132

15. Félix, E., Dalke, A., Landrum, G., Bushuiev, R.: chembl/FPSim2: 0.5.0 (2023). https://doi.org/10.5281/ZENODO.10041218, https://zenodo.org/doi/10.5281/zenodo.10041218

16. Gahbauer, S., et al.: Iterative computational design and crystallographic screening identifies potent inhibitors targeting the Nsp3 macrodomain of SARS-CoV-2. Proc. Nat. Acad. Sci. 120(2), e2212931120 (2023). https://doi.org/10.1073/pnas.2212931120, https://www.pnas.org/doi/abs/10.1073/pnas.2212931120

17. Grygorenko, O.O.: Enamine Ltd.: the science and business of organic chemistry and beyond. Eur. J. Org. Chem. 2021(47), 6474–6477 (2021). https://doi.org/10.1002/ejoc.202101210, https://onlinelibrary.wiley.com/doi/10.1002/ejoc.202101210

18. Gupta, D., Loane, R., Gayen, S., Demner-Fushman, D.: Medical image retrieval via nearest neighbor search on pre-trained image features (arXiv:2210.02401). https://doi.org/10.48550/arXiv.2210.02401, http://arxiv.org/abs/2210.02401, version: 1

19. Johnson, J., Douze, M., Jégou, H.: Billion-scale similarity search with GPUs. IEEE Trans. Big Data 7(3), 535–547 (2019)

20. Johnson, W.B., Lindenstrauss, J., Schechtman, G.: Extensions of lipschitz maps into banach spaces 54(2), 129–138. https://doi.org/10.1007/BF02764938

21. Jolliffe, I.: Principal Component Analysis. In: Balakrishnan, N., Colton, T., Everitt, B., Piegorsch, W., Ruggeri, F., Teugels, J.L. (eds.) Wiley StatsRef: Statistics Reference Online. Wiley, first edn. (2014). https://doi.org/10.1002/9781118445112.stat06472

22. Karlova, A., Dehaen, W., Svozil, D.: Molecular Fingerprint VAE p. 6

23. Keogh, E., Chakrabarti, K., Pazzani, M., Mehrotra, S.: Dimensionality reduction for fast similarity search in large time series databases 3(3), 263–286. https://doi.org/10.1007/PL00011669

24. Kirchoff, K.E., Maxfield, T., Tropsha, A., Gomez, S.M.: SALSA: Semantically-aware latent space autoencoder (2023)

25. Korn, P., Sidiropoulos, N.D., Faloutsos, C., Siegel, E.L., Protopapas, Z.: Fast and effective similarity search in medical tumor databases using morphology: Multimedia storage and archiving systems **2916**, 116–129. https://doi.org/10.1117/12.257282

26. Lavecchia, A., Giovanni, C.D.: Virtual screening strategies in drug discovery: A critical review **20**(23), 2839–2860. https://www.eurekaselect.com/article/53238

27. Lyu, J., Irwin, J.J., Shoichet, B.K.: Modeling the expansion of virtual screening libraries **19**(6), 712–718. https://doi.org/10.1038/s41589-022-01234-w, https://www.nature.com/articles/s41589-022-01234-w, number: 6 Publisher: Nature Publishing Group

28. Maggiora, G., Vogt, M., Stumpfe, D., Bajorath, J.: Molecular similarity in medicinal chemistry: Miniperspective. J. Med. Chem. **57**(8), 3186–3204 (2014). https://doi.org/10.1021/jm401411z

29. Mahdi, M.A., Hosny, K.M., Elhenawy, I.: Scalable clustering algorithms for big data: A review **9**, 80015–80027. https://doi.org/10.1109/ACCESS.2021.3084057, conference Name: IEEE Access

30. McInnes, L., Healy, J., Melville, J.: UMAP: Uniform Manifold Approximation and Projection for Dimension Reduction (2018). https://doi.org/10.48550/ARXIV.1802.03426

31. Nasser, M., et al.: Feature reduction for molecular similarity searching based on autoencoder deep learning. Biomolecules **12**(4), 508 (2022). https://doi.org/10.3390/biom12040508

32. Nikolova, N., Jaworska, J.: Approaches to measure chemical similarity - a Review. QSAR Comb. Sci. **22**(9–10), 1006–1026 (2003). https://doi.org/10.1002/qsar.200330831

33. Omohundro, S.: Five Balltree Construction Algorithms. Tech. Rep. TR-89-063, International Computer Science Institute (1989)

34. Riesen, K., Bunke, H.: Approximate graph edit distance computation by means of bipartite graph matching **27**(7), 950–959. https://doi.org/10.1016/j.imavis.2008.04.004, https://www.sciencedirect.com/science/article/pii/S026288560800084X

35. Riniker, S., Landrum, G.A.: Open-source platform to benchmark fingerprints for ligand-based virtual screening. J. Cheminform. **5**(1), 26 (2013)

36. Rogers, D., Hahn, M.: Extended-Connectivity Fingerprints. J. Chem. Inf. Model. **50**(5), 742–754 (2010). https://doi.org/10.1021/ci100050t

37. Sanfeliu, A., Fu, K.S.: A distance measure between attributed relational graphs for pattern recognition **SMC-13**(3), 353–362. https://doi.org/10.1109/TSMC.1983.6313167, conference Name: IEEE Transactions on Systems, Man, and Cybernetics

38. Sankaranarayanan, J., Samet, H., Varshney, A.: A fast all nearest neighbor algorithm for applications involving large point-clouds. Comput. Graph. **31**(2), 157–174 (2007). https://doi.org/10.1016/j.cag.2006.11.011

39. Swamidass, S.J., Baldi, P.: Bounds and algorithms for fast exact searches of chemical fingerprints in linear and sublinear time **47**(2), 302–317. https://doi.org/10.1021/ci600358f

40. Swamidass, S.J., Baldi, P.: Bounds and algorithms for fast exact searches of chemical fingerprints in linear and sublinear time. J. Chem. Inf. Model. **47**(2), 302–317 (2007). https://doi.org/10.1021/ci600358f, pMID: 17326616

41. Tanimoto, T.T.: An elementary mathematical theory of classification and prediction by T.T. Tanimoto. International Business Machines Corporation New York (1958)

42. Tingle, B.I., et al.: ZINC-22-A free multi-billion-scale database of tangible compounds for ligand discovery. J. Chem. Inf. Model. **63**(4), 1166–1176 (2023). https://doi.org/10.1021/acs.jcim.2c01253
43. Toth, C.D., O'Rourke, J., Goodman, J.E.: Handbook of Discrete and Computational Geometry. CRC Press, google-Books-ID: 9mlQDwAAQBAJ
44. van der Maaten, L., Hinton, G.: Visualizing data using t-SNE. J. Mach. Learn. Res. **9**(86), 2579–2605 (2008)
45. Wald, I., Havran, V.: On building fast kd-Trees for Ray Tracing, and on doing that in O(N log N). In: 2006 IEEE Symposium on Interactive Ray Tracing, pp. 61–69. IEEE, Salt Lake City, UT, USA (2006). https://doi.org/10.1109/RT.2006.280216
46. Wang, J., Shen, H.T., Song, J., Ji, J.: Hashing for similarity search: A survey (arXiv:1408.2927). https://doi.org/10.48550/arXiv.1408.2927, http://arxiv.org/abs/1408.2927
47. Wang, Y., Wang, J., Cao, Z., Farimani, A.B.: MolCLR: Molecular contrastive learning of representations via graph neural networks. CoRR abs/2102.10056 (2021). https://arxiv.org/abs/2102.10056
48. Weber, R., Schek, H.J., Blott, S.: Analysis and Performance Study for Methods in High-Dimensional Spaces p. 12
49. Weininger, D.: SMILES, a chemical language and information system. 1. introduction to methodology and encoding rules **28**(1), 31–36. https://doi.org/10.1021/ci00057a005, https://doi.org/10.1021/ci00057a005, publisher: American Chemical Society
50. Weiss, Y., Torralba, A., Fergus, R.: Spectral hashing 21
51. Wigh, D.S., Goodman, J.M., Lapkin, A.A.: A review of molecular representation in the age of machine learning. WIREs Comput. Mol. Sci. **12**(5), e1603 (2022). https://doi.org/10.1002/wcms.1603
52. Yang, J., Cai, Y., Zhao, K., Xie, H., Chen, X.: Concepts and applications of chemical fingerprint for hit and lead screening **27**(11), 103356. https://doi.org/10.1016/j.drudis.2022.103356
53. Zeng, Z., Tung, A.K.H., Wang, J., Feng, J., Zhou, L.: Comparing stars: on approximating graph edit distance **2**(1), 25–36. https://doi.org/10.14778/1687627.1687631

Translate-Distill: Learning Cross-Language Dense Retrieval by Translation and Distillation

Eugene Yang[1]([envelope])[ID], Dawn Lawrie[1][ID], James Mayfield[1][ID], Douglas W. Oard[1,2][ID], and Scott Miller[3][ID]

[1] HLTCOE. Johns Hopkins University, Baltimore 21211, USA
{eugene.yang,lawrie,mayfield}@jhu.edu, oard@umd.edu
[2] University of Maryland, College Park 20742, USA
[3] Information Sciences Institute, University of Southern California, Los Angeles 90292, USA
smiller@isi.edu

Abstract. Prior work on English monolingual retrieval has shown that a cross-encoder trained using a large number of relevance judgments for query-document pairs can be used as a teacher to train more efficient, but similarly effective, dual-encoder student models. Applying a similar knowledge distillation approach to training an efficient dual-encoder model for Cross-Language Information Retrieval (CLIR), where queries and documents are in different languages, is challenging due to the lack of a sufficiently large training collection when the query and document languages differ. The state of the art for CLIR thus relies on translating queries, documents, or both from the large English MS MARCO training set, an approach called *Translate-Train*. This paper proposes an alternative, *Translate-Distill*, in which knowledge distillation from either a monolingual cross-encoder or a CLIR cross-encoder is used to train a dual-encoder CLIR student model. This richer design space enables the teacher model to perform inference in an optimized setting, while training the student model directly for CLIR. Trained models and artifacts are publicly available on Huggingface.

Keywords: CLIR · Dense retrieval · Knowledge distillation · Translate-Train

1 Introduction

Cross-language information retrieval (CLIR) enables users to express queries in one language and search for content in another. Matching potentially ambiguous queries to documents is already challenging monolingually; matching them across languages adds the additional complexity of translation. With recent improvements in machine translation, translating queries, documents, or both into the same language and then performing monolingual retrieval is a viable approach. However, machine translation is computationally expensive. Query translation is particularly challenging because queries are often short, writing styles for queries differ from the styles of documents (for which machine translation systems are typically trained), and tight query latency budgets provide little time for complex processing. However, the alternative of translating very large document collections may not be feasible in cost-constrained applications. Therefore, this work—and a good deal of the work on CLIR—focuses on creating

© The Author(s), under exclusive license to Springer Nature Switzerland AG 2024
N. Goharian et al. (Eds.): ECIR 2024, LNCS 14609, pp. 50–65, 2024.
https://doi.org/10.1007/978-3-031-56060-6_4

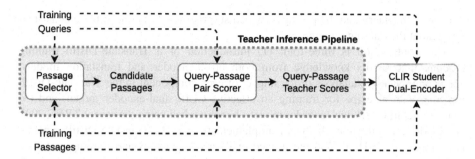

Fig. 1. Translate-Distill training pipeline. The white boxes with blue borders are the fixed teacher models. The hatched green box is the trainable student model. Dashed arrows indicate the optional machine translation middle step, i.e., the input text the model receives can either be original or translated, with different translation decisions made for each dashed arrow. (Color figure online)

CLIR systems that do not require full neural machine translation during either indexing or query processing.

CLIR dual-encoders, such as DPR-X [42] (also known as mDPR [2]), ColBERT-X [25] and BLADE [26], use multilingual pretrained language models (e.g., multilingual BERT [8] or XLM-RoBERTa [4]) to match queries to documents across languages without using machine translation at indexing or retrieval time. However, the zero-shot approach of using English MS MARCO query-passage pairs to train a ColBERT dual-encoder using XLM-RoBERTa has been shown suboptimal [25]. Instead, a more effective approach is to translate the MS MARCO pairs to match the CLIR setting of the final task before fine-tuning to the task, an approach called *Translate-Train*. This Translate-Train approach allows the model to simultaneously learn both the retrieval task and the details of the translation task for the specific language pair. Translate-Train is the current state of the art for optimizing retrieval effectiveness in single-stage (i.e., "end-to-end") CLIR for large document collections [18].

It is, however, possible to exceed the effectiveness of the single-stage approach by using cross-encoders as rerankers [21,29]. CLIR cross-encoders, such as MonoT5 with mT5XXL [12], are computationally expensive, limiting their application in cost-constrained tasks. This paper addresses that limitation by using effective but inefficient cross-encoders at training time, as teacher models for training efficient CLIR dual-encoder models. Specifically, we propose *Translate-Distill* (summarized in Fig. 1), an approach that distills ranking knowledge from cross-encoders through translations of the training data to create an effective CLIR dual-encoder model. While using distillation to train a monolingual dual-encoder retrieval model is not new [31,37], generalizing the technique to train CLIR models is not straightforward, both because the CLIR task has a larger design space for cross-encoder teacher models, and because the choice of the teacher model can substantially influence the effectiveness of the resulting student model. In this work, we explore a suite of design options for Translate-Distill pipelines that use translation in different ways. Of particular importance are the languages used to express the query and the passages, which surprisingly do not need to be consistent throughout the training pipeline.

Our contribution is three-fold: (1) introduction of a Translate-Distill training pipeline that distills knowledge from both cross-encoder and translation models; (2) a comprehensive analysis of the impact of each component in the Translate-Distill pipeline, with a recipe for training an effective CLIR dual-encoder model; and (3) state-of-the-art CLIR dual-encoder models, benchmarked using the recent TREC 2022 NeuCLIR test collection. A Python implementation of the Translate-Distill pipeline is available on GitHub[1], and the trained dual-encoder models and training materials (including the teacher models and scores) are available on Huggingface.[2]

2   Background

Early work on CLIR explored the use of parallel corpora [17], comparable corpora [38], and bilingual lexicons [30] to cross the language barrier. However, early work with shared semantic spaces using Cross-Language Latent Semantic Indexing (CL-LSI) [35] was not widely adopted, largely for efficiency reasons. Using full neural document translation to reduce CLIR to a monolingual task is now substantially more effective than all of these earlier techniques, although translation costs can limit its use on large collections [26]. Query translation is typically less effective, however, as language use in queries differs in important ways from that in documents, and training resources that are specialized for query translation are not widely available. The research frontier is thus currently focused on balancing effectiveness and efficiency, seeking to develop approaches that are as effective as, and considerably more efficient at indexing time than, full neural machine translation [18].

Cross-encoders, such as MonoBERT [29] or MonoT5 [28], are a family of retrieval models that use neural pretrained language models (PLM) such as BERT [8], RoBERTa [23], or T5 [32] by concatenating the query and document as a single input text sequence; this allows cross-attention between the query and document representations. Although very effective, such models require that the query be available before each document can be processed. This prevents document representations from being indexed in advance, thus limiting the number of documents to which the technique can be applied. Cross-encoders are thus primarily used as rerankers in a retrieval pipeline, following an efficient retrieval model such as BM25; this is known as a *Retrieve-and-Rerank* pipeline [29].

On the other hand, dual-encoders (e.g., DPR [14], SPLADE [10], and Col-BERT [15]) are a family of model architectures that encode queries and documents separately, using a PLM as the encoder. This is a modern neural realization of LSI's shared semantic space, with the additional benefit of the non-linearity that neural models make possible (LSI was limited to lossy linear transformations on the term-document matrix). These dual encoder models enable offline document preprocessing before any query has arrived, thus (when used with efficient approximate nearest neighbor techniques) making it possible to respond to queries quickly, even for large collections.

---

[1] https://github.com/hltcoe/ColBERT-X/tree/plaid-x.

[2] https://huggingface.co/collections/hltcoe/translate-distill-659a11e0a7f2d2491780a6bb.

English neural retrieval models use collections with millions of judged query-passage pairs, such as MS MARCO [27] and Natural Questions [16], as training data. Until similar cross-language training data becomes available for a broad range of language pairs, training neural CLIR models will remain more complicated than training monolingual models. While it is possible to train a model that uses an mPLM with English query-passage pairs zero-shot, prior work has found that providing cross-language supervision during training leads to more effective CLIR models [25,26,42]. Such supervision can be in the form of pretraining, such as by using XLM-Align [3] as the mPLM [42], continued pretraining of an mPLM with parallel or comparable text for a specific language pair [26,42], or fine-tuning with queries and documents that have been translated into the languages of the final CLIR task [25]. These training strategies strengthen the model's ability to bridge the query/document language barrier.

The top-ranked run for each document language in the TREC 2022 NeuCLIR track [18] used a cross-encoder reranker at the end of a retrieval pipeline [12]. Each used MonoT5 with the mT5XXL mPLM, trained from mMARCO [2] and document-language queries produced by machine translation to rerank the top-ranked documents. We refer to this reranker as *Mono-mT5XXL*. Although effective, Mono-mT5XXL has 13 billion parameters, requiring GPUs with large memory or special model-sharding approaches. To exploit the effectiveness of cross-encoder rerankers and the efficiency of dual-encoders, this paper explores distilling ranking knowledge from an effective cross-encoder, such as Mono-mT5XXL, to train a CLIR dual-encoder model.

Prior work on English retrieval has explored distillation for training monolingual retrieval models [9,22,31,37,43]. Such approaches use an English cross-encoder as a teacher model to score training query-passage pairs, using those scores to optimize a student dual-encoder model with KL-divergence loss. Li et al. [20] jointly trained a cross-language alignment model (a form of translation model) with a distillation objective from an English dual-encoder as teacher and a CLIR ColBERT model as student. While effective, this training requires a parallel corpus with IR relevance judgments, which are rarer than unicorns. Therefore, they used XOR-TyDi [1], a synthetic multilingual retrieval collection built from Wikipedia, for supervision on alignment and ranking. This is similar to prior work on continued pretraining with comparable text [39,42]. We explore an alternative approach.

Given that Translate-Train has been shown to be an effective training strategy and that it only requires translation of the training corpora, such as MS MARCO, we propose a unified framework *Translate-Distill*. Our framework combines Translate-Train with cross-encoder distillation to train a CLIR dual-encoder model without additional data resources. In the next section, we introduce the Translate-Distill framework, and we identify a series of design choices in a Translate-Distill pipeline.

3   Translate-Distill

Illustrated in Fig. 1, our proposed Translate-Distill pipeline consists of two inference steps using teacher models and a training step for the student CLIR dual-encoder model using knowledge distillation. We first introduce the distillation training framework for CLIR and then discuss the pipeline for generating the training material. Of particular importance is the languages of the queries and passages, which is not necessarily consistent throughout the pipeline.

3.1 Notation

Let $\mathcal{L}_{\mathbb{Q}}$ and $\mathcal{L}_{\mathbb{D}}$ be the query and the document languages of the final CLIR task. In this work, we focus on CLIR passage retrieval, limiting the retrieved items to a fixed length, due to the limitations of current transformer models. We do, however, map those passages back to the documents they came from when evaluating using the NeuCLIR 2022 test collection. We follow the convention in the CLIR literature of using the name "document language" to refer to the language of the passages we retrieve. We assume that the original training collection is monolingual, containing a set of training queries $\mathcal{Q}$ and passages $\mathcal{P}$ in the same language $\mathcal{L}_T$.

We use $\mathcal{L}$ to denote machine translation into language $\mathcal{L}$. We place this notation as a superscript to indicate the language into which text has been translated. For example, machine-translated queries that have been translated into the document language $\mathcal{L}_{\mathbb{D}}$ are denoted as $\mathcal{Q}^{\widetilde{\mathcal{L}_{\mathbb{D}}}}$. We also use this notation in compound superscripts to denote the languages on which functions operate. For example, a function denoted $f^{\mathcal{L}_a, \mathcal{L}_b}$ accepts a pair of texts in language $\mathcal{L}_a$ and $\mathcal{L}_b$. For convenience, we define $\mathcal{Q}^{\widetilde{\mathcal{L}_T}} = \mathcal{Q}^{\mathcal{L}_T}$ and $\mathcal{D}^{\widetilde{\mathcal{L}_T}} = \mathcal{D}^{\mathcal{L}_T}$ since the text in $\mathcal{Q}$ and $\mathcal{P}$ are already written in language $\mathcal{L}_T$.

3.2 Cross-Language Knowledge Distillation

To select the training passages for each query q, we retrieve the top k passages from $\mathcal{P}$. Specifically, given a passage selector[3] accepting queries in language $\mathcal{L}_a$ and passages in language $\mathcal{L}_b$ as $PS^{\mathcal{L}_a, \mathcal{L}_b}(\cdot) \to \mathbb{R}$, we define the set of candidate passages $\mathcal{C}_q \subset P$ of size k for query $q \in \mathcal{Q}$ such that

$$PS^{\mathcal{L}_a, \mathcal{L}_b}(q^{\widetilde{\mathcal{L}_a}}, p_{in}^{\widetilde{\mathcal{L}_b}}) \geq PS^{(\mathcal{L}_a, \mathcal{L}_b)}(q^{\widetilde{\mathcal{L}_a}}, p_{out}^{\widetilde{\mathcal{L}_b}}) \quad \forall p_{in} \in \mathcal{C}_q \text{ and } p_{out} \notin \mathcal{C}_q$$

Passages in $\mathcal{C}_q$ are similar to the "hard negatives" for training IR models. However, in our approach, these candidate passages may be relevant to the queries. $\mathcal{C}_q$ forms the "teaching material" for the subsequent CLIR student dual-encoder model.

For a trainable CLIR student dual-encoder model $DE^{\mathcal{L}_{\mathbb{Q}}, \mathcal{L}_{\mathbb{D}}}(\cdot) \to \mathbb{R}$, we construct a distillation loss using KL-divergence against a fixed query-passage pair scorer $QP^{\mathcal{L}_c, \mathcal{L}_d}(\cdot) \to \mathbb{R}$, typically an effective cross-encoder model. Specifically, we define the loss function for Translate-Distill L_{TD} based on the predicted query-passage scores from the teacher and student as

$$L_{TD}(q) = D_{KL}\left(DE^{\mathcal{L}_{\mathbb{Q}}, \mathcal{L}_{\mathbb{D}}} \| QP^{\mathcal{L}_c, \mathcal{L}_d}\right)$$

$$= \sum_{p \in \mathcal{C}_q} DE^{\mathcal{L}_{\mathbb{Q}}, \mathcal{L}_{\mathbb{D}}}(q^{\widetilde{\mathcal{L}_{\mathbb{Q}}}}, p^{\widetilde{\mathcal{L}_{\mathbb{D}}}}) \log\left(\frac{DE^{\mathcal{L}_{\mathbb{Q}}, \mathcal{L}_{\mathbb{D}}}(q^{\widetilde{\mathcal{L}_{\mathbb{Q}}}}, p^{\widetilde{\mathcal{L}_{\mathbb{D}}}})}{QP^{\mathcal{L}_c, \mathcal{L}_d}(q^{\widetilde{\mathcal{L}_c}}, p^{\widetilde{\mathcal{L}_d}})}\right) \quad (1)$$

---

[3] What we call a passage selector assigns scores to passages that are the basis for selection. We call it a selector rather than a scorer to avoid confusion with the query-passage scorer.

Table 1. Example input text query-passage language pairs for each module in the Translate-Distill Pipeline. These examples assume that the final CLIR task requires English (Eng) queries and Persian (Fas) documents, and the training resources are all in English.

	Teacher Inference		Training	Retrieval
	Passage	Query-Passage	CLIR Student	CLIR Student
Availability of Key Modules	Selector	Pair Scorer	Dual-encoder	Dual-encoder
Eng Training Query & Passages	Eng-Eng	Eng-Eng	Eng-Eng	Eng-Fas
+ Eng-Fas MT model	Eng-Eng	Eng-Eng	Eng-Fas	Eng-Fas
+ CLIR cross-encoder	Eng-Eng	Eng-Fas	Eng-Fas	Eng-Fas
+ Multilingual Cross-encoder	Eng-Eng	Fas-Fas	Eng-Fas	Eng-Fas
+ E2E CLIR Retrieval System	Eng-Fas	Fas-Fas	Eng-Fas	Eng-Fas

The pair scorer $QP^{\mathcal{L}_c,\mathcal{L}_d}$ provides supervision to the student dual-encoder $DE^{\mathcal{L}_Q,\mathcal{L}_D}$ based on $\mathcal{C}_q$. Therefore, if the $\mathcal{C}_q$ is not informative or sufficiently hard, the cross-encoder cannot adequately distill its ranking knowledge.

Constructing the training loss by Equation (1) has the important benefit of decoupling the input languages of the three modules in the pipeline. We can optimize the effectiveness of each teacher model by choosing languages that matches its training, avoiding language mismatches between training and inference time on a module-by-module basis. Similarly, we can choose the languages of the input text for the student model based on how we intend to use that model at inference time. Such decoupling avoids unnecessary language transfer, and enables each module to operate or train in the languages for which it was designed.

The input languages for the teacher model's passage selector PS ($\mathcal{L}_a$ and $\mathcal{L}_b$) and query-passage pair scorer QP ($\mathcal{L}_c$, and $\mathcal{L}_d$) could also be different. However, in practice, they are often chosen to be among $\mathcal{L}_Q$, $\mathcal{L}_D$, and $\mathcal{L}_T$ to align with either the final CLIR task or the available monolingual training resource. In the rest of this section, we discuss language selection given varying constraints.

3.3 Training Pipeline

The Translate-Distill pipeline is illustrated in Fig. 1. We pre-construct (1) the translations of $\mathcal{Q}$ and $\mathcal{P}$; (2) the candidate passages $\mathcal{C}_q$ for each query q; and (3) the teacher scores $s_{q,p} = QP^{\mathcal{L}_c,\mathcal{L}_d}(q^{\widetilde{\mathcal{L}_c}}, p^{\widetilde{\mathcal{L}_d}})$ before training. Although all these resources can be generated on-the-fly during the dual-encoder training, pre-computing them limits peak computational resource requirements during training. Training queries and passages are first fed into the passage selector to select $\mathcal{C}_q$ in the language pairs that the selector accepts. For each $p \in \mathcal{C}_q$, we use the query-passage pair scorer to produce the teacher score for each (q, p) pair. These scores are stored and later loaded back into memory when training the CLIR student dual-encoder model.

The first row in Table 1 is an English training setup; the remaining four rows illustrate a series of possible input language choices of the teacher and student models for an English-Persian CLIR task. The Translate-Distill training pipeline requires the availability of machine translation models that can translate the training queries and passages from $\mathcal{L}_T$ to $\mathcal{L}_\mathbb{Q}$ and $\mathcal{L}_\mathbb{D}$. More precisely, it requires the existence of $Q^{\widetilde{\mathcal{L}_\mathbb{Q}}}$ and $\mathcal{D}^{\widetilde{\mathcal{L}_\mathbb{D}}}$. For example, the publicly available neuMARCO [18] and mMARCO [2] collections provide machine translation of the popular training collection MS MARCO [27]. Such requirements are identical to those of Translate-Train.

If effective cross-encoders for the document language are available, either using CLIR or using queries in that same language, we can use teacher models that more closely match the final CLIR setting. When the original language of the training passages is English rather than Persian, a CLIR or a multilingual cross-encoder could directly produce teacher scores using translated Persian passages. To see why this could be helpful, consider the case of a passage that is relevant in its original English version, but that becomes non-relevant after translation due to translation errors. A teacher cross-encoder that directly (and correctly) scores Persian passages would reflect such shifts in its scores. This approach reduces the mismatch of the input text between the teacher and student models.

Similarly, if a CLIR system already exists, we could use this CLIR system to select the candidate passages $\mathcal{C}_q$ for each query q. This approach would provide in-language hard negatives, helping the student model learn to distinguish relevant and non-relevant passages as they are expressed in the document language.

Of course, the benefits to be gained from such approaches depend on the quality of the multilingual or CLIR systems used for those purposes; Sect. 5 demonstrates how using them can improve distillation results.

4 Experiments

This section introduces our evaluation collections and metrics, and discusses resources and models used in our Translate-Distill pipeline experiments.

4.1 Evaluation Collections and Metrics

We evaluate Translate-Distill on the TREC 2022 NeuCLIR track [18] and HC3 [19] test collections. Collection statistics are summarized in Table 2. Both of these test collections model three CLIR tasks: searching Chinese, Persian, or Russian documents using English queries. The NeuCLIR collections consist of news articles, a genre for which many machine translation models are optimized, extracted from Common Crawl. HC3, by contrast, consists of short informal Twitter conversations, which can pose challenges for machine translation. We report nDCG@20 (the primary evaluation metric for NeuCLIR 2022) as our effectiveness score. We use the topic title concatenated with the description as the query. Note that prior work has found that reversing the order of the title and description in the query (i.e., description followed by title) is more effective for Chinese [12]. However, for experimental consistency we retain the conventional order for all languages and collections.

Table 2. Collection statistics.

Collection	NeuCLIR 2022			HC3		
Language	zho	fas	rus	zho	fas	rus
# Documents	3.2M	2.2M	4.6M	5.6M	7.3M	26.8M
# Passages	19.8M	14.0M	25.1M	6.3M	0.4M	27.0M
# Topics	49	46	45	50	50	69

4.2   Training Queries and Passages

We use MS MARCOv1 training queries and passages [27] as our training set. Passage translations were provided by the NeuCLIR organizers and released along with the test collection with the name *neuMARCO* [18] on `ir-datasets` [24].[4] These translations are generated by Sockeye v2 trained with general domain parallel text. MS MARCO training query translations were obtained from mMARCO [2]. These translations are generated using the Google Translate service. Since Persian is not included in mMARCO, we used Google Translate ourselves to generate Persian translations of the queries.[5] For consistency, although mMARCO also provides translation for Chinese and Russian passages, we use the neuMARCO version in our experiments whenever passage translation is needed.

4.3   Teacher and Student Models

We use publicly-available models as passage selector and teacher query-passage pair scorer. In the Translate-Distill pipeline, we select the top 50 passages from the MS MARCOv1 passage collection for each training query.

Passage Selector. For our main experiments in Sect. 5.1 and Table 3 we use the Col-BERTv2 [37] model released by the ColBERT authors as the passage selector. This is a monolingual English retrieval model; thus, the input query and passages are not translated, but kept in English. To assess training robustness, we also experiment using different retrieval systems as passage selectors in contrastive experiments. As one alternative, for English, we experiment with CoCondenser [11], a single-vector dual-encoder model.[6] Results using these alternatives appear in Table 4. Assuming the existence of an end-to-end CLIR system, we experiment with using an existing Translate-Train ColBERT-X model [25] to select candidate passages.

---

[4] https://ir-datasets.com/neumarco.html.

[5] https://huggingface.co/datasets/hltcoe/tdist-msmarco-scores/blob/main/msmarco.train.query.fas.tsv.gz.

[6] We use the hard negatives from `co-condenser-margin_mse-sym_mnrl-mean-v1`, which are publicly available on Huggingface Datasets released by the Sentence-Transformers: https://huggingface.co/datasets/sentence-transformers/msmarco-hard-negatives.

Query-Passage Pair Scorers. We experimented with MiniLM [41], MonoT5-3b [28], Mono-mT5XXL [12], and four cross-encoders (one English-Trained and three Translate-Trained models on Chinese, Persian, and Russian passages) that we trained ourselves. MiniLM (a lightweight model distilled from a large BERT cross-encoder) and MonoT5-3b are commonly used cross-encoder rerankers that have very different resource requirements; Mono-mT5XXL is the state-of-the-art cross-encoder for CLIR (according to the NeuCLIR 2022 evaluation results [18]). MonoT5-3b and Mono-mT5XXL are executed on 2 NVidia 40GB A100 GPUs with DeepSpeed [34] ZeRO-3 [33] model sharding.

To understand the effect of the input language pair for the teacher scorer, we fixed the size of the teacher cross-encoders using the XLM-RoBERTa-large model with the released MS MARCO small training triples using English training and Translate-Train. The classification head is attached to the last hidden state of the starter <s> token. We concatenate the training query and passage as the input text sequence and use cross-entropy loss to train the model. For the English-Trained (ET) cross-encoder, we provide the native MS MARCO training queries and passages in English; for Translate-Train (TT), we translate the passages into the document language of the final CLIR task (Chinese, Persian, or Russian) and use those with English queries, resulting in three TT cross-encoder models.

CLIR Student Dual-Encoder. We use ColBERT-X, a CLIR variant of the Col-BERT [15] retrieval architecture, as the dual-encoder model. Prior work has shown that single-vector dual-encoder models, such as DPR-X [2,42], and learned-sparse models, such as BLADE [26], are substantially less effective than multi-vector dense dual-encoders [42]. Therefore, we evaluate Translate-Distill with ColBERT-X as the student model. Trained models are available on Huggingface Models.

We use the PLAID [36] implementation for ColBERT-X retrieval. Both evaluation queries and passages are processed in their native languages without any translation. The number of residual bits is set to one for each dimension of the passage representation. Based on our preliminary studies, the number of residual bits has a substantial impact on index size and query latency but not on retrieval effectiveness. In some cases, using only one bit may result in numerically better retrieval results than using two or four bits.

The student CLIR dual-encoder models are trained with KL-divergence loss on the predicted query-passage scores, on a batch of 64 queries distributed across eight NVidia V100 GPUs (with the DGX platform). Each query is associated with six passages randomly sampled from the candidate set at every epoch. Such random selections enable larger batches while stochastically allowing the models to see all candidate passages. We use the AdamW optimizer with a learning rate of 5×10^{-6} with 16-bit floating point.

Since documents in the evaluation collections are generally longer than the input length of XLM-RoBERTa, following prior work in neural IR, we create passages from the documents by using a sliding window of size 180 tokens with a stride of 90 [25]. At retrieval time, we aggregate the passage scores using MaxP [6] to form the document score. The number of generated passages is shown in Table 2.

4.4 Baselines

We compare the models trained against Translate-Train, a predecessor to our proposed Translate-Distill training pipeline. We also report the effectiveness of the Patapsco [5] implementation of BM25 with RM3 query expansion, using document translation into English. This retrieval system was a baseline run for the TREC 2022 NeuCLIR track, and was designated as the standard initial retrieval result for the reranking task [18]. For the NeuCLIR collection, we use the document translations released by TREC. We translate the HC3 documents with the same set of translation models that was used to translate the NeuCLIR collection.

5 Results and Analysis

This section presents our results for distillation, passage selection, and comparison to a Retrieve-and-Rerank pipeline.

5.1 Distillation with a Different Query-Passage Scorer

We used the ColBERTv2 English passage selector for these experiments. As shown in Table 3, ColBERT-X models trained with knowledge distilled from Mono-mT5XXL are substantially more effective than the Translate-Train baseline (statistically significant with 95% confidence, using paired t-tests and Bonferroni correction for three tests). The models trained with Mono-mT5XXL as the teacher scorer achieve state-of-the-art nDCG@20 performance for CLIR systems on both HC3 and NeuCLIR 2022 collections, and they do so with a single-stage model.

As Table 3 also shows, the query-passage teacher scorer greatly affects retrieval effectiveness. Using publicly available English rerankers (MiniLM and MonoT5-3b) as teacher scorers does not yield more effective student dual-encoders. Instead, using an XLMR cross-encoder (CE) trained with English MS MARCO as the teacher scorer in English (the E-E setting), in contrast to using it as a reranker in a Retrieve-and-Rerank pipeline [29], provides better supervision for training the student dual-encoders. While the difference between XLMR CE with ET and MonoT5-3b is not statistically significant, the fact that an English-trained XLMR cross-encoder distills to the student ColBERT-X model as well as the MonoT5-3b suggests that the model size of the teacher scorer is not a dominant factor in the training pipeline. Among all teacher scorers operating in monolingual English, using Mono-mT5XXL as the teacher scorer results in statistically significantly more effective student ColBERT-X models (Row 8) than using either the MonoT5-3b teacher (Row 5) or the XLMR English-trained CE (Row 6) We hypothesize that the Mono-mT5XXL model is particularly well-trained and in this case the additional power does not come from the model size. However, we cannot eliminate the possibility that growing the model size even further would eventually lead to improvements based solely on size. We leave this hypothesis for future investigation.

The languages of the input query and passages for the teacher models also substantially affect retrieval effectiveness. For both XLMR cross-encoder and MonoT5XXL

Table 3. nDCG@20 and Judged@20 (J@20) using different teacher query-passage pair scorers. The "Scorer Lang." columns indicate the input query (Q) and passage (P) languages of the query-passage pair scorer, respectively, where "E" and "*L*" indicate English and the document language of the final CLIR task (Chinese, Persian or Russian depending on the collection), respectively. All models other than BM25+RM3 (which indexes MT results) index documents in their native language, and all accept English queries at inference time. All Translate-Distill models use ColBERTv2 as the passage selector. The average columns report the micro-average of all 309 topics across the six collections. Subscript numbers indicate statistically significantly better than the system in the corresponding row with 95% confidence using a paired *t*-test on all 309 topics.

Query-Passage Pair Scorer	Scorer Lang. Q P	nDCG@20						Micro Avg.	J@20 Micro Avg.
		HC3			NeuCLIR 22				
		zho	fas	rus	zho	fas	rus		
Baselines									
1 BM25+RM3 using DT		0.262	0.261	0.136	0.340	0.355	0.292	0.301	0.656
2 BLADE		–	–	–	0.330	0.341	0.347	0.339[†]	0.702[†]
3 ColBERT-X with TT		0.333	0.411	0.303	0.441	0.438	0.470	0.392	0.547
ColBERT-X Student Models with Translate-Distill using Different Teacher Scorers									
4 MiniLM	E E	0.429	0.450	0.318	0.415	0.419	0.474	0.410	0.520
5 MonoT5-3b	E E	0.391	0.510	0.310	0.479	0.467	0.497	0.433[3]	0.544
6 XLMR CE (ET)	E E	0.465	0.501	0.309	0.452	0.466	0.503	0.440[3]	0.540
7 XLMR CE (TT)	E *L*	0.438	0.479	0.328	0.450	0.441	0.493	0.430[3]	0.535
8 Mono-mT5XXL	E E	**0.473**	**0.539**	**0.355**	0.492	**0.484**	**0.522**	**0.469**[356]	0.560
9 Mono-mT5XXL	E *L*	0.468	0.508	0.332	0.471	0.475	**0.522**	0.453[35]	0.548
10 Mono-mT5XXL	*L* *L*	0.453	0.524	0.342	**0.493**	0.469	0.503	0.456[356]	0.550

[†] Micro-average values for BLADE are computed over only the three NeuCLIR 2022 collections, and thus are not comparable to the other micro-averages.

as the teacher scorer, training student models with teacher scorers operating in the language of the training corpus (English in this case) is more effective than using document languages. Keeping the input MS MARCO query-passage pairs for the teacher scorer in English (the E-E setting) when using the English-trained (ET) XLMR cross-encoder (CE) produces a more effective subsequent ColBERT-X model than when using its Translate-Trained (TT) counterparts (the E-L setting). Using MonoT5XXL as the teacher scorer shows a similar trend. Such results indicate that the teacher scorer should be trained using the original text without translation (English for MS MARCO in this case).

Nevertheless, these observations suggest that language knowledge can be passed down through the teacher scores for MS MARCO training queries and passages. While the *vessels* for the knowledge transfer (the training queries and passages) are the same (because the same passage selector was used), the *cargo* carried by those vessels (the scores) differ; this can have a large effect on what the student model learns. In the next section, we explore the effects of changing the *vessel*.

Table 4. nDCG@20 and R@1000 on NeuCLIR 22 Chinese using different passage selectors and query-passage pair scorers. The tested passage selectors include English ColBERTv2 (CBv2), English CoCondenser (CoC), and English-Chinese Translate-Trained ColBERT-X (CB-TT).

Pair Scorer	Scorer Lang. Q	P	nDCG@20 CBv2	CoC	CB-TT	R@1000 CBv2	CoC	CB-TT
MiniLM	E	E	**0.415**	0.408	0.397	0.840	0.819	**0.853**
Mono-mT5XXL	*L*	*L*	**0.493**	0.483	0.460	0.867	0.867	**0.884**
Average			**0.454**	0.446	0.429	0.853	0.843	**0.869**

5.2 Passage Selector

The previous section focused on the effect of the query-passage pair scorers when using ColBERTv2 as the passage selector. The following experiments vary the passage selector, using Chinese as an example. Table 4 shows the results of those experiments. There was no statistically significant difference between the two English selectors (ColBERTv2 and CoCondenser) for nDCG@20 or Recall@1000 (R@1000), although models trained with passages selected by the ColBERTv2 model achieved numerically higher results by both measures. Interestingly, using a Translate-Trained ColBERT-X as the passage selector does not lead to a more effective student model when measured by nDCG@20, but it does result in a more effective student model when measured by R@1000. We hypothesize that the Translate-Trained ColBERT-X model is not effective enough to retrieve sufficiently hard (and thus informative) candidate passages. Since it selects passages using their translated text, the candidate sets contain a larger diversity in the document language (Chinese, in this set of experiments), resulting in student models with higher Recall@1000.

Given these observations, we conclude that both the query-passage pair scorer and the passage selector affect final CLIR retrieval effectiveness. The recipe for creating an effective CLIR dual-encoder model requires that the selected passages are sufficiently hard, for which we can use monolingual English neural retrieval models. Also, the subsequent query-passage pair scorer should be selected carefully to be both effective and aligned with the native language of the training corpus (English for the MS MARCO).

5.3 Comparison to Retrieve-and-Rerank Pipeline

Finally, we compare two ways of using a cross-encoder: reranking and distillation. To facilitate our comparison, the cross-encoder we use as teacher for our student dual-encoders is also used as the reranker of the top 200 first-stage results in the Retrieve-and-Rerank pipeline (R&R). To eliminate the effect of translation and potential translationese (language artifacts from the machine translation models), none of our retrieval pipelines uses query or document translation. Instead, we use probabilistic structured queries (PSQ) with HMM [7,40] as the first-stage retriever, which is a strong sparse CLIR baseline model that does not require one-best neural machine translation. PSQ indexes documents in the query language tokens (English in our experiments) by

Table 5. nDCG@20 on NeuCLIR 2022 without query or document translation in the retrieval pipeline. R&R indicates the retrieve-and-rerank pipeline using PSQ as the first-stage retriever followed by the cross-language cross-encoder reranker specified in the first column. T-D indicates the score of the ColBERT-X model trained with Translate-Distill using the cross-encoder specified in the first column as the teacher scorer.

	zho			fas			rus		
	R&R	T-D	△	R&R	T-D	△	R&R	T-D	△
First Stage PSQ	0.329	–	–	0.358	–	–	0.330	–	–
XLMR CE (ET)	0.299	**0.452**	151%	0.306	**0.466**	152%	0.355	**0.503**	142%
XLMR CE (TT)	0.371	**0.450**	121%	0.362	**0.441**	122%	0.405	**0.493**	122%
Mono-mT5XXL	0.459	**0.493**	107%	**0.501**	0.469	94%	**0.503**	0.503	100%

translating the document tokens probabilistically using an alignment matrix. To be more specific, each token in the original document is translated into a bag of query language tokens, where the weight of each resulting token is the product of the original weight multiplied by the translation probability. Therefore, each translated document is a bag-of-token with probabilistic weights in the query language that can be indexed by sparse retrieval systems such as Lucene.

The rerankers are also restricted to those capable of accepting queries and documents in different languages. Note that the official TREC 2022 NeuCLIR Track first-stage retrieval results used BM25 with document translation, which differs from the setting here. However, to compare systems under identical conditions, we use the PSQ model in the pipeline, which is a slightly weaker but much more efficient CLIR system [26].

As summarized in Table 5, models trained with Translate-Distill provide at least 94% of their teachers' retrieval effectiveness. For XLMR cross-encoders, the distilled ColBERT-X models even outperform the use of their cross-encoder teacher models as rerankers. While a Retrieve-and-Rerank pipeline using Mono-mT5XXL is no better than student dual-encoders trained with its knowledge, using the cross-encoders for reranking is much more computationally expensive at retrieval time, and thus query latency could be much longer than for an end-to-end ColBERT-X model.

6   Conclusions and Future Work

This work introduces Translate-Distill, a training pipeline that distills ranking knowledge from an effective cross-encoder to train a CLIR student dual-encoder model. Evaluated on two CLIR collections, each with three language pairs, we have shown that this training pipeline produces statistically significantly more effective CLIR dual-encoders than the earlier Translate-Train approach. Dual-encoders trained with Mono-mT5XXL as the query-passage pair scorer achieve state-of-the-art effectiveness for CLIR end-to-end neural retrieval on the TREC 2022 NeuCLIR benchmark.

We expect that Translate-Distill may help not just with dual-encoders, but also with other neural retrieval models. For example, work on BLADE [26] (a CLIR variant of

SPLADE [10]) has shown Translate-Train to be beneficial; perhaps it may also benefit from Translate-Distill. Distilling from even larger models such as GPT-4, which is known to be a multilingual model [13], may further improve the student CLIR models.

References

1. Asai, A., Kasai, J., Clark, J., Lee, K., Choi, E., Hajishirzi, H.: XOR QA: cross-lingual open-retrieval question answering. In: Proceedings of the 2021 Conference of the North American Chapter of the Association for Computational Linguistics: Human Language Technologies, pp. 547–564, Association for Computational Linguistics, Online (2021). https://doi.org/10.18653/v1/2021.naacl-main.46
2. Bonifacio, L., et al.: mMARCO: A multilingual version of the MS MARCO passage ranking dataset (2021). arXiv preprint arXiv:2108.13897
3. Chi, Z., et al.: Improving pretrained cross-lingual language models via self-labeled word alignment. In: Proceedings of the 59th Annual Meeting of the Association for Computational Linguistics and the 11th International Joint Conference on Natural Language Processing (Volume 1: Long Papers), pp. 3418–3430, Association for Computational Linguistics, Online (2021). https://doi.org/10.18653/v1/2021.acl-long.265
4. Conneau, A., et al.: Unsupervised cross-lingual representation learning at scale. In: Proceedings of the 58th Annual Meeting of the Association for Computational Linguistics, pp. 8440–8451, Association for Computational Linguistics, Online (2020)
5. Costello, C., Yang, E., Lawrie, D., Mayfield, J.: Patapsco: a python framework for cross-language information retrieval experiments. In: Proceedings of the 44th European Conference on Information Retrieval (ECIR) (2022)
6. Dai, Z., Callan, J.: Deeper text understanding for IR with contextual neural language modeling. In: Proceedings of the 42nd International ACM SIGIR Conference on Research and Development in Information Retrieval, pp. 985–988 (2019)
7. Darwish, K., Oard, D.W.: Probabilistic structured query methods. In: Proceedings of the 26th Annual International ACM SIGIR Conference on Research and Development in Information Retrieval, pp. 338–344 (2003)
8. Devlin, J., Chang, M.W., Lee, K., Toutanova, K.: BERT: Pre-training of deep bidirectional transformers for language understanding. In: Proceedings of the 2019 Conference of the North American Chapter of the Association for Computational Linguistics: Human Language Technologies, Volume 1 (Long and Short Papers), pp. 4171–4186, Association for Computational Linguistics, Minneapolis, Minnesota (2019)
9. Formal, T., Lassance, C., Piwowarski, B., Clinchant, S.: From distillation to hard negative sampling: Making sparse neural IR models more effective. In: Proceedings of the 45th International ACM SIGIR Conference on Research and Development in Information Retrieval, pp. 2353–2359 (2022)
10. Formal, T., Piwowarski, B., Clinchant, S.: SPLADE: sparse lexical and expansion model for first stage ranking. In: Proceedings of the 44th International ACM SIGIR Conference on Research and Development in Information Retrieval, pp. 2288–2292 (2021)
11. Gao, L., Callan, J.: Unsupervised corpus aware language model pre-training for dense passage retrieval. In: Proceedings of the 60th Annual Meeting of the Association for Computational Linguistics (Volume 1: Long Papers), pp. 2843–2853, Association for Computational Linguistics, Dublin, Ireland (2022)
12. Jeronymo, V., Lotufo, R., Nogueira, R.: NeuralMind-UNICAMP at 2022 TREC NeuCLIR: Large boring rerankers for cross-lingual retrieval. arXiv preprint arXiv:2303.16145 (2023)

13. Jiao, W., Wang, W., Huang, J., Wang, X., Tu, Z.: Is ChatGPT a good translator? Yes with GPT-4 as the engine (2023). arXiv preprint arXiv:2301.08745

14. Karpukhin, V., et al.: Dense passage retrieval for open-domain question answering (2020). arXiv preprint arXiv:2004.04906

15. Khattab, O., Zaharia, M.: ColBERT: efficient and effective passage search via contextualized late interaction over BERT. In: Proceedings of the 43rd International ACM SIGIR conference on research and development in Information Retrieval, pp. 39–48 (2020)

16. Kwiatkowski, T., et al.: Natural questions: a benchmark for question answering research. Trans. Assoc. Comput. Linguist. **7**, 453–466 (2019)

17. Landauer, T.K., Littman, M.L.: A statistical method for language-independent representation of the topical content of text segments. In: Proceedings of the Eleventh International Conference: Expert Systems and Their Applications, vol. 8, p. 85, Citeseer (1991)

18. Lawrie, D., et al.: Overview of the TREC 2022 NeuCLIR track (2023)

19. Lawrie, D., Mayfield, J., Oard, D.W., Yang, E., Nair, S., Galuščáková, P.: HC3: A suite of test collections for CLIR evaluation over informal text. In: Proceedings of the 46th International ACM SIGIR Conference on Research and Development in Information Retrieval (Taipei, Taiwan) (SIGIR 2023). (2023)

20. Li, Y., Franz, M., Sultan, M.A., Iyer, B., Lee, Y.S., Sil, A.: Learning cross-lingual IR from an English retriever. In: Proceedings of the 2022 Conference of the North American Chapter of the Association for Computational Linguistics: Human Language Technologies, pp. 4428–4436 (2022)

21. Lin, J., Nogueira, R., Yates, A.: Pretrained transformers for text ranking: BERT and beyond. Springer Nature (2022). https://doi.org/10.1007/978-3-031-02181-7

22. Lin, Z., et al.: PROD: progressive distillation for dense retrieval. In: Proceedings of the ACM Web Conference 2023, pp. 3299–3308 (2023)

23. Liu, Y., et al.: RoBERTa: A robustly optimized BERT pretraining approach (2019)

24. MacAvaney, S., Yates, A., Feldman, S., Downey, D., Cohan, A., Goharian, N.: Simplified data wrangling with IR_datasets. In: Proceedings of the 44th International ACM SIGIR Conference on Research and Development in Information Retrieval, pp. 2429–2436 (2021)

25. Nair, S., et al.: Transfer learning approaches for building cross-language dense retrieval models. In: Hagen, M., et al. (eds.) ECIR 2022. LNCS, vol. 13185, pp. 382–396. Springer, Cham (2022). https://doi.org/10.1007/978-3-030-99736-6_26

26. Nair, S., Yang, E., Lawrie, D., Mayfield, J., Oard, D.W.: BLADE: combining vocabulary pruning and intermediate pretraining for scaleable neural CLIR. In: Proceedings of the 46th International ACM SIGIR Conference on Research and Development in Information Retrieval, pp. 1219–1229, SIGIR 2023, Association for Computing Machinery, New York, NY, USA (2023). ISBN 9781450394086

27. Nguyen, T., et al.: MS MARCO: A human generated machine reading comprehension dataset. arXiv preprint arXiv:1611.09268 (2016). http://arxiv.org/abs/1611.09268

28. Nogueira, R., Jiang, Z., Pradeep, R., Lin, J.: Document ranking with a pretrained sequence-to-sequence model. In: Findings of the Association for Computational Linguistics: EMNLP 2020, pp. 708–718, Association for Computational Linguistics, Online (2020). https://doi.org/10.18653/v1/2020.findings-emnlp.63

29. Nogueira, R., Yang, W., Cho, K., Lin, J.: Multi-stage document ranking with BERT (2019). arXiv preprint arXiv:1910.14424

30. Pirkola, A.: The effects of query structure and dictionary setups in dictionary-based cross-language information retrieval. In: Proceedings of the 21st Annual International ACM SIGIR Conference on Research and Development in Information Retrieval, pp. 55–63 (1998)

31. Qu, Y., et al.: RocketQA: An optimized training approach to dense passage retrieval for open-domain question answering. In: Proceedings of the 2021 Conference of the North American

Chapter of the Association for Computational Linguistics: Human Language Technologies, pp. 5835–5847, Association for Computational Linguistics, Online (2021)

32. Raffel, C., et al.: Exploring the limits of transfer learning with a unified text-to-text transformer. J. Mach. Learn. Res. **21**(140), 1–67 (2020). http://jmlr.org/papers/v21/20-074.html

33. Rajbhandari, S., Ruwase, O., Rasley, J., Smith, S., He, Y.: Zero-infinity: breaking the GPU memory wall for extreme scale deep learning. In: Proceedings of the International Conference for High Performance Computing, Networking, Storage and Analysis, pp. 1–14 (2021)

34. Rasley, J., Rajbhandari, S., Ruwase, O., He, Y.: DeepSpeed: system optimizations enable training deep learning models with over 100 billion parameters. In: Proceedings of the 26th ACM SIGKDD International Conference on Knowledge Discovery & Data Mining, pp. 3505–3506 (2020)

35. Rehder, B., Littman, M.L.: Automatic 3-language cross-language information retrieval. In: Information Technology: The Sixth Text REtrieval Conference (TREC-6), vol. 500, p. 233, National Institute of Standards and Technology (1998)

36. Santhanam, K., Khattab, O., Potts, C., Zaharia, M.: PLAID: an efficient engine for late interaction retrieval. In: Proceedings of the 31st ACM International Conference on Information & Knowledge Management, pp. 1747–1756 (2022)

37. Santhanam, K., Khattab, O., Saad-Falcon, J., Potts, C., Zaharia, M.: ColBERTv2: effective and efficient retrieval via lightweight late interaction. In: Proceedings of the 2022 Conference of the North American Chapter of the Association for Computational Linguistics: Human Language Technologies, pp. 3715–3734, Association for Computational Linguistics, Seattle, United States (Jul 2022)

38. Sheridan, P., Ballerini, J.P.: Experiments in multilingual information retrieval using the SPIDER system. In: Proceedings of the 19th Annual International ACM SIGIR Conference on Research and Development in Information Retrieval, pp. 58–65 (1996)

39. Sun, S., Duh, K.: CLIRMatrix: a massively large collection of bilingual and multilingual datasets for cross-lingual information retrieval. In: Proceedings of the 2020 Conference on Empirical Methods in Natural Language Processing (EMNLP), pp. 4160–4170 (2020)

40. Wang, J., Oard, D.W.: Matching meaning for cross-language information retrieval. Inf. Process. Manag. **48**(4), 631–653 (2012). ISSN 0306–4573, https://doi.org/10.1016/j.ipm.2011.09.003

41. Wang, W., Wei, F., Dong, L., Bao, H., Yang, N., Zhou, M.: MiniLM: deep self-attention distillation for task-agnostic compression of pre-trained transformers. In: Proceedings of the 34th International Conference on Neural Information Processing Systems, NIPS 2020, Curran Associates Inc., Red Hook, NY, USA (2020), ISBN 9781713829546

42. Yang, E., Nair, S., Chandradevan, R., Iglesias-Flores, R., Oard, D.W.: C3: continued pretraining with contrastive weak supervision for cross language ad-hoc retrieval. In: Proceedings of the 45th International ACM SIGIR Conference on Research and Development in Information Retrieval, pp. 2507–2512, SIGIR 2022, Association for Computing Machinery, New York, NY, USA (2022). https://doi.org/10.1145/3477495.3531886

43. Zeng, H., Zamani, H., Vinay, V.: Curriculum learning for dense retrieval distillation. In: Proceedings of the 45th International ACM SIGIR Conference on Research and Development in Information Retrieval, pp. 1979–1983 (2022)

Prompt-Based Generative News Recommendation (PGNR): Accuracy and Controllability

Xinyi Li[1]([⊠]) [iD], Yongfeng Zhang[2] [iD], and Edward C. Malthouse[1] [iD]

[1] Northwestern University, Evanston, IL, USA
XINYILI2024@u.northwestern.edu, ecm@northwestern.edu
[2] Rutgers University, Piscataway, NJ, USA
yongfeng.zhang@rutgers.edu

Abstract. Online news platforms often use personalized news recommendation methods to help users discover articles that align with their interests. These methods typically predict a matching score between a user and a candidate article to reflect the user's preference for the article. Given that articles contain rich textual information, current news recommendation systems (RS) leverage natural language processing (NLP) techniques, including the attention mechanism, to capture users' interests based on their historical behaviors and comprehend article content. However, these existing model architectures are usually task-specific and require redesign to adapt to additional features or new tasks. Motivated by the substantial progress in pre-trained large language models for semantic understanding and prompt learning, which involves guiding output generation using pre-trained language models, this paper proposes *Prompt-based Generative News Recommendation* (PGNR). This approach treats personalized news recommendation as a text-to-text generation task and designs personalized prompts to adapt to the pre-trained language model, taking the generative training and inference paradigm that directly generates the answer for recommendation. Experimental studies using the Microsoft News dataset show that PGNR is capable of making accurate recommendations by taking into account various lengths of past behaviors of different users. It can also easily integrate new features without changing the model architecture and the training loss function. Additionally, PGNR can make recommendations based on users' specific requirements, allowing more straightforward human-computer interaction for news recommendation.

Keywords: Large Language Model · Recommender Systems · Natural Language Processing · Information Retrieval

1 Introduction

The newspaper industry has experienced a steady and steep decline over the past decade. There have been widespread layoffs and closures, resulting in 'ghost

N. Goharian et al. (Eds.): ECIR 2024, LNCS 14609, pp. 66–79, 2024.
https://doi.org/10.1007/978-3-031-56060-6_5

newspapers' and 'news deserts', where almost 200 out of 3,143 counties in the U.S have been left with no daily newspaper and 1,540 counties with only one weekly newspaper [1]. The demise of local newspapers is not only a commercial problem, but also a public and social problem. Communities without news organizations have seen an increase in government spending due to a lack of accountability [8]. Citizens who consume less news are unable to evaluate elected officials and are less likely to participate in voting. Reading news is one way for people to gain knowledge and to become more open-minded. Online platforms such as Google News and Microsoft News attract users to read news online [27]. However, in the current information-overloaded society, it is difficult for users to find news articles of interest from the massive set of articles published each day [14]. Therefore, it is important to design news RS to find articles of interest for users.

News RS typically involve three fundamental tasks: analyzing users' interests based on their past behaviors, comprehending news content by considering its contextual information, and predicting a user's preferences for candidate articles for personalized ranking [25]. News articles contain rich textual information, including their titles, bodies, and topics, making NLP techniques like Gated Recurrent Unit (GRU) [5], Long-short Term Memory (LSTM) [10], Convolutional Neural Network (CNN) [4], and attention mechanisms [21] popular choices for modeling users' interests and comprehending article contents [2,23,26].

There have recently been significant developments in pre-trained language models that can be used across various language tasks. These models can transfer knowledge from one task to another without extensive additional training, making them useful for fine-tuning for specific domains with little data compared to training a model from scratch. T5 [19], GPT-3 [3], BERT [7], and RoBERTa [15], are popular pre-trained language models that demonstrate impressive performance on NLP tasks. However, these models are usually large and complex, making it difficult to modify their structures or re-train them. To address this limitation and leverage the pre-trained language models, prompt learning [12], which provides specific prompts to guide the output generation, has been introduced. Prompt learning makes it possible to generate outputs that adapt to the input and has been an effective approach for various NLP tasks. Though some recent works attempted to explore pre-trained language models and prompt learning for news RS, they are mostly based on the slot filling paradigm using the BERT architecture or its variants [29], while the effectiveness of direct generative modeling based on large language models (LLMs) is still left unexplored.

Motivated by the power of LLMs and prompt learning, this paper presents a novel news recommendation model, PGNR (Prompt-based Generative News Recommendation), that treats the personalized news recommendation task as a text-to-text language generation task. In summary, the key contributions are:

– We introduce PGNR, a novel approach that predicts a user's preference for an article by applying personalized prompts that model the user's past behavior and article information. Unlike existing deep neural news RS, PGNR takes a generative recommendation paradigm and meanwhile allows various history lengths for different users throughout the training process.

- We incorporate language generation loss and ranking loss during model training to enhance the model's performance on the recommendation task.
- We demonstrate PGNR's flexibility in incorporating additional article features to improve recommendation performance without any need to modify its model architecture and training loss function.
- We investigate the potential for controlling news recommendations based on individual user requirements, which can enhance user experiences, improve human-computer interaction, and improve the interpretability of news RS with the help of personalized prompt learning. This is also the main advantage that distinguishes PGNR from existing news RS.

2   Related Works

Sequential News Recommendation. Since news articles are items with rich textual information, NLP techniques are often utilized to extract useful information from news contexts and understand users' interests [2,23,26]. Okura et al. [17] propose using a denoising autoencoder to study news representations and use a GRU network to model users' interests. An et al. [2] adopt CNN and the attention mechanism to learn a news representation from its title, topic and subtopic; learn a user's short-term representation using a GRU network; and learn a user's long-term representation using his/her ID embedding. The NRMS model proposed by Wu et al. [24] studies news representation from its title using a word-level, multi-head, self-attention and additive word-attention network, and studies a user's interest using a multi-head, self-attention network with the given historical clicked news sequence by a user. Wu et al. [22] also propose a neural news RS approach that studies news representation using an attentive multi-view network. Besides adopting various language models to better represent users and articles, An et al. [2] suggest focusing not only on users' short-term interests from their past behavior but also their long-term interests from their ID embedding, and Wu et al. [26] suggest being aware of temporal diversity when modeling the match between a user and an article. All of these deep news RS highly rely on the thriving of language techniques, but they are typically trained from scratch. In contrast, PGNR directly treats news RS as a text-to-text language generative task, utilizing prompt learning to generatively fine-tune the T5 language model [19] for news recommendation. Moreover, if further features are available, existing news RS models may require architectural modifications. However, PGNR can simply integrate them into its prompts without any need to modify its model architecture.

Pre-trained Language Models and RS. Motivated by the effectiveness of pre-trained language models and prompt learning techniques, RS researchers tend to formulate recommendation as a language task. Zhang et al. [28] convert the item-based recommendation to a text-based cloze task by modeling a user's historical interactions to a text inquiry. Li et al. [13] design personalized prompt learning for explainable recommendation by treating user and item IDs as prompts. Cui *et al.* [6] propose M6-Rec, which converts a user's behavior to a text inquiry using general textual descriptions. Inspired by the T5 model [19] that studies a unified text-to-text generation model, Geng *et al.* [9] design a flexible and unified text-to-text paradigm called 'Pretrain, Personalized Prompt, and Predict Paradigm' (P5) for RS. Similar to P5 [9], our PGNR is also an encoder-decoder transformer that uses T5 as a backbone. However, different from P5, which relies on user IDs and item IDs [9] and may encounter challenges due to discrepancies between the semantic space of these IDs and that of the pre-trained language models, PGNR describes users' behaviors and news textually in the designed prompts. Zhang *et al.* [29] employ prompt learning for news recommendation by formulating it as a slot filling task for the [MASK] prediction. In contrast, PGNR formulates news recommendation as a direct generative recommendation task. Furthermore, to enhance the language model's performance on the recommendation task, PGNR innovatively integrates the ranking loss into the language generation loss throughout the training.

3   Methodology

Our objective is to estimate user u's preference $\hat{r}_{ui}$ for candidate article i by analyzing the user's interests based on previously read articles. This section describes PGNR and the loss function it employs to train the model parameters. The codes, dataset, and prompts are released at Github.[1]

3.1   Model Architecture

The PGNR employs the pre-trained T5 [19] as its backbone, utilizing transformer [21] blocks to build an encoder-decoder framework. As illustrated in Fig. 1, each user's sequential behavior is converted into a textual input sequence. The encoder processes this input sequence by summing the raw token embedding $X = \{x_1, x_2, \ldots, x_n\}$ with an additional position embedding to capture the token positional information.

Both the encoder and decoder consist of a stack of H identical layers. At the ℓ-th layer of the encoder, the output $X^{\ell-1}$ from the previous layer undergoes a multi-head self-attention mechanism, which generates a set of attention weights for each token based on its interactions with all other tokens in the sequence, resulting in:

$$MH(X^{\ell-1}) = [head_1, \ldots, head_h]W^O$$

where

$$head_i = Attention(X^{\ell-1}W_i^Q, X^{\ell-1}W_i^K, X^{\ell-1}W_i^V).$$

The attention applies the scaled dot product attention

$$Attention(Q, K, V) = softmax\left(\frac{QK^T}{\sqrt{d}}\right)V,$$

---

[1] https://github.com/imrecommender/PGNR.

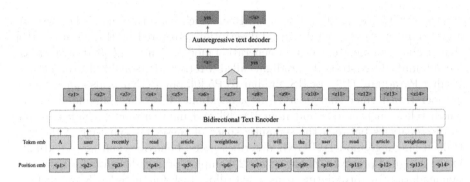

Fig. 1. PGNR utilizes an encoder-decoder framework, where a user's historical behavior is converted into a text inquiry and each news article is described textually, and then PGNR generates the answer to indicate a user's preference to a candidate article through an auto-regressive decoder.

where Q, V and K represent query, value and key of dimension d respectively. The output then undergoes a residual connection and layer normalization, resulting in O^ℓ, which is further passed through a position-wise feed-forward neural network to get

$$ReLU(O^\ell W_1 + b_1)W_2 + b_2.$$

The output of this feed-forward network is added to the original input tokens using a residual connection, and the resulting sequence is normalized using layer normalization to get X^ℓ. In the aforementioned formulas, $\theta = \{W^O, W_i^Q, W_i^K, W_i^V, W_1, W_2, b_1, b_2\}$ are model parameters.

The output of the encoder is a sequence of continuous representations denoted as $Z = \{z_1, z_2, \ldots, z_n\}$. Subsequently, given Z, the decoder engages in auto-regressive generation to produce an output sequence $Y = \{y_1, y_2, \ldots, y_m\}$. The decoder employs a linear transformation and a softmax layer to obtain a probability distribution over all tokens during the generation.

3.2   Training Loss Function

PGNR treats the news RS as a text-to-text language generation task, thus the language generation loss function (i.e., negative log-likelihood (NLL)) is applied to estimate the model parameters θ for auto-regressive model

$$L_{NLL} = -\sum_t \log P_\theta(y_t | y_{<t}, X),$$

where L_{NLL} measures how well the language model can generate the observed output sequence. However, RS often care about how well a model ranks items for a given user, so pair-wise or list-wise training are often applied to maximize the margin between a user u's preference for a clicked positive sample $\hat{r}_{u,pos}$ and that for an unclicked negative sample $\hat{r}_{u,neg}$. To improve the language model's performance in news recommendation task, we incorporate L_{NLL} and Bayesian Personalized Ranking (BPR) loss L_{BPR} [18]

$$L = (1 - \lambda)L_{NLL} + \lambda L_{BPR},$$

where λ is a positive hyper-parameter, and L_{BPR} is defined as

$$L_{BPR} = - \sum_{(u,pos,neg)} \log(\sigma((\hat{r}_{u,pos} - \hat{r}_{u,neg}))),$$

considering all pair of clicked positive and unclicked negative items for users in training. Here, $\sigma(\cdot)$ denotes the sigmoid function.

4 Experiments

We aim to investigate several research questions about the performance of PGNR:

- **RQ1**: How does PGNR perform compared to other baselines in the task of sequential news recommendation?
- **RQ2**: Does PGNR possess the adaptability to incorporate additional article features to enhance recommendation performance?
- **RQ3**: Is it possible for PGNR to produce personalized recommendations according to the specific needs of users?
- **RQ4**: How does the ranking loss L_{BPR} impact the performance of PGNR?
- **RQ5**: How does the definition of additional article features, such as its diversity, have on the performance of PGNR?

4.1 Dataset

We conduct experiments over the Microsoft News dataset (MIND) [27], which is the only well-established benchmark for researchers in the field of news RS [29]. Following common settings [27], the impressions collected from November 9, 2019 to November 14, 2019 are used for model training, while the impressions on November 15, 2019 are used for validation and testing. A summary of the statistical details of the used dataset is provided in Table 1.

4.2 Baseline Methods

We compare the performance of PGNR with some representative baselines:

- **MostPop** and **RecentPop** [11]: recommend the top-K popular articles based on the number of real-time news clicks each article receives, with **RecentPop** considers the clicks in the past 24 h.
- **LSTUR** [2]: captures users' interests by modeling both their long- and short-term preferences.

Table 1. Statistics of MIND used for model evaluations.

#users	#news	#impressions	avg. history length	avg. click rate (%)	avg. title length	#category
141,935	71,671	297,715	23.56	0.10	10.77	18

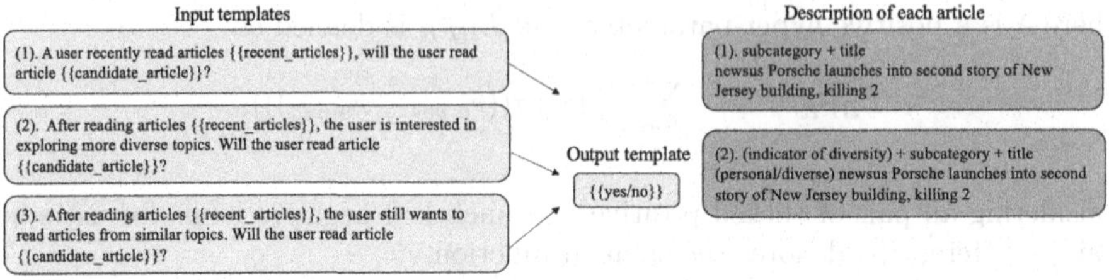

Fig. 2. The personalized prompts are created by designing input-target templates, wherein the relevant fields in the prompts are replaced with corresponding information from the raw data. In this study, the model denoted by PGNR $(i-j)$ employs input template (i) and article description (j). The phrasing of these prompts are common phrasing for recommendations [9,29].

– **TANR** [23]: employs a topic-aware news encoder, and utilizes an attention network to select essential words from the news title and important news from the user's past behavior.
– **NRMS** [24]: models users' and articles' representations via a multi-head self-attention network.
– **NAML** [26]: models users' and articles' representations via multi-view self-attention.
– **Prompt4NR** [29]: approaches news recommendation by formulating it as a slot filling task. For a fair comparison, we opt for its discrete action prompt because it closely corresponds to our designed prompts.

4.3 Implementation Details

The personalized prompts used are presented in Fig. 2. The personalization is from the depiction of a user's past behaviors and the detailed description of each article in input templates. For creating personalized recommendations, input template (1) is utilized. Meanwhile, input templates (2) and (3) are employed to evaluate the controllability of PGNR based on specific user requests, such as exploring more topics or reading articles from similar topics next. To maintain clarity and uniformity, a standardized target template of {yes/no} is adopted.

Different from existing deep neural news recommendations, where a user's preference is calculated through the dot-product of the news embedding and the user embedding, the user's preference for an article $\hat{r}_{ui}$ used for personalized ranking is estimated as the probability that the output from the auto-regressive decoder is 'yes'. During the inference stage, constrained text generation is used, given the prior knowledge that the target output is limited to either 'yes' or 'no'.

The T5 pre-trained checkpoint [19] serves as PGNR's backbone, consisting of 6 layers in both the encoder and decoder components, with a dimension size of 512 and an 8-headed attention mechanism. The SentencePiece [20] tokenizer is used with a vocabulary size of 32,128 sub-word units. A batch size of 16 is employed during training. To incorporate the ranking loss L_{BPR} for each user, a pair of positive and negative sample is generated every time, resulting in the generation of 32 input-target templates for each batch. The peak learning rate is set to 10^{-3}, and the maximum length of input tokens is restricted to 512. The warmup strategy is applied with the warmup stage set to the first 5% of all iterations to adjust the learning rate during training. PGNR is trained for 10 epochs with AdamW optimization on four NVIDIA RTX A6000 GPUs. For training with negative sampling, we used a positive-to-negative sample ratio of approximately 1:4.

5   Performance Evaluations

5.1   Sequential News Recommendations (RQ1)

We assess the effectiveness of PGNR in sequential news recommendations and ensure a fair comparison with other baseline methods by incorporating information on the subcategory and title of the news articles for all methods. For users with a history length shorter than the setting of the history length, existing deep neural news RS add vector embeddings to fill in the remaining articles to a fixed length while PGNR simply adds padding tokens at the end of the input template. As a result, PGNR is capable of considering different lengths of history throughout the training compared to deep neural baselines LSTUR, TANR, NRMS, and NAML.

The experimental results shown in Table 2 provide several insights about personalized news RS. Firstly, we see the methods based on popularity turn out to very competitive baselines for news recommendation. This could be attributed to users' tendency to favor popular articles. Furthermore, the measurement of news popularity could also be influenced by impression bias. Secondly, the performance of our approach without tuning the hyper-parameter λ in the training objective, PGNR (1–1), is comparable to other deep neural baselines. This is because our approach also considers users' interests from their historical behaviors, attempts to understand the articles read by a user before, and employs the attention mechanism. Lastly, adjusting the hyper-parameter λ leads to superior performance of PGNR (1–1)* over Prompt4NR, which adopts a slot filling paradigm for news recommendation. This observation emphasizes the effectiveness of treating news recommendation with a direct generative language model and underscores the potency of our methodology in introducing ranking loss and utilizing paired data to enhance the language model's performance in the news recommendation task. Overall, the performance indicates that treating personalized news RS as a language generation task and utilizing constrained text generation with the assistance of prompt learning is an effective approach for sequential news RS.

Table 2. Experimental results on Mean Reciprocal Rank (MRR), Normalized Discounted Cumulative Gain (NDCG@k), and Hit Ratio (HR@k). * indicates that the hyper-parameter λ in the training objective is adjusted; otherwise, $\lambda = 0$. Bold numbers represent best performance utilizing solely subcategory and title information of articles. ↑ indicates the improved performance when additional article feature is included. The statistical significance was assessed using the Student's t-test, with a significance level of $p < 0.1$.

Methods	MRR	HR@5	NDCG@5	NDCG@10
MostPop	0.2699	0.4899	0.2906	0.3510
RecentPop	0.2704	0.4939	0.2924	0.3519
LSTUR	0.2522	0.4715	0.2712	0.3352
TANR	0.2918	0.5519	0.3241	0.3876
NRMS	0.2847	0.5253	0.3101	0.3763
NAML	0.2943	0.5426	0.3235	0.3870
Prompt4NR	0.2997	0.5487	0.3249	0.3880
PGNR (1–1)	0.2924	0.5450	0.3218	0.3862
PGNR (1–1)*	**0.3084**	**0.5574**	**0.3387**	**0.4012**
PGNR (1–2)*	0.3168 ↑	0.5688 ↑	0.3454↑	0.4068↑

5.2 Incorporate Additional Article Features (RQ2)

Following a comprehensive examination of the dataset, it was discovered that 54% of the 297,715 impressions are associated with articles covering distinct topics compared to those read by users before. Interestingly, users still click on articles within these new topics. This discovery motivated us to improve recommendation accuracy by enhancing article descriptions through the introduction of a diversity signal.

For each user, an article within the impression is categorized as 'diverse' if its topic differs from those of the T most recently read articles by the user. Otherwise, it is classified as 'personal'. With PGNR, all that is required to include the diversity signal is to add this signal to the description using description (2) from Fig. 2. The results of PGNR (1–2)* from Table 2 indicate that PGNR can readily accommodate diversity signals, leading to enhanced recommendation performance without modifying the model architecture or training loss function.

5.3 Controllability of PGNR (RQ3)

We define a RS as controllable when it can tailor recommendations to individual users' preferences. This control is crucial because users might want to explore different topics after reading some articles, and their reading habits vary. Therefore, there should be a mechanism for users to express their preferences to the RS, enabling it to generate personalized recommendations aligned with their specific interests. Current news RS may not recognize this preference and keep suggesting articles with similar content. We now test how our model can consider users' requests to enhance the news RS accordingly. In particular, if a user expresses a preference for exploring articles from a new topic, we expect PGNR to recommend such articles to the user. Conversely, if the user wants to read content similar to what they have previously engaged with, we anticipate the model to provide such recommendations.

To achieve this, we employ input template (2) and article description (2) as shown in Fig. 2 to assess the controllability of PGNR in recommending articles that are tagged as 'diverse' (i.e., from a new topic for the user). Similarly, we use input template (3) and article description (2) to evaluate whether PGNR can provide personal recommendations when necessary. To demonstrate PGNR's effectiveness in considering users' preferences in generating recommendations, we test its performance on three groups of testing impressions: (1) all impressions in the test dataset, (2) diverse impressions in the test set where all clicked articles are labeled as 'diverse', and (3) personal impressions in the test set where all clicked articles are labeled as 'personal'.

Figure 3 compares the performances among different prompts for news recommendation. Subfigure (a) demonstrates that PGNR (1–1)* and PGNR (1–2)* perform similarly in providing sequential news recommendations, while PGNR (2–2)* and PGNR (3–2)*, which aim to recommend articles based on users' preferences for topic diversity, perform worse than PGNR (1–1)* and PGNR (1–2)*. However, PGNR (2–2)*, which targets articles labeled as 'diverse', performs better than all other models in terms of diverse impressions, while PGNR (3–2)*, which targets articles labeled as 'personal', performs better than all other models in terms of personal impressions. Subfigure (b) also presents the number of recommended articles labeled as 'diverse' among the top-10 recommendations. As expected, PGNR (2–2)* suggests a more varied range of topics, while PGNR (3–2)* recommends a limited range. These findings confirm the controllability of PGNR in terms of enabling readers to tailor it to provide either personal or diverse recommendations based on readers' preferences, which is beyond the capability of most currently existing news RS.

5.4   Ablation Study on Ranking Loss (RQ4)

This section describes an ablation study of the training objective function to assess the impact of jointly training the ranking loss L_{BPR} and the language generation loss L_{NLL} in the training process. The results are presented in Fig. 4. We find that not considering L_{BPR} results in sub-optimal recommendations. If λ is too small, the model fails to utilize the benefits of adopting L_{BPR}. Conversely, if λ is too large, the performance of the language model in generating responses may be overlooked, leading to a decline in overall performance. This observation highlights the significance of jointly considering and carefully balancing the ranking loss and the language generation loss during training to enhance the language model's performance on recommendation task.

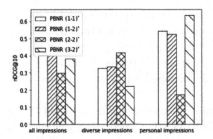
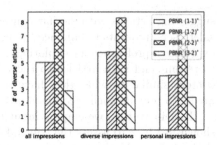
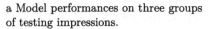

a Model performances on three groups of testing impressions.

b The number of 'diverse' articles within the top-10 recommendations.

Fig. 3. Evaluation of PGNR's controllability to make recommendations based on individual user requirements. PGNR $(2$–$2)^*$ and PGNR $(3$–$2)^*$ are omitted from Table 2 as they focus on evaluating the model's controllability through resampled training data based on requirements.

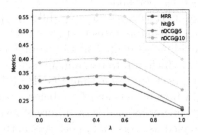

Fig. 4. PGNR performance on sequential recommendation with different λ values – weight on ranking loss.

5.5 Influence of Definition of Diversity (RQ5)

We have shown that the PGNR can effectively incorporate extra article feature, such as whether an article is 'diverse' or 'personal' for the user, to improve its performance. In this section, we present our experimental analysis of how the definition of the diversity of articles affects the model's recommendations.

Figure 5a illustrates the performance of PGNR's recommendations with different threshold values, T, used to classify articles as either 'diverse' or 'personalized'. It shows a general trend that the performance decreases as T either increases or decreases, and we observe that $T = 4$ is an appropriate choice for defining articles' diversity to achieve the best recommendation performance. One underlying reason for the findings is the memorization ability of the large language model [16]. To assess the influence of an article's diversity, we analyzed the proportion of articles labeled as 'diverse' versus 'personal' that were clicked on, denoted as clicked diverse/clicked personal. Based on Fig. 5b, our results demonstrate that when T equals 4, the proportion of clicked articles labeled as 'diverse' is approximately equal to those labeled as 'personal', indicating no dominant label during the training process. This observation implies that the

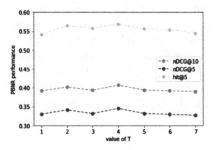

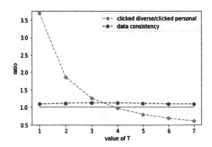

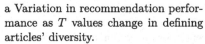

a Variation in recommendation performance as T values change in defining articles' diversity.

b Visualization of training data samples and the consistency between the training data and the testing data with different T values.

Fig. 5. Evaluation of PGNR performance using different threshold value T in defining articles' diversity. For each user, an article within the impression is categorized as 'diverse' if its topic differs from those of the T most recently read articles by the user.

language model may memorize the 'diverse' signal when generating the output sequence for the testing data. Optimal performance of PGNR is achieved when the memorization capability of the large language model is reduced. Since language models have the capability of memorization, it is crucial to carefully define these additional features when incorporating additional features to enhance the model's performance.

6   Conclusion

This work introduces a novel generative news recommendation approach called PGNR that capitalizes on the strengths of pre-trained language models and prompt learning. Rather than considering news recommendation as a conventional task, we treat it as a text-to-text language generation task. To enhance the language model's performance in the recommendation domain, we incorporate both ranking and language generation losses during model training. Our experimental findings show that PGNR outperforms existing baselines in recommendation accuracy and does not require a fixed length of history for all users throughout the training process. This improvement can be attributed to the enhanced language understanding capabilities of pre-trained language models. Unlike other baselines that may necessitate a change in the model's architecture to integrate additional article features, PGNR remains unchanged in structure and training loss function, allowing easy integration of extra features through prompt design. PGNR also stands out from the existing news RS methods in its ability to produce personalized recommendations to meet users' specific needs, improving the human-computer interaction in the domain of news RS through the memorization capabilities of LLMs.

   In the future, we will consider incorporating more and multimodal news information in prompts for news recommendation. Additionally, while our study employs manually designed personalized prompts, future research could explore automated approaches to prompt design, which would allow the system to design prompts more efficiently and independently. The current token limit of 512 may pose challenges in handling long user news interactions, suggesting a potential avenue for future investigation. Furthermore, we aim to extend our exploration to leverage LLMs for enhancing other recommendation metrics, such as the recommendation diversity, in contrast to the emphasis on recommendation accuracy in the current work.

References

1. Abernathy, P.: The expanding news desert, center for innovation and sustainability in local media (2018)
2. An, M., Wu, F., Wu, C., Zhang, K., Liu, Z., Xie, X.: Neural news recommendation with long-and short-term user representations. In: Proceedings of the 57th Annual Meeting of the Association for Computational Linguistics, pp. 336–345 (2019)
3. Brown, T., et al.: Language models are few-shot learners. Adv. Neural. Inf. Process. Syst. **33**, 1877–1901 (2020)
4. Chen, Y.: Convolutional neural network for sentence classification. Master's thesis, University of Waterloo (2015)
5. Cho, K., et al.: Learning phrase representations using RNN encoder-decoder for statistical machine translation. arXiv preprint arXiv:1406.1078 (2014)
6. Cui, Z., Ma, J., Zhou, C., Zhou, J., Yang, H.: M6-Rec: generative pre-trained language models are open-ended recommender systems. arXiv preprint arXiv:2205.08084 (2022)
7. Devlin, J., Chang, M.W., Lee, K., Toutanova, K.: BERT: pre-training of deep bidirectional transformers for language understanding. arXiv preprint arXiv:1810.04805 (2018)
8. Gao, P., Lee, C., Murphy, D.: Municipal borrowing costs and state policies for distressed municipalities. J. Financ. Econ. **132**(2), 404–426 (2019)
9. Geng, S., Liu, S., Fu, Z., Ge, Y., Zhang, Y.: Recommendation as language processing (RLP): a unified pretrain, personalized prompt & predict paradigm (P5). arXiv preprint arXiv:2203.13366 (2022)
10. Graves, A., Graves, A.: Supervised Sequence Labelling with Recurrent Neural Networks, Long Short-term Memory, pp. 37–45. Springer, Heidelberg (2012). https://doi.org/10.1007/978-3-642-24797-2
11. Ji, Y., Sun, A., Zhang, J., Li, C.: A re-visit of the popularity baseline in recommender systems. In: Proceedings of the 43rd International ACM SIGIR Conference on Research and Development in Information Retrieval, pp. 1749–1752 (2020)
12. Jin, W., Cheng, Y., Shen, Y., Chen, W., Ren, X.: A good prompt is worth millions of parameters? Low-resource prompt-based learning for vision-language models. arXiv preprint arXiv:2110.08484 (2021)
13. Li, L., Zhang, Y., Chen, L.: Personalized prompt learning for explainable recommendation. arXiv preprint arXiv:2202.07371 (2022)
14. Lian, J., Zhang, F., Xie, X., Sun, G.: Towards better representation learning for personalized news recommendation: a multi-channel deep fusion approach. In: IJCAI, pp. 3805–3811 (2018)
15. Liu, Y., et al.: RoBERTa: a robustly optimized BERT pretraining approach. arXiv preprint arXiv:1907.11692 (2019)

16. Mireshghallah, F., Uniyal, A., Wang, T., Evans, D.K., Berg-Kirkpatrick, T.: An empirical analysis of memorization in fine-tuned autoregressive language models. In: Proceedings of the 2022 Conference on Empirical Methods in Natural Language Processing, pp. 1816–1826 (2022)

17. Okura, S., Tagami, Y., Ono, S., Tajima, A.: Embedding-based news recommendation for millions of users. In: Proceedings of the 23rd ACM SIGKDD International Conference on Knowledge Discovery and Data Mining, pp. 1933–1942 (2017)

18. Qi, T., Wu, F., Wu, C., Huang, Y.: PP-Rec: news recommendation with personalized user interest and time-aware news popularity. arXiv preprint arXiv:2106.01300 (2021)

19. Raffel, C., et al.: Exploring the limits of transfer learning with a unified text-to-text transformer. J. Mach. Learn. Res. **21**(1), 5485–5551 (2020)

20. Sennrich, R., Haddow, B., Birch, A.: Neural machine translation of rare words with subword units. arXiv preprint arXiv:1508.07909 (2015)

21. Vaswani, A., et al.: Attention is all you need. In: Advances in Neural Information Processing Systems, vol. 30 (2017)

22. Wu, C., Wu, F., An, M., Huang, J., Huang, Y., Xie, X.: Neural news recommendation with attentive multi-view learning. arXiv preprint arXiv:1907.05576 (2019)

23. Wu, C., Wu, F., An, M., Huang, Y., Xie, X.: Neural news recommendation with topic-aware news representation. In: Proceedings of the 57th Annual Meeting of the Association for Computational Linguistics, pp. 1154–1159 (2019)

24. Wu, C., Wu, F., Ge, S., Qi, T., Huang, Y., Xie, X.: Neural news recommendation with multi-head self-attention. In: Proceedings of the 2019 Conference on Empirical Methods in Natural Language Processing and the 9th International Joint Conference on Natural Language Processing (EMNLP-IJCNLP), pp. 6389–6394 (2019)

25. Wu, C., Wu, F., Huang, Y., Xie, X.: Personalized news recommendation: methods and challenges. ACM Trans. Inform. Syst. **41**(1), 1–50 (2023)

26. Wu, C., Wu, F., Qi, T., Li, C., Huang, Y.: Is news recommendation a sequential recommendation task?. In: Proceedings of the 45th International ACM SIGIR Conference on Research and Development in Information Retrieval, pp. 2382–2386 (2022)

27. Wu, F., et al.: MIND: a large-scale dataset for news recommendation. In: Proceedings of the 58th Annual Meeting of the Association for Computational Linguistics, pp. 3597–3606 (2020)

28. Zhang, Y., et al.: Language models as recommender systems: evaluations and limitations. In: I (Still) Can't Believe It's Not Better! NeurIPS 2021 Workshop (2021)

29. Zhang, Z., Wang, B.: Prompt learning for news recommendation. arXiv preprint arXiv:2304.05263 (2023)

CaseGNN: Graph Neural Networks for Legal Case Retrieval with Text-Attributed Graphs

Yanran Tang[(✉)][ID], Ruihong Qiu[ID], Yilun Liu[ID], Xue Li[ID], and Zi Huang[ID]

The University of Queensland, Brisbane, Australia
{yanran.tang,r.qiu,yilun.liu,helen.huang}@uq.edu.au, xueli@eecs.uq.edu.au

Abstract. Legal case retrieval is an information retrieval task in the legal domain, which aims to retrieve relevant cases with a given query case. Recent research of legal case retrieval mainly relies on traditional bag-of-words models and neural language models. Although these methods have achieved significant improvement in retrieval accuracy, there are still two challenges: (1) **Legal structural information neglect**. Previous neural legal case retrieval models mostly encode the unstructured raw text of case into a case representation, which causes the lack of important legal structural information in a case and leads to poor case representation; (2) **Lengthy legal text limitation**. When using the powerful BERT-based models, there is a limit of input text lengths, which inevitably requires to shorten the input via truncation or division with a loss of legal context information. In this paper, a graph neural networks-based legal case retrieval model, CaseGNN, is developed to tackle these challenges. To effectively utilise the legal structural information during encoding, a case is firstly converted into a Text-Attributed Case Graph (TACG), followed by a designed Edge Graph Attention Layer and a Readout function to obtain the case graph representation. The CaseGNN model is optimised with a carefully devised contrastive loss with both easy and hard negative samples. Since the text attributes in the case graph come from individual sentences, the restriction of using language models is further avoided without losing the legal context. Extensive experiments have been conducted on two benchmarks from COLIEE 2022 and COLIEE 2023, which demonstrates that CaseGNN outperforms other state-of-the-art legal case retrieval methods. The code has been released on https://github.com/yanran-tang/CaseGNN.

Keywords: Legal Case Retrieval · Graph Neural Networks

1 Introduction

Legal case retrieval (LCR) is a specialised and indispensable retrieval task that focuses on retrieving relevant cases given a query case. For legal practitioners such as judges and lawyers, using retrieval models is more efficient than manually

N. Goharian et al. (Eds.): ECIR 2024, LNCS 14609, pp. 80–95, 2024.
https://doi.org/10.1007/978-3-031-56060-6_6

finding relevant cases by looking into thousands of legal documents. It is said that 59% of lawyers in the US are using web-based software to get technical services and solution suggestions[1]. LCR also greatly helps a broader community who has legal questions but does not want to spend money on expensive consultation fees.

Existing LCR models can be categorised into two streams: statistical models and language models (LM). Statistical models [16,33,41] focus on measuring term frequency as the case similarity while LMs [1,2,7–9,20,22,27,38,42,49,50, 52,56] conduct nearest neighbour search with case representations from language models. Among the LMs for LCR task, there are different case text matching strategies that focus on sentence [55], paragraph [42] and whole-case [20] levels.

Although the powerful LMs have achieved higher accuracy performance compared to traditional statistical models in LCR, two critical challenges still remain unsolved. (1) **Legal structural information neglect**. Under the context of legal domain, the case structural information typically refers to the relationship among different elements in a legal case, such as parties, crime activities and evidences. Recent LMs for LCR are trying to encode the unstructured raw text of a legal case into a high dimensional representation to measure the similarity with different text matching strategies [20,42,55]. However, only using bag-of-words statistical retrieval models or sequence LMs will restrict the interactions between different elements of cases, which will cause a significant loss of the useful structural information of legal cases. (2) **Lengthy legal text limitation**. As in the study of the LCR benchmark, COLIEE2023 [13], the average length of a case is 5,566 tokens [45], which exceeds most input limitations of LMs, such as 512-token limit of BERT-based models [11]. Therefore, most existing researches rely on truncation [20] or division [42] to shorten the input text. These preprocessing methods for adapting the LMs' input length will lead to incomplete legal case text and finally cause the loss of legal information from a global view.

To address the above two challenges, a novel CaseGNN framework is proposed in this paper. Firstly, for each case, the informative and useful case structural information that refers to the parties, crime activities or evidences of cases will be extracted by Named Entity Recognition and Relation Extraction tools to construct a case graph. Furthermore, to collaboratively utilise the textual and structural information in legal cases, the extracted structural information will be represented by LMs to transform the case graph into a Text-Attributed Case Graph (TACG). Secondly, to effectively obtain a case representation from the TACG, a CaseGNN framework is proposed by utilising the Edge Graph Attention Layer (EdgeGAT) and a Readout function to obtain a graph level representation for retrieval. Finally, to train CaseGNN, a contrastive loss is designed to incorporate effective easy and hard negative samples. Empirical experiments are conducted on two benchmark datasets COLIEE 2022 [12] and COLIEE 2023 [13], which demonstrates that the proposed case structural information and CaseGNN can achieve state-of-the-art performance on LCR by effectively leveraging the

---

[1] https://www.clio.com/blog/lawyer-statistics/.

structural information. The main contributions of this paper are summarised as follows:

- A CaseGNN framework is proposed for LCR to tackle the challenges on incorporating legal structural information and avoiding overlong input text.
- A Text-Attributed Case Graph (TACG) is developed to transform the format of unstructured case text into structural and textual format.
- A GNN layer called edge graph attention layer (EdgeGAT) is designed to learn the representation in the TACG.
- Extensive experiments conducted on two benchmark datasets demonstrate the state-of-the-art performance of CaseGNN.

2   Related Work

2.1   Legal Case Retrieval Models

As a specialised information retrieval (IR) task, the methods of LCR task can be categorised into two streams as IR task: statistical retrieval models [16,33,41] and language models [17,32,34,40]. In LCR task, TF-IDF [16], BM25 [41] and LMIR [33] are the statistical models that also frequently used, which are all based on calculating the text matching score by utilising term frequency and inverse document frequency of words in legal case. General LMs [10,11,26,31] are highly used for LCR task by using LMs to encode the case into representative embeddings for the powerful language understanding ability of LMs [1,2,4–9,22, 23,27,38,44,46,49,50,52,54–57]. In the state-of-the-art research, to tackle the long text problem in legal domain, BERT-PLI [42] divides cases into paragraphs and calculates the similarity between two paragraphs while SAILER [20] directly truncates the case text to cope with the input limit of LM.

2.2   Graph Neural Networks

GNN models can effectively capture the structural information from graph data [14,19,24,25,35–37,48]. To further utilise the edge information, SCENE [30] proposed a GNN layer that can deal with the edge weights. Recently, text-attributed graph are widely used for the capacity of combining both the text understanding ability of LMs and the structural information of graphs, such as TAPE [15], G2P2 [51] and TAG [21].

There are two existing graph-based legal understanding methods, LegalGNN for legal case recommendation [53] and SLR for LCR [29]. Both methods utilise an external legal knowledge database, such as legal concepts and charges, to construct a knowledge graph with human knowledge while encoding legal cases with general LMs. Our proposed CaseGNN is different from these two methods that there is no external knowledge and the encoding of a case actually uses the structural information from the case itself.

3   Preliminary

In the following, a bold lowercase letter denotes a vector, a bold uppercase letter denotes a matrix, a lowercase letter denotes a scalar or a sequence of words, and a scripted uppercase letter denotes a set.

3.1   Task Definition

In legal case retrieval, given the query case q, and the set of n candidate cases $D = \{d_1, d_2, ..., d_n\}$, our task is to retrieve a set of relevant cases $D^* = \{d_i^* | d_i^* \in D \wedge relevant(d_i^*, q)\}$, where $relevant(d_i^*, q)$ denotes that d_i^* is a relevant case of the query case q. The relevant cases are called precedents in legal domain, which refer to the historical cases that can support the judgement of the query case.

3.2   Graph Neural Networks

Graph: A graph is denoted as $G = (V, E)$, where a node v with feature $\mathbf{x}_v \in \mathbb{R}^d$ for $v \in V$, and an edge with feature $\mathbf{e}_{uv} \in \mathbb{R}^d$ for $e \in E$ between node u and v.

Graph Neural Networks: GNNs utilise node features, edge features, and the graph structure to learn representations for nodes, edges and the graph. Most GNNs use iterative neighbourhood aggregation to calculate the representations. After $l - 1$ iterations of aggregation, the output features of a node v after l-th layer is:

$$\mathbf{h}_v^l = \text{Map}^l(\mathbf{h}_v^{l-1}, \text{Agg}(\mathbf{h}_v^{l-1}, \mathbf{h}_u^{l-1}, \mathbf{h}_{e_{uv}})) : u \in N(v))), \tag{1}$$

where $\mathbf{h}_v^l \in \mathbb{R}^d$ is the node representation of v at lth layer, $\mathbf{h}_{e_{uv}} \in \mathbb{R}^d$ is the edge representation between node u and v, and $N(v)$ is the neighbour node set of node v. Specially, the input of the first layer is initialised as $\mathbf{h}_v^0 = \mathbf{x}_v$. Agg and Map are two functions that can be formed in different ways, where Agg performs aggregation to the neighbour node features and edge features while Map utilises the node self features and the neighbour features together for mapping node v to a new feature vector. To generate a graph representation $\mathbf{h}_G$, a Readout function is used to transform all the node features:

$$\mathbf{h}_G = \text{Readout}(G). \tag{2}$$

4   Method

4.1   Text-Attributed Case Graph

Text-Attributed Case Graph (TACG) aims to convert the unstructured case text into a graph. To construct a TACG, the structure and the features of the graph will be obtained by using information extraction tools and language models.

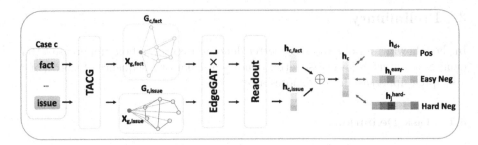

Fig. 1. The framework of CaseGNN. Given legal case c, the legal fact and the legal issue sections are converted into TACG based on information extraction and text encodings. The TACG is processed by L layers of EdgeGAT and a Readout function to obtain an overall case graph representation. The whole framework is trained with the contrastive loss with positive and negative samples.

Information Extraction. To leverage the legal structural information for graph construction, named entity recognition tool and relation extraction tool are used for information extraction. From the legal perspective, the determining factor of relevant cases is the alignment of *legal fact* and *legal issue* [45]. Specifically, legal fact is a basic part of a case that describe "who, when, what, where and why" while legal issue is the legal disputes between parties of a case and need to be settled by judges [45]. The details of generating legal fact and legal issue can be found in PromptCase [45]. Therefore, in this paper, the important legal structural information refers to the relation triplets that generated from legal fact and legal issue, which are extracted from a case text using the PromptCase framework [45]. For example, in COLIEE datasets collected from the federal court of Canada [12,13], a triplet example is extracted as *(applicant, is, Canadian)* from a sentence "The applicant is a Canadian." in legal fact of a case, where *applicant* denotes the "who" and *is, Canadian* refers to the "what" in the case. After conducting information extraction, a set of triplets $R = \{(h, r, t)_{i=1:n}\}$ can be obtained, where h is the head entity, t is the tail entity, r is the relation between h and t, and n is the number of triplets in a case.

Graph Construction. With the extracted set of triplets, the graph construction is to convert these triplets into a case graph. For a legal case c and its triplet set $R_{c,\text{fact}}$ and $R_{c,\text{issue}}$ specifically for its *legal fact* and *legal issue*, the TACG is constructed as $G_{c,\text{fact}} = (V_{c,\text{fact}}, E_{c,\text{fact}})$ and $G_{c,\text{issue}} = (V_{c,\text{issue}}, E_{c,\text{issue}})$. For $G_{c,\text{fact}}$, $V_{c,\text{fact}}$ includes the

Fig. 2. The constructed legal fact graph, $G_{c,\text{fact}}$ and legal issue graph, $G_{c,\text{issue}}$ of case c.

set of nodes of all head and tail entities h and t in $R_{c,\text{fact}}$ as v_h and v_t, and $E_{c,\text{fact}}$ includes the set of edges corresponding to the relations r from head entity h to tail entity t in $R_{c,\text{fact}}$ as $e_{v_h v_t}$. The same construction process is applied to the $G_{c,\text{issue}}$ for the legal issue. Additionally, for both the legal fact graph and the legal issue graph of a case, two virtual node $v_{g,\text{ fact}}$ and $v_{g,\text{ issue}}$ that representing the global textual semantics are added to fact graph and issue graph respectively. To help propagate the global information in the graph representation learning, this virtual global node is connected to every node in the graph. The detailed illustration of the TACG is demonstrated in Fig. 2.

Text Attribute. In the case graph above with the extracted entities as nodes and relations as edges, the node and edge features are obtained by using language models encoding of the text in nodes and edges. For node u, node v, and edge e_{uv} in the case, the text attribute encoding is processed as:

$$\mathbf{x}_u = \text{LM}(t_u); \quad \mathbf{x}_v = \text{LM}(t_v); \quad \mathbf{x}_{e_{uv}} = \text{LM}(t_{e_{uv}}), \tag{3}$$

where $t_u, t_v, t_{e_{uv}}$ is the text of node u, node v and edge e_{uv} respectively, and LM is a pre-trained language model, such as BERT [11], SAILER [20] or Prompt-Case [45]. $\mathbf{x}_u \in \mathbb{R}^d$, $\mathbf{x}_v \in \mathbb{R}^d$, and $\mathbf{x}_{e_{uv}} \in \mathbb{R}^d$ are the output of text-attributed encoding and serve as the feature vector of node u, v and edge e_{uv}. For the virtual global nodes $v_{g,\text{fact}}$ and $v_{g,\text{issue}}$, the node feature is the whole text encodings of the legal fact and legal issue extracted by using PromptCase [45] from a case respectively. The feature of the edge between the virtual global node and other nodes such as entity u will directly reuse the feature of other nodes u in the TACG to simplify the feature extraction:

$$\begin{aligned} \mathbf{x}_{v_{g,\text{fact}}} &= \text{LM}(t_{\text{fact}}); \quad \mathbf{x}_{e_{uv_{g,\text{fact}}}} = \mathbf{x}_u; \\ \mathbf{x}_{v_{g,\text{issue}}} &= \text{LM}(t_{\text{issue}}); \quad \mathbf{x}_{e_{uv_{g,\text{issue}}}} = \mathbf{x}_u. \end{aligned} \tag{4}$$

4.2   Edge Graph Attention Layer

After obtaining the text-attributed features of nodes and edges, a self-attention module will be used to aggregate the nodes and its neighbour nodes and edges information to an informative representation. Moreover, to avoid over-smoothing, a residual connection is added. As previous study shows, multi-head attention has better performance than original attention [47]. According to multi-head attention mechanism, the update of node v with K attention heads is defined as:

$$\mathbf{h}'_v = \mathbf{W}_s \cdot \mathbf{h}_v + \underset{k=1:K}{\text{Avg}} \left(\sum_{u \in N(v)} \alpha^k (\mathbf{W}_n^k \cdot \mathbf{h}_u + \mathbf{W}_e^k \cdot \mathbf{h}_{e_{uv}}) \right), \tag{5}$$

where $\mathbf{h}'_v \in \mathbb{R}^{d'}$ is the updated node feature, and Avg means the average of the output vectors of K heads. Specially, the input of the first EdgeGAT layer is initialised as $\mathbf{h}_v = \mathbf{x}_v$, $\mathbf{h}_u = \mathbf{x}_u$ and $\mathbf{h}_{e_{uv}} = \mathbf{x}_{e_{uv}}$. All the weight matrices $\mathbf{W}$

are in $\mathbb{R}^{d \times d'}$. Specifically, $\mathbf{W}_s$ is the node self update weight matrix, $\mathbf{W}_n$ is the neighbour node update weight matrix and $\mathbf{W}_e$ is the edge update weight matrix respectively. α^k is the attention weight in the attention layer as:

$$\alpha^k = \text{Softmax}(\text{LeakyReLU}(\mathbf{w}_{\text{att}}^k{}^T [\mathbf{W}_n^k \cdot \mathbf{h}_v \parallel \mathbf{W}_n^k \cdot \mathbf{h}_u \parallel \mathbf{W}_e^k \cdot \mathbf{h}_{e_{uv}}])), \quad (6)$$

where Softmax is the softmax function, LeakyReLU is the non-linear function, $\mathbf{w}_{\text{att}}^k \in \mathbb{R}^{3d'}$ is the weight vector of attention layer and $\parallel$ denotes concatenation of vectors. Specifically, the same edge features are reused to make the EdgeGAT simpler. Further model development of updating edge can be designed.

4.3 Readout Function

With the updated node and edge representations, a graph Readout function is designed to obtain the case graph representation for the case c:

$$\mathbf{h}_{c,\text{fact}} = \text{Readout}(G_{c,\text{fact}}), \quad \mathbf{h}_{c,\text{issue}} = \text{Readout}(G_{c,\text{issue}}), \quad (7)$$

where Readout is an aggregation function to output an overall representation in the graph level. One example is the average pooling of all the node embeddings:

$$\mathbf{h}_{c,\text{fact}} = \text{Avg}(\mathbf{h}_{v_i}|v_i \in V_{c,\text{fact}}); \quad \mathbf{h}_{c,\text{issue}} = \text{Avg}(\mathbf{h}_{v_i}|v_i \in V_{c,\text{issue}}). \quad (8)$$

In addition, since the virtual global node has already been in TACG, the updated virtual global node vector $\mathbf{h}_g$ can be considered as the final representation of the case graph:

$$\mathbf{h}_{c,\text{fact}} = \mathbf{h}_{g,\text{fact}}; \quad \mathbf{h}_{c,\text{issue}} = \mathbf{h}_{g,\text{issue}}. \quad (9)$$

For the experiments in this paper, the final virtual global node vector is used as the graph representation.

For case c, the fact graph feature $\mathbf{h}_{c,\text{fact}}$ and issue graph feature $\mathbf{h}_{c,\text{issue}}$ are generated by using $\mathbf{h}_{c,\text{fact}}$ and $\mathbf{h}_{c,\text{issue}}$ respectively. Therefore, the case graph representation $\mathbf{h}_c \in \mathbb{R}^{2d'}$ is the concatenation of $\mathbf{h}_{c,\text{fact}}$ and $\mathbf{h}_{c,\text{issue}}$:

$$\mathbf{h}_c = \mathbf{h}_{c,\text{fact}} \parallel \mathbf{h}_{c,\text{issue}}. \quad (10)$$

4.4 Objective Function

To train the CaseGNN model for the LCR task, it is required to distinguish the relevant cases from irrelevant cases given a large candidate case pool. To provide the training signal, contrastive learning is a tool that aims at pulling the positive samples closer while pushing the negative samples far away used in retrieval tasks [20,50,56]. In this paper, given a query case q and a set of candidate case D that includes both relevant cases d^+ and irrelevant cases d^-, the objective function is defined as a contrastive loss:

$$\ell = -\log \frac{e^{(s(\mathbf{h}_q, \mathbf{h}_{d^+}))/\tau}}{e^{(s(\mathbf{h}_q, \mathbf{h}_{d^+}))/\tau} + \sum_{i=1}^n e^{(s(\mathbf{h}_q, \mathbf{h}_{d_i^{easy-}}))/\tau} + \sum_{j=1}^m e^{(s(\mathbf{h}_q, \mathbf{h}_{d_j^{hard-}}))/\tau}}, \quad (11)$$

where s is the similarity metric such as dot product or cosine similarity, n is the number of easy negative samples, m is the number of hard negative samples, and τ is the temperature coefficient. During training, the positive samples are given by the ground truth from the dataset. The easy negative samples are randomly sampled from the whole candidate pool as well as using the in-batch samples from other queries. For the hard negative samples, it is designed to make use of harder samples to effectively guide the training. Therefore, hard negative samples are sampled based on the BM25 relevance score. If a candidate case has a high score from BM25 yet it is not a positive case, such a case is considered as a hard negative case because it has a high textual similarity to the query case while it is still not a positive case. The overall pipeline is detailed in Fig. 1.

5   Experiments

5.1   Setup

Datasets. To evaluate the proposed CaseGNN, the experiments are conducted on two benchmark LCR datasets, COLIEE2022 [12] and COLIEE2023 [13] from the Competition on Legal Information Extraction/Entailment (COLIEE), where

Table 1. Statistics of datasets.

Datasets	COLIEE2022		COLIEE2023	
	train	test	train	test
# Query	898	300	959	319
# Candidates	4415	1563	4400	1335
# Avg. relevant cases	4.68	4.21	4.68	2.69
Avg. length (# token)	6724	6785	6532	5566
Largest length (# token)	127934	85136	127934	61965

the cases are collected from the federal court of Canada. Given a query case, relevant cases are retrieved from the entire candidate pool. The difference between two datasets are: (1) Although the training sets have overlap, the test sets are totally different; (2) As shown in Table 1, the average relevant cases numbers per query are different, leading to different difficulties in finding relevant cases. These datasets focus on the most widely used English legal case retrieval benchmarks and CaseGNN can be easily extended to different languages with the corresponding information extraction tools and LMs.

Metrics. In this experiment, the metric of precision (P), recall (R), Micro F1 (Mi-F1), Macro F1 (Ma-F1), Mean Reciprocal Rank (MRR), Mean Average Precision (MAP) and normalized discounted cumulative gain (NDCG) are used for evaluation. According to the previous LCR works [20,28,45], top 5 ranking results are evaluated. All metrics are the higher the better.

Baselines. According to the recent research [20,45], 5 popular and state-of-the-art methods are compared as well as the competition winners:

- **BM25** [41]: a strong retrieval benchmark that leverages both term frequency and inverse document frequency for retrieval tasks.
- **LEGAL-BERT** [8]: a legal LM that is pre-trained on large English corpus.

Table 2. Overall performance on COLIEE2022 and COLIEE2023 (%). Underlined numbers indicate the best baselines. Bold numbers indicate the best performance of all methods. Both one-stage and two-stage results are reported.

Methods	COLIEE2022							COLIEE2023						
	P@5	R@5	Mi-F1	Ma-F1	MRR@5	MAP	NDCG@5	P@5	R@5	Mi-F1	Ma-F1	MRR@5	MAP	NDCG@5
One-stage														
BM25	17.9	21.2	19.4	21.4	23.6	25.4	33.6	16.5	30.6	21.4	22.2	23.1	20.4	23.7
LEGAL-BERT	4.47	5.30	4.85	5.38	7.42	7.47	10.9	4.64	8.61	6.03	6.03	11.4	11.3	13.6
MonoT5	0.71	0.65	0.60	0.79	1.39	1.41	1.73	0.38	0.70	0.49	0.47	1.17	1.33	0.61
SAILER	16.6	15.2	14.0	16.8	17.2	18.5	25.1	12.8	23.7	16.6	17.0	25.9	25.3	29.3
PromptCase	17.1	20.3	18.5	20.5	35.1	33.9	38.7	16.0	29.7	20.8	21.5	32.7	32.0	36.2
CaseGNN (Ours)	35.5±0.2	42.1±0.2	38.4±0.3	42.4±0.1	66.8±0.8	64.4±0.9	69.3±0.8	17.7±0.7	32.8±0.7	23.0±0.5	23.6±0.5	38.9±1.1	37.7±0.8	42.8±0.7
Two-stage														
SAILER	23.8	25.7	24.7	25.2	43.9	42.7	48.4	19.6	32.6	24.5	23.5	37.3	36.1	40.8
PromptCase	23.5	25.3	24.4	30.3	41.2	39.6	45.1	21.8	36.3	27.2	26.5	39.9	38.7	44.0
CaseGNN (Ours)	22.9±0.1	27.2±0.1	24.9±0.1	27.0±0.1	54.9±0.4	54.0±0.5	57.3±0.6	20.2±0.2	37.6±0.5	26.3±0.3	27.3±0.2	45.8±0.9	44.4±0.8	49.6±0.8

- **MonoT5** [31]: a pre-trained LM that utilises T5 [39] architecture for document ranking tasks.
- **SAILER** [20]: a pre-trained legal structure-aware LM that obtains competitive performance on both datasets.
- **PromptCase** [45]: an input reformulation method that works on LM for LCR, which achieves sate-of-the-art performance on COLIEE2023 [13] dataset. Two-stage usage of PromptCase with BM25 is evaluated as well.

Implementation. The French text in both datasets are removed. The spaCy[2], Stanford OpenIE [3] and LexNLP[3] packages are used for information extraction. Two-stage experiment uses the top 10 retrieved cases by BM25 as the first stage result. The embedding size are set to 768 based on BERT. The number of EdgeGAT layers are set to 2 and the number of EdgeGAT heads are chosen from {1, 2, 4}. The training batch sizes are chosen from {16, 32, 64, 128}. The Dropout [43] rate of every layer's representation is chosen from {0.1, 0.2, 0.3, 0.4, 0.5}. Adam [18] is applied as optimiser with the learning rate chosen from {0.00001, 0.00005, 0.0001, 0.0005, 0.000005} and weight decay from {0.00001, 0.0001, 0.001, 0.01}. For every query during training, the number of positive sample is set to 1; the number of randomly chosen easy negative sample is set to 1; the number of hard negative samples is chosen from {1, 5, 10, 30}. The in-batch samples from other queries are also employed as easy negative samples. SAILER [20] is chosen as the LM model to generate the text attribute of nodes and edges, which is a BERT-based model that pre-trained and fine-tuned on large corpus of legal cases.

5.2   Overall Performance

In this experiments, the overall performance of CaseGNN is evaluated on COLIEE2022 and COLIEE2023 by comparing with state-of-the-art models, as shown in Table 2. According to the results, CaseGNN achieves the best performance

---

[2] https://spacy.io/.

[3] https://github.com/LexPredict/lexpredict-lexnlp.

Table 3. Ablation study. (%)

Methods	COLIEE2022							COLIEE2023						
	P@5	R@5	Mi-F1	Ma-F1	MRR@5	MAP	NDCG@5	P@5	R@5	Mi-F1	Ma-F1	MRR@5	MAP	NDCG@5
PromptCase	17.1	20.3	18.5	20.5	35.1	33.9	38.7	16.0	29.7	20.8	21.5	32.7	32.0	36.2
w/o v_g	1.6 ± 0.1	2.9 ± 0.1	2.1 ± 0.1	2.2 ± 0.1	4 ± 0.1	4.0 ± 0.1	4.8 ± 0.2	1.6 ± 0.1	2.9 ± 0.1	2.1 ± 0.1	2.2 ± 0.1	4.0 ± 0.1	2.9 ± 0.1	4.8 ± 0.1
Avg Readout	30.5 ± 0.5	36.2 ± 0.6	33.1 ± 0.5	36.6 ± 0.5	61.3 ± 0.4	59.0 ± 0.1	64.6 ± 0.5	17.6 ± 0.4	32.6 ± 0.7	22.8 ± 0.5	23.6 ± 0.4	37.7 ± 1.0	36.6 ± 0.7	41.7 ± 0.5
CaseGNN	35.5 ± 0.2	42.1 ± 0.2	38.4 ± 0.3	42.4 ± 0.1	66.8 ± 0.8	64.4 ± 0.9	69.3 ± 0.8	17.7 ± 0.7	32.8 ± 0.7	23.0 ± 0.5	23.6 ± 0.5	38.9 ± 1.1	37.7 ± 0.8	42.8 ± 0.7

compared with all the baseline models by a large margin in both one-stage and two-stage settings. In COLIEE2022, one-stage CaseGNN has a much higher performance than other two-stage methods, and in COLIEE2023, one-stage CaseGNN has a comparable results with two-stage methods.

For one-stage retrieval setting, compared to the state-of-the-art performance on COLIEE2022 and COLIEE2023, CaseGNN significantly improved the LCR performance by utilising the important legal structural information with graph neural network. Compared with the strong baseline of traditional retrieval model BM25, CaseGNN achieves outstanding performance. The reason of inferior performance of BM25 is because only using the term frequency will ignore the important legal semantics of a case. The performance of CaseGNN also outperforms LEGAL-BERT, a legal corpus pre-trained LCR model, which indicates that only using legal corpus to simply pre-trained on BERT-based model is not enough for the difficult LCR task. The performances of MonoT5 model on two datasets are the poorest in the experiment, which may for the reason that MonoT5 is pre-trained for text-to-text tasks instead of information retrieval tasks. Although SAILER model uses the structure-aware architecture, the performances are not that good as CaseGNN, which shows that the combining of both TACG and GNN model can largely improve the learning and understanding ability of model. Since CaseGNN model uses the fact and issue format to construct the TACG to encode a graph representation, the worse performances of PromptCase indicates the importance of transforming fact and issue into TACG and applying to GNN to learn an expressive case graph representation for LCR.

For two-stage retrieval setting, all methods use a BM25 top10 results as the first stage retrieval and conduct the re-ranking based on these ten retrieved cases. SAILER and PromptCase are compared since these two methods have a comparable one-stage retrieval performance. CaseGNN outperforms both methods in most metrics, especially those related to ranking results, such as MRR@5, MRR and NDCG@5. In COLIEE2022, although CaseGNN has a much higher performance than other baselines, CaseGNN actually cannot benefit from a two-stage retrieval since BM25 cannot provide a higher and useful first stage ranking result compared with CaseGNN itself. In COLIEE2023, all methods can benefit from the two-stage retrieval and CaseGNN can further improve the performance over the one-stage setting.

Table 4. Effectiveness of GNNs. (%)

Methods	COLIEE2022							COLIEE2023						
	P@5	R@5	Mi-F1	Ma-F1	MRR@5	MAP	NDCG@5	P@5	R@5	Mi-F1	Ma-F1	MRR@5	MAP	NDCG@5
GCN	21.3 ± 0.3	25.3 ± 0.4	23.2 ± 0.2	26.0 ± 0.4	46.0 ± 0.2	44.4 ± 0.1	49.7 ± 0.3	12.8 ± 0.2	23.7 ± 0.3	16.5 ± 0.1	16.9 ± 0.2	27.8 ± 0.8	26.8 ± 0.6	31.6 ± 0.7
GAT	29.3 ± 0.1	34.8 ± 0.3	31.8 ± 0.1	35.3 ± 0.2	59.2 ± 0.5	56.9 ± 0.3	62.3 ± 0.7	17.4 ± 0.3	32.2 ± 0.5	22.5 ± 0.3	23.1 ± 0.5	37.5 ± 0.4	36.4 ± 0.4	41.4 ± 0.4
EdgeGAT	35.5 ± 0.2	42.1 ± 0.2	38.4 ± 0.3	42.4 ± 0.1	66.8 ± 0.8	64.4 ± 0.9	69.3 ± 0.8	17.7 ± 0.7	32.8 ± 0.7	23.0 ± 0.5	23.6 ± 0.5	38.9 ± 1.1	37.7 ± 0.8	42.8 ± 0.7

5.3 Ablation Study

The ablation study is conducted to verify the effectiveness of the graph components of CaseGNN: (1) not using any graph, which is equivalent to PromptCase; (2) using TACG without the virtual global node (w/o v_g); and (3) using the average of updated node features as case graph representation (Avg Readout). The experiments are conducted on both datasets and measured under all metrics. Results are reported in Table 3. Only one-stage experiments are considered.

As shown in Table 3, the CaseGNN framework with all the proposed components can significantly outperform the other variants for both datasets. Prompt-Case utilises the text encodings of the legal fact and the legal issue to obtain the case representation, which serves as a strong baseline for LCR. The virtual global node in TACG comes from the encoding of the legal fact and the legal issue. For the w/o v_g variant, the performance is the worst that the model almost cannot learn any useful information because without the proper text encodings, the overall semantics are not effectively encoded. For the Avg Readout variant, the Readout function of CaseGNN is set to using the average node embeddings as the case graph representation. This variant is outperformed by CaseGNN because in this variant, the Readout function ignores the information in the edge features. Nevertheless, Avg Readout has a better result compared with PromptCase, which verifies that the graph structure in the case can provide useful information.

5.4 Effectiveness of GNNs

To validate the effectiveness of the EdgeGAT layer, CaseGNN is compared with variants of substituting EdgeGAT with GCN [19] and GAT [48]. The experiments are conducted on both datasets and evaluated with all metrics. The results are shown in Table 4. Only one-stage experiments are considered. For a fair comparison, all variants will be trained with the proposed TACG.

As shown in the experimental results, EdgeGAT has the highest performance compared with the widely used GNN models GCN and GAT. The outstanding results of EdgeGAT on LCR tasks is because in the proposed CaseGNN method, there is a TACG module including both node and edge features in the case graph. Correspondingly, EdgeGAT has a novel design to incorporate the edge features into the case representation calculation. These edge features are important in terms of the legal information contained in the relations between different entities extracted from the legal case. For both GCN and GAT, since these general GNN layers do not have the capability to utilise the edge information, only the node information encoded from the entities are used in the calculation, which leads to information loss of the legal case. More specifically, GAT has a better performance compared with GCN. This phenomenon is aligned with the performance gap between GAT and GCN on other general graph learning tasks because of the graph learning ability difference between GAT and GCN.

5.5 Parameter Sensitivity

In this experiment, the temperature coefficient τ and the number of hard negative samples in the contrastive loss in Eq. (11) are investigated for their parameter sensitivity. The results are presented in Fig. 3 and Fig. 4 respectively.

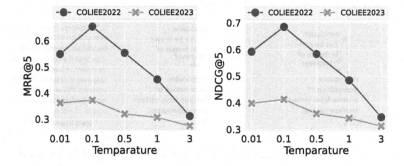

Fig. 3. Parameter sensitivity for the temperature τ in the contrastive loss.

Temperature Coefficient. As shown in Fig. 3, temperature is chosen from {0.01, 0.1, 0.5, 1, 3}. For both datasets, τ set to 0.1 achieves the best performance. When the temperature is too large, the similarity score will be flatten and it cannot provide sufficient training signal to the model via the contrastive loss. In the contrast, when τ is too small, it will extremely sharpen the similarity distribution, which will make the objective function neglect the less significant prediction in the output.

Number of Hard Negative Samples. According to Fig. 4, the choice of number of hard negative samples are from {0, 1, 5, 10}. The hard negative samples are sampled from highly rank irrelevant cases by BM25. Different numbers of hard negative samples in Eq. (11) have different impacts to the final model.

When there is no hard negative samples in the training objective, the model will be only trained with easy negative samples by random sampling. Model trained without hard negative samples will have an inferior performance compared with a proper selected number of hard negative samples. This is because hard negative samples can provide a more strict supervision signal to train the CaseGNN. The performance decreases when there are too many hard negative samples, which is because the training task becomes extremely difficult and the model can barely obtain useful information from the training signal.

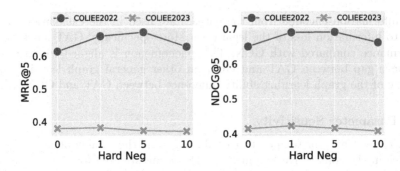

Fig. 4. Parameter sensitivity for the number of hard negative samples in the contrastive loss.

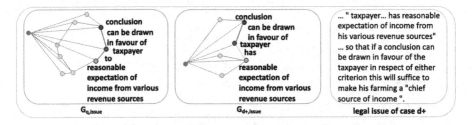

Fig. 5. TACG of cases successfully retrieved by CaseGNN but not by PromptCase.

5.6   Case Study

To further demonstrate how the graph structure helps with the LCR task in CaseGNN, the constructed TACG is visualised in Fig. 5, where CaseGNN successfully performs retrieval while LM-based PromptCase fails. In this visualisation, the constructed TACG of the legal issue in both the query case q and the ground truth candidate case $d+$, $G_{q,\text{issue}}$ and $G_{d+,\text{issue}}$ are presented. From this visualisation, it clear that the graph structure can bring multiple entities together to create a candidate graph that is similar to the query graph. On the contrary, from the case text of the candidate $d+$ on the right of Fig. 5, the corresponding language of these entities and relationships are far from each other, which leads to the unsuccessful retrieval of PromptCase for only using sequential LM without case structural information.

6   Conclusion

This paper identifies two challenges remaining in recent LM-based LCR models about the legal structural information neglect and the lengthy legal text limitation. To overcome the problems from these challenges, this paper proposes a novel framework, CaseGNN, with specific design to utilise the graph neural networks. A Text-Attributed Case Graph (TACG) is developed to transform the unstructured case text into structural graph data. To learn an effective case representation, an Edge Graph Attention Layer (EdgeGAT) is developed. Extensive experiments conducted on two benchmark datasets verify the state-of-the-art performance and the effectiveness of CaseGNN.

Acknowledgements. This work is supported by Australian Research Council CE200100025 and DP230101196.

References

1. Abolghasemi, A., Verberne, S., Azzopardi, L.: Improving BERT-based query-by-document retrieval with multi-task optimization. In: ECIR (2022)
2. Althammer, S., Askari, A., Verberne, S., Hanbury, A.: DoSSIER@COLIEE 2021: leveraging dense retrieval and summarization-based re-ranking for case law retrieval. CoRR abs/2108.03937 (2021)
3. Angeli, G., Premkumar, M.J.J., Manning, C.D.: Leveraging linguistic structure for open domain information extraction. In: ACL (2015)
4. Askari, A., Abolghasemi, A., Pasi, G., Kraaij, W., Verberne, S.: Injecting the BM25 score as text improves BERT-based re-rankers. In: ECIR (2023)
5. Askari, A., Peikos, G., Pasi, G., Verberne, S.: LeiBi@COLIEE 2022: aggregating tuned lexical models with a cluster-driven BERT-based model for case law retrieval. CoRR abs/2205.13351 (2022)
6. Askari, A., Verberne, S.: Combining lexical and neural retrieval with longformer-based summarization for effective case law retrieval. In: DESIRES. CEUR (2021)
7. Askari, A., Verberne, S., Abolghasemi, A., Kraaij, W., Pasi, G.: Retrieval for extremely long queries and documents with RPRS: a highly efficient and effective transformer-based re-ranker. CoRR abs/2303.01200 (2023)
8. Chalkidis, I., Fergadiotis, M., Malakasiotis, P., Aletras, N., Androutsopoulos, I.: LEGAL-BERT: the Muppets straight out of law school. CoRR abs/2010.02559 (2020)
9. Chalkidis, I., Kampas, D.: Deep learning in law: early adaptation and legal word embeddings trained on large corpora. Artif. Intell. Law **27**(2), 171–198 (2019)
10. Dai, Z., Callan, J.: Context-aware sentence/passage term importance estimation for first stage retrieval. CoRR abs/1910.10687 (2019)
11. Devlin, J., Chang, M., Lee, K., Toutanova, K.: BERT: pre-training of deep bidirectional transformers for language understanding. In: NAACL-HLT (2019)
12. Goebel, R., et al.: Competition on legal information extraction/entailment (COLIEE) (2022)
13. Goebel, R., et al.: Competition on legal information extraction/entailment (COLIEE) (2023)
14. Hamilton, W.L., Ying, Z., Leskovec, J.: Inductive representation learning on large graphs. In: NeurIPS (2017)
15. He, X., Bresson, X., Laurent, T., Hooi, B.: Explanations as features: LLM-based features for text-attributed graphs. CoRR abs/2305.19523 (2023)
16. Jones, K.S.: A statistical interpretation of term specificity and its application in retrieval. J. Documentation **60**(5), 493–502 (2004)
17. Khattab, O., Zaharia, M.: ColBERT: efficient and effective passage search via contextualized late interaction over BERT. In: SIGIR (2020)
18. Kingma, D.P., Ba, J.: Adam: a method for stochastic optimization. In: ICLR (2015)

19. Kipf, T.N., Welling, M.: Semi-supervised classification with graph convolutional networks. In: ICLR (2017)
20. Li, H., et al.: SAILER: structure-aware pre-trained language model for legal case retrieval. CoRR abs/2304.11370 (2023)
21. Li, Y., Hooi, B.: Prompt-based zero- and few-shot node classification: a multimodal approach. CoRR abs/2307.11572 (2023)
22. Liu, B., et al.: Investigating conversational agent action in legal case retrieval. In: ECIR (2023)
23. Liu, B., et al.: Query generation and buffer mechanism: towards a better conversational agent for legal case retrieval. Inf. Process. Manag. **59**, 103051 (2022)
24. Liu, Y., Qiu, R., Huang, Z.: CaT: balanced continual graph learning with graph condensation. CoRR abs/2309.09455 (2023)
25. Liu, Y., Qiu, R., Tang, Y., Yin, H., Huang, Z.: PUMA: efficient continual graph learning with graph condensation, vol. abs/2312.14439 (2023)
26. Liu, Y., et al.: RoBERTa: a robustly optimized BERT pretraining approach. CoRR abs/1907.11692 (2019)
27. Ma, Y., et al.: Incorporating retrieval information into the truncation of ranking lists for better legal search. In: SIGIR (2022)
28. Ma, Y., et al.: LeCaRD: a legal case retrieval dataset for Chinese law system. In: SIGIR (2021)
29. Ma, Y., et al.: Incorporating structural information into legal case retrieval. ACM Trans. Inf. Syst. **42**, 1–28 (2023)
30. Monninger, T., et al.: SCENE: reasoning about traffic scenes using heterogeneous graph neural networks. IEEE Robot. Autom. Lett. **8**(3), 1531–1538 (2023)
31. Nogueira, R., Jiang, Z., Pradeep, R., Lin, J.: Document ranking with a pretrained sequence-to-sequence model. In: EMNLP (2020)
32. Nogueira, R.F., Yang, W., Lin, J., Cho, K.: Document expansion by query prediction. CoRR abs/1904.08375 (2019)
33. Ponte, J.M., Croft, W.B.: A language modeling approach to information retrieval. In: SIGIR (2017)
34. Qiao, Y., Xiong, C., Liu, Z., Liu, Z.: Understanding the behaviors of BERT in ranking. CoRR abs/1904.07531 (2019)
35. Qiu, R., Huang, Z., Li, J., Yin, H.: Exploiting cross-session information for session-based recommendation with graph neural networks. ACM Trans. Inf. Syst. **38**, 1–23 (2020)
36. Qiu, R., Li, J., Huang, Z., Yin, H.: Rethinking the item order in session-based recommendation with graph neural networks. In: CIKM (2019)
37. Qiu, R., Yin, H., Huang, Z., Chen, T.: GAG: global attributed graph neural network for streaming session-based recommendation. In: SIGIR (2020)
38. Rabelo, J., Kim, M., Goebel, R.: Semantic-based classification of relevant case law. In: JURISIN (2022)
39. Raffel, C., et al.: Exploring the limits of transfer learning with a unified text-to-text transformer. J. Mach. Learn. Res. **21**, 5485–5551 (2020)
40. Reimers, N., Gurevych, I.: Sentence-BERT: sentence embeddings using Siamese BERT-networks. In: EMNLP-IJCNLP (2019)
41. Robertson, S.E., Walker, S.: Some simple effective approximations to the 2-Poisson model for probabilistic weighted retrieval. In: SIGIR (1994)
42. Shao, Y., et al.: BERT-PLI: modeling paragraph-level interactions for legal case retrieval. In: IJCAI (2020)

43. Srivastava, N., Hinton, G.E., Krizhevsky, A., Sutskever, I., Salakhutdinov, R.: Dropout: a simple way to prevent neural networks from overfitting. J. Mach. Learn. Res. **15**, 1929–1958 (2014)
44. Sun, Z., Xu, J., Zhang, X., Dong, Z., Wen, J.: Law article-enhanced legal case matching: a model-agnostic causal learning approach. CoRR abs/2210.11012 (2022)
45. Tang, Y., Qiu, R., Li, X.: Prompt-based effective input reformulation for legal case retrieval. CoRR abs/2309.02962 (2023)
46. Tran, V.D., Nguyen, M.L., Satoh, K.: Building legal case retrieval systems with lexical matching and summarization using a pre-trained phrase scoring model. In: ICAIL (2019)
47. Vaswani, A., et al.: Attention is all you need. In: NeurIPS, pp. 5998–6008 (2017)
48. Velickovic, P., Cucurull, G., Casanova, A., Romero, A., Liò, P., Bengio, Y.: Graph attention networks. In: ICLR (2018)
49. Vuong, T., Nguyen, H., Nguyen, T., Nguyen, H., Nguyen, T., Nguyen, H.: NOWJ at COLIEE 2023 - multi-task and ensemble approaches in legal information processing. CoRR abs/2306.04903 (2023)
50. Wang, Z.: Legal element-oriented modeling with multi-view contrastive learning for legal case retrieval. In: IJCNN (2022)
51. Wen, Z., Fang, Y.: Augmenting low-resource text classification with graph-grounded pre-training and prompting. In: SIGIR (2023)
52. Xiao, C., Hu, X., Liu, Z., Tu, C., Sun, M.: Lawformer: a pre-trained language model for Chinese legal long documents. AI Open **2**, 79–84 (2021)
53. Yang, J., Ma, W., Zhang, M., Zhou, X., Liu, Y., Ma, S.: LegalGNN: legal information enhanced graph neural network for recommendation. ACM Trans. Inf. Syst. **40**, 33 (2022)
54. Yao, F., et al.: LEVEN: a large-scale Chinese legal event detection dataset. In: ACL (2022)
55. Yu, W., et al.: Explainable legal case matching via inverse optimal transport-based rationale extraction. In: SIGIR (2022)
56. Zhang, H., Dou, Z., Zhu, Y., Wen, J.R.: Contrastive learning for legal judgment prediction. ACM Trans. Inf. Syst. **41**(4), 25 (2023)
57. Zhong, H., Wang, Y., Tu, C., Zhang, T., Liu, Z., Sun, M.: Iteratively questioning and answering for interpretable legal judgment prediction. In: AAAI (2020)

Ranking Heterogeneous Search Result Pages Using the Interactive Probability Ranking Principle

Kanaad Pathak[(✉)][iD], Leif Azzopardi[(✉)][iD], and Martin Halvey[(✉)][iD]

University of Strathclyde, Glasgow, UK
{kanaad.pathak,leif.azzopardi,martin.halvey}@strath.ac.uk

Abstract. The Probability Ranking Principle (PRP) ranks search results based on their expected utility derived solely from document contents, often overlooking the nuances of presentation and user interaction. However, with the evolution of Search Engine Result Pages (SERPs), now comprising a variety of result cards, the manner in which these results are presented is pivotal in influencing user engagement and satisfaction. This shift prompts the question: How does the PRP and its user-centric counterpart, the Interactive Probability Ranking Principle (iPRP), compare in the context of these heterogeneous SERPs? Our study draws a comparison between the PRP and the iPRP, revealing significant differences in their output. The iPRP, accounting for item-specific costs and interaction probabilities to determine the "Expected Perceived Utility" (EPU), yields different result orderings compared to the PRP. We evaluate the effect of the EPU on the ordering of results by observing changes in the ranking within a heterogeneous SERP compared to the traditional "ten blue links". We find that changing the presentation affects the ranking of items according to the (iPRP) by up to 48% (with respect to DCG, TBG and RBO) in ad-hoc search tasks on the TREC WaPo Collection. This work suggests that the iPRP should be employed when ranking heterogeneous SERPs to provide a user-centric ranking that adapts the ordering based on the presentation and user engagement.

Keywords: Probability Ranking Principle · Interactive Probability Ranking Principle · Expected Utility · Expected Perceived Utility

1 Introduction

The primary aim of search engines is to help users find information relevant to their specific needs. This process often involves issuing multiple queries, scanning numerous documents, and evaluating the relevance of retrieved documents [11]. To enhance this search experience, results should be presented in a manner that allows users to efficiently identify relevant information [9]. Traditional result cards, consisting of a title, link, and summary, have been designed with this goal in mind. However, today's SERPs feature a variety of card types, including

© The Author(s), under exclusive license to Springer Nature Switzerland AG 2024
N. Goharian et al. (Eds.): ECIR 2024, LNCS 14609, pp. 96–110, 2024.
https://doi.org/10.1007/978-3-031-56060-6_7

images, data, and suggestions, among others [1,3]. Existing user-centric research suggests that both the design of result cards and the layout of SERPs significantly influence user interactions and their satisfaction and success in search tasks [4,5,17].

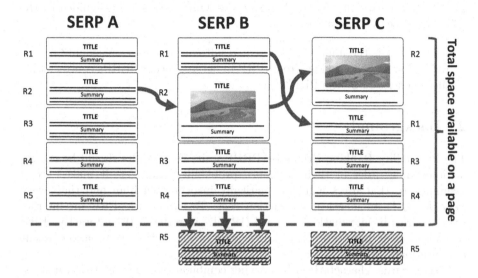

Fig. 1. Compared to SERP A, only four cards can be shown above the fold (dotted horizontal blue line) on SERP B and C. However, changing the card type (e.g., TS to TIS) may also lead to changes in the ranking under the iPRP. (Color figure online)

For example, in Fig. 1, we can see three different SERP layouts: A, B, and C. In SERP A, all results {R1, R2, R3, R4 and R5} are presented using a title and summary (TS). So changing R2 to be presented with a title, image, and summary (TIS) means that it may:

- attract more (or less) clicks, thus changing its interaction probabilities,
- take more (or less) time for a user to decide if they want to click the result, or skip over it, thus changing its cost of interaction, and,
- occupy more (or less) screen space, resulting in a different number of results being displayed above the fold.

The Probability Ranking Principle (PRP) ranks results based purely on the decreasing order of their relevance [13]. Whereas, the Interactive Probability Ranking Principle (iPRP) incorporates interaction probabilities and the cost of processing each result card [7]. Such considerations might cause R2 to rank higher than R1 in terms of "Expected Perceived Utility" (EPU) under the iPRP, as demonstrated in SERP C. Moreover, the type of result cards can significantly influence the overall utility presented to users. Due to space constraints, different card types can alter the number of results displayed on the results page or above the fold, as exemplified by the 5 results in SERP A versus the 4 in SERP B and C. Consequently, adjusting the combination and type of result cards within a SERP introduces trade-offs between EPU, overall utility, and the number of results shown. In this study, we delve into the potential variations in rankings and performance outcomes when using the iPRP, especially within heterogeneous SERPs that feature diverse result card types. Our investigation is guided by the following Research Questions (RQs):

- RQ1: What is the impact of different result cards on user behaviour?
- RQ2: How do the rankings from iPRP in heterogeneous SERPs contrast with those generated by the PRP?

2   Background

The Probability Ranking Principle (PRP) posits that results should be ranked in descending order of their relevance probability [13]. Yet, it makes broad assumptions: that users will systematically inspect each item and that every item demands an identical interaction cost. To extend the PRP for interactive scenarios in information retrieval (IR), [7] introduced the interactive PRP (iPRP), considering both costs and benefits of interacting with various result presentations.

Result cards, characterized by varying combinations such as, titles, images, URLs, and summaries, dictate distinct interaction costs and processing times for users. A substantial body of empirical research has delved into these aspects. For instance, while [12] found negligible differences in processing times between list and tabular displays featuring titles and summaries, [15] observed consistent interaction metrics, such as page clicks, across different card types. Yet, variations in user satisfaction were significant. The accuracy of decisions was enhanced when users were presented results with titles complemented by images [6]. Meanwhile, [9] determined that user satisfaction remained stable across the different search result cards, although the efficacy of summaries was notably task-specific. Direct comparisons between title-only and title + summary displays revealed that the latter had a pronounced positive effect on relevance assessments [16]. Additionally, the length of the summary emerged as a key factor in task performance. While short summaries aligned better with navigational tasks, longer summaries were found to be more beneficial for informational tasks [5,11].

These observations highlight a crucial point: the way results are displayed significantly affects how users interact with them and their satisfaction levels. With the ongoing shift from the traditional "ten blue links" approach to more heterogeneous SERPs, an interesting question emerges: How do different card formats affect the way results are ranked?

The iPRP theorises that such differences will lead to different results rankings. One such instantiation of the iPRP is the Card Model [19] which conceptualizes the interaction process as a cooperative game between two participants:

the system and the user. The goal of the game is to maximize the information gain, while trying to minimize the user effort. The model estimates the utility of a displayed card by considering both its presentation cost and the resulting user benefit (which we will refer to as the expected perceived utility).

Consider a given result item **i** and a specified result card "**card**". In this context, the user can perform actions, denoted as $\mathbf{A_{i,j}}$, within an action space $\mathcal{A}$.

Here, j signifies the type of action: for example clicking, skipping scrolling etc.

Additionally $\mathbf{R_i}$, signifies the relevance of the result item, for example in a relevance space $\mathcal{R}$ of graded relevance, this can be relevant, partially relevant and non-relevant. These actions allow the user to transition to subsequent sets of choices within the retrieved results. Each action $\mathbf{A_{i,j}}$ undertaken by the user within this space comes with an associated expected benefit $\mathbf{B(A_{i,j})}$ and a corresponding expected cost $\mathbf{C(A_{i,j})}$, incurred from performing that specific action, considering the relevance $\mathbf{R_i}$ of the item. The Expected Perceived Utility (EPU) of a result card, for a result item **i**, is thus generally formulated as:

$$EPU_{\text{card}}(i) = \sum_{R_i \in \mathcal{R}} \sum_{A_{i,j} \in \mathcal{A}} P(A_{i,j}|R_i)P(R_i)\Big(B(A_{i,j}|R_i) - C(A_{i,j}|R_i)\Big) \quad (1)$$

This utility can be extended for a result list as $\mathbf{L}$ [19]:

$$EPU\,(\mathbf{L}) = \sum_{i=1}^{n} \left(\prod_{j=1}^{i-1}\left(1 - P(R)_j\right)\right) EPU_{\text{card}}(i)$$
$$\text{subject to } 1 \leq W \leq M \quad (2)$$

Here, $P(R)_j$ represents the relevance probability of the result item, $\mathbf{W}$ is the space occupied by the result card and $\mathbf{M}$ is the total units of screen space available, and **n** is the total number of results in the list L. While the Card Model/iPRP are underpinned by strong theoretical principles, applying them in real-world situations presents challenges related to the estimation of the different costs and benefits. In the following section, we discuss a methodology for estimating the associated costs and benefits, aiming to offer a renewed vision for the Card Model's (specifically the "Plain Card" model) practical execution and evaluate its influence on result ordering.

3   Proposed Method

We begin by considering a common search scenario, where a user is presented with a list items on different result cards. The user examines each result card sequentially, choosing either to: (1) click on the result card, or (2) skip the result card and proceed to the next one. This is the user model assumed by the interactive Probability Ranking Principle [7] and implemented in the Card Model [19]. In this section, we propose a method to estimate the costs and benefits, to compute the EPU for this setting. We use the following key variables:

- $\mathcal{A}$: Action space, containing actions: clicking, c and skipping, s.
- $\mathcal{R}$: Relevance space containing R and $\bar{R}$.
- $\mathbf{R_i}$: Relevance of an item i: relevant, R or non-relevant, $\bar{R}$.
- $\mathbf{P(R_i)}$: Probability of the relevance of item i.
- $\mathbf{P(A_{i,j}|R_i)}$: Probability of taking action $A_{i,j}$ given R_i.
- $\mathbf{B(A_{i,j}|R_i)}$: Benefit of taking action $A_{i,j}$ given R_i.
- $\mathbf{C(A_{i,j}|R_i)}$: Cost of taking action $A_{i,j}$ given R_i.

To estimate the EPU, we must first define the costs and benefits associated with each action, given the result card and associated document. In this work, we adopt the suggestion of [7], who proposed using *time* to represent both the benefit (time saved) and the cost (time spent). The rationale is that users invest their time to find relevant result items (a cost), and discovering relevant result item saves them time as they do not need to keep searching for the required information. The time taken for various actions is influenced by the relevance of the result items and the presentation of the result cards. This is just one method in which benefits can be computed. We can also incorporate other heuristics beyond dwell time such as mouse position and scroll behaviour and incorporate them into our action space for estimating the EPU [8], however, we leave that to be incorporated in the future. Thus, we can calculate the expected cost and benefit based on the summation of the item's relevance, $\mathbf{R}$ (relevant) or $\bar{\mathbf{R}}$ (non-relevant), given an action $\mathbf{A_{i,j}}$. The expected benefit of an action can be written as:

$$\mathbf{B(A_{i,j})} = P(A_{i,j}|R)B(A_{i,j}|R) + P(A_{i,j}|\bar{R})B(A_{i,j}|\bar{R}) \tag{3}$$

and the expected cost of an action can be written as:

$$\mathbf{C(A_{i,j})} = P(A_{i,j}|R)C(A_{i,j}|R) + P(A_{i,j}|\bar{R})C(A_{i,j}|\bar{R}) \tag{4}$$

Given our expressions for the expected cost and benefit, we can re-write the EPU of a card from Eq. 1, where the action space is limited to clicking and skipping and the relevance is binary; for a given item as follows:

$$EPU_{\text{card}}(i) = \sum_{R_i \in \{R, \bar{R}\}} \sum_{A_{i,j} \in \{c,s\}} P(A_{i,j}|R_i)P(R_i)\Big(B(A_{i,j}|R_i) - C(A_{i,j}|R_i)\Big) \tag{5}$$

An open question is: how to meaningfully estimate these costs and benefits in terms of time? The remainder of this paper is divided into several parts to address this. Firstly, we will derive estimations for these variables from the experimental data. Subsequently, we will investigate how different result cards affect the EPU. Lastly, we will evaluate whether incorporating various result cards within the same ranked list has the potential to alter the ranking of results under the card model/iPRP.

4   Experimental Methodology

To explore the impact of the iPRP on ranking heterogeneous search engine result pages, we experimented with four different result card types (see Figs. 2, 3, 4 and 5). These cards represent typical variations on SERPs.

To ground our analysis, we gathered timing data and click data on these result cards across three topics from the TREC WaPo collection, employing 150 annotators. Following this, we utilized the annotation data to estimate interaction probabilities and timing components of EPU, leading to the determination of rankings under the iPRP using the Card Model, as previously detailed.

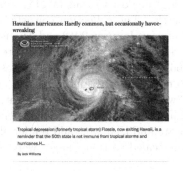

Fig. 2. Title+Image+Summary (TIS)

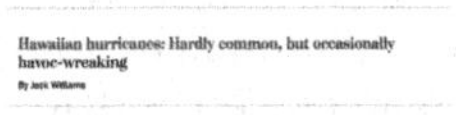

Fig. 3. Title Only (T)

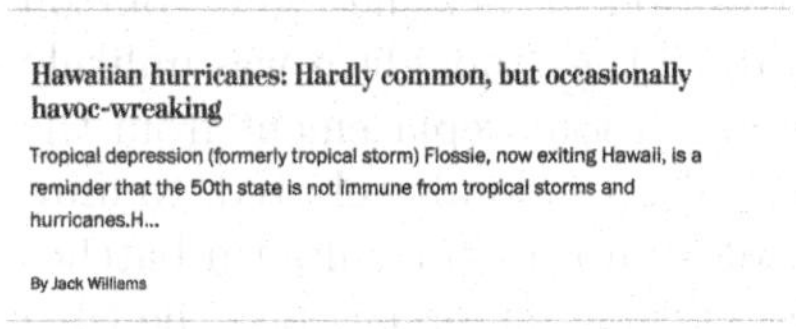

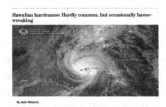

Fig. 4. Title+Summary (TS)

Fig. 5. Title+Image (TI)

The four result card types used were:

(a) **TIS**: These cards displayed the article's Title, Image and Summary (representing the most attractive and informative result card, often used for promoted news articles).

(b) **TI**: These cards displayed the article's Title along with the Image (representing a similar result card to what is used on Google News).

(c) **TS**: These cards displayed the title and summary (representing the default result cards used by the Washington Post).

(d) **T**: These cards only displayed the title of the article (representing the sufficient headlines result card).

To ensure consistency across cards we used the Washington Post's: (1) style sheets to give them all the same look and feel, and (2) summaries which consisted of the lead sentence. The width of the cards was the same (6 columns out of 12 – using Bootstrap), however, because the different cards housed different elements the heights varied, but in a controlled way, where one TIS card used six rows, one TI card used four rows, one TS card used three rows, and the T card used one row. So, for example, if the SERP had 12 rows of vertical space it could hold either two TIS cards, three TI cards, four TS cards, 12 T cards, or some combination of.

Collection: This study used the TREC Washington Post Corpus (WaPo) collection from the TREC Common Core 2018 track[1]. The collection consists of 608,180 news articles and blog posts published between January 2012 and August 2017 categorized into 50 topics for information retrieval tasks. We selected three topics for annotation (341: Airport Security, 363: Transportation Tunnel Disaster, and 408: Tropical Storms). These topics were selected because they had at least 100 TREC judgments, of which at least 50 were judged relevant documents, for which there were downloadable images. This ensured we had a sufficient mixture of relevant/non-relevant items to annotate and that we could render all card types. The images were downloaded and re-scaled so that images were of the same height and width.

Annotations: An interface was developed to collect annotations. Given a description of the topic, the annotators were shown results styled as one of the different result cards. They were then given the option to click the "view" button (if they thought it was relevant), or skip the card (if they thought it was unlikely to be relevant). Results were randomly selected without replacement from the pool of TREC documents and result cards were also randomly selected, to minimize any order effects. Participants could annotate up to 50 results (in batches of 10). We recorded participants' actions (e.g., clicking, skipping) and the time taken to perform these actions. We recruited 150 participants from the United Kingdom and the United States on Prolific[2]. Participants were required to use a desktop or laptop computer and have English as their first language. Pre-screening checks on Prolific ensured they met these criteria. After reading the on-screen information sheet and providing consent, participants completed a set of practice annotations to familiarize themselves with the task and setup. Ethics approval (no. 1643) was granted via the departmental ethics committee at the University of Strathclyde for this task, and participants were compensated in line with national working wage guidelines.

---

[1] https://trec-core.github.io/2018/.
[2] https://www.prolific.co.

Participant Demographics: : In total, we collected 6,052 annotations from 150 participants (approx. 40 annotations per participant, approx. 10 annotations per result card per topic per participant.) The study sample comprised a near-equal gender distribution of 77 males, 71 females and two participants preferring not to identify with either gender. Participants ranged in age from 21 to 75 years, capturing a broad spectrum of adult age groups. Within this cohort, a minority of 14 individuals (9.3%) identified as students, while the majority, 136 participants (90.7%), were non-students.

A significant proportion of the participants, 83 individuals (55.3%), were engaged in either full-time or part-time employment. The remaining 67 participants (44.7%) were not involved in paid employment at the time of the study, which includes groups such as homemakers, retired, or disabled individuals.

4.1 Estimating the EPU

Before we can exactly instantiate the iPRP via the Card Model, we still need to define how we estimate the costs and benefits of clicks and skips, as well as how we estimated the probability of relevance.

For the costs, we used the time spent performing each of the different actions denoted by T i.e.:

- Cost to click a relevant $C(\mathbf{A}_{i,j} = c | \mathbf{R}_i = \mathbf{R}) = \mathbf{T}(c | \mathbf{R})$ and non-relevant item $\mathbf{C}(\mathbf{A}_{i,j} = c | \mathbf{R}_i = \bar{\mathbf{R}}) = \mathbf{T}(c | \bar{\mathbf{R}})$, and,
- Cost to skip a relevant item $C(\mathbf{A}_{i,j} = s | \mathbf{R}_i = \mathbf{R}) = \mathbf{T}(s | \mathbf{R})$ and non-relevant item $\mathbf{C}(\mathbf{A}_{i,j} = s | \mathbf{R}_i = \bar{\mathbf{R}}) = \mathbf{T}(s | \bar{\mathbf{R}})$.

For the benefits, we need to map the gain from a result, to be in the same units as the cost (i.e., in units of time). First we assume that users derive no benefit from choosing to "view" a non-relevant result, or skipping over a result (relevant or not).

So we can represent the benefit for these as: $B(\mathbf{A}_{i,j} = c | \mathbf{R}_i = \bar{\mathbf{R}}) = 0$, $B(\mathbf{A}_{i,j} = s | \mathbf{R}_i = \mathbf{R}) = 0$ and $B(\mathbf{A}_{i,j} = s | \mathbf{R}_i = \bar{\mathbf{R}}) = 0$ for all i that are not-relevant, i.e. they get no benefit from these actions. This leaves the final case when a user clicks "view" on a relevant result. We consider that the time spent reading a relevant result $T(read | R)$ facilitates information acquisition, and thus aligns with the concept of time well spent [14].

We define our benefit from a relevant click to be the time required to read the result $\mathbf{B}(\mathbf{A}_{i,j} = c | \mathbf{R}) = \mathbf{T}(\text{read} | \mathbf{R})$ (we describe below how to estimate this from our measurements) There are potentially other ways to map the gain of information to time or vice versa (e.g., [2,10]), however, we leave exploring such avenues for future work.

Estimation of Time: For each of the card types, given the relevance, we calculated the average time (in seconds) to click or skip the card.

- $\mathbf{T}(c | \mathbf{R})$ and $\mathbf{T}(s | \mathbf{R})$: For each result presented in the annotation interface, we measured the time from when the result appeared until the user either clicked the "view" or "skipped" button next to the result, respectively. We then computed the average time across all results and users for each result card type, taking into account its relevance.

- **T(read|R)**: For each relevant result, given a card type; we measured the time spent reading the result. We then recorded the maximum reading time for each card type and user to compute the average reading time across all users for a given card type. This approach ensures that spending less time on a relevant result does not negatively impact the utility value. For example, if different results are presented in the same card type, depending on the density of the information in the result it may take longer or shorter amounts of time to read it. This could potentially mean that longer reading times would give more benefits. Therefore, if we cap and fix the benefit per card type to the maximum time to read the result, quickly reading a relevant result will not give a small benefit to one card type or vice versa. Thus, in our benefit computation, we account for different reading speeds, lengths and comprehension of information in the result by taking the maximum time to read the result per card type.

Estimation of Probabilities: To estimate the interaction probabilities, we counted the number of times an item was shown and how often it was clicked or skipped, given its relevance score and card type. We used maximum likelihood estimation to calculate the probability of each type of interaction occurring for each card type.

To estimate the probability of relevance in our analysis, we employed the BM25 retrieval function with $\beta = 0.75$. Given that BM25 yields an unbounded retrieval score, it was necessary to convert it to a probability. Following the approach from a related study [11], we used a set of previously submitted queries used on this test collection and issued them to our retrieval engine. For every query variation, the top 50 documents were selected. Then, across all the documents retrieved, the BM25 scores were normalized using z-normalization, and subsequently mapped to a range of [0–1] through a logistic curve transformation. A regression model was then constructed, with an R^2 value of 0.791, to predict the probability of relevance based on a BM25 score. However, it's worth noting that perfecting this model is not the primary focus of this paper.

Given our estimates of the different components, we calculated the expected perceived utility for a given result list (**EPU(L)**) (see Eq. 2) for a list layout using different results card types for each result to determine how the rankings between the iPRP and PRP vary.

5   Results

Table 1 presents the timings, probabilities, and Expected Perceived Utility (EPU) for each card type, with timings measured in seconds. From this data, we crafted a typical user profile, grounding it in the observed timings and interaction probabilities. For every TREC topic, we used the topic title as a query to fetch the top 20 results using BM25 ($\beta = 0.75$). We then calculated the EPU values for each card type, which are displayed in the table as reflective of our average user profile. To pinpoint variations among the card types, used one-way ANOVA tests. Post-hoc analysis was done with Tukey's IISD test. In the table, the mean $\pm$ standard deviation of the timings and probabilities are shown and significant differences ($p < 0.05$) are emphasized using superscripts.

Table 1. Components of the Utility Function, Probabilities, and Expected Perceived Utility (EPU) for All Card Types. Significant differences in values between the card types are indicated by a, b, c or d in superscript.

Card Type	$T(s\|\bar{R})$	$T(c\|\bar{R})$	$T(c\|R)$	$T(s\|R)$	$P(s\|\bar{R})$	$P(c\|R)$	EPU_{card}
a. TS	$4.63 \pm 2.42^{c,d}$	4.41 ± 3.31	4.13 ± 2.32^{d}	5.49 ± 2.81	0.69 ± 0.27	0.81 ± 0.25	$11.78 \pm 4.64^{c,d}$
b. TIS	4.40 ± 2.19^{d}	$5.15 \pm 3.61^{c,d}$	$4.38 \pm 2.70^{c,d}$	$5.86 \pm 3.53^{c,d}$	0.73 ± 0.27	0.82 ± 0.23	$11.77 \pm 4.66^{c,d}$
c. T	3.58 ± 1.62	3.86 ± 2.42	3.64 ± 2.05	4.48 ± 2.23	0.68 ± 0.30	0.80 ± 0.25	10.90 ± 4.14^{d}
d. TI	3.80 ± 1.67	3.72 ± 1.95	3.42 ± 1.64	4.43 ± 2.21	0.73 ± 0.25	0.78 ± 0.26	6.40 ± 2.89

RQ1: What is the impact of different result cards on user behaviour?

Our observations indicate that there are distinct variations in timings and probabilities associated with different card types, and these differences are statistically significant. Referring to Table 1, we can see that integrating an image with a T card diminishes its EPU (T: 10.90, TI: 6.4). This trend implies that the addition of images can potentially divert or mislead users, thereby compromising their ability to swiftly discern relevant details. The interaction probabilities bolster this argument as they display a decreased probability of interacting with a relevant item when an image is included ($P(c\|R)$ for TI: 0.78 ± 0.26).

Conversely, augmenting a T card with a summary enhances its EPU (T: 10.9, TS: 11.78). Summaries, especially when coupled with titles and images, potentially supply crucial context, enabling users to better assess the accompanying image. They also present an overview of the result's content, which aids users in determining its relevance and deciding about further engagement. Hence, while T cards enriched with summaries do incur a slightly higher processing time (T: 3.71 ± 1.95, TS: 4.23 ± 2.36), they demonstrate a reduced probability of mistakes, corroborated by the heightened $P(c\|R)$ values (T: 0.80 ± 0.25, TS: 0.81 ± 0.25).

In the context of Expected Perceived Utility (EPU), we discerned variations in the average EPU across card types. Explicitly, TIS and TS cards exhibit a superior EPU compared to T and TI cards. A one-way ANOVA was conducted, unveiling a statistically significant difference in EPU ($F(3, 3996) = 406.33, p < 0.05$). Delving deeper via Tukey's HSD Test, it became evident that TIS, TS, and TI cards possess an EPU surpassing that of T cards. Moreover, TIS and TS cards outperformed TI cards in terms of EPU. However, the distinction between TS and TIS in EPU was not statistically significant (p = 0.129).

While our study observed differences in EPU, timings, and interaction probabilities across card types, the probability of a user clicking or skipping a relevant item, intrinsic to their behavior, remained unaffected by the card type. This finding is consistent with [15], which reported no variance in click behaviour across diverse interfaces. However, user satisfaction did differ, highlighting individual differences in satisfaction preferences. Although the card type does not consistently alter click probabilities across all users, the EPU can encapsulate these individual variations by incorporating additional context such as the time required to process and read items. For instance, even if a particular card type inherently takes longer to process, it could still be more effective for some users due to their personal preferences or cognitive strengths. Such advantages, like lower error rates (higher $P(c|R)$), could counterbalance the longer processing times, resulting in a higher EPU for specific card designs, such as TS cards, for certain users. Given that these individual card features play a role, an overarching metric, the Expected Perceived Utility (EPU), presents a more holistic view. Our findings underscore that the card type significantly affects user interactions with search results. This raises a subsequent question of how mixing card types on a search results page influences the overall rankings, since changing the card type can change its EPU.

RQ2: How do the rankings obtained from heterogeneous SERPs differ compared to the PRP (in terms of performance)?

Table 2. Comparison of RBO, DCG of Page, and TBG for different card type combinations. Results show a statistically significant difference in RBO between different groups of combinations after running a one-way ANOVA of ($F(7,31841) = 2517.66$, $p < 0.001$). "∼" shows that there is no statistically significant difference with that row.

Combination Type	RBO	DCG of Page	TBG of Page
a. Baseline	1.000 ± 0.000	3.137 ± 1.625	3.073 ± 0.095
b. T or TI	$0.952 \pm 0.135^{\sim c}$	2.437 ± 1.405	1.960 ± 0.482
c. TIS or TS	0.951 ± 0.136	$2.437 \pm 1.407^{\sim g,b}$	$1.962 \pm 0.478^{\sim b}$
d. TIS or T	0.762 ± 0.251	$2.614 \pm 1.649^{\sim g}$	2.381 ± 1.130
e. TS or T	0.741 ± 0.222	$\mathbf{3.588 \pm 2.029}$	$\mathbf{4.363 \pm 0.916}$
f. Random	0.637 ± 0.291	$2.640 \pm 1.646^{\sim d}$	$2.413 \pm 0.784^{\sim d}$
g. TS or TI	$0.505 \pm 0.321^{\sim h}$	$2.525 \pm 1.636^{\sim b}$	2.318 ± 0.215
h. TIS or TI	0.501 ± 0.385	2.024 ± 1.384	1.595 ± 0.249

To explore the differences between ranking results by EPU and by EU (ordering with the PRP), we ran a simulation using all 50 TREC WaPo topics. In this simulation, we used the EU from retrieved results with BM25 ($\beta = 0.75$) as our baseline. We assumed that the default result card type was 'TS' and that a page could display up to 12 rows. Thereby creating a baseline ranked order similar to "n blue links". The core of our simulation involved altering this baseline according to the space constraint. Specifically, we selected every result in the list and changed its card type randomly to one of two possibilities, as illustrated in

Table 2. For example, the first result might change to TIS, the second to TS, and so on. Since the result page is constrained to 12 rows, a page containing TIS and TS cards can have cards in the following combinations – TIS, TS, TS or TIS, TIS or TS, TS, TS etc. We repeated this random alteration 100 times for each result list combination type to observe how such changes impacted the ranking order. After applying these changes, we re-ranked the documents in the altered result list in decreasing order of EPU and then compared this new order with our baseline EU ranking. We used the Rank Biased Overlap (RBO) metric [18] to measure any changes in ranking order. Additionally, we looked at the DCG and TBG metrics ($h = 224$) to see how different SERP layouts affected search result effectiveness.

Our results, presented in Table 2, show that adjusting the presentation of results via different card types to construct heterogeneous SERPs can change document ordering. The RBO metric can quantify this change, however, we acknowledge that RBO is opaque in the sense that it cannot tell us if the change was positive or negative. We leave the exploration of this to future studies that will collect user satisfaction scores to quantify this.

In analyzing DCG scores for our altered result pages, we found that some SERP layouts influenced both RBO and DCG scores similarly. Post-hoc tests using Tukey's HSD Test revealed no significant difference in RBO for certain combinations of card layouts such as T, TI and TIS, TS or TS, TI and TIS or TI. Notably, for DCG, there wasn't a significant difference among several combinations of card layouts, despite the differences in card type mixes.

Time Biased Gain (TBG), accounts for the time spent by the users and their attention on retrieved results [14]. We can observe with the TBG how the costs associated with reading each item in the result list affects the gain of the page. For example the TBG of Page is significantly higher when we combine T cards with TS cards for a result list, where as combining TIS cards with TI cards has a significantly lower TBG compared to the baseline. These results emphasize the role of card types and their arrangement in influencing search result effectiveness and how the time spent assessing the results will affect users' gain. This underscores the need to carefully consider both the presentation and number of search results to optimize user experience (space-utility trade-off).

Our observations show how the alteration of presentation influences the order of ranked list for the iPRP compared to the PRP. In our implementation of TBG, we have implemented a simplistic user model that assumes linear browsing, like the iPRP. In further work, we aim to explore how changing the presentation affects other complex browsing models.

6   Discussion and Future Work

Our study examined whether the Interactive Probability Ranking Principle (iPRP), instantiated via the Card Model, significantly affects the ranking of search result pages based on the relevance of items and their presentation. We aimed to understand the impact of presentation when ranking heterogeneous result pages with four common types of result cards under the iPRP. We framed the iPRP/Card Model as producing the expected perceived utility of each result presented, factoring in different interaction probabilities and decision-making times for various result card types. This approach contrasts with the original PRP, which only considers item relevance for ranking. Our research focused on two main questions, exploring the Expected Perceived Utility (EPU) of different result card types, and the impact of ranking results by EPU on performance with respect to Rank Biased Overlap (RBO), Discounted Cumulative Gain (DCG) and Time Biased Gain (TBG).

Our findings indicate that in the context of ad-hoc news search for the TREC WaPo dataset, result cards using a title, image, and summary (TIS) or title and summary (TS) yield the highest EPU, which is in line with previous research that finds users tend to be more satisfied with a Title and Summary or a Title and Image [6,9,12,15,16]. However, these card types also limit the number of cards that can be displayed on the screen, creating a trade-off between space and utility. We found that this trade-off is crucial, as the choice of result card type can significantly affect SERP effectiveness through the re-ordering of documents at higher ranks, as evidenced by RBO, DCG and TBG measurements.

Moreover, we show how altering the result card type on a SERP changes the ranking of items on the SERP (and also the DCG and TBG of the page) compared to a homogeneous result card format. This suggests that when ranking heterogeneous result pages, it may be possible to manipulate the presentation of results to demote or promote items in the ranking, given the differences in how people engage with different card types. This can raise some ethical concerns as manipulating the presentation can be used to bias users toward specific results. Diving into more detail about these ethical considerations is currently out of the scope of this paper.

In conclusion, our study underscores the importance of considering the presentation of search results when designing ranking algorithms. The perceived relevance of items can change the ranking of documents depending on the presentation of results. We have established that presentation matters when ranking, and that presentation effects can be encoded within a theoretical framework to estimate the expected "perceived" utility.

However, due to the highly controlled nature of our study, the applicability of our observations on other domains such as e-commerce or travel search remains to be explored. Other future work will also investigate different ways to estimate the costs and benefits of interaction and different result card types/styles, and how they impact user performance and satisfaction when interacting with heterogeneous search engine result pages. The findings from our study sets up the foundation for instantiating the benefits and costs and estimating the EPU with a simple model for an ad-hoc search task for the TREC WaPo dataset.

Using this new understanding of EPU, we can think about adapting and optimizing heterogenous SERPs according to user preferences. We should thus theoretically be able to create SERP layouts that increase user satisfaction and make the user more efficient at finding relevant information.

Acknowledgements. We want to thank the reviewers for their insightful suggestions and feedback and all the participants who took part in the study. The work reported here is funded by the DoSSIER project under European Union's Horizon 2020 research and innovation program, Marie Skłodowska-Curie grant agreement No. 860721.

References

1. Azzopardi, L., Thomas, P., Craswell, N.: Measuring the utility of search engine result pages: an information foraging based measure. In: 41st International ACM SIGIR Conference on Research and Development in Information Retrieval, SIGIR 2018, pp. 605–614, June 2018. https://doi.org/10.1145/3209978.3210027, https://dl.acm.org/doi/10.1145/3209978.3210027

2. Azzopardi, L., Zuccon, G.: An analysis of theories of search and search behavior. In: ICTIR 2015 - Proceedings of the 2015 ACM SIGIR International Conference on the Theory of Information Retrieval (2015). https://doi.org/10.1145/2808194.2809447

3. Bota, H., Zhou, K., Jose, J.M.: Playing your cards right: the effect of entity cards on search behaviour and workload. In: CHIIR 2016 - Proceedings of the 2016 ACM Conference on Human Information Interaction and Retrieval, pp. 131–140, March 2016. https://doi.org/10.1145/2854946.2854967, https://dl.acm.org/doi/10.1145/2854946.2854967

4. Chierichetti, F., Kumar, R., Raghavan, P.: Optimizing two-dimensional search results presentation. In: Proceedings of the 4th ACM International Conference on Web Search and Data Mining, WSDM 2011, pp. 257–266 (2011). https://doi.org/10.1145/1935826.1935873

5. Cutrell, E., Guan, Z.: What are you looking for? An eye-tracking study of information usage in Web search. In: Conference on Human Factors in Computing Systems - Proceedings (2007). https://doi.org/10.1145/1240624.1240690

6. Dziadosz, S., Chandrasekar, R.: Do thumbnail previews help users make better relevance decisions about web search results? In: SIGIR Forum (ACM Special Interest Group on Information Retrieval) (2002). https://doi.org/10.1145/564437.564446

7. Fuhr, N.: A probability ranking principle for interactive information retrieval. Inf. Retrieval **11**(3), 251–265 (2008). https://doi.org/10.1007/s10791-008-9045-0

8. Guo, Q., Agichtein, E.: Beyond dwell time: estimating document relevance from cursor movements and other post-click searcher behavior. In: WWW 2012 - Proceedings of the 21st Annual Conference on World Wide Web, pp. 569–578 (2012). https://doi.org/10.1145/2187836.2187914, https://dl.acm.org/doi/10.1145/2187836.2187914

9. Joho, H., Jose, J.M.: A comparative study of the effectiveness of search result presentation on the Web. In: Lalmas, M., MacFarlane, A., Rüger, S., Tombros, A., Tsikrika, T., Yavlinsky, A. (eds.) Advances in Information Retrieval, ECIR 2006. LNCS, vol. 3936. Springer, Heidelberg (2006). https://doi.org/10.1007/11735106_27

10. Kim, Y., Hassan, A., White, R.W., Zitouni, I.: Modeling dwell time to predict click-level satisfaction. In: WSDM 2014 - Proceedings of the 7th ACM International Conference on Web Search and Data Mining, pp. 193–202 (2014). https://doi.org/10.1145/2556195.2556220, https://dl.acm.org/doi/10.1145/2556195.2556220

11. Maxwell, D., Azzopardi, L., Moshfeghi, Y.: A study of snippet length and informativeness behaviour, performance and user experience. In: SIGIR 2017 - Proceedings of the 40th International ACM SIGIR Conference on Research and Development in Information Retrieval, pp. 135–144, August 2017. https://doi.org/10.1145/3077136.3080824

12. Rele, R.S., Duchowski, A.T.: Using eye tracking to evaluate alternative search results interfaces. In: Proceedings of the Human Factors and Ergonomics Society (2005). https://doi.org/10.1177/154193120504901508

13. Robertson, S.E.: The probability ranking principle in IR (1977). https://doi.org/10.1108/eb026647

14. Smucker, M.D., Clarke, C.L.A.: Time-based calibration of effectiveness measures. In: SIGIR 2012 - Proceedings of the International ACM SIGIR Conference on Research and Development in Information Retrieval, pp. 95–104 (2012). https://doi.org/10.1145/2348283.2348300

15. Teevan, J., et al.: Visual snippets: summarizing web pages for search and revisitation. In: Conference on Human Factors in Computing Systems - Proceedings (2009). https://doi.org/10.1145/1518701.1519008

16. Tombros, A., Sanderson, M.: Advantages of query biased summaries in information retrieval. In: SIGIR Forum (ACM Special Interest Group on Information Retrieval) (1998). https://doi.org/10.1145/290941.290947

17. Wang, Y., et al.: Beyond ranking: optimizing whole-page presentation. In: WSDM 2016 - Proceedings of the 9th ACM International Conference on Web Search and Data Mining, pp. 103–112, February 2016. https://doi.org/10.1145/2835776.2835824, https://app.litmaps.com

18. Webber, W., Moffat, A., Zobel, J.: A similarity measure for indefinite rankings. ACM Trans. Inf. Syst. **28**(4), 1–38 (2010). https://doi.org/10.1145/1852102.1852106

19. Zhang, Y., Zhai, C.: Information retrieval as card playing: a formal model for optimizing interactive retrieval interface. In: SIGIR 2015 - Proceedings of the 38th International ACM SIGIR Conference on Research and Development in Information Retrieval (2015). https://doi.org/10.1145/2766462.2767761

DESIRE-ME: Domain-Enhanced Supervised Information Retrieval Using Mixture-of-Experts

Pranav Kasela[1,2(✉)] [ID], Gabriella Pasi[1] [ID], Raffaele Perego[2] [ID],
and Nicola Tonellotto[2,3] [ID]

[1] University of Milano-Biocca, Milan, Italy
gabriella.pasi@unimib.it
[2] ISTI-CNR, Pisa, Italy
pranav.kasela@unimib.it, raffaele.perego@isti.cnr.it,
nicola.tonellotto@unipi.it
[3] University of Pisa, Pisa, Italy

Abstract. Open-domain question answering requires retrieval systems able to cope with the diverse and varied nature of questions, providing accurate answers across a broad spectrum of query types and topics. To deal with such topic heterogeneity through a unique model, we propose DESIRE-ME, a neural information retrieval model that leverages the Mixture-of-Experts framework to combine multiple specialized neural models. We rely on Wikipedia data to train an effective neural gating mechanism that classifies the incoming query and that weighs the predictions of the different domain-specific experts correspondingly. This allows DESIRE-ME to specialize adaptively in multiple domains. Through extensive experiments on publicly available datasets, we show that our proposal can effectively generalize domain-enhanced neural models. DESIRE-ME excels in handling open-domain questions adaptively, boosting by up to 12% in NDCG@10 and 22% in P@1, the underlying state-of-the-art dense retrieval model.

Keywords: Open-domain Q&A · Mixture-of-Experts · Domain Specialization

1 Introduction

The Information Retrieval (IR) research landscape has been fundamentally reshaped by the rapid adoption and emergence of neural models, generating a new paradigm known as Neural Information Retrieval (NIR). Within this transformation, one prominent application of neural models within IR systems is achieved through dense retrieval techniques that have shown promising results in situations where understanding the semantic context of queries and documents is crucial for accurate retrieval. In contrast to their traditional counterparts, which heavily rely on lexical similarities captured by scoring functions

N. Goharian et al. (Eds.): ECIR 2024, LNCS 14609, pp. 111–125, 2024.
https://doi.org/10.1007/978-3-031-56060-6_8

such as TF-IDF or BM25, dense retrieval techniques naturally capture query and document semantics and can be easily adapted to handle multi-modal data and cross-lingual retrieval [19]. However, their training requires large labeled datasets, and the resulting models are typically highly specialized to the task they are trained on and do not generalize well to a new task or domain without additional fine-tuning.

Numerous efforts have been directed towards creating a single neural model that can generalize across many domains, but achieving this goal has proven challenging [26]. In attaining this objective, we must also consider that the queries in many IR tasks are often brief and concise, sometimes lacking sufficient information for comprehensive semantic matching. Moreover, users typically do not explicitly specify the domain of their query, so, if necessary, the system must deduce it in a latent manner.

A sub-field of neural IR is open-domain Q&A, where the questions are posed in natural language and the answer is retrieved from an extensive collection of documents. In this work, to address the above issues, we propose DESIRE-ME, a model for open-domain Q&A that can specialize in multiple domains without changing the underlying pre-trained language model. This specialization is achieved by adaptively focusing the retrieval on the current query domain by leveraging the Mixture-of-Experts (MoE) framework [14]. The MoE framework provides a machine learning architecture combining multiple specialized models, called "specializers" or "experts", to collectively solve a task, such as Q&A. Each specializer within the framework is designed to excel in a specific topical subdomain or under certain conditions, and the MoE model dynamically selects and combines these specializers to make predictions tailored to the input data. A gating mechanism determines which specializer(s) to use for a given input. This gating mechanism is a trained neural network that takes the input query and assigns an importance weight to each expert. The weights indicate the relevance of each specializer for the current input and determine their contribution to the final prediction. The DESIRE-ME approach applied to a complex and faceted task such as open-domain Q&A permits learning a robust and adaptive MoE model that handles the heterogeneity of questions better than state-of-the-art monolithic dense retrievers. To summarize, our research contributions are:

- A modular MoE framework for open-domain Q&A integrated into a dense retrieval system that significantly boosts the performance of the underlying model by exploiting domain specialization;
- A supervised gating method able to understand the query topic and correspondingly weighting the domain contextualization computed by the various MoE specializers;
- A novel experimental framework exploiting the folksonomy of Wikipedia to derive automatically the domains of documents and queries used to train the supervised gating mechanisms;

We evaluate our proposal against state-of-the-art baselines with reproducible experiments on three different datasets[1]. The results of the experiments show

---

[1] The code is available at this link: https://github.com/pkasela/DESIRE-ME.

that DESIRE-ME consistently improves the performance of the underlying dense retriever with an increase of up to 12% in NDCG@10 and 22% in P@1, outlining the potential of the proposed model for the open-domain Q&A task. Furthermore, we utilize a fourth dataset having similar characteristics to investigate the generalization capabilities of DESIRE-ME in a zero-shot scenario. Even in this case, we observe a significant performance boost over the underlying dense retriever.

The paper is organized as follows. Section 2 discusses the relevant related work. Section 3 formally introduces the DESIRE-ME architecture and methodology while Sect. 4 discusses the results of our experimental analysis on public datasets. Finally, Sect. 5 concludes the work and drafts some future work.

2 Related Work

2.1 Open Domain Q&A

Models most commonly used for open-domain Q&A in IR can be broadly classified into five different families based on their architecture: Lexical models, Neural Sparse models, Late-interaction models, Re-ranking models, and Dense retrieval models. Lexical models include all adaptations to open-domain Q&A of classical IR models, such as BM25 [23], that do lexical matching. Neural Sparse models leverage deep neural networks to enhance and overcome some of the limitations of the lexical models, e.g. query-document vocabulary mismatch. They include models such as docT5query [21] that uses sequence-to-sequence models to expand document terms by generating possible queries for which the document would be relevant. Late-interaction models rely on a bi-encoder architecture to encode the query and documents at a token level. The relevance is assessed by computing the similarity between the representations of queries terms and document terms. Late-interaction models allow the pre-computation of documents' representation by delaying the interaction between the query and document representations. A notable example is ColBERT [16], which computes contextualized token-level embeddings for both documents and queries and uses them at retrieval and scoring time. Re-ranking models employ a computationally expensive neural model to re-rank documents retrieved by a fast first-stage ranker. The best-performing re-ranking model in a zero-shot retrieval scenario is currently based on a MonoT5 cross-encoder and utilizes BM25 as the first stage ranker [24]. Dense retrieval models project the query and the documents (or passages) in a common semantic dense vector space and leverage similarity functions to score the documents according to a given query. Many different dense models have been recently proposed because they empirically perform better than lexical and sparse models in many tasks while not being computationally expensive like cross-encoder re-ranking models. Two dense models, namely COCO-DR [29] and Contriever [13], are specifically attractive in this regard for open-domain Q&A as they generalize very well to new domains without the need for labeled data. They are currently among the best performing dense retrieval models on the BEIR benchmarks[2].

---

[2] Official BEIR performance spreadsheet [Deprecated since Jan 10, 2023].

Both models rely on *contrastive learning*, a method that uses pairs of positive and negative examples to learn meaningful and discriminative representations for queries and passages. This is generally done using a synthetic dataset pseudo-labeled in a self-supervised fashion using the target domain corpus.

2.2   Mixture-of-Experts

In this work we employ COCO-DR and contriever in a MoE [14] framework for open-domain Q&A. MoE has been used in many different contexts by the machine learning community [3,7,22]. Shazeer et al. [25] introduced MoE in natural language processing. Their proposal routes a token-level representation through a fixed number of experts. Many works later used MoE in NLP [5,8,9]. MoE models have also been applied in the field of IR for various tasks, for example, for question answering in the biomedical domain [4], visual question answering [20], and for rank fusion for multi-task dense retrieval [18].

MoE allows the creation of expert sub-networks that specialize in an unsupervised manner and improve performance. Even though COCO-DR and Contriever perform exceptionally well on the BEIR benchmark, the domain knowledge is not explicitly leveraged in their training. Due to the high domain specialization of neural networks in NLP tasks, we argue that enforcing specialized MoE IR models should yield better performance. In this work, we rely on these pre-trained dense retrieval models and focus on improving their performance by injecting domain specialization based on a supervised variant of MoE.

3   DESIRE-ME

In this section, we introduce the DESIRE-ME model: in Sect. 3.1, we give an overview of the MoE models; in Sect. 3.2 we describe DESIRE-ME, detailing its components and the training procedure, along with the differences from the classical MoE models.

3.1   MoE Background

Mixture-of-Experts [14] (MoE) is an ensemble learning model that relies on the collective information provided by multiple expert models, which we will also call *domain specializer*, or simply *specializer* from hereon. Each of these specializers is dedicated to a specific topical domain or to a specific sub-task within a broader problem domain. One of the most remarkable aspects of MoEs is their versatility as they can be employed for various types of data and tasks [3,18,20]. In the context of MoEs, a key issue is determining which specializer(s) to rely on for a specific input. This decision process is managed by a gating function, a significant component of a MoE model, which aims to determine the contribution of each specializer in producing the final outcome for a given input. The gating function is trained alongside the specializers to ensure that the gating mechanism and the specializers work together to improve the overall model's performance. For example, let us assume to tackle a complex primary task; MoE can be employed to learn to divide it into M sub-tasks, each handled by a distinct specializer. The gating mechanism learns to predict which sub-task the input will likely belong to and select the appropriate specializer accordingly.

MoE operates as an ensemble model, aggregating the outputs of each specializer in a final pooling stage. Let $\mathbf{x}$ be the vector encoding the input item and $f_i(\mathbf{x})$ the output of the function, f_i, learned by the i-th specializer. Moreover, let $g_i(\mathbf{x})$ be the weight of the i-th specializer computed by the gating mechanism for input $\mathbf{x}$. Various pooling methods have been proposed in the literature to aggregate the output of the specializers. The simplest pooling stage proposed in [30], often referred to as *Top-1 gating*, is a trivial decision model that always chooses the output of the specializer with the highest weight, i.e.:

$$m = \arg\max_{i=1,...,M}(g_i(\mathbf{x}))$$

$$\mathbf{y} = f_m(\mathbf{x})$$

Alternatively, probability scores can be derived from the gating function's output values, possibly using a *softmax* normalization [15]. The resulting probability distribution indicates the likelihood of a specializer being the most appropriate for a given input. In this case, the pooling method makes use of the probability values from the above probability distribution as weights to compute the weighted sum of the M specializers' outputs:

$$\mathbf{y} = \sum_{i=1}^{M} f_i(\mathbf{x}) \cdot g_i(\mathbf{x}) \tag{1}$$

3.2 The DESIRE-ME Model

The overall structure of DESIRE-ME is very similar to that of the underlying bi-encoder dense retrieval model: we have a *query encoder*, which computes the query representation, and a *document encoder*, which computes the document representation. A scoring function, e.g., the dot product or cosine similarity, is used to compute the similarity between the dense vectors representing the query and the document. For efficiency purposes, the embeddings of all the documents in the collection are computed offline using the document encoder and indexed for fast retrieval. In addition to the components of the underlying dense retriever, we introduce in DESIRE-ME a MoE module acting on the query representation only. Such a component inputs the embedding computed by the query encoder and outputs a modified representation of the query having the same dimensionality. The transformation is made utilizing the DESIRE-ME MoE specializers detailed in the following. Since the documents are encoded and indexed offline for fast retrieval, DESIRE-ME applies the MoE only to the query representation that is typically computed online; document representations are not modified based on the specific query processed.

The DESIRE-ME MoE is detailed in Fig. 1. The component has three major modules: the *gating function*, the *specializers*, and the *pooling module*.

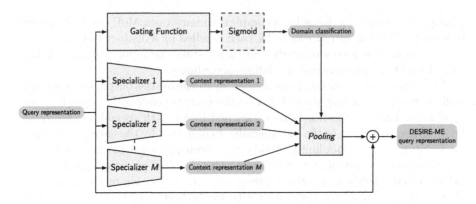

Fig. 1. The MoE module of the proposed model.

The Gating Function. It has the primary purpose of computing the likelihood for the query to belong to any of M predefined domains. Our gating mechanism differs from classical MoE gating functions in several ways. Firstly, it relies on a multi-label domain classifier. Using a classifier as a gating function is not entirely novel in MoE; for example, in [10] a Bayes posterior probability model is used to compute the output values of the gating function. Instead, we do not make the assumption of mutual exclusivity of labels, and we allow an input to belong to multiple domains. To handle multiple labels per query, we enforce that each domain is classified independently by applying a sigmoid function to the gating function output, as opposed to the commonly used softmax function. The use of softmax could compel the model to specialize even for out-of-domain queries, potentially resulting in unexpected outcomes. Another difference from the classical MoE models, where the gating function and the specializers' representation are trained together, is that we train end-to-end the gating function and the specializers using two distinct loss functions. While the multi-label classifier is trained using binary cross-entropy, the MoE specializers rely on the contrastive loss computed on query-document similarity, i.e., the same loss function employed for training the underlying dense retrieval bi-encoder architecture. The multi-label classifier used and the process followed for generating the query labels and training it are detailed in Sect. 4.

The Specializers. They are very similar to those proposed in [14]. Each of the M specializers focuses on tuning the input query representation for the corresponding domain. At training time they learn via the contrastive loss function how to contextualize the query for the specific domain.

The Pooling Module. Finally we have the *pooling module* that merges the domain context representations computed by the specializers on the basis of the domain likelihood estimated by the gating function in the form of a normalized vector of M weights. Merging is accomplished by simply weighting and summing up the outputs of the specializers, as shown in Eq. 1 and depicted in Fig. 1.

We note that a consequence of the enforced domain independence condition is that an input query can be classified by our gating function as not belonging to any of the predefined domains. This is the reason why DESIRE-ME model has a skip connection for the input query representation that is updated with the domain context representation computed by the previous modules. Thanks to such a skip connection, when DESIRE-ME encounters an out-of-domain query, it outputs the unmodified representation of the query not benefiting from specialization.

4   Experimental Analysis

In the following we detail the extensive experiments conducted to answer the following research questions:

RQ1: Can DESIRE-ME enhance the effectiveness of state-of-the-art dense retrieval models for open-domain Q&A?
RQ2: Does a DESIRE-ME model trained on a dataset generalize to datasets having similar characteristics in a zero-shot scenario?

4.1   Experimental Settings

In this Section, we detail the characteristics of the datasets used for the experiments; we then discuss how the datasets are used to train and test DESIRE-ME.

Datasets. In our experiments, we use four datasets included in BEIR (BEnchmarking IR [26]), a valuable resource for tackling the issue of models' generalization. The datasets are: NaturalQuestion [17], HotpotQA [28], FEVER [27], and Climate-FEVER [6]. The main characteristics of the four datasets are resumed in Table 1. They all rely on a corpus based on Wikipedia, and provide binary relevance assessments for query-document pairs:

- NaturalQuestion (NQ) contains queries submitted to the Google search engine and their answers drawn from Wikipedia articles. The passages within the Wikipedia articles that provide satisfactory answers to the questions have been identified by human annotators.
- HotpotQA focuses on complex questions that a single span of text might not answer and could involve reasoning over multiple documents. Queries and relevance labels have been generated with crowd-sourcing.
- FEVER is a resource proposed to tackle fact-checking and verification claims. It encompasses queries and documents from various domains and relies, as the previous datasets, on a Wikipedia-based corpus.
- Climate-FEVER is a dataset for verifying climate change-related claims. It includes ~1500 test queries (no training data). The corpus is the same as FEVER, with the addition of 25 more documents unavailable in FEVER.

Table 1. Characteristics of the datasets used. Labeled queries and the average number of labels per query refer to training queries only.

Dataset	#Docs	#Training	#Validation	#Test	Labeled docs	Labeled queries	Avg labels
NaturalQuestions [17]	2,681,468	132,803	-	3,452	97.1%	97.8%	2.04
HotpotQA [28]	5,233,329	85,000	5,447	7,405	95.45%	99.9%	3.62
FEVER [27]	5,416,568	109,810	6,666	6,666	91.96%	99.1%	2.28
Climate-FEVER [6]	5,416,593	-	-	1,535	91.95%	-	-

Query-Domain Labels. As discussed in the previous section, the DESIRE-ME gating function is trained in a supervised way by exploiting domain labels available for the training queries. We automatically generated such labels for all the questions in the first three datasets by resorting to the category assigned by contributors to their Wikipedia articles[3]. For example, the page on *Eleventh Amendment to the United States Constitution* belongs to the category Law. In contrast, the page on *Chinese New Year* belongs to categories Human behavior, Culture, Society, and Religion. The straightforward approach we employ to create query labels involves assigning to each query the category of the corresponding Wikipedia article containing the relevant passage. However, this basic methodology proved inadequate in specific situations, necessitating the implementation of more specific actions. The first issue arises when the relevant Wikipedia article lists specific subcategories without mentioning the main category. In such instances, starting from each subcategory, we navigate the Wikipedia category graph backward in a breadth-first manner until we reach the category to which the subcategory belongs. The second scenario occurs when the relevant article pertains to multiple categories and/or two or more Wikipedia pages are pertinent to the same query. In such cases, we identify the categories for each page and simply label the query with all the categories of all relevant pages.

By following this approach, we successfully label the vast majority of questions in the datasets. The percentage of labeled documents and queries and the average number of per-query labels are reported in Table 1 for the three datasets having training queries. The labels per query are not equally distributed: for instance, in FEVER there are ~5000 queries in the category *Life*, meanwhile only ~500 queries belong to the category *Mathematics*.

MoE Specializers and Training Hyperparameters. Since in DESIRE-ME each specializer focuses on a specific query category, we employ 37 distinct MoE specializers, a number equal to the number of distinct query categories in the datasets. DESIRE-ME specializers feature a simple architecture: they consist of a down-projection layer using a feed-forward network (FFN) that reduces the input dimension by half. The output layer comprises an up-projection FFN layer, which restores the vector dimension to match the input dimension. This design draws inspiration from the adapter layer proposed in [12]. However, we opted not to use that complete adapter layer in our setup, as the skip connection is

---

[3] https://en.wikipedia.org/wiki/Wikipedia:FAQ/Categories.

already introduced within the MoE module. The gating function classifier has two up-projection layers, which increase the vector dimension to 2× and 4×, respectively. The output layer is a down-projection FFN with the same size as the number of categories, i.e., 37 in our case. We set the training batch size to 512, the learning rate to 10^{-5}, and train for 60 epochs. We use 5% of the training set for validation and keep only the checkpoint with the lowest validation loss.

Metrics and Baselines. We assess the results of the experiments using: MAP@100, MRR@100, R@100, NDCG@10, NDCG@3 and P@1. While NDCG@10 and R@100 are commonly used on BEIR benchmarks, the additional metrics allow us to have a deeper understanding of the potential improvement of DESIRE-ME at small cutoffs. We also report statistically significant differences according to a Bonferroni corrected two-sided paired Student's t-tests with p-value < 0.001. We rely on the *ranx* library [1] for evaluation. To simplify comparative evaluations and to give the possibility of computing other evaluation metrics, all the runs are made publicly available on *ranxhub*[4] [2]. We compare DESIRE-ME variants integrated within the following different state-of-the-art dense retrieval models[5]: COCO-DR, COCO-DR$_{XL}$, and Contriever against the following baselines, for each dense retrieval model:

- *Base.* The original dense retrieval model without MoE in a zero-shot scenario.
- *Fine-tuned.* We fine-tuned the base models with the training data with a batch size of 32 and a learning rate of 10^{-6} for 10 epochs. All the other training hyper-parameters are taken from their original settings.
- *Random_gating (RND-G).* We use randomly generated weights to merge specializers' outputs. This baseline is introduced to assess the benefits of our supervised gating function. All other DESIRE-ME settings are unchanged.

4.2   Results and Discussion

Answering RQ1. To answer RQ1, we conduct multiple experiments using the NQ, HotpotQA, and FEVER datasets to assess DESIRE-ME capability to enhance the effectiveness of the underlying dense retrieval model. The results on the three datasets are reported in Table 2, Table 3, and Table 4, respectively.

Table 2 reports the results of the experiments conducted with the NQ dataset. The figures reported in the table show that fine-tuning the base model using the training data does not yield any benefit and that the integration of DESIRE-ME into the different dense retrieval systems always results in a remarkable improvement of the performances. Irrespective of the metrics considered and the dense retriever used, our solution boosts the base models of a statistically significant margin. The Contriever relative improvement reaches an astonishing 12% in NDCG@10 and 22% in P@1 over the base model. This indicates that DESIRE-ME contributes significantly to enhancing the ranking quality of retrieved documents, particularly in the top positions. Furthermore, it is also worth noting

---

[4] https://amenra.github.io/ranxhub.

[5] Available on HuggingFace: COCO-DR, COCO-DR$_{XL}$ and Contriever.

Table 2. Results on the NQ dataset. In *italic* the best results per model, in **boldface** the best results overall. Symbol * indicates a statistically significant difference over Base, Fine-tuned and RND-G.

Retriever	Variant	MAP@100	MRR@100	R@100	NDCG@10	P@1	NDCG@3
BM25	-	0.292	0.295	0.758	0.339	0.198	0.268
COCO-DR	Base	0.441	0.455	0.923	0.504	0.325	0.424
	Fine-tuned	0.433	0.446	0.942	0.501	0.310	0.411
	RND-G	0.434	0.448	0.926	0.499	0.313	0.417
	DESIRE-ME	*0.463**	*0.477**	*0.941*	*0.526**	*0.339**	*0.448**
Contriever	Base	0.432	0.446	0.927	0.498	0.311	0.414
	Fine-tuned	0.427	0.438	0.940	0.497	0.295	0.406
	RND-G	0.441	0.457	0.928	0.510	0.320	0.426
	DESIRE-ME	*0.493**	*0.511**	*0.941*	*0.559**	*0.379**	*0.480**
COCO-DR$_{XL}$	Base	0.480	0.495	0.937	0.546	0.359	0.465
	Fine-tuned	0.465	0.478	***0.955***	0.537	0.331	0.447
	RND-G	0.488	0.503	0.939	0.553	0.371	0.473
	DESIRE-ME	*0.510**	*0.527**	*0.951*	***0.577****	***0.390***	*0.497**

that the RND-G model, which relies on a random gating mechanism, does not improve substantially the base model. This observation, which holds also for the experiments presented in the following, proves that our gating mechanism is an important factor contributing to improved retrieval performance.

In Table 3, we report the results on the HotpotQA dataset. In this case, fine-tuning the base model improves model performance, especially for R@100. For COCO-DR and COCO-DR$_{XL}$ DESIRE-ME improves the performance over the baselines across all three models. The improvements are consistently statistically significant for NDCG@3. For the other metrics, except R@100, we observe a slight improvement, but not always statistically significant. The relative performance improvement over the base model on HotpotQA is lower than that measured on NQ, reaching a margin of 3% in MAP@100 and 2% in NDCG@10. For Contriever, instead, the fine-tuned model outperforms DESIRE-ME in terms of R@100 and NDCG@10; meanwhile, for the other metrics DESIRE-ME performs slightly better than all baselines but not statistically significantly.

Table 4 shows the performance achieved on the FEVER dataset. FEVER presents a unique set of challenges compared to the other two datasets: the queries in FEVER are not questions but statements, and the relevant documents support or refute the claim made in the query statement. On this dataset, fine-tuning the base model, surprisingly, deteriorates the model performances, while BM25 performs very well, showing that the statement and the relevant documents share a similar vocabulary. As in the previous cases, DESIRE-ME improves over the COCO-DR and Contriever retrievers baselines, with a relative margin of 6% and 9% in NDCG@10 and P@1, respectively.

It is crucial to outline that while we could replicate the COCO-DR and COCO-DR$_{XL}$ results on the NQ dataset, our results diverged slightly from those

Table 3. Results on the HotpotQA dataset. In *italic* the best results per model, in **boldface** the best results overall. Symbol * indicates a statistically significant difference over Base, Fine-tuned and RND-G.

Retriever	Variant	MAP@100	MRR@100	R@100	NDCG@10	P@1	NDCG@3
BM25	-	0.521	0.770	0.740	0.603	0.707	0.558
COCO-DR	Base	0.519	0.795	0.727	0.604	*0.737*	0.563
	Fine-tuned	0.527	0.753	*0.805*	0.608	0.678	0.553
	RND-G	0.523	0.794	0.742	0.607	0.734	0.566
	DESIRE-ME	*0.530*	*0.795*	0.753	*0.614*	0.734	*0.571**
Contriever	Base	0.553	0.819	0.777	0.638	0.758	0.592
	Fine-tuned	**0.575**	0.799	**0.848**	**0.657**	0.728	0.600
	RND-G	0.552	0.817	0.780	0.636	0.757	0.592
	DESIRE-ME	0.567	**0.824**	0.787	0.648	**0.767**	**0.606**
COCO-DR$_{XL}$	Base	0.549	0.819	0.756	0.633	0.763	0.592
	Fine-tuned	0.542	0.757	*0.831*	0.622	0.681	0.563
	RND-G	0.555	0.819	0.767	0.637	0.763	0.595
	DESIRE-ME	*0.564**	*0.821*	0.780	*0.646**	*0.767*	*0.602**

reported in the original paper [29] for FEVER and HotpotQA. The Contriever results, instead, align exactly with those reported in the original article [13].

In summary, independently of these minor differences, our experiments on the three datasets demonstrate a consistent and significant improvement in retrieval performance obtained by integrating DESIRE-ME into the respective dense retrieval models. We can thus definitely answer positively RQ1.

Answering RQ2. We evaluate DESIRE-ME trained on FEVER in a zero-shot scenario on Climate-FEVER. This experiments aims to assess the generalization power of DESIRE-ME on a similar yet distinct dataset. Climate-FEVER and FEVER share a substantial portion of their corpus. However, an important distinction lies in the queries: Climate-FEVER relies on real-world user queries, while FEVER employs synthetic queries. We report in Table 5 the results of the experiments conducted using the DESIRE-ME models trained on the FEVER on the questions of Climate-FEVER. Despite the difference in query types, we notice improvements over the baselines across all models, similar to the previous three experiments. Specifically, the improvements over the respective base models are statistically significant for all the metrics measured with both COCO-DR retrievers. The relative margin in terms of NDCG@10 reaches 9%. These results outlines the capacity of DESIRE-ME to adapt to incoming queries that differs substantially from the ones seen at training time. We can thus answer positively also the second research question (RQ2) even if further experiments involving also other corpora are needed to undoubtedly assess the generalization power of DESIRE-ME across totally different Q&A scenarios.

Table 4. Results on the FEVER dataset. In *italic* the best results per model, in **boldface** the best results overall. Symbol * indicates a statistically significant difference over Base, Fine-tuned and RND-G.

Retriever	Variant	MAP@100	MRR@100	R@100	NDCG@10	P@1	NDCG@3
BM25	-	0.707	0.744	0.931	0.753	0.646	0.719
COCO-DR	Base	0.660	0.698	0.935	0.715	0.586	0.670
	Fine-tuned	0.544	0.568	0.928	0.607	0.431	0.544
	RND-G	0.652	0.690	0.937	0.710	0.565	0.666
	DESIRE-ME	*0.696**	*0.736**	*0.945**	*0.749**	*0.623**	*0.712**
Contriever	Base	0.708	0.749	*0.949*	0.758	0.642	0.724
	Fine-tuned	0.466	0.483	0.920	0.531	0.343	0.458
	RND-G	0.709	0.749	0.947	0.761	0.640	0.725
	DESIRE-ME	*0.722**	*0.764**	0.948	*0.772**	*0.655**	*0.739**
COCO-DR$_{XL}$	Base	0.699	0.740	0.946	0.749	0.633	0.713
	Fine-tuned	0.421	0.434	0.916	0.487	0.296	0.406
	RND-G	0.716	0.759	0.948	0.765	0.654	0.733
	DESIRE-ME	***0.745****	***0.789****	**0.952**	***0.792****	***0.691****	***0.762****

5   Conclusions

In this work we introduced DESIRE-ME, a new retrieval model for open-domain Q&A task that leverages the Mixture-of-Experts (MoE) framework to improve the performance of state-of-the-art dense retrieval models. The proposed MoE component uses supervised methods in the gating mechanism and predicts the likelihood of a query belonging to predefined domains, while the specializer modules focus on contextualizing the query vector for specific domains. We conducted extensive experiments across multiple datasets to investigate two research questions. For the first experiment, we chose three diverse datasets. Our experiments show that integrating the DESIRE-ME model into dense retrieval models leads to significant improvements in various retrieval metrics, answering positively the **RQ1**. These findings highlight the robustness and adaptability of DESIRE-ME. In response to the **RQ2**, the experiment performed on the Climate-FEVER dataset, using a model trained on FEVER shows that MoE can generalize to new datasets in a zero-shot scenario. This also shows the potential of leveraging knowledge from a similar corpus and encourages further exploration of techniques, such as transfer learning in the open-domain Q&A tasks.

Limitations and Future Work. Our primary focus was understanding the improvements achieved by using domain specialization in open-domain Q&A; we did not concentrate on optimizing the underlying neural architectures for the specializers and gating mechanism. The main limitation of this work is the assumption of having query domain information, which might not be true in most IR tasks. In our experiments, we relied on Wikipedia corpora and categories; our labeling process is however not exportable to other cases. Consequently,

Table 5. Results on the Climate-FEVER dataset using models trained on FEVER. In *italic* the best results per model, in **boldface** the best results overall. Symbol * indicates a statistically significant difference over Base and RND-G.

Retriever	Variant	MAP@100	MRR@100	R@100	NDCG@10	P@1	NDCG@3
BM25	-	0.162	0.293	0.436	0.213	0.205	0.179
COCO-DR	Base	0.164	0.290	0.514	0.210	0.201	0.171
	RND-G	0.170	0.298	0.536	0.218	0.207	0.176
	DESIRE-ME	*0.178**	*0.312**	*0.544*	*0.228**	*0.219*	*0.185**
Contriever	Base	0.184	0.317	0.574	0.237	0.216	0.189
	RND-G	*0.205*	0.351	*0.609*	0.264	0.241	0.211
	DESIRE-ME	*0.205*	*0.358*	0.600	*0.268*	*0.250*	*0.213*
COCO-DR$_{XL}$	Base	0.180	0.322	0.547	0.231	0.227	0.189
	RND-G	0.182	0.325	0.564	0.234	0.229	0.188
	DESIRE-ME	*0.191**	*0.343**	*0.573*	*0.247**	*0.243*	*0.199**

given the diversity in real-world queries and documents our insights could be not directly generalizable to other settings. Future research could address this issue by evaluating DESIRE-ME on more diverse and extensive datasets. This would require extensive user studies and crowd-sourcing to label query domains or topics. Another option would be using LLMs to create soft labels for queries [11]. Another future research topic is query augmentation, which can be addressed by adapting the DESIRE-ME specializer modules to domain-specific query expansion modules. This way, the query expansion would occur by using models that can leverage domain-specific vocabularies.

Acknowledgements. Funding for this research has been provided by spoke "FutureHPC & BigData" of the ICSC - Centro Nazionale di Ricerca in High-Performance Computing, Big Data and Quantum Computing, Spoke "Human-centered AI" of the M4C2 - Investimento 1.3, Partenariato Esteso PE00000013 - "FAIR - Future Artificial Intelligence Research", the FoReLab project (Departments of Excellence), the NEREO and CAMEO PRIN projects funded by the Italian Ministry of Education and Research Grant no. 2022AEFHAZ and 2022ZLL7MW, and the EFRA project funded by the European Commission under the NextGeneration EU programme grant agreement n. 101093026. However, the views and opinions expressed are those of the authors only and do not necessarily reflect those of the EU or European Commission-EU. Neither the EU nor the granting authority can be held responsible for them.

References

1. Bassani, E.: ranx: a blazing-fast python library for ranking evaluation and comparison. In: Hagen, M., et al. (eds.) ECIR 2022, Part II. LNCS, vol. 13186, pp. 259–264. Springer, Cham (2022). https://doi.org/10.1007/978-3-030-99739-7_30
2. Bassani, E.: Ranxhub: an online repository for information retrieval runs. In: Proceedings of the 46th International ACM SIGIR Conference on Research and Development in Information Retrieval, SIGIR 2023, pp. 3210–3214. Association for Computing Machinery, New York (2023). https://doi.org/10.1145/3539618.3591823

3. Collobert, R., Bengio, S., Bengio, Y.: A parallel mixture of SVMs for very large scale problems. In: Dietterich, T., Becker, S., Ghahramani, Z. (eds.) Advances in Neural Information Processing Systems, vol. 14. MIT Press (2001). https://proceedings.neurips.cc/paper_files/paper/2001/file/36ac8e558ac7690b6f44e2cb5ef93322-Paper.pdf

4. Dai, D., et al.: Mixture of experts for biomedical question answering. arXiv abs/2204.07469 (2022). https://api.semanticscholar.org/CorpusID:248218762

5. Dauphin, Y.N., Fan, A., Auli, M., Grangier, D.: Language modeling with gated convolutional networks. In: Precup, D., Teh, Y.W. (eds.) Proceedings of the 34th International Conference on Machine Learning. Proceedings of Machine Learning Research, vol. 70, pp. 933–941. PMLR (2017). https://proceedings.mlr.press/v70/dauphin17a.html

6. Diggelmann, T., Boyd-Graber, J., Bulian, J., Ciaramita, M., Leippold, M.: Climate-fever: a dataset for verification of real-world climate claims (2021)

7. Eigen, D., Ranzato, M., Sutskever, I.: Learning factored representations in a deep mixture of experts (2014)

8. Fedus, W., Zoph, B., Shazeer, N.: Switch transformers: scaling to trillion parameter models with simple and efficient sparsity. J. Mach. Learn. Res. **23**(1), 5232–5270 (2022)

9. Gaur, N., et al.: Mixture of informed experts for multilingual speech recognition. In: ICASSP 2021-2021 IEEE International Conference on Acoustics, Speech and Signal Processing (ICASSP), pp. 6234–6238 (2021). https://doi.org/10.1109/ICASSP39728.2021.9414379

10. Gururangan, S., Lewis, M., Holtzman, A., Smith, N.A., Zettlemoyer, L.: Demix layers: disentangling domains for modular language modeling (2021)

11. Hashemi, H., Zhuang, Y., Kothur, S.S.R., Prasad, S., Meij, E., Croft, W.B.: Dense retrieval adaptation using target domain description (2023)

12. Houlsby, N., et al.: Parameter-efficient transfer learning for NLP. In: Proceedings of the 36th International Conference on Machine Learning (2019)

13. Izacard, G., et al.: Unsupervised dense information retrieval with contrastive learning (2021). https://doi.org/10.48550/ARXIV.2112.09118. https://arxiv.org/abs/2112.09118

14. Jacobs, R.A., Jordan, M.I., Nowlan, S.J., Hinton, G.E.: Adaptive mixtures of local experts. Neural Comput. **3**(1), 79–87 (1991). https://doi.org/10.1162/neco.1991.3.1.79

15. Jordan, M.I., Jacobs, R.A.: Hierarchical mixtures of experts and the EM algorithm. Neural Comput. **6**(2), 181–214 (1994). https://doi.org/10.1162/neco.1994.6.2.181

16. Khattab, O., Zaharia, M.: Colbert: efficient and effective passage search via contextualized late interaction over bert. In: Proceedings of the 43rd International ACM SIGIR Conference on Research and Development in Information Retrieval, SIGIR 2020, pp. 39–48. Association for Computing Machinery, New York (2020). https://doi.org/10.1145/3397271.3401075

17. Kwiatkowski, T., et al.: Natural questions: a benchmark for question answering research. Trans. Assoc. Comput. Linguist. **7**, 453–466 (2019)

18. Li, M., Li, M., Xiong, K., Lin, J.: Multi-task dense retrieval via model uncertainty fusion for open-domain question answering. In: Findings of the Association for Computational Linguistics: EMNLP 2021, Punta Cana, Dominican Republic, pp. 274–287. Association for Computational Linguistics (2021). https://doi.org/10.18653/v1/2021.findings-emnlp.26. https://aclanthology.org/2021.findings-emnlp.26

19. Mitra, B., Craswell, N.: An introduction to neural information retrieval. Found. Trends® Inf. Retrieval **13**(1), 1–126 (2018). https://doi.org/10.1561/1500000061

20. Mun, J., Lee, K., Shin, J., Han, B.: Learning to specialize with knowledge distillation for visual question answering. In: Bengio, S., Wallach, H., Larochelle, H., Grauman, K., Cesa-Bianchi, N., Garnett, R. (eds.) Advances in Neural Information Processing Systems, vol. 31. Curran Associates, Inc. (2018). https://proceedings.neurips.cc/paper_files/paper/2018/file/0f2818101a7ac4b96ceeba38de4b934c-Paper.pdf

21. Nogueira, R., Yang, W., Lin, J., Cho, K.: Document expansion by query prediction (2019)

22. Puigcerver, J., Riquelme, C., Mustafa, B., Houlsby, N.: From sparse to soft mixtures of experts (2023)

23. Robertson, S.E., Walker, S., Jones, S., Hancock-Beaulieu, M., Gatford, M.: Okapi at TREC-3. In: Harman, D.K. (ed.) Proceedings of the Third Text REtrieval Conference, TREC 1994, Gaithersburg, Maryland, USA, 2–4 November 1994. NIST Special Publication, vol. 500–225, pp. 109–126. National Institute of Standards and Technology (NIST) (1994). http://trec.nist.gov/pubs/trec3/papers/city.ps.gz

24. Rosa, G.M., et al.: No parameter left behind: How distillation and model size affect zero-shot retrieval (2022)

25. Shazeer, N., et al.: Outrageously large neural networks: the sparsely-gated mixture-of-experts layer (2017)

26. Thakur, N., Reimers, N., Rücklé, A., Srivastava, A., Gurevych, I.: BEIR: a heterogeneous benchmark for zero-shot evaluation of information retrieval models. In: Thirty-Fifth Conference on Neural Information Processing Systems Datasets and Benchmarks Track (Round 2) (2021). https://openreview.net/forum?id=wCu6T5xFjeJ

27. Thorne, J., Vlachos, A., Christodoulopoulos, C., Mittal, A.: FEVER: a large-scale dataset for fact extraction and VERification. In: Proceedings of the 2018 Conference of the North American Chapter of the Association for Computational Linguistics: Human Language Technologies, Volume 1 (Long Papers), New Orleans, Louisiana, pp. 809–819. Association for Computational Linguistics (2018). https://doi.org/10.18653/v1/N18-1074. https://aclanthology.org/N18-1074

28. Yang, Z., et al.: HotpotQA: a dataset for diverse, explainable multi-hop question answering. In: Proceedings of the 2018 Conference on Empirical Methods in Natural Language Processing, Brussels, Belgium, pp. 2369–2380. Association for Computational Linguistics (2018). https://doi.org/10.18653/v1/D18-1259. https://aclanthology.org/D18-1259

29. Yu, Y., Xiong, C., Sun, S., Zhang, C., Overwijk, A.: COCO-DR: combating distribution shifts in zero-shot dense retrieval with contrastive and distributionally robust learning. In: Proceedings of the 2022 Conference on Empirical Methods in Natural Language Processing, pp. 1462–1479 (2022)

30. Zhou, Y., et al.: Mixture-of-experts with expert choice routing. Adv. Neural. Inf. Process. Syst. **35**, 7103–7114 (2022)

Probing Pretrained Language Models with Hierarchy Properties

Jesús Lovón-Melgarejo[1(✉)] [ID], Jose G. Moreno[1] [ID], Romaric Besançon[2] [ID],
Olivier Ferret[2] [ID], and Lynda Tamine[1] [ID]

[1] Université Paul Sabatier, IRIT, Toulouse, France
{jesus.lovon,jose.moreno,tamine}@irit.fr
[2] Université Paris-Saclay, CEA, List, Palaiseau, France
{romaric.besancon,olivier.ferret}@cea.fr

Abstract. Since Pretrained Language Models (PLMs) are the corner-stone of the most recent Information Retrieval models, the way they encode semantic knowledge is particularly important. However, little attention has been given to studying the PLMs' capability to capture hierarchical semantic knowledge. Traditionally, evaluating such knowledge encoded in PLMs relies on their performance on task-dependent evaluations based on proxy tasks, such as hypernymy detection. Unfortunately, this approach potentially ignores other implicit and complex taxonomic relations. In this work, we propose a task-agnostic evaluation method able to evaluate to what extent PLMs can capture complex taxonomy relations, such as ancestors and siblings. This evaluation, based on intrinsic properties capturing these relations, shows that the lexico-semantic knowledge implicitly encoded in PLMs does not always capture hierarchical relations. We further demonstrate that the proposed properties can be injected into PLMs to improve their understanding of hierarchy. Through evaluations on taxonomy reconstruction, hypernym discovery and reading comprehension tasks, we show that knowledge about hierarchy is moderately but not systematically transferable across tasks.

Keywords: Neural language models · Taxonomic relations · Evaluation

1 Introduction

The hierarchical representation of concepts is a fundamental aspect of human cognition and essential in performing numerous Information Retrieval (IR) (e.g., Web search [45]) and Natural Language Processing (NLP) tasks (e.g., hypernym discovery [13]). Therefore, works in this research direction have explored

This work has been supported by the MEERQAT project (ANR-19-CE23-0028), granted by ANR. This work was also granted access to the HPC resources of IDRIS under the allocation 2023-AD011012638R2 made by GENCI.

incorporating hierarchical knowledge extracted from knowledge graphs and taxonomies to refine models [16,26,35,38]. In particular, prior studies on static word embeddings showed that designing dedicated space representations to encode the hierarchy benefits the performance of various downstream tasks [38,50], giving rise to the importance of encoding concept hierarchy in these representations.

Most of previous studies have adopted task-dependent evaluations to assess the extent to which models capture hierarchical linguistic knowledge. These evaluations are based on the premise that the model's performance on downstream tasks sheds light on whether the model captures hierarchy. Specifically, such evaluations target tasks that require applying hierarchical knowledge to some degree, such as taxonomy reconstruction and hypernymy detection [12,38]. However, existing task-dependent evaluations have two main limitations. First, they heavily rely on detecting a single relation type (hypernymy) and overlook the implicit taxonomy structure by ignoring other related relations, such as ancestors and siblings [37]. As a result, none of the state-of-the-art (SOTA) evaluation methodologies are able to reveal this implicit yet essential hierarchical information [2]. Second, particularly in the context of Pretrained Language Models (PLMs), task-dependent evaluations might conflate the model's understanding of a given target task and the model's understanding of hierarchy per se. Hence, there is a need for more comprehensive evaluation methods and datasets that consider complex hierarchical relations and are able to reveal how well models capture those relations and apply them to downstream tasks.

The present work addresses the two aforementioned limitations. We propose a task-agnostic methodology to evaluate language models' understanding of hierarchical knowledge considering implicit and more complex taxonomic relations beyond the direct hypernymy (parent, ancestors, and siblings). In particular, our evaluation focuses on PLMs, such as BERT [18], considering their ability to encode lexical knowledge, as well as their outstanding performance on different tasks [40]. To the best of our knowledge, no task-agnostic methodology exists so far that enables us to reveal to what extent PLMs encode hierarchy relations intrinsically. Following previous work on information retrieval [14] and lexical semantics [48,53], we use a *probe*-based methodology [39] to evaluate whether SOTA models capture task-agnostic linguistic knowledge.

The article is structured as follows. We first introduce a set of intrinsic properties of a hierarchy based on edge-distance observations in a taxonomy. Then, we design probes consisting of triplets of entities, named *ternaries*, that encode these properties and are used to evaluate and teach PLMs a notion of hierarchy.

We aim to answer three research questions. Firstly, **(RQ1)** *To which extent do PLM representations encode hierarchy w.r.t. hierarchy properties?* To address this question, we analyze PLM representations with our task-agnostic evaluation based on taxonomic relations between concepts. Secondly, **(RQ2)** *Does injecting hierarchy properties into PLMs using a task-agnostic methodology impact their representations?* where we fine-tuned hierarchy-enhanced PLMs with probes built upon the hierarchy properties and evaluated them. Lastly, **(RQ3)** *Can hierarchy-enhanced PLM representations be transferred to downstream tasks?*

where we evaluated the performance impact of hierarchy-enhanced PLMs on Hypernym Discovery, Taxonomy Reconstruction, and Reading Comprehension.

Our main findings regarding these questions are: 1) the evaluated PLMs struggle to capture hierarchical relations, such as siblings and ancestor; 2) PLMs can enhance their representations by learning the hierarchy properties, which improve their performance on our evaluation; 3) there appears to be a gap in understanding between hierarchy knowledge and task, making it difficult to achieve a clear trend of performance increase of enhanced PLMs across downstream tasks.

2   Related Work and Background

2.1   Hierarchical Representations: Learning and Evaluation

Learning knowledge representations with an underlying hierarchical structure is an active research topic [44]. Approaches include learning embedding representations from symbolic knowledge sources (e.g., knowledge graphs (KGs)) such as TransE [9] and TransR [29] or constructing dedicated vector spaces based on hyperbolic spaces [38] and order embeddings [50]. Recently, with the emergence of PLMs, different works explored infusing factual knowledge from KGs. The infusion is typically done by injecting triplets embeddings to capture entity and relation features [35,56]. However, current approaches are limited to the triplet representation and risk only to capture partial graph structure information [55]. In our work, we use a *ternary* of entities to represent three entities implicitly linked between them with a hierarchy-based relation, aiming to build more hierarchy-aware representations.

To assess the quality of model representations, a frequent evaluation task is Taxonomy Reconstruction [12,38]. Multiple frameworks extended this evaluation with different features of a taxonomy. For instance, evaluations are based on the granularity and generalization paths [23] or structural features such as cycles, connected components, and intermediate nodes [7,8]. Similarly, recent work studied PLMs' adaptation to Taxonomy Reconstruction [27] and Hypernymy Detection tasks via sequence classification [16]. Additionally, Chen et al. [15] explored the syntactic and semantic hierarchies at the term-level of BERT within the hyperbolic space tailored to the sentiment classification task. However, our methodology differs from this approach. We conceive and analyze the notion of hierarchy between concepts, translating it into a set of properties that provide an intuitive and robust basis for evaluation. These properties, based on SOTA semantic similarity metrics [11,42], impose constraints on the relations between concepts. Finally, while non-Euclidean spaces for hierarchy representation have garnered recent attention [21,46,49,58], they fall beyond our current scope.

2.2   Probes on PLMs

We consider the probes as challenging datasets, not conceived to provide new knowledge but to assess what a model already knows about a task, typically under a light fine-tuning or zero-shot setup [43]. In the context of PLMs, different efforts are proposed to unveil what kind of knowledge these models encode using these probes. Recent work has explored probes for information retrieval [14], natural language inference [17], symbolic reasoning [48], and lexical semantics [53]. Other approaches [43,48] developed evaluation methodologies, including probe evaluations before and after fine-tuning these models on target data samples. Under this approach, it is observable how adaptable a model is to the learning probe format. In our work, we apply probes in a zero-shot setup to study the default model PLM representations.

3   A Task-Agnostic Methodology: Design and Evaluation

This section presents our task-agnostic methodology for probing hierarchical knowledge of PLMs. First, we define a set of intrinsic properties for a hierarchy (§3.1). Then, we describe how to design probes upon these properties and how to fine-tune PLMs to identify these properties by learning these probes (§3.2).

3.1   Intrinsic Hierarchy Properties

We characterize a hierarchical structure by a set of intrinsic properties that mirror the distribution of concepts within a given taxonomy. We rely on evaluating the taxonomic similarity to study the hierarchical relationship between words. From a linguistic perspective, this can be understood as evaluating the hierarchical relation based on the *paradigmatic* knowledge encoded by PLMs [30,33]. To validate this similarity between two concepts, we represent concepts as nodes and use edge-based approaches, such as the shortest path between two nodes [11,57]. Please note that this differs from a *syntagmatic* approach, where only co-occurrence is considered.

Concretely, we define a set of *relations* and *properties* from edge-based observations. Then, we evaluate these properties empirically considering semantic distance methods, inspired by [11][1].

Notably, for a fixed node n, we define four basic relations –*parent, ancestor, sibling*, and *far relative*– in a taxonomy in the following way. A *parent* node p is a direct hypernym at a one-edge distance from n, while an *ancestor* node is an indirect hypernym at a two-edge distance. A *sibling* node shares a parent with a two-edge distance, and a *far relative* shares the ancestor but not the parent.

We use these relations and the corresponding edge-based distances to define hierarchy properties. Table 1 presents the six hierarchy properties considered in this work based on the four defined relations for all possible combinations. In

---

[1] We consider distance as the complement metric for similarity.

order to verify these properties, we formulate them as inequalities that examine the distance between two distinct relations for a fixed node, as shown in Table 2. Our motivation for considering three nodes (*ternary*) in these evaluations comes from recent research on pattern-based relation extraction [25,34], where the inclusion of a third "anchor" node has proven useful in capturing various relation types, including hypernymy and co-hyponymy.

For ease of reading, we adopt a naming schema of properties that highlights the two used relations, i.e. a **R**elation-**R**elation format. The relation identifier, denoted by **R**, can take one of four possible values: **P**arent, **A**ncestor, **S**ibling, and **F**ar relative. Furthermore, we split the initial properties into three *groups*, P-*, A-*, and S-*, based on the left relation in the inequality. These *groups* serve as an aggregated representation of the properties for each type of relations.

Table 1. The six hierarchy properties with their names and textual descriptions.

Property	Description
P-A	*"A node is closer to its parent than its ancestor."* The similarity between a pair of concepts is proportional to their path length under an edge-based approach [42].
P-S	*"The distance between 'siblings' should be longer than between 'father' and 'son'."*. Similarly, it is a straightforward application of edge-counting, also found in other approaches such as scaling the taxonomy [57].
P-F	*"The parent is the closest element to a node."* It generalizes P-S by comparing to further relations in a taxonomy. If a model does not satisfy this property, it struggles to differentiate the hypernym relation from others.
A-S	*"A node is closer to its ancestor than to its sibling."* We did not find an expected behavior about this property in the literature. We empirically choose the path-level evaluation proposed by [23], favoring the correct edges in the hypernym path rather than adding incorrect elements that could cause a cascade of generalization errors.
A-F	*"The ancestor is the closest term for a node, except for the father."* This property generalizes the ancestor relationship evaluated in further relationships in a taxonomy.
S-F	*"The sibling is the closest element for a node, except for the father and the ancestor."* Based on edge-counting approaches, we should find a sibling node closer to other relations beyond the ancestor in a hierarchy

3.2 Probing Hierarchical Representations

Designing probes. For this purpose, we align the properties into a single form. Given a taxonomy $T = (V, E)$, with V representing a set of nodes and E a set of edges, we define a *ternary* as $t = (x_n, x_l, x_r)$ encoding a hierarchy property $\mathbf{R}_l$-$\mathbf{R}_r$ where x_n represents a fixed node, x_l the node of the left Relation ($\mathbf{R}_l$), and x_r the node of the right Relation ($\mathbf{R}_r$). Note that each of the tuples (x_n, x_l) and (x_n, x_r) corresponds to one of the defined relations (parent **P**, ancestor **A**, sibling **S** or far relative **F**) and the taxonomic distances (edge-based) must satisfy the associated property $dist(x_n, x_l) < dist(x_n, x_r)$.

Table 2. The hierarchy properties along with their distance-based definitions, and the three participant (colored) nodes in the taxonomy. Three *groups* are identified based on the left relation in the inequality: P-* (regrouping P-A, P-S, and P-F), A-* (regrouping A-S and A-F), and S-* (with only S-F).

Property	Definition	Property	Definition
P-A	$dist(n,p) < dist(n,a)$	A-S	$dist(n,a) < dist(n,s)$
P-S	$dist(n,p) < dist(n,s)$	A-F	$dist(n,a) < dist(n,f)$
P-F	$dist(n,p) < dist(n,f)$	S-F	$dist(n,s) < dist(n,f)$

Each node within the ternary tuple is converted into textual representations through predefined phrases (prompts) (§4.3). These concept representations are used to compute a model representation within a given language model. Consequently, for a given model, we obtain a representation in the form $(\hat{x}_n, \hat{x}_l, \hat{x}_r)$ for each ternary (x_n, x_l, x_r). Finally, we use a distance method to evaluate the inequality $d(\hat{x}_n, \hat{x}_l) < d(\hat{x}_n, \hat{x}_r)$.

Teaching hierarchy properties to a PLM. To teach the hierarchy properties to PLMs, we leverage the concept representations derived from the probes, $(\hat{x}_n, \hat{x}_l, \hat{x}_r)$, and train a model to satisfy the inequality $d(\hat{x}_n, \hat{x}_l) < d(\hat{x}_n, \hat{x}_r)$. We assume that a model's performance on these probes indicates how well its representations align to a hierarchy-like distribution. We employ the Sentence Transformer framework [41] with a triplet network architecture and a pooling operation to the output of the PLM to generate the embedding of our concepts. This approach achieved promising results on lexical knowledge evaluations [52].

Given the inner contrastive nature of our probes in a triplet form, we adopt a contrastive loss, specifically the Triplet loss [20]. This loss fine-tunes the network to minimize the distance between related inputs, $(\hat{x}_n, \hat{x}_l)$, while maximizing the distance for unrelated inputs, $(\hat{x}_n, \hat{x}_r)$, by minimizing the following loss:

$$\mathcal{L}(x_n, x_l, x_r) = -max(\|\hat{x}_n - \hat{x}_l\| - \|\hat{x}_n - \hat{x}_r\| + \alpha, 0) \tag{1}$$

4 Experimental Setup

4.1 Datasets and Metrics

We used the Bansal et al. dataset to sample our probes, consisting of medium-sized taxonomies generated from WordNet subtrees [4]. This dataset comprises subtrees of height 3 (i.e., the longest path from the root to the leaf is 4 nodes) containing between 10 and 50 terms. Specifically, we used the version extended with WordNet definitions [16]. We generated ternaries for each property, splitted into train/dev partitions to fine-tune PLMs with the hierarchy properties. We generated approximately 20,000/4,000/14,000 for each property's train/dev/test split, except for *P-A* with 6,300/1,400/1,400 ternaries. As mentioned before, we do not consider the property *A-S* for training due to its absence in the literature. For evaluation, we use the accuracy metric defined as the number of correct predictions where the ternary inequality was satisfied, divided by the total number of ternaries in the test set.

4.2   Baselines and Models

As baselines, we use three groups of models:

Group 1. Comprises random and non-contextualized models as a lower bound performance of the PLMs following previous work [48,52]: a) **Random:** we generated symmetrical random distances for all node pairs and report the average of ten random runs; b) **FastText:** to compare static word embeddings in a fair setup, i.e. avoiding the out-of-vocabulary problem, we used the character-based version of **FastText** embeddings [6] trained on Wikipedia (**FT-wiki**).

Group 2. Comprises a total of five PLMs available in the HuggingFace library: a) **BERT** (*bert-base-cased*); b) **BERT-L** (*bert-large-cased*); c) **RoB-L** (*roberta-large*); d) **S-RoB** (*all-distilroberta-v1*); and e) **S-MPNet** (*all-mpnet-base-v2*). We notice that the first three models are cross-encoders and the last two models (**S-RoB** and **S-MPNet**) are bi-encoders. The bi-encoders are trained using a dual-encoder network and oriented towards semantic textual similarity tasks on multiple datasets. These characteristics give some advantages in the final results of our probes compared to classical PLMs. However, we evaluated these models to obtain insights on the most robust ones.

Group 3. Comprises two SOTA knowledge enhanced PLMs: **ERNIE** [47], a BERT-based trained on multiple tasks to capture lexical, syntactic and semantic aspects of information, and **CTP** [16], a RoBERTa large model trained to perform hypernym prediction.

4.3   Concept and Ternary Representations

Each concept in a ternary is represented textually by its name and definition from a knowledge source to provide context information. We consider two vector-based methods, namely **cls** and **avg**, and prompt-based methods to generate the representation of a ternary to text with a PLM. For vector-based methods, we use the last layer output of the PLM and represent each concept as a list of tokens with the form: *'[CLS] [concept name] is defined as [definition] [SEP]'*. The **cls** method uses the special token *[CLS]* and the **avg** method computes the average of all subtokens. We compute the distance between these representations based on cosinus (**cos**) and euclidean (**L2**) distances. In contrast to vector-based methods, the prompt-based method condenses the ternary representation into a single textual representation. We use the LMScorer method [28] to score sentences for factual accuracy. LMScorer computes a pseudo-likelihood

Table 3. Accuracy for Property *P-A* with different representation methods. LMScorer is not computed for bi-encoder models since they are not adapted for this task [54].

Model Representation	Distance Method	BERT	BERT-L	RoB-L	S-RoB	S-MPNet
cls	cos	72.6	68.2	58.5	83.1	83.2
	L2	72.4	68.7	58.5	79.1	83.1
avg	cos	**76.0**	**78.9**	**75.0**	**85.2**	**85.6**
	L2	75.7	78.0	74.8	80.6	**85.6**
LMScorer	–	50.2	52.6	54.6	–	–

score for each token in a sequence by iteratively masking it, considering past and future tokens: $LMScorer_{PLM}(\mathbf{W}) = \exp\left(\sum_{i=1}^{|\mathbf{W}|} logP_{PLM}(w_i|\mathbf{W}_{\backslash i})\right)$. We experiment with different templates for converting these ternaries and report the best-scoring template,[2] different from the one for vector-based methods. For static word embeddings, we use only the **avg** method, as there is no *[CLS]* token.

5   Results and Analysis

In this section, we report the results to answer our research questions. To simplify our analysis, we use the *groups* presented in Table 2 and report the average values. We only report the best configuration for each model due to space limitations, but we ensure we obtain the best configuration for each model.[3]

5.1   Evaluation of Hierarchy Properties in PLMs

First, we carry out preliminary experiments to explore the best performing settings in terms of model representation and distance methods. Table 3 shows our results based only on the representative property *P-A* using various methods on all PLMs. Our findings indicate that the vector-based representations (75.0–85.6) outperform all the explored prompts. We also observe that the **avg** representation is always superior to **cls**. Moreover, the **avg** representation with **cos** distance frequently obtains slightly better scores (76.0) than **L2** (75.7). Under this evidence, we adopted the **avg** representation with **cos** distance for our remaining evaluations.

We now answer our first research question **RQ1**: *To which extent do PLM representations encode hierarchy w.r.t. hierarchy properties?* Our analysis focuses on the model's overall performance in our evaluation (named *All*, as the average score from all properties) and provides insights into the property groups (*P-**, *S-**, and *F-**) captured by *Group 1* and *Group 2* models.

---

[2] Template: "A is a B. C is a B. [definition a][definition b] [definition c].", where A, B, and C are nodes.

[3] Resources available at: https://github.com/jeslev/hierarchy_properties_plms.

Considering the model's overall performance (*All*), we first test the quality of our probes from our baselines in *Group 1*. Table 4 shows that **Fast-Text (FT-wiki)** embeddings perform better (60.3) than the **Random** approach (50.2). Aligned with previous work [22] claiming that static embeddings encode hypernymy-like relations to some degree, we argue that our probes help to capture these hierarchical relations. Similarly, all *Group 2* models obtained *All* scores higher than **Random**. In particular, the **S-RoB** model obtained the highest score (70.3), closely followed by **S-MPNet** (70.0). Moreover, only these models outperformed the **FT-wiki** static embeddings. From *Group 3* models, we can surprisingly see that **ERNIE** outperforms **CPT**, while the latter is a RoBERTa large model specifically trained on hypernym classification task with a classification layer on top of the PLM. By comparing these models against their vanilla version, we can interestingly observe that while **ERNIE** showed constant improvement in all properties, with +2 score points on *All* (56.2) w.r.t. **BERT**, our probes suggest that **CTP**, trained on a task-dependent evaluation (37.7), degrades its initial representations w.r.t **RoB-L** (53.3). Moreover, **CTP** is the only PLM-based model with lower performance than **Random**. Overall, these results provide insights on the conflation problem faced with task-dependent evaluation regarding task understanding and hierarchy understanding of PLM's.

Now, we analyze the results at *groups* level. Considering the semantic similarity approach in pre-training, we expect that *sibling* relations (*S-**), akin to analogies, will exhibit better representations than other hierarchical relations. Across all models, we can observe that the *siblings* relation (*S-**), followed by the *parent* relation (*P-**), obtained higher performances than the *ancestor* relation (*A-**). Our results capture the preference of the semantic similarity of these

Table 4. Accuracy on each hierarchy property on the test dataset with method *avg(cos)*. Best results in **bold** for each probe per group. * shows 1 sample t-test > 0.05.

Model	P-A	P-S	P-F	A-S	A-F	S-F	P-*	A-*	S-*	All
	Group 1: Baselines									
Random	50.5	49.8	50.1	50.1	50.4	50.1	50.1	50.3	50.1	50.2
FT-wiki	79.7	49.9	81.6	24.1	48.5	78.2	**70.4**	36.3	**78.2**	60.3
	Group 2: Vanilla PLMs									
BERT	75.8	37.9	71.8	22.3	43.0	74.5	61.8	32.6	74.5	54.2
BERT-L	78.8	42.0	77.6	24.6	47.6	76.1	66.1	36.1	76.1	57.8
RoB-L	73.2	37.2	71.1	24.6	40.1	73.8	60.5	32.4	73.8	53.3
S-RoB	85.2	68.6	90.4	33.7	64.8	78.8	**81.4**	**49.2**	78.8	**70.3**
S-MPNet	85.6	70.0	88.3	29.6	65.9	80.6	81.3	47.8	**80.6**	70.0
	Group 3: KG-Enhanced PLMs									
ERNIE	77.0	42.7	74.2	23.1	44.9	75.6	**64.6**	34.0	**75.6**	**56.2**
CTP	59.9	22.4	36.2	23.9	22.8	61.0	39.5	23.4	61.0	37.7
	Group 4: Hierarchy-Enhanced PLMs									
BERT$_{hp}$	79.9	56.2	82.6	29.2	58.4	75.2	72.9	43.8	75.2	64.0*
BERT-L$_{hp}$	85.7	69.1	91.0	31.6	67.6	79.3	81.9	49.6	79.3	70.3*
RoB-L$_{hp}$	82.0	64.3	86.0	35.1	61.9	76.1	77.4	48.5	76.1	67.4*
S-RoB$_{hp}$	87.7	74.9	93.9	34.5	73.0	81.6	85.5	53.8	**81.6**	**73.6***
S-MPNet$_{hp}$	84.0	80.0	92.9	35.5	72.3	80.4	**85.6**	**53.9**	80.4	73.3*

models, particularly when compared to other *far* relations, because of the high values in *S-F* (*min*: 73.8 − *max*: 80.6). Similarly, most *ancestor* representations are further than *parents* (73.2−85.6 in *P-A*). Based on property *P-S* (37.2−42.0), PLMs such as **BERT**, **BERT-L**, and **RoB-L** tend to represent *siblings* closer (73.8 − 76.1) than the *parent* (60.5 − 66.1). For the *ancestor* relation, higher scores on property *A-F* (40.1 − 65.9) than on *A-S* (22.3 − 33.7) imply that *far relatives* are easier than *siblings*. However, these scores are generally lower than those obtained with other properties. Besides, when comparing the *P-F* and *A-F* columns, we observe higher scores with the *parent*, one-edge distance, than the *ancestor*, two edges away (similarly, for the scores between *P-S* and *A-S*). These trends indicate that *ancestor* representations are not as clearly defined as *parents*.

Overall, our evaluation reveals that the analyzed PLMs struggle to capture some hierarchical relationships such as *sibling* and *ancestor*. Besides, the results confirm that the performance of PLM specifically trained on a hierarchy-aware task (e.g., **CTP**) is not a salient signal of PLM's understanding of hierarchy.

Discussion About Probing Methodology Our probing methodology requires a textual definition of concepts for evaluation. Hence, the resulting representations' quality can be impacted by the definition's length [10] or absence. In the latter scenario, alternative approaches must be considered to obtain a suitable vector representation [53]. Additionally, the taxonomy's granularity can impact the representations' quality. Our methodology has been tested on low domain-specific levels (for example, a concept of "presenile dementia", with parent "dementia" and sibling "insanity"). However, coarse-grained taxonomies, with more abstract concepts in proportion, may not respect our target properties since relations between abstract concepts tend to be different from relations between more specific concepts, as suggested by [51].

5.2   Enhancing PLMs with Hierarchy Properties

In the following, we answer **RQ2:** *Does injecting hierarchy properties into PLMs using a task-agnostic methodology impact their representations?* Our underlying objective is to investigate the feasibility of our task-agnostic evaluation of PLMs w.r.t to hierarchy through the set of defined properties (§3.1). To this end, we fine-tuned (§3.2) and evaluated the models on the probes following the best evaluation setup for PLMs (discussed in §5.1). These fine-tuned models are referenced as hierarchy-enhanced PLMs (denoted PLM_{hp}) belonging to *Group 4* in Table 4, where we report the average accuracy after fine-tuning each model with 5 different random seeds and optimal hyper-parameters. Similar to RQ1, we consider the *All* for comparison and we take as reference the corresponding vanilla model for our analysis.

Our results showed improvement in all models for the *All* column w.r.t the original PLM. For instance, the models **BERT-L**$_{hp}$, **S-RoB**$_{hp}$, and

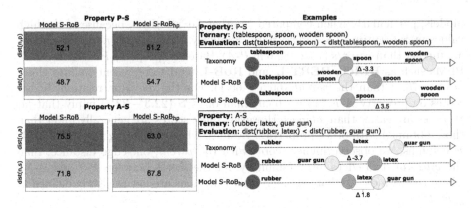

Fig. 1. Avg. distance in the property evaluation for properties *P-S* and *A-S* (left). One example for each property, showing the distances of different concepts for a fixed node with models S-RoB and S-RoB$_{hp}$. We indicate the Δ distance between concepts (right).

S-MPNet$_{hp}$ are improved by 12.5, 3.3, and 3.3 points, respectively in comparison to their vanilla counterparts. In particular, the property *A-S*, not included in the training, shows a constant improvement for all models. These findings align with our hypothesis that *A-S* complements other property criteria without compromising initial performance. In contrast, the trained property *S-F* shows a more modest improvement (+1.8 on average), implying that while our fine-tuning approach yields promising results, it still struggles to differentiate *sibling* and *far relation* representations effectively.

We further deepen our understanding of the impact of considering hierarchy on PLM's representations by comparing between **S-RoB** and **S-RoB**$_{hp}$. We particularly analyze the failures, in terms of wrong predictions of **S-RoB** using properties *P-S* and *A-S*. Figure 1 (left) shows that both *ancestor* and *parent* relations initially exhibit greater distances, dist(n,p)= 52.1 and dist(n,a)= 75.5, for **S-RoB**, which subsequently decrease to 51.2, 63.0, respectively, after fine-tuning, for **S-RoB**$_{hp}$. Figure 1 (right) shows hand-picked examples where we observe how the enhanced PLMs accurately reverse the trend of the distances between representations w.r.t. the evaluated properties.

To sum up, our experiments confirmed the feasibility of injecting hierarchy properties into PLMs, particularly for **BERT-L**, **S-RoB** and **S-MPNet** models, with overall performance higher than all the evaluated vanilla PLMs.

5.3 Analyzing the Transfer of Hierarchy Knowledge of PLMs to Downstream Tasks

Finally, we answer **RQ3:** *Can hierarchy-enhanced PLMs representations be transferred to downstream tasks?* We investigate in a sequential mode whether our hierarchy-enhanced PLMs using probes retain and leverage their knowledge

Table 5. Results for vanilla and enhanced PLMs (left) and SOTA models (right). Hypernymy Discovery: report MAP and P@5. Taxonomy Reconstruction: report Precision (P), Recall(R), and F1. Reading Comprehension: report accuracy. Results are average of three runs and best values in **bold**.

| Model | Hypernym Discovery | | | | | | Taxonomy Reconstruct. | | | Reading Comprehension | | |
| | FT-wiki | | BM25+RM3 | | Oracle | | | | | | | |
	MAP	P@5	MAP	P@5	MAP	P@5	P	R	F1	Middle	High	All
BERT-L	5.9	4.6	7.0	5.5	42.6	37.2	15.5	48.0	22.9	75.0	66.3	68.8
RoB-L	2.0	1.3	1.8	1.2	18.3	15.7	17.3	46.7	24.8	86.3	80.0	81.9
S-RoB	10.4	8.7	11.6	9.8	45.5	39.8	17.2	33.6	22.6	57.9	51.4	53.3
S-MPNet	9.9	7.8	10.6	8.6	46.1	39.6	9.4	41.9	15.1	74.4	67.8	69.7
BERT-L$_{hp}$	**11.3**	**9.7**	**12.4**	**11.0**	**53.6**	**48.5**	16.2	45.0	23.3	74.3	63.4	66.6
RoB-L$_{hp}$	5.0	4.1	5.5	4.5	32.2	27.0	15.9	**49.4**	23.6	**87.8**	**81.4**	**83.2**
S-RoB$_{hp}$	9.2	7.8	10.5	9.1	45.6	39.9	18.3	32.8	23.3	57.3	51.0	52.8
S-MPNet$_{hp}$	9.4	7.7	10.5	8.8	47.9	42.0	**41.4**	27.1	**32.3**	46.2	42.2	43.4

Hyper. Disc.	MAP	P@5	
BERT-L$_{hp}$	12.4	11.0	
CRIM [5]	19.8	19.0	
RMM [3]	27.1	23.4	
Taxonomy Reconst.	**P**	**R**	**F1**
S-MPNet$_{hp}$	41.4	27.1	32.3
TaxoRL [37]	35.1	35.1	35.1
CTP [16]	26.3	25.9	26.1
LMScorer [27]	29.8	28.6	29.1
Reading Compr.	**Middle**	**High**	**All**
RoB-L$_{hp}$	87.8	81.4	83.2
RoBERTa [36]	86.5	81.8	82.8
DeBERTa [24]	90.5	86.8	87.5
CoLISA[19]	90.8	86.9	87.9

when fine-tuned to perform a given downstream task. In the following, we introduce each downstream task and report our results.

Hypernym Discovery. This task evaluates an input term as a query and retrieves its suitable hypernyms from a target corpus. We used the SemEval-2018 Task 9 benchmark [13]. Specifically, we consider the English subtask 1A which includes a vocabulary with all valid hypernyms, and a training/test set of annotated hypernym pairs. Due to the size of the corpus and queries, we employed a re-ranking approach involving two steps. Firstly, we considered two different models as first rankers to retrieve the top 1000 most relevant candidates: a semantic-retrieval approach with FT-wiki embeddings and cosine similarity, and BM25 with RM3 [1]. Then, we re-ranked these filtered results using our hierarchy-enhanced and vanilla PLMs, which were fine-tuned with a binary classification layer using the CTP architecture (for details, please refer to [16]). We further explored an Oracle first ranker which extends the first ranker with missing golden candidates and acts as an upper bound of the performance of PLMs. We report MAP and P@5 ranking measures as proposed in [13].

Taxonomy Reconstruction. This task aims to construct a hierarchical taxonomy from a given set of words. We used the SemEval *TexEval-II* [8] dataset[4] to compare with previous work [16,27]. The dataset consists of an evaluation set with edge-based accuracy as the metric. We followed CTP approach by training the vanilla and our hierarchy-enhanced PLMs on the Bansal et al. dataset, omitting overlapping terms. We report standard Precision, Recall, and F1.

Reading Comprehension. We use the RACE [32] dataset, consisting of English exams for middle and high school Chinese students with up to four possible answers. The questions are split into Middle and High sets, where the High set is the most difficult. Reported values are given in terms of accuracy.

---

[4] We used the English-language version of the taxonomies *environment* and *science*.

We examine the transferability of the learned hierarchical representations to downstream tasks. Table 5 (left) shows results for all tasks. For Hypernym Discovery, our initial MAP scores were 2.2, 3.4, and 3.7 for FT-wiki, BM25, and BM25+RM3, respectively. Additionally, re-rankers **BERT-L**$_{hp}$ and **RoB-L**$_{hp}$ improved +5.4 and +5.1 w.r.t. vanilla versions for MAP and P@5 based on BM25+RM3 initial ranking. However, we noticed a slight degradation in performance for **S-RoB**$_{hp}$ and **S-MPNet**$_{hp}$ (similarly for FT-wiki column). We also observed that all models improved their results for the Oracle results, with smaller margin (+0.1) for **S-RoB**$_{hp}$ and **S-MPNet**$_{hp}$. These results motivate us to explore better first rankers. Moreover, considering the fact that this particular task heavily relies on hypernym relations (*group P-**), we assume that the enhanced representations are easily transferable to this task. For Taxonomy Reconstruction, we report the average metrics from both taxonomies. Three models, **BERT-L**$_{hp}$, **S-RoB**$_{hp}$, and **S-MPNet**$_{hp}$, showed improvements in F1 scores w.r.t their vanilla version (+0.4, +0.7, and +17.2, respectively), while **RoB-L**$_{hp}$ degraded its performance by −1.2. The improvements were primarily driven by better precision $(0.7 - 32.0)$, with a smaller penalization on the recall $(0.8 - 14.8)$. Considering higher precision as an indication of better quality in the concept representations, we argue that the enhanced representations from *groups P-**, *A-**, and *S-** are transferable for this task. For Reading Comprehension, learned representations may not be fully transferable. Specifically, we observed improvements for **RoB-L**$_{hp}$, but not for other models. This suggests that hierarchy knowledge is either forgotten [31] or penalizing for this task. Thus, an appropriate setup is called since a sequential fine-tuning might lead to the hierarchy representations drift for this task.

Table 5 (right) shows our results w.r.t SOTA models. For the tasks of Taxonomy Reconstruction and Reading Comprehension[5], **S-MPNet**$_{hp}$ and **RoB-L**$_{hp}$ achieve competitive performances. However, the main trend is that specialized models, such as RRM, TaxoRL, and CoLISA, outperform fine-tuned enhanced models, which sheds light to the desirable room for improvement that a suitable transfer learning of hierarchy knowledge to downstream tasks, would achieve. Overall, our empirical findings suggest that the enhanced PLM representations are moderately transferable in a sequential mode of probe then fine-tune for hierarchy-aware tasks such as Hypernymy Discovery and Taxonomy Reconstruction, but detrimental for Reading Comprehension.

6   Conclusion and Future Work

In this paper, we proposed a task-agnostic methodology for probing the capability of PLMs to capture hierarchy. We first identified hierarchy properties from taxonomies and then, constructed probes encoding these properties to train hierarchy-enhanced PLMs. Our experiments showed that cross-encoder models like BERT struggle to capture hierarchy and that hierarchy properties could be

---

[5] All scores recalculated for RoBERTa as suggested in the fairseq GitHub repository.

readily injected into PLMs regardless of specific tasks. Evaluation of hierarchy-aware PLMs in downstream tasks reveals that a kind of catastrophic forgetting can occur leading to performance results under upper-bound performance of PLMs trained specifically on hierarchy-aware tasks.

Future work will explore the use of transfer learning methods, such as adding hierarchy-aware auxiliary losses, that are able to learn robust features of a wide-range of hierarchy properties in a cross-task setting. This work is generalizable to other properties beyond hierarchy, leading to foster future research on the design of interpretable and effective PLMs for IR and NLP tasks.

References

1. Abdul-Jaleel, N., et al.: UMass at TREC 2004: Novelty and HARD. Computer Science Department Faculty Publication Series, p. 189 (2004)
2. Alsuhaibani, M., Maehara, T., Bollegala, D.: Joint learning of hierarchical word embeddings from a corpus and a taxonomy. In: Automated Knowledge Base Construction (AKBC) (2019)
3. Bai, Y., Zhang, R., Kong, F., Chen, J., Mao, Y.: Hypernym discovery via a recurrent mapping model. In: Findings of the Association for Computational Linguistics: ACL-IJCNLP 2021, pp. 2912–2921. Online (2021)
4. Bansal, M., Burkett, D., de Melo, G., Klein, D.: Structured learning for taxonomy induction with belief propagation. In: ACL 2014, pp. 1041–1051 (2014)
5. Bernier-Colborne, G., Barrière, C.: CRIM at SemEval-2018 task 9: a hybrid approach to hypernym discovery. In: Proceedings of the 12th International Workshop on Semantic Evaluation, pp. 725–731. New Orleans, Louisiana (2018)
6. Bojanowski, P., Grave, E., Joulin, A., Mikolov, T.: Enriching Word Vectors with Subword Information. TACL **2017**(5), 135–146 (2017)
7. Bordea, G., Buitelaar, P., Faralli, S., Navigli, R.: SemEval-2015 Task 17: Taxonomy Extraction Evaluation (TExEval). In: SemEval-2015, pp. 902–910 (2015)
8. Bordea, G., Lefever, E., Buitelaar, P.: SemEval-2016 Task 13: Taxonomy Extraction Evaluation (TExEval-2). In: SemEval-2016, pp. 1081–1091 (2016)
9. Bordes, A., Usunier, N., Garcia-Duran, A., Weston, J., Yakhnenko, O.: Translating embeddings for modeling multi-relational data. In: Advances in Neural Information Processing Systems (NIPS). vol. 26 (2013)
10. Bouraoui, Z., Camacho-Collados, J., Schockaert, S.: Inducing relational knowledge from BERT. In: Proceedings of the AAAI Conference on Artificial Intelligence. vol. 34, pp. 7456–7463 (2020)
11. Budanitsky, A., Hirst, G.: Evaluating WordNet-based measures of lexical semantic relatedness. Comput. Linguist. **32**(1), 13–47 (2006)
12. Camacho-Collados, J.: Why we have switched from building full-fledged taxonomies to simply detecting hypernymy relations (2017). arXiv preprint arXiv:1703.04178
13. Camacho-Collados, J., et al.: SemEval-2018 task 9: Hypernym discovery. In: Proceedings of the 12th International Workshop on Semantic Evaluation, pp. 712–724. Association for Computational Linguistics, New Orleans, Louisiana (2018)
14. Câmara, A., Hauff, C.: Diagnosing BERT with retrieval heuristics. In: Jose, J.M., et al. (eds.) ECIR 2020. LNCS, vol. 12035, pp. 605–618. Springer, Cham (2020). https://doi.org/10.1007/978-3-030-45439-5_40
15. Chen, B., et al.: Probing BERT in hyperbolic spaces. In: International Conference on Learning Representations (2020)

16. Chen, C., Lin, K., Klein, D.: Constructing taxonomies from pretrained language models. In: Proceedings of the 2021 Conference of the North American Chapter of the Association for Computational Linguistics: Human Language Technologies (NAACL), pp. 4687–4700. Online (2021)

17. Clark, P., Richardson, K., Tafjord, O.: Transformers as soft reasoners over language. In: IJCAI 2020. vol. 4, pp. 3882–3890 (2020)

18. Devlin, J., Chang, M.W., Lee, K., Toutanova, K.: BERT: pre-training of deep bidirectional transformers for language understanding. In: NAACL 2019, pp. 4171–4186 (2019)

19. Dong, M., Zou, B., Li, Y., Hong, Y.: CoLISA: inner interaction via contrastive learning for multi-choice reading comprehension. In: Kamps, J., et al. Advances in Information Retrieval. ECIR 2023. LNCS, vol. 13980. Springer, Cham (2023). https://doi.org/10.1007/978-3-031-28244-7_17

20. Dong, X., Shen, J.: Triplet Loss in Siamese Network for Object Tracking. In: ECCV 2018, pp. 459–474 (2018)

21. Du, Y., Zhu, X., Chen, L., Zheng, B., Gao, Y.: HAKG: hierarchy-aware knowledge gated network for recommendation. In: Proceedings of the 45th International ACM SIGIR Conference on Research and Development in Information Retrieval, pp. 1390–1400 (2022)

22. Fu, R., Guo, J., Qin, B., Che, W., Wang, H., Liu, T.: Learning Semantic Hierarchies via Word Embeddings. In: ACL 2014, pp. 1199–1209 (2014)

23. Gupta, A., Piccinno, F., Kozhevnikov, M., Paşca, M., Pighin, D.: Revisiting taxonomy induction over Wikipedia. In: COLING 2016, pp. 2300–2309 (2016)

24. He, P., Gao, J., Chen, W.: DeBERTav3: improving deBERTa using ELECTRA-style pre-training with gradient-disentangled embedding sharing. In: The Eleventh International Conference on Learning Representations (2023). https://openreview.net/forum?id=sE7-XhLxHA

25. Hovy, E., Kozareva, Z., Riloff, E.: Toward completeness in concept extraction and classification. In: Proceedings of the 2009 Conference on Empirical Methods in Natural Language Processing (EMNLP), pp. 948–957 (2009)

26. Hu, Z., Huang, P., Deng, Y., Gao, Y., Xing, E.: Entity Hierarchy Embedding. In: IJCNLP 2015, pp. 1292–1300 (2015)

27. Jain, D., Espinosa Anke, L.: Distilling hypernymy relations from language models: on the effectiveness of zero-shot taxonomy induction. In: Nastase, V., Pavlick, E., Pilehvar, M.T., Camacho-Collados, J., Raganato, A. (eds.) Proceedings of the 11th Joint Conference on Lexical and Computational Semantics, pp. 151–156. Association for Computational Linguistics, Seattle, Washington (2022). https://doi.org/10.18653/v1/2022.starsem-1.13, https://aclanthology.org/2022.starsem-1.13

28. Jain, D., Espinosa Anke, L.: Distilling hypernymy relations from language models: on the effectiveness of zero-shot taxonomy induction. In: Proceedings of the 11th Joint Conference on Lexical and Computational Semantics, pp. 151–156. Seattle, Washington (2022)

29. Ji, G., He, S., Xu, L., Liu, K., Zhao, J.: Knowledge graph embedding via dynamic mapping matrix. In: Proceedings of the 53rd Annual Meeting of the Association for Computational Linguistics and the 7th International Joint Conference on Natural Language Processing (NAACL), pp. 687–696. Beijing, China (2015)

30. Kacmajor, M., Kelleher, J.D.: Capturing and measuring thematic relatedness. Lang. Resour. Eval. **54**(3), 645–682 (2020)

31. Kirkpatrick, J., et al.: Overcoming catastrophic forgetting in neural networks. Proc. Nat. Acad. Sci. **114**(13), 3521–3526 (2017). https://doi.org/10.1073/pnas.1611835114, https://www.pnas.org/doi/abs/10.1073/pnas.1611835114

32. Lai, G., Xie, Q., Liu, H., Yang, Y., Hovy, E.: RACE: Large-scale ReAding Comprehension Dataset From Examinations. In: EMNLP 2017, pp. 785–794 (2017)
33. Lapesa, G., Evert, S., Schulte im Walde, S.: Contrasting syntagmatic and paradigmatic relations: insights from distributional semantic models. In: Bos, J., Frank, A., Navigli, R. (eds.) Proceedings of the Third Joint Conference on Lexical and Computational Semantics (*SEM 2014), pp. 160–170. Association for Computational Linguistics and Dublin City University, Dublin, Ireland (2014). https://doi.org/10.3115/v1/S14-1020, https://aclanthology.org/S14-1020
34. Liu, C., Cohn, T., Frermann, L.: Seeking Clozure: robust Hypernym extraction from BERT with anchored prompts. In: Proceedings of the The 12th Joint Conference on Lexical and Computational Semantics (* SEM 2023), pp. 193–206 (2023)
35. Liu, W., et al.: K-BERT: enabling language representation with knowledge graph. In: Proceedings of the AAAI Conference on Artificial Intelligence. vol. 34, pp. 2901–2908 (2020)
36. Liu, Y., et al.: RoBERTa: A Robustly Optimized BERT Pretraining Approach (2019). arXiv:1907.11692 [cs]
37. Mao, Y., Ren, X., Shen, J., Gu, X., Han, J.: End-to-end reinforcement learning for automatic taxonomy induction. In: Proceedings of the 56th Annual Meeting of the Association for Computational Linguistics (ACL), pp. 2462–2472. Melbourne, Australia (2018)
38. Nickel, M., Kiela, D.: Poincaré Embeddings for Learning Hierarchical Representations. In: NIPS 2017. vol. 30 (2017)
39. Pimentel, T., Valvoda, J., Hall Maudslay, R., Zmigrod, R., Williams, A., Cotterell, R.: Information-theoretic probing for linguistic structure. In: Proceedings of the 58th Annual Meeting of the Association for Computational Linguistics (ACL) (2020)
40. Qiu, X., Sun, T., Xu, Y., Shao, Y., Dai, N., Huang, X.: Pre-trained models for natural language processing: a survey. Sci. China Technol. Sci. 63(10), 1872–1897 (2020)
41. Reimers, N., Gurevych, I.: Sentence-BERT: Sentence Embeddings using Siamese BERT-Networks. In: EMNLP-IJCNLP (2019)
42. Resnik, P.: Using information content to evaluate semantic similarity in a taxonomy. In: IJCAI 1995, pp. 448–453 (1995)
43. Richardson, K., Sabharwal, A.: What does my QA model know? Devising controlled probes using expert. In: TACL 2020 vol. 8, pp. 572–588 (2020)
44. Rossi, A., Barbosa, D., Firmani, D., Matinata, A., Merialdo, P.: Knowledge graph embedding for link prediction: a comparative analysis. ACM Trans. Knowl. Discov. Data 15(2), 1–49 (2021)
45. Sieg, A., Mobasher, B., Lytinen, S., Burke, R.: Using concept hierarchies to enhance user queries in web-based information retrieval. Artificial Intelligence and Applications (AIA) (2004)
46. Song, M., Feng, Y., Jing, L.: Hyperbolic relevance matching for neural keyphrase extraction. In: Proceedings of the 2022 Conference of the North American Chapter of the Association for Computational Linguistics: Human Language Technologies, pp. 5710–5720 (2022)
47. Sun, Y., et al.: ERNIE 2.0: a continual pre-training framework for language understanding. Proc. AAAI Conf. Artif. Intell. 34(05), 8968–8975 (2020). https://doi.org/10.1609/aaai.v34i05.6428, https://ojs.aaai.org/index.php/AAAI/article/view/6428
48. Talmor, A., Elazar, Y., Goldberg, Y., Berant, J.: oLMpics-On what language model pre-training captures. TACL 2020(8), 743–758 (2020)

49. Tifrea, A., Becigneul, G., Ganea, O.E.: Poincare Glove: Hyperbolic Word Embeddings. In: International Conference on Learning Representations (2019). https://openreview.net/forum?id=Ske5r3AqK7

50. Vendrov, I., Kiros, R., Fidler, S., Urtasun, R.: Order-embeddings of images and language. In: 4th International Conference on Learning Representations, ICLR 2016, San Juan, Puerto Rico, 2016, Conference Track Proceedings (2016)

51. Vulić, I., Gerz, D., Kiela, D., Hill, F., Korhonen, A.: HyperLex: a large-scale evaluation of graded lexical entailment. Comput. Linguist. **43**(4), 781–835 (2017)

52. Vulić, I., Ponti, E.M., Korhonen, A., Glavaš, G.: LexFit: lexical fine-tuning of pretrained language models. In: ACL 2021, pp. 5269–5283 (2021)

53. Vulić, I., Ponti, E.M., Litschko, R., Glavaš, G., Korhonen, A.: Probing pretrained language models for lexical semantics. In: EMNLP 2020, pp. 7222–7240 (2020)

54. Wang, K., Reimers, N., Gurevych, I.: TSDAE: using transformer-based sequential denoising auto-encoderfor unsupervised sentence embedding learning. In: Findings of the Association for Computational Linguistics: EMNLP 2021, pp. 671–688. Association for Computational Linguistics, Punta Cana, Dominican Republic (2021)

55. Yang, J., et al.: A survey of knowledge enhanced pre-trained models (2022)

56. Zhang, Z., Han, X., Liu, Z., Jiang, X., Sun, M., Liu, Q.: ERNIE: enhanced Language Representation with Informative Entities. In: ACL 2019, pp. 1441–1451 (2019)

57. Zhong, J., Zhu, H., Li, J., Yu, Y.: Conceptual graph matching for semantic search. In: ICCS, pp. 92–106 (2002)

58. Zhu, Y., Zhou, D., Xiao, J., Jiang, X., Chen, X., Liu, Q.: HyperText: Endowing FastText with hyperbolic geometry. In: Cohn, T., He, Y., Liu, Y. (eds.) Findings of the Association for Computational Linguistics: EMNLP 2020, pp. 1166–1171. Association for Computational Linguistics, Online (2020). https://doi.org/10.18653/v1/2020.findings-emnlp.104, https://aclanthology.org/2020.findings-emnlp.104

Query Exposure Prediction for Groups of Documents in Rankings

Thomas Jaenich[✉], Graham McDonald, and Iadh Ounis

University of Glasgow, Glasgow, UK
t.jaenich.1@research.gla.ac.uk,
{graham.mcdonald,iadh.ounis}@glasgow.ac.uk

Abstract. The main objective of an Information Retrieval (IR) system is to provide a user with the most relevant documents to the user's query. To do this, modern IR systems typically deploy a re-ranking pipeline in which a set of documents is retrieved by a lightweight first-stage retrieval process and then re-ranked by a more effective but expensive model. However, the success of a re-ranking pipeline is heavily dependent on the performance of the first stage retrieval, since new documents are not usually identified during the re-ranking stage. Moreover, this can impact the amount of exposure that a particular group of documents, such as documents from a particular demographic group, can receive in the final ranking. For example, the fair allocation of exposure becomes more challenging or impossible if the first stage retrieval returns too few documents from certain groups, since the number of group documents in the ranking affects the exposure more than the documents' positions. With this in mind, it is beneficial to predict the amount of exposure that a group of documents is likely to receive in the results of the first stage retrieval process, in order to ensure that there are a sufficient number of documents included from each of the groups. In this paper, we introduce the novel task of query exposure prediction (QEP). Specifically, we propose the first approach for predicting the distribution of exposure that groups of documents will receive for a given query. Our new approach, called GEP, uses lexical information from individual groups of documents to estimate the exposure the groups will receive in a ranking. Our experiments on the TREC 2021 and 2022 Fair Ranking Track test collections show that our proposed GEP approach results in exposure predictions that are up to ∼40% more accurate than the predictions of suitably adapted existing query performance prediction (QPP) and resource allocation approaches.

1 Introduction

The main objective of an information retrieval (IR) system is to provide the users with the documents that are the most relevant to their search query. Recently, deep neural ranking models that use contextualised language representations, such as BERT [12], have been shown to effectively score documents and determine their relevance to a user's query [27]. However, while effective, applying these neural models is computationally expensive and is often infeasible over large collections [23]. Therefore, deep neural ranking models are usually deployed in re-ranking pipelines. In a re-ranking pipeline, an inexpensive ranking model, such as BM25 [34] or the more recent SPLADE [18], is used to create an initial pool of documents in a first-pass retrieval process. The documents from this pool are then re-scored by deep neural ranking models such as ELECTRA [31] or T5 [32]. The re-scored documents are then presented to the user in decreasing order of relevance. However, the amount of exposure to a user that a document is likely to receive is dependent on the document's position in the ranking, since according to the well-known position bias model [10], lower-ranked documents have a lower probability of being exposed to, or examined by, a user. This can lead to an unintended bias against particular *groups* of documents. For example, if relevant documents that are associated with a demographic group, such as race or gender, are systematically ranked at lower positions, then the group will receive less exposure to the user than the groups that have documents ranked at higher positions [37]. In re-ranking pipelines, the exposure distribution in the final ranking is typically limited by the recall of the initial candidate pool, i.e., documents that are not retrieved by the first-pass retrieval process are not scored by the re-ranker. If the initial pool of documents contains only a few or no relevant documents from a group, a fair distribution of exposure in the final ranking might become infeasible. For example, Bower et al. [7] have shown that inequality in the initial pool of documents in a multi-stage ranking process can hinder the effectiveness of existing fairness mitigation methods, such as randomisation [13]. Therefore, predicting the exposure of a group before deploying the initial ranker is both important and useful. Indeed, this would, for example, allow the system to adjust the weights of different groups in the first-pass retrieval model, so as to have enough documents from each group in the candidate pool and a fair chance of achieving the desired exposure in the search results when a re-ranker is applied.

In this work, we introduce the new task of Query Exposure Prediction (QEP), i.e., predicting how the available exposure in search results will be distributed among groups of (related) documents in response to a user's query. Developing exposure distribution prediction models will enable system designers to choose an appropriate initial ranker and deploy fairness strategies on a case-by-case basis, i.e., when the predicted exposure distribution is not closely correlated with a desired, or target, distribution. Specifically, we propose the first QEP approach, called Group Exposure Predictor (GEP). Our new approach is inspired by pre-retrieval Query Performance Prediction (QPP) [21] approaches that aim to estimate how difficult it will be for an IR system to provide relevant documents in response to a particular query. However, differently from pre-retrieval QPP approaches, our proposed GEP approach leverages statistics about how a query's terms are distributed in the indexed documents from each group to estimate how the available exposure will be distributed among the different groups in a ranking. We evaluate the performance of GEP in two scenarios. One where the query is passed to the predictor and executed as it is, and another where the query is expanded by Pseudo-Relevance Feedback (PRF) methods before being passed to the predictor. Our extensive experiments on the TREC 2021 & 2022 Fair Ranking

Track [15, 16] test collections show that our proposed exposure prediction model GEP is able to predict the distribution of exposure better than traditional performance predictors as well as an established federated information retrieval approach. Specifically, our predictor shows statistically significant improvements over the baselines for the majority of observed exposure distributions, with decreases in the prediction error of up to ∼40% compared to the next best baseline.

2   Related Work

A fair allocation of document exposure in a ranking is increasingly recognised as a key objective besides relevance when developing search systems [14]. Specifically, in the fairness literature, many prior works [6, 22, 24, 26, 29, 30, 36–38, 41] have proposed approaches to avoid unfair distributions of exposure in rankings. Such approaches have typically focused on fairly distributing the exposure over individual documents or over predefined groups of documents [42]. In this work, we are concerned with groups of documents in a ranking. The definition of the groups of documents in a ranking usually depends on the search task at hand. In many cases, the documents are grouped by an associated protected characteristic. The protected characteristic, sometimes also referred to as a protected attribute, is defined as an associated label to a given document, which should not be used as the basis of decision-making. Some prominent examples of protected attributes include geographic locations, ethnicity, and gender. Among the many studies that propose mitigation approaches, several works have suggested how to measure fairness in terms of exposure [13, 33, 40]. However, we are not aware of any previous work on predicting the exposure of groups in a ranking. Therefore, we aim to fill this research gap by proposing the first query performance predictor for measuring the exposure of groups of documents. Query Performance Prediction (QPP) is traditionally applied to predict the effectiveness of the ranking produced by an IR system in response to a query without using human-relevance judgements. This is useful since the performance of IR systems varies over different queries. The QPP predictors are usually separated in *post-retrieval* and *pre-retrieval* methods. Post-retrieval methods are applied after an initial ranking was created by the system and therefore have access to the estimated relevance scores, which can be used for the performance prediction [9]. However, post-retrieval methods are not applicable in our scenario, since after the initial retrieval stage, the exposure can be directly calculated, rendering the prediction redundant. On the other hand, pre-retrieval query performance prediction methods traditionally use properties from the underlying document corpus or information obtained from the query terms to make their predictions. For example, average inverse document frequency, AvIDF [11], approximates the average specificity of query terms by calculating the inverse document frequency (idf) for each query term. Related to this, He et al. [20] have introduced the average inverse collection term frequency (AvICTF). AvICTF is similar to AvIDF but uses the term frequency in a collection instead of the document frequency in the calculation of IDF. Further extending the concept of AvICTF, He et al. proposed to use the Simplified Clarity Score (SCS). SCS measures how ambiguous a query is. SCS is calculated using the sum of the AvICTF and a maximum likelihood estimate on how often a term appears in a query. These predictors serve as baselines in our experiments. Differently from predictors that solely use lexical information of the query terms from the collection statistics, the term-relatedness predictor, Averaged Pointwise Mutual Information (AvPMI) [19], calculates the probability of two terms appearing together in a document. AvPMI has been shown to be a robust predictor due to its ability to exploit co-occurrence statistics in a collection. Therefore, we include it as a baseline in our experiments, and evaluate how well the query performance predictors AvIDF, AvICTF, SCS and AvPMI can estimate the exposure of groups of documents in a ranking.

In a related manner, resource selection approaches such as the prominent CORI [8] algorithm from federated search, predict the suitability of different data sources for retrieving content for a given query. Conceptually similar to the task we are tackling, the CORI algorithm uses a probability denoted as *belief*, which is calculated from the collection statistics and predicts how suitable a data source is for a given query. We adapt this algorithm to our task and use it as one of our baselines. In this work, we particularly focus on and adapt QPP methods based on lexical information. Unlike neural approaches, lexical QPP predictors rely on clear and interpretable features derived from lexical information. This makes it easier to capture the underlying mechanisms contributing to the prediction as well as developing and evaluating a new predictor for our new query exposure prediction task. Next, before the introduction of our proposed predictor, we first describe in more detail the exposure distribution problem.

3   Exposure in Rankings

In this work, we aim to predict how much exposure to a user a group of documents will receive when the documents are ranked by their relevance to the user's query. The expected exposure of a given document depends on its position in the ranking, since a user usually starts from the top and examines each document in order with a certain probability of stopping based on the number of relevant (or partially relevant) documents they have seen. Therefore, the further down in the ranking the document is, the less likely it will be exposed to the user (often referred to as the position bias [10]). There have been a number of user models, which have been proposed in the literature that estimate the likelihood of a user examining a document at a particular rank position, for example [3,10,39]. In this work, we estimate the available document exposure at a given position in the ranking using the standard exposure drop-off (position bias) from the well-known Discounted Cumulative Gain (DCG) [25] metric, defined as: $\frac{1}{(log_2(p)+1)}$ where p is a position in the ranking. Figure 1(a) illustrates the exposure distribution that is available for a ranking of length $k=100$ under the DCG position bias/exposure model. Figure 1(a) shows that the available exposure drops sharply in rank positions 1..10 and then decreases more gradually in positions 11..100. Figure 1(b) shows the number of possible orderings for rankings of size $k=1..100$. As can be seen from Fig. 1(b), there are 10^{152} different possible orderings of documents in a ranking of size $k=100$.

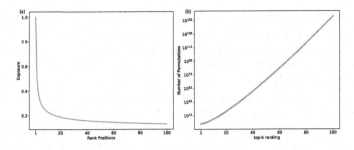

Fig. 1. Plot (a) shows the amount of available exposure for each position in a ranking of size $k{=}100$ according to the position bias model from the well-known DCG metric [25]. Plot (b) shows the number of possible orderings of documents for rankings of size $k{=}1...100$. The y-axis of Plot (b) is in log scale.

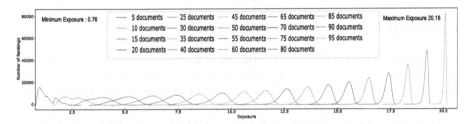

Fig. 2. The number of rankings that can achieve a particular amount of exposure for a given number of documents in a single group for a ranking of size $k{=}100$.

The amount of exposure that a group of documents receives is calculated by summing the exposure values for each position in the ranking where a document from the group appears. For example, if a group has two documents in a ranking, with the first document placed at position 1 (i.e., the top rank position) and the second document placed in the 5th position, the first document receives an exposure of 1 and the second document receives an exposure of 0.39. In such a scenario, the group will receive an exposure of $1 + 0.39 = 1.39$. Such an exposure model only considers the rank of a document to determine its amount of exposure. In this case, the amount of exposure that a group of documents will receive is dependent on (1) the size of the ranking, (2) the number of documents from the group that are in the ranking, and (3) the positions in the ranking where the documents from the group appear. However, there are multiple possible rankings (or orderings) that produce the same amount of exposure for a group. Returning to our previous example of two documents from a group in rank positions 1 and 5, if the positions of the two documents in the ranking are swapped then the group's exposure is still 1.39.

For a ranking of size $k{=}100$, Fig. 2 presents (1) the range of exposure values that are achievable for a single group depending on the number of documents from the group that are included in the ranking, and (2) the number of possible rankings that achieve that particular amount of exposure. From Fig. 2, we can see that the range of achievable exposure is larger when there is a relatively small number of

documents from a group in the ranking, but that the amount of achievable exposure increases as more documents from the group are added to the ranking. Moreover, the number of documents in the ranking affects the specific range of achievable exposure. Furthermore, certain exposure values are more likely to be achieved
given the number of documents in the ranking. Hence, adding more documents
from a group to the ranking can notably increase the group's exposure, much
more than just moving the existing documents to higher ranks. This highlights the
importance of the first-pass retrieval and the impact of low recall on the re-ranking
pipeline since the exposure that a group receives will be much more enhanced by
adding more documents from a group in the first-pass retrieval than re-ranking
the documents in the existing ranking. Moreover, this suggests that enhancing
recall using techniques such as pseudo-relevance feedback (PRF) [2,4,5,17] has
the potential to notably impact the distribution of exposure.

4   Group Exposure Prediction - GEP

Our proposed GEP predictor estimates how the available exposure will be distributed over groups, $g \in G$, of documents, in a ranking r of size k in response
to a user's query Q with query terms $t_0..t_n$. To predict how the available exposure will be distributed over the groups, for each group, $g \in G$, we first create
a vector representation S_g, where $|S_g| = n$. The vector S_g contains one score, s,
per query term, $t \in Q$, where s_i is the mean of the k largest tf-idf scores for the
documents in g and the query term t_i. To calculate s we first calculate the tf-idf
scores for every document within a group $d_{g,i}$ for a term t_i, where the inverse
document frequency is calculated as: $idf(d_g) = log(\frac{||d_g|| - df_g + 0.5}{df_g + 0.5})$, and $||d_g||$ is
the number of documents that exist for a group g in the collection, df_g is the
document frequency, i.e., the number of group documents the term appears in.
In the next step, we collect the term frequencies tf, i.e., the number of term
occurrences in each document $d_{g,i}$ and combine it with the idf score tf-idf $=$
$tf \cdot idf$. To obtain the score s from the calculated tf-idf values, we use only the
k largest tf-idf scores, where k refers to the length of the ranking r for which
we predict the exposure distribution. In other words, the maximum number of
group documents that could be included in the ranking. If a term appears in less
than k documents, we consider the missing scores to be zero. We only consider
the k highest scores because the ranking length is essential for the exposure the
documents receive. For example, let us assume that for a considered group, only
one document contains the query term t_i. We obtain only one tf-idf score, which
is potentially very high. However, a group can only occupy a single high position
in the final ranking with a single document, so the overall exposure is still low.
Therefore, to avoid any under or over-estimation of s we use the average of the
top-k scores, calculating $i = \frac{1}{k} \sum_{i=1}^{k} tf$-idf. For every query term, we store s in
the vector S. Conceptually, S is a suitable representation of the group indicating
which query terms are most important. To use S to make a performance prediction for the group, we need to check its similarity to the original query Q. We
use the traditional way of calculating tf-idf for each t in Q. However, instead of

calculating the scores only over documents from one group we calculate them over the whole collection. This is necessary since the query will be executed over all of the documents. To compute the similarity e_g for each group to the query we calculate the dot product between S and the query $Q : e_g = \mathbf{I_g} \cdot \mathbf{Q}$. A higher similarity indicates a higher predicted exposure. However, since the exposure is relative to the other groups we need to calculate the similarity for every group $g \in G$. We calculate our final prediction score GEP_g by comparing the individual scores e_g and normalise them to be between 0 and 1. This results in a predictor score per group that can be used to estimate the exposure a group will obtain.

5 Experimental Setup

To evaluate our proposed GEP predictor in the task of query exposure prediction, we formulate the following two research questions:

1. **RQ1:** Is our proposed GEP approach effective for predicting the exposure that groups will receive when an inexpensive ranking model is deployed?
2. **RQ2:** Is GEP effective for predicting the exposure that groups receive in a ranking, when the query is expanded using pseudo-relevance feedback?

To conduct our experiments and answer our research questions, we use the PyTerrier [28] Information Retrieval Framework.

Test Collections: In our experiments, we use the TREC 2021 and 2022 Fair Ranking Track test collections [15,16]. Both test collections consist of English language Wikipedia articles. These test collections were created to encourage research on how editors of Wikipedia can be presented with a fair ranking of articles that need editing, in relation to a topic of their interest. In the TREC Fair Ranking Track, fairness is measured in terms of exposure. Hence, the corresponding test collections are well suited for investigating our research questions. The test collections consist of a set of queries as well as fairness group category labels for the documents in the underlying collection. The labels categorise the different features of documents such as their popularity or creation date. Both TREC collections were indexed using a Porter Stemmer and stop-word removal. In the next paragraph, we explain the categories.

Categories: Figure 1 provides an overview of the available categories and their corresponding groups for which we aim to predict the exposure. In the TREC 2021 Fair Ranking Track collection, only the geographic location category has labels for all of the documents in the collection. Therefore, we use the geographic category in our experiments. For the TREC 2022 collection, we use the categories proposed by the organisers with group labels for every document in the collection. The first category is *Age of an article*, which denotes the length of time the article (or document) has existed. The age is mapped into four discrete categories shown in Figure 1. We also use the categories: *Age of Topic, Alphabetical* and *Popularity*. Age of Topic denotes the age of the main topic that is

Table 1. Categories and their groups.

Geo-Location
Unknown, Europe, Africa, Northern America, Asia, Oceania, Latin America and the Caribbean, Antarctica
Age of Article
2001–2006, 2007–2011, 2012–2016, 2017–2022
Popularity
Low, Medium-Low, Medium-High, High
Alphabetical
a-d, e-k, l-r, and s-z
Age of the topic
Unk, Pre-1900 s, 20th century, 21st century

discussed in an article. Alphabetical categorises articles by the initial letter of an article's title. The *Popularity* category labels articles based on their viewing frequency in February 2022.

Queries: We use all of the evaluation queries from the two Fair Ranking Track collections. Specifically, we use the 49 evaluation queries in the TREC 2021 Fair Ranking Track collection and the 47 evaluation queries from the 2022 collection. All queries consist of keywords relating to a Wikipedia topic, e.g., a query can be a collection of words such as "museum baroque librarian architecture library".

Retrieval models: In our experiments, we use a selection of prominent first-pass ranking models commonly used in re-ranking pipelines. We evaluate our predictor on BM25 [34] (with default parameters), TF-IDF [35] and SPLADE [18].

PRF: To evaluate the impact of pseudo-relevance feedback on the prediction performance of GEP, we use the RM3 relevance language model [2], and KLQueryExpansion [4] (KLQ) in our experiments, with the number of feedback documents set to 3 and the number of expansion terms set to 10.

Baselines: As baselines, we use the CORI [8] federated IR algorithm as well as several pre-retrieval Query Performance Predictors. We use Average Inverse Collection Term Frequency (AvICTF) [21], Average Inverse Document Frequency (AvIDF) [11], Averaged Pointwise Mutual Information (AvPMI) [19] and Simplified Clarity Score (SCS) [20] as baselines. To have a fair comparison with our proposed approach, we create a prediction score for each group and then normalise these group scores to be between 0 and 1. The normalised scores are used as a way to measure how much exposure a group will receive in the ranking. We assume that if a query is predicted to perform better on documents of one group, compared to the documents of another group, then the first group is likely to receive more exposure in the ranking. We normalise the prediction scores, since the predictions between groups are dependent i.e., if all prediction scores between groups are high, then the difference in exposure will likely be low. If the predicted performance of a query is high over one group of documents but low on the others, then it is likely that the high-performing group will receive most of the exposure in the ranking. As introduced in the related work section, the CORI algorithm contains a belief component, which depicts how likely a group of documents contains relevant information to a query. In this work, we choose to use an average of the belief values and a value of 0.4 for the belief parameter in the algorithm, as suggested in the original paper [8].

Measures: To evaluate our predictions, we calculate the actual exposure the groups receive in a ranking and compare the exposure to our predicted values. Following the TREC Fair Ranking Track, we use the Jensen-Shannon Distance for the calculation. Jensen-Shannon Distance calculates the distance between two probability distributions. We use it to calculate the distance between the distributions of our predicted exposure P and the actual Exposure E:

$$JSD(P\|E) = \sqrt{\frac{1}{2}\text{KL}(P\|M) + \frac{1}{2}\text{KL}(E\|M)} \qquad (1)$$

where $JSD(P\|E)$ is the distance between the predicted exposure P and the actual exposure E. $KL(P\|M)$ is the Kullback-Leibler (KL) divergence between P and the average distribution M. $KL(E\|M)$ is the Kullback-Leibler divergence between E and M. M denotes the midpoint distribution between P and E. $M = \frac{1}{2}(P + E)$. If there are no differences between the actual exposure and the predicted value, the distance is 0. In our experiments, we consider a predictor to be better when it produces values closer to 0 compared to another predictor. In practice, when choosing a predictor for deployment, we usually set a target threshold to ascertain its usefulness. For example, a predictor should show a high degree of similarity (e.g., <0.2) with the real values of a test set before being deployed. However, the choice of threshold depends on the predictor's application and the fairness definition in that context. Hence, we leave such a study to future work. When analysing our experiments, we test the results of GEP for statistically significant differences to the baselines using a paired t-test (p<0.01) with Bonferroni correction.

6   Results

In this section, we answer both our research questions and draw further insights into the problem of exposure prediction.

Results for RQ1: To answer RQ1, we evaluate the performance of our GEP predictor in terms of its ability to predict the exposure that groups of documents will receive in a ranking of size $k=100$. We generate rankings with different retrieval models namely BM25, TF-IDF and SPLADE. We calculate the actual exposure values the documents receive and compare these with the predictions

Table 2. Evaluation of the predictors on common first-pass retrieval rankers, for rankings with $k=100$. * indicates significant differences (t-test, with Bonferroni correction, p<0.01) in prediction performance between GEP and the baselines.

Approach	Age of Topic	Popularity	Age of the Article	Alphabetical	Geo-Location (TREC 21)
SPLADE					
GEP	**0.2624**	0.2013	**0.2129**	**0.2041**	**0.3543**
SCS	0.3738*	0.2243	0.2782*	0.2438*	0.5261*
AvIDF	0.3995*	0.2128	0.2889*	0.2471*	0.5820*
AvICTF	0.4088*	**0.1884**	0.2714*	0.2469*	0.5695*
AvPMI	0.3713*	0.2083	0.2758*	0.2414*	0.5474*
CORI	0.4062*	0.1888	0.2735*	0.2467*	0.5631*
BM25					
GEP	**0.2880**	**0.1567**	**0.1759**	**0.1335**	**0.2449**
SCS	0.3505*	0.1620	0.2336*	0.1407	0.4134*
AvIDF	0.3726*	0.1621	0.2443*	0.1436	0.4844*
AvICTF	0.3773*	0.1706*	0.2305*	0.1431	0.4668*
AvPMI	0.3562*	0.1645	0.2302*	0.1413	0.4388*
CORI	0.3735*	0.1703*	0.2338*	0.1422	0.4582*
TF-IDF					
GEP	**0.2886**	**0.1583**	**0.1759**	**0.1336**	**0.2454**
SCS	0.3512*	0.1628	0.2339*	0.1420	0.4138*
AvIDF	0.3734*	0.1629	0.2447*	0.1449	0.4847*
AvICTF	0.3780*	0.1716*	0.2307*	0.1445	0.4671*
AvPMI	0.3568*	0.1652	0.2304*	0.1423	0.4390*
CORI	0.3726*	0.1708*	0.2331*	0.1435	0.4586*

made by our proposed approach and the baselines for each of the categories (where a category, such as Geo-Location, is a set of related groups, such as Europe, Asia, etc.).

Table 2 shows the results of the evaluation when a first-pass retrieval model is used for ranking. The table reports the Jensen-Shannon distance between the prediction and the actual exposure of each of the groups. The optimal prediction perfectly estimates the actual exposure and would result in a distance of 0. From the table, it is apparent that our GEP approach achieves the best results on all the groups when BM25 or TF-IDF is used as a ranker. GEP significantly outperforms (paired t-test, with Bonferroni correction, p<0.01) the baselines on three out of the five categories. The differences are significant for the Geographic Location, the Age of Topic, as well as the Age of Article categories. Analysing the Popularity category, we note that our predictor is significantly better in predicting the exposure compared to the AvICTF and the CORI baselines. The biggest differences between the predictors can be observed in the geographic location category, on which our prediction outperforms the next best predictor by up to ~40%. Geographic location is the category with the highest number of groups. When SPLADE is used as a retrieval model, we observe that our

Table 3. Evaluation of the predictors over common first-pass retrieval rankers with different Query Expansion models applied for rankings with k=100. * indicates significant differences (t-test, with Bonferroni correction, p<0.01) in prediction performance between GEP and the baselines.

Approach	Age of Topic	Popularity	Age of the Article	Alphabetical	Geo-Location (TREC 21)
BM25 + RM3					
GEP	**0.3134**	**0.1722**	**0.2030**	**0.1456**	**0.2723**
SCS	0.3772*	0.1763	0.2507*	0.1548	0.4364*
AvIDF	0.3941*	0.1746	0.2603*	0.1571	0.5038*
AvICTF	0.3986*	0.1835	0.2493*	0.1567	0.4876*
AvPMI	0.3823*	0.1747	0.2481*	0.1561	0.4613*
CORI	0.3940*	0.1836	0.2514*	0.1557	0.4796*
BM25 + KLQ					
GEP	**0.2962**	**0.1625**	**0.1911**	**0.1430**	**0.2614**
SCS	0.3624*	0.1706	0.2491*	0.1489	0.4226*
AvIDF	0.3813*	0.1670	0.2598*	0.1513	0.4915*
AvICTF	0.3864*	0.1708	0.2467*	0.1510	0.4750*
AvPMI	0.3669*	0.1651	0.2461*	0.1505	0.4479*
CORI	0.3815*	0.1708	0.2487*	0.1500	0.4667*
SPLADE + RM3					
GEP	**0.1912**	**0.2870**	**0.2549**	**0.1574**	**0.3511**
SCS	0.2365*	0.3300*	0.2871*	0.1670	0.5149*
AvIDF	0.2438*	0.4150*	0.2898*	0.1637	0.5665*
AvICTF	0.2293*	0.4241*	0.2895*	0.1576	0.5564*
AvPMI	0.2344*	0.3857*	0.2830*	0.1610	0.5321*
CORI	0.2315*	0.4216*	0.2895*	0.1572	0.5508*
SPLADE + KLQ					
GEP	0.2291	**0.2410**	**0.2662**	**0.1777**	**0.3251**
SCS	0.2200	0.3300*	0.2980*	0.1738	0.4961*
AvIDF	0.2161	0.3538*	0.3009*	0.1847*	0.5516*
AvICTF	**0.2120***	0.3613*	0.3009*	0.2054*	0.5395*
AvPMI	0.2230	0.3245*	0.2933*	0.1932*	0.5163*
CORI	0.2120*	0.3615*	0.3010*	0.2045*	0.5335*

GEP predictor gives the best prediction on four out of five categories. In all of these four categories, our predictor significantly outperforms the baselines. GEP is only outperformed in the Popularity category. However, there is no observed statistical significance. These results show that our approach is overall the most effective predictor compared to all other evaluated baselines. The TF-IDF retrieval model also leads to the same trends as with BM25.

To further investigate the behaviour of the predictors, we consider how dispersed the exposure is distributed over the groups in the actual rankings. We have identified the coefficient of variation (CV) [1], which is the ratio of the standard deviation of a distribution σ to the mean μ of the distribution to be a suitable measure for this, where: $CV = \frac{\sigma}{\mu}$. We express the CV as a percentage. A lower CV indicates a flatter distribution, e.g., if the actual exposure is distributed $[0.25, 0.25, 0.25, 0.25]$ over four groups then $CV = 0$.

Figure 3 shows the distribution of the CV values over all queries per category. Analysing the figure, it is clear that the categories Popularity and Alphabetical have the lowest median (orange line) and mean (blue star) CV. The category Geographic Location has the highest median and its interquartile range contains the highest CV values. Notably, all categories on which we significantly outperform the baselines show a higher variability in the distribution of the actual exposure scores in the ranking than the other categories, which suggests that our predictor is better than the baselines when the exposure is not uniformly distributed. We use statistical testing to ascertain whether the differences between the distributions of the two categories with the lowest CV values and the other categories are significant (t-test, $p<0.05$). We compare the category Age of Article (which has the lowest CV mean of the groups on which our predictor outperforms the baselines) against the Popularity and Alphabetical categories (for which we observe visible but non-significant improvements in Table 2). The results of our statistical tests show a significant difference, denoted as †, between Age of Article and the two other categories in terms of the distributions of CV-values. This might indicate that the coefficient of variation plays an important role in predicting the exposure over groups since all of our baselines provide better estimates for categories with more flat distributions. We leave the investigation of this interesting finding to future work. To answer RQ1, we conclude that our proposed predictor can outperform all of the used baselines and provide promising results independent of the deviations in the exposure distribution.

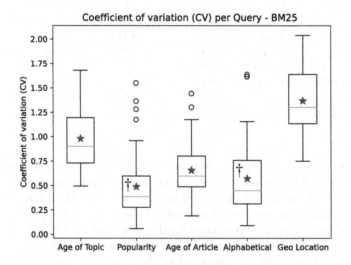

Fig. 3. Dispersion of the exposure in the categories per query measured by Coefficient of Variation. We use † to indicate significant differences (t-test, $p<0.05$) to the Age of Article category values.

Results RQ2: To answer RQ2, we investigate how the prediction performance is affected when a query is altered through Pseudo-Relevance Feedback. The first two sections of Table 3 show the results for the predictors when BM25 is combined with different PRF strategies (RM3, KLQ). We observe that for both PRF approaches over all categories, our predictor GEP outperforms the baselines. On three categories, our approach provides significant improvements over the baselines. This is the same trend we have observed when analysing the standard retrieval approaches. However, the results in Tables 2 and 3 show that for all of the categories and approaches, the prediction quality drops compared to the standard retrieval rankings. This is to be expected since the predictors have no knowledge of how the query might be changed. Nevertheless, the drop in performances is marginal, indicating that the effect of PRF on the exposure is very limited and that our predictor is still useful even when the query is altered. When SPLADE is combined with RM3, our predictor outperforms the baselines on all categories with significant differences on four out of five categories. Analysing the ranking with the KLQ expansion, we observe statistical differences to all baselines on three out of five categories, with GEP providing the lowest numbers. On the age of topic category, the CORI and the AvICTF baseline produce the best prediction, outperforming our approach. On the Alphabetical category, however, GEP achieves the best scores. Overall, we observe that our predictor remains robust with only small drops in the prediction quality even when PRF is applied in the pipeline. A possible explanation for the robustness of our predictor when PRF is applied and unknown query terms are added can be found by revisiting Fig. 2. First let us assume a scenario where the PRF component leads to only a re-ranking of documents, and does not add any new documents to the ranking. Assuming such a ranking, we can observe from Fig. 2 how the exposure can change per group by looking at the individual bell curves. If the number of documents in a group stays the same, an individual bell curve shows the range of how much exposure a group can obtain. From the figure, it is clear that the range of possible exposure values is very restricted and narrows down even more for an increasing number of documents. Explanations for a second scenario where the exposure changes with variations in the number of documents per group due to alterations of the query can also be found in Fig. 2. The figure shows that substantial changes in the exposure distribution only occur if a relatively large number of documents (e.g., +5) from a group are added to the ranking. Moreover, as can be seen from Fig. 1(a), if the documents are added in lower ranking positions, then relatively little additional exposure is accumulated, as previously discussed in Sect. 3. To answer RQ2, we can conclude that even when PRF is applied, our proposed predictor is still effective. Moreover, we have outlined two possible explanations, for why PRF interventions might only lead to limited effects on the exposure among groups. We have also conducted additional PRF experiments using the Bo1 model from the Divergence from Randomness framework [5], and axiomatic query expansion (AEQ) [17] and have observed that all the observations from RQ2 also hold for these. Due to page limitations, we do not present these results in this paper.

7   Conclusions

In this paper, we have introduced the novel task of query exposure prediction. We have conducted a detailed investigation of the exposure allocation problem and have proposed a new predictor, GEP, which has been explicitly designed for query exposure prediction. Our extensive experiments on two standard test collections of the TREC Fair Ranking Track showed that our predictor is a suitable and effective choice for the task of query exposure prediction. GEP outperforms the used baselines across multiple retrieval models, even when the knowledge of the query is limited, i.e., when PRF is applied. Our evaluation shows that our predictor achieves improvements of up to ~40% in prediction accuracy over the next best performing baseline. The main aim of this work is to present this important task to the community and to encourage the development of further exposure predictors. In future work, we aim to develop end-to-end predictors across the entire pipeline, exploring potential avenues such as the development of a supervised learning approach trained on exemplary rankings.

References

1. Abdi, H.: Coefficient of variation. Encycl. Res. Des. **1**(5), 169–171 (2010)
2. Abdul-Jaleel, N., et al.: UMass at TREC 2004: Novelty and HARD. Computer Science Department Faculty Publication Series, p. 189 (2004)
3. Agarwal, A., Zaitsev, I., Wang, X., Li, C., Najork, M., Joachims, T.: Estimating position bias without intrusive interventions. In: Proceedings of ICWSM (2019)
4. Amati, G.: Probabilistic Models for Information Retrieval based on Divergence from Randomness. University of Glasgow, UK. Ph.D. thesis, PhD Thesis (2003)
5. Amati, G., Van Rijsbergen, C.J.: Probabilistic models of information retrieval based on measuring the divergence from randomness. ACM Trans. Inf. Syst. (TOIS) **20**(4), 357–389 (2002)
6. Biega, A.J., Gummadi, K.P., Weikum, G.: Equity of attention: amortizing individual fairness in rankings. In: Proceedings of SIGIR, pp. 405–414 (2018)
7. Bower, A., Lum, K., Lazovich, T., Yee, K., Belli, L.: Random isn't always fair: Candidate set imbalance and exposure inequality in recommender systems (2022). arXiv preprint arXiv:2209.05000
8. Callan, J.: Distributed information retrieval. In: Croft, W.B. (eds.) Advances in Information Retrieval. The Information Retrieval Series, vol. 7. Springer, Boston, MA (2002). https://doi.org/10.1007/0-306-47019-5_5
9. Carmel, D., Yom-Tov, E.: Estimating the query difficulty for information retrieval. Synth. Lect. Inf. Concepts Retrieval Serv. **2**(1), 1–89 (2010)
10. Craswell, N., Zoeter, O., Taylor, M., Ramsey, B.: An experimental comparison of click position-bias models. In: Proceedings of WSDM (2008)
11. Cronen-Townsend, S., Zhou, Y., Croft, W.B.: Predicting query performance. In: Proceedings of SIGIR, pp. 299–306 (2002)
12. Devlin, J., Chang, M.W., Lee, K., Toutanova, K.: BERT: Pre-training of deep bidirectional transformers for language understanding (2018). arXiv preprint arXiv:1810.04805
13. Diaz, F., Mitra, B., Ekstrand, M.D., Biega, A.J., Carterette, B.: Evaluating stochastic rankings with expected exposure. In: Proceedings of CIKM, pp. 275–284 (2020)

14. Ekstrand, M.D., Burke, R., Diaz, F.: Fairness and discrimination in retrieval and recommendation. In: Proceedings of SIGIR, pp. 1403–1404 (2019)
15. Ekstrand, M.D., McDonald, G., Raj, A., Johnson, I.: Overview of the TREC 2021 fair ranking track. In: Proceedings of TREC (2021)
16. Ekstrand, M.D., McDonald, G., Raj, A., Johnson, I.: Overview of the TREC 2022 fair ranking track. In: Proceedings of TREC 2022 (2022)
17. Fang, H., Zhai, C.: Semantic term matching in axiomatic approaches to information retrieval. In: Proceedings of SIGIR, pp. 115–122 (2006)
18. Formal, T., Piwowarski, B., Clinchant, S.: SPLADE: sparse lexical and expansion model for first stage ranking. In: Proceedings of SIGIR, pp. 2288–2292 (2021)
19. Hauff, C., Azzopardi, L., Hiemstra, D., de Jong, F.: Query performance prediction: evaluation contrasted with effectiveness. In: Gurrin, C., et al. (eds.) ECIR 2010. LNCS, vol. 5993, pp. 204–216. Springer, Heidelberg (2010). https://doi.org/10.1007/978-3-642-12275-0_20
20. He, B., Ounis, I.: Inferring query performance using pre-retrieval predictors. In: Apostolico, A., Melucci, M. (eds.) SPIRE 2004. LNCS, vol. 3246, pp. 43–54. Springer, Heidelberg (2004). https://doi.org/10.1007/978-3-540-30213-1_5
21. He, B., Ounis, I.: Query performance prediction. Inf. Syst. **31**(7), 585–594 (2006)
22. Heuss, M., Sarvi, F., de Rijke, M.: Fairness of exposure in light of incomplete exposure estimation. In: Proceedings of SIGIR, pp. 759–769 (2022)
23. Hofstätter, S., Hanbury, A.: Let's measure run time! Extending the IR replicability infrastructure to include performance aspects (2019). arXiv preprint arXiv:1907.04614
24. Jaenich, T., McDonald, G., Ounis, I.: ColBERT-FairPRF: towards fair pseudo-relevance feedback in dense retrieval. In: Kamps, J., et al. Advances in Information Retrieval. ECIR 2023. LNCS, vol. 13981. Springer, Cham (2023). https://doi.org/10.1007/978-3-031-28238-6_36
25. Järvelin, K., Kekäläinen, J.: Cumulated gain-based evaluation of IR techniques. ACM Trans. Inf. Syst. (TOIS) **20**(4), 422–446 (2002)
26. Kletti, T., Renders, J.M., Loiseau, P.: Pareto-optimal fairness-utility amortizations in rankings with a DBN exposure model. In: Proceedings of SIGIR, pp. 748–758 (2022)
27. Lin, J., Nogueira, R., Yates, A.: Pretrained transformers for text ranking: BERT and beyond. Springer Nature (2022). https://doi.org/10.1007/978-3-031-02181-7
28. Macdonald, C., Tonellotto, N.: Declarative experimentation in information retrieval using PyTerrier. In: Proceedings of ICTIR (2020)
29. McDonald, G., Macdonald, C., Ounis, I.: Search results diversification for effective fair ranking in academic search. Inf. Retrieval J. **25**(1), 1–26 (2022)
30. Morik, M., Singh, A., Hong, J., Joachims, T.: Controlling fairness and bias in dynamic learning-to-rank. In: Proceedings of SIGIR, pp. 429–438 (2020)
31. Pradeep, R., Liu, Y., Zhang, X., Li, Y., Yates, A., Lin, J.: Squeezing water from a stone: a bag of tricks for further improving cross-encoder effectiveness for reranking. In: Hagen, M., et al. (eds.) ECIR 2022. LNCS, vol. 13185, pp. 655–670. Springer, Cham (2022). https://doi.org/10.1007/978-3-030-99736-6_44
32. Pradeep, R., Nogueira, R., Lin, J.: The expando-mono-duo design pattern for text ranking with pretrained sequence-to-sequence models (2021). arXiv preprint arXiv:2101.05667
33. Raj, A., Wood, C., Montoly, A., Ekstrand, M.D.: Comparing fair ranking metrics (2020). arXiv preprint arXiv:2009.01311
34. Robertson, S.E., Walker, S., Jones, S., Hancock-Beaulieu, M.M., Gatford, M., et al.: Okapi at TREC-3. NIST Special Publication SP **109**, 109 (1995)

35. Salton, G., Buckley, C.: Term-weighting approaches in automatic text retrieval. Inf. Process. Manag. **24**(5), 513–523 (1988)
36. Sarvi, F., Heuss, M., Aliannejadi, M., Schelter, S., de Rijke, M.: Understanding and mitigating the effect of outliers in fair ranking. In: Proceedings of WSDM, pp. 861–869 (2022)
37. Singh, A., Joachims, T.: Fairness of exposure in rankings. In: Proceedings of KDD (2018)
38. Usunier, N., Do, V., Dohmatob, E.: Fast online ranking with fairness of exposure. In: Proceedings of FAccT, pp. 2157–2167 (2022)
39. Wang, X., Golbandi, N., Bendersky, M., Metzler, D., Najork, M.: Position bias estimation for unbiased learning to rank in personal search. In: Proceedings of ICWSM (2018)
40. Wu, H., Mitra, B., Ma, C., Diaz, F., Liu, X.: Joint multisided exposure fairness for recommendation. In: Proceedings of SIGIR, pp. 703–714 (2022)
41. Zehlike, M., Castillo, C.: Reducing disparate exposure in ranking: a learning to rank approach. In: Proceedings of The Web Conference, pp. 2849–2855 (2020)
42. Zehlike, M., Yang, K., Stoyanovich, J.: Fairness in ranking: A survey (2021). arXiv preprint arXiv:2103.14000

ProMap: Product Mapping Datasets

Kateřina Macková[iD] and Martin Pilát[(✉)][iD]

Faculty of Mathematics and Physics, Charles University, Malostranské náměstí 25,
Prague 1, 118 00, Czech Republic
{Katerina.Mackova,Martin.Pilat}@mff.cuni.cz

Abstract. The goal of product mapping is to decide, whether two listings from two different e-shops describe the same products. Existing datasets of matching and non-matching pairs of products, however, often suffer from incomplete product information or contain only very distant non-matching products. In this paper, we introduce two new datasets for product mapping: ProMapCz consisting of 1,495 Czech product pairs and ProMapEn consisting of 1,555 English product pairs of matching and non-matching products manually scraped from two pairs of e-shops. The datasets contain both images and textual descriptions of the products, including their specifications, making them one of the most complete datasets for product mapping. Additionally, we divide the non-matching products into two different categories – close non-matches and medium non-matches, based on how similar the products are to each other. Even the medium non-matches are, however, pairs of products that are much more similar than non-matches in other datasets – for example, they still need to have the same brand and similar name and price. Finally, we train a number of product matching models on these datasets to demonstrate the advantages of having these two types of non-matches for the analysis of these models.

Keywords: Product Mapping · Product Matching · Similarity Computation · Machine Learning

1 Introduction

Product mapping or product matching (PM) is the process of matching identical products from different e-shops, where each product can be described by different graphical and textual data. It has an important application in e-commerce as it allows for general marketplace analysis and price comparison among different e-shops. Product mapping is challenging as there is no general identification of products available on all the websites. Therefore, models for measuring similarity of the products need to be trained to identify matching pairs of products based on as much textual and image data describing each product as available. Existing freely available datasets [2–5] are quite limited as they often do not provide all data describing each product. Moreover, they consist only of very distant non-matches. Therefore, training the predictive models to distinguish matches and

© The Author(s), under exclusive license to Springer Nature Switzerland AG 2024
N. Goharian et al. (Eds.): ECIR 2024, LNCS 14609, pp. 159–172, 2024.
https://doi.org/10.1007/978-3-031-56060-6_11

non-matches is very simple. To fill this gap, we create a new group of freely available datasets for product matching by manually scraping selected e-shops and searching for matching and non-matching pairs of products that create a good challenge for further research. In this paper, we present a dataset for Czech product mapping - ProMapCz and a dataset for English product mapping - ProMapEn.

Both of the datasets contain approximately 1,500 product pairs with roughly equal distribution of matches, and close and medium non-matches. Each of the products in these datasets contains all the available information, including name, price, images, long and short description, specification, and URL. More detailed information about the datasets can be found in Sect. 3.1. Both datasets - raw versions and precomputed similarities are available at https://github.com/kackamac/Product-Mapping-Datasets.

In order to demonstrate the advantages of having much closer non-matches in the data, we preprocessed these datasets and trained several machine-learning models to solve product mapping tasks on them. We did the same for two of the existing datasets – Amazon-Walmart [3] and Amazon-Google [2,5]. The best models on ProMapCz and ProMapEn reach F1 scores of 0.777 and 0.706 respectively, while obtaining F1 scores of 0.93 and 0.99 on Amazon-Walmart and Amazon-Google respectively, thus showing that the newly proposed datasets are indeed more challenging and provide a better benchmark for product mapping models.

2   Existing Datasets

The existing datasets all suffer from some deficiencies, most importantly, they only contain pairs of matching products, or the included pairs of non-matching pairs are only very distant, non-similar products. Some of the datasets also contain only limited information about the products, such as the Web Data Commons (WDC) Dataset [4]. While it is the largest dataset for product mapping, containing 26 million product offers in many languages (out of which 16 million are in English), it contains only the names of the products, completely omitting any other available information.

The Abt-Buy dataset [2,5] includes more information, such as product description and product price. It contains 1,081 entities from Abt.com and 1,092 entities from Buy.com as well as 1,097 matching product pairs between these two data sources. Similarly, the Amazon-Google [2,5] contains the product names, descriptions, manufacturer and prices of 1,363 products from Amazon and 3,226 products from Google with 1,363 matches defined between them.

The Amazon-Walmart [3] is more complete, containing 24,583 individual products with detailed information such as title, brand, category, price, short description, long description, technical details, image URLs, etc. Among these products 1,154 matching pairs were identified.

All the datasets discussed above contain only matches, and non-matches have to be generated from the rest of the products. While this is simple to do in principle, the generated non-matches are typically very distant non-matches (i.e. pairs of completely different products). We believe that such non-matches are not particularly useful for the evaluation of product matching models, as we would expect the models to be used mainly for similar products (e.g. such that they have the same manufacturer and are in the same category). Therefore, in this paper we provide two datasets, each containing over 2,000 products and around 1,500 matches, close non-matches and medium non-matches defined between them. We believe that such a dataset can be used for better evaluation of product matching models.

3   ProMap Datasets

We created our datasets in several stages: first, we selected the URLs of the products on the source website, then we let the annotators find the matching URLs, close non-matching and medium non-matching URLs on the target website, and we manually verified the correctness of the data to ensure their quality. For ProMapCz the source website is Alza.cz and the target is Mall.cz. For ProMapEn the source is Walmart.com and the target is Amazon.com.

We created both product mapping datasets with several different categories. We scraped all data from the source e-shop, and we selected the 10 most common product categories intending to cover a wide range of different product categories in ProMapCz. In ProMapEn, we also selected the categories according to the most common products but also regarding the selected categories in ProMapCz to cover similar categories for further language comparisons between Czech and English language. From each category, we randomly sampled 100 products.

For each source product, we searched for three products on the target website: match, close non-match, and medium non-match. A *match* is the identical product on the target website, the only difference we allowed was the product color. A *close non-match* is defined as the product from the target website which is very similar but not the same as the source product. It was supposed to have the same brand, similar name, similar price (at a maximum 20% difference from the source price) and almost the same attributes. A *medium non-match* is a product that is different but still has a number of similar attributes – it is supposed to still have the same brand and similar name and price, but the specifications and prices can be more different than for the close non-matches. If there were multiple suitable candidate products, the annotator was supposed to select one randomly. If the target product was not found, the particular source product was left empty. We prohibited repeating target products for different source products.

The inclusion of different types of non-matches in this dataset allows us to study product matching models in more detail. This also makes the dataset harder, as even medium non-matches are still products from the same category and with the same brand as opposed to being completely different products. This is more realistic as we do not expect to run product matching on completely different products – these can be often easily identified, for example, by having a different brand or being in a different category.

After the annotators created the datasets, we performed manual control of created data by randomly selecting 20 products in each category and checking corresponding products to ensure data quality and annotators' correctness and to avoid and correct misunderstandings and errors.

After having pairs of URLs of corresponding products from both websites, we automatically scraped all possible product data about all the products from the URLs. Specifically, we scraped names, long descriptions, short descriptions (often stated immediately under the product names), specifications containing technical information, images and prices. The following types of columns and types are stored in the dataset: id1, id2 (URL), name1, name2, short_description1, short_description2, long_description1, long_description2, specification1, specification2 (text), price1, price2 (number), images1, images2 (list of URLs) and match (binary label). Product information extracted from the source URLs are suffixed by 1 and those from the target URLs are suffixed by 2.

3.1   Datasets Description

In the end, the ProMapCz dataset contains 1,409 unique products from Mall.cz and 706 unique products from Alza.cz, these are organized as 1,495 pairs of matching and non-matching products. Out of these 504 are matches, 456 are close non-matches are 535 are medium non-matches. The ProMapEn dataset contains 1,555 unique products from Walmart.com and 751 unique products from Amazon.com. These form 1,555 pairs of products divided into 509 matches, 509 close non-matches and 537 medium non-matches. The distribution of the products, matches and both types of non-matches by category, as well as the list of categories is presented in Table 1. In most categories, we were able to find around 50 matching products, however, some categories were harder than others. For example, in the ProMapCZ dataset in the Laptops category we found only 20 matching products. This is mostly given by the fact that Alza.cz carries much wider range of electronics with many more different laptop models than Mall.cz.

4   Dataset Preprocessing and Similarity Computations

We preprocessed both datasets to train multiple baseline models for the product mapping task. The images and text attributes of each pair of products are preprocessed and converted to numerical vectors and the cosine distance between each pair of corresponding attributes is computed. Additionally, we performed keyword detection in textual attributes and computed the ratio of matching keywords between products for each textual column separately. In this manner, we created a vector of 34 features representing the distances of all attributes between each product pair. Along with the label whether the pair is matching or not, we used these vectors to train several machine learning models to predict corresponding pairs of products between the two e-shops. The feature extraction process is described in more details below.

Table 1. Number of matching (=), close non-matching (c≠) and medium non-matching (m≠) products in each category of ProMapCz and ProMapEn

ProMapCz Categories	=	c≠	m≠	ProMapEn Categories	=	c≠	m≠
Pet Supplies	30	24	30	Pet Supplies	48	30	29
Backpacks&Bags	35	20	44	Backpacks&Accessories	43	51	43
Hobby&Garden	68	51	50	Patio&Garden	47	52	52
Appliances	61	37	46	Kitchen Appliances	83	75	81
Mobile phones	50	29	44	Mobile Phones	45	46	53
Household Supplies	46	53	59	Home Improvement	48	52	52
Laptops	20	50	49	Laptops	40	58	59
TVs	71	68	71	Toys	63	40	59
Headphones	65	63	85	Sports&Clothes	39	38	39
Fridges	58	61	57	Health&Beauty	48	67	70

4.1 Image Preprocessing and Similarity Computations

Each product is represented by several images. We preprocessed all these images to obtain their numerical representation using a perceptual hashing method. Having these hashes, we computed the similarities of images between each product pair to obtain one number representing the products' image similarity.

Image Preprocessing. The images of the products often vary in sizes, colors, rotations and centering. The first step of the process is thus resizing the image into a maximal width and height to preserve memory and increase speed. We experimentally set the size to 1024 × 1024. We added a white border around the image to allow easier object detection, and we converted the image into grayscale. Then, we created a black and white mask of the object and applied canny edge detection and we found contours using methods from the OpenCV[1] library. Finally, we selected the largest object in the picture and cropped the image to its bounding box to obtain the final preprocessed image. In preliminary experiments, we also tried to replace object detection with simpler techniques such cropping of the white borders to the closest edge of the product or by using contour detection, but these did not work that well due to the presence of various logos and other elements added by the e-shops around the images.

Image Hash Creation. After having the images preprocessed, we created the image hashes preserving the main features and taking much less memory. To this end, we used the Image Perceptual Hashing method [7]. We used the implementation from image-hash Node.js library[2]. This technique works by splitting the array of pixels into several blocks and performing kernel operations for finding the main features and creating a hash representing the image and preserving the most important information from it. We used hashes of size 8 with 8 blocks

---

[1] https://docs.opencv.org/.

[2] https://www.npmjs.com/package/image-hash.

thus creating 64-bit hashes. We selected these parameters experimentally as smaller sizes did not provide enough information to compute the similarity of the images and larger sizes unnecessarily slowed down the run of the algorithm without improving the results.

Image Hashes Similarity Computation. Having a set of image hashes characterizing each product, we compared the hashes of images between corresponding product pairs to obtain overall image similarity. For each image in the source product image set, we computed the Hamming distance to all images in the target product image set and we selected the most similar image. We kept only images having a similarity higher than a given threshold to filter out images that are present in only one of the image sets. The threshold was set experimentally to 0.9 concerning image hashes sizes meaning that images having similarities above 90 % are corresponding to each other. By finding the most similar image and filtering away too-distant images, we obtained the closest image from the second product image set for each image from the first product image set. Finally, we computed the overall similarity of images by summing up all these precomputed similarities. This summing means that products with a greater number of more similar images are more similar. The number obtained in this way is referred to as *hash_similarity* in the rest of the paper.

4.2 Text Preprocessing and Similarity Computations

Each product is represented by several textual attributes: name, long and short description, and specification. We preprocessed all these attributes in the same way to create numerical vectors. We also extracted several main keywords in attributes to create additional vectors. In the end, we computed the cosine distance of corresponding attributes between both products in the given product pair to obtain additional similarity measures.

Text Preprocessing. We preprocessed every attribute containing some text in every product by the same procedure. First, we deleted useless characters from the text such as additional tabs, spaces, brackets etc. Afterwards, we detected the cases of units and values not separated by space (e.g. 15GB or 4") and we separated them using a space. We separated the whole text into single words and we lemmatized them using the morphological analyzer Majka [6] to avoid problems with declination and conjugation which are quite common in some languages such as Czech. The last step in the preprocessing pipeline is lowercasing the text.

Additionally, we added an *all_texts* meta-attribute consisting of the concatenation of the name, long and short descriptions, and specification. This allows us to compute features not only for each separate attribute but also for all the texts at once.

Text Similarity Computations. After having all the texts of each product preprocessed, the similarity of each attribute in each product pair needs to be computed. The final similarity number is computed by converting preprocessed texts into numerical vectors using the simplest baseline methods tf.idf and computing the cosine similarity of both vectors This creates three additional similarity measures: *short_ description_ cos*, *name_ cos*, *long_ description_ cos*. Note that we do not use cosine similarity of the specifications of the products as this field has a different structure and its similarity is computed differently.

Keywords Detection and Similarity Computations. The textual fields can contain a number of words with special meaning, therefore, we decided to perform detection and comparison of such keywords – IDs, brands, numbers, and descriptive words. The similarity of detected keyword sets in a product pair is computed as the ratio of the same keywords and the number of all keywords for every detected keyword type, i.e. as the Jaccard similarity of the two sets: $J(A, B) = |A \cap B|/|A \cup B|$, where A and B are the two sets of keywords.

The detection of each of the types of keywords is described below:

ID detection is performed by selecting unique words longer than five characters that are not included in the vocabularies of English and Czech words created based on the ParaCrawl [1] corpus. ID detection is important as some of the products may have a unique identification that can facilitate the identification of matching products between e-shops. IDs are detected and compared in the name, short description and all text attributes giving us: *name_ id*, *short_ description_ id*, and *all_ texts_ id* similarities.

Brand detection is based on our vocabulary created by scraping all brands on the source and target websites. It is important as brands in names are another important identification of matching products. Brands are detected and compared in name, short description and all texts attributes, giving us *name_ brand*, *short_ description_ brand*, *all_ texts_ brand* similarities.

Numbers Detection is based on detecting numbers in the text and searching for units around them. If no units are found near the number, the number is detected as a *free number*. Such numbers can contain for example model numbers or other crucial information. Free numbers are detected and compared in every attribute giving us the features: *name_ numbers*, *short_ description_ numbers*, *long_ description_ numbers*, *specification_ text_ numbers*, and *all_ texts_ numbers*.

Descriptive Words are a set of the most characterising words for each attribute of the product. These words are selected as the top k words that occur in a maximum of p per cent of documents (in our case among all textual attributes of all products). We experimentally set the k to 50 and p to 50 % to eliminate the most common words such as 'and', 'or' etc. having no important meaning but to preserve words characterising each attribute of the products. Descriptive words are also detected and compared in every attribute, giving us *name_ descriptives*, *short_ description_ descriptives*, *long_ description_ descriptives*, and *all_ texts_ descriptives* similarities.

Units Detection is based on the extraction of numbers followed by units from each attribute that can specify the product in detail. Units are extracted and compared in all attributes, giving us *name_ units*, *short_ description_ units*, *long_ description_ units*, *specification_ text_ units*, and *all_ texts_ units* similarities.

Words are a ratio of the same words taking all words from corresponding attributes of two products. Words are computed only between names, short descriptions and, all texts, giving us *name_words, short_description_words, all_texts_words* similarities.

All Detected Keywords Comparisons. We created one list from each detected units, IDs, numbers and brands for each product and we compared the ratio of matching values in those lists between two compared products giving us *all_units_list, all_ids_list, all_numbers_list, all_brands_list* similarity numbers allowing to compare detected keywords across all text attributes. We computed the similarity of the detected keyword using the ratio of the same keywords as in other texts.

Specification Preprocessing. As the product specification is often in a specific format containing parameter names and their values, we also performed preprocessing of the specification based on the comparisons of these parameters. We extracted numbers followed by units from each attribute and compared them between two products to obtain parameters and their similarity as they can specify the product in detail. As the parameter values can be in different units, we transformed all values to metric units in basic form without any prefixes. We computed the ratio of corresponding parameter names *specification_key_matches*, and ratio of corresponding parameter names and values *specification_key_value*. While comparing the values, 5% deviation is allowed to account for inaccuracies during conversions.

5   Experiments and Results

After processing text and image attributes, keywords detection and similarity computations, we obtained the overall similarity of two products characterized by a vector containing 34 features. All the features are listed in Table 2. Along with the *match* label representing either matching or non-matching products, we trained logistic regression (LR), support vector machines (SVM), decision trees (DT), random forests (RF), and neural network (NN) classifiers with different parameter setups to predict matching and non-matching pairs, see Table 3. To preserve objectivity and stability in the results, in all following experiments the resulting scores are an average of five runs.

Train-Test Data Preparation. We split ProMap datasets into train-test data with a ratio of 80:20 giving us 1,196 vectors for training and 299 for testing in ProMapCz and 1,244 vectors for training and 311 for testing in ProMapEn. We release this train-test split to enable the training of other machine learning models and to compare their results in future research.

Training and Parameters Tuning. We trained linear regression, support vector machines, decision trees, random forests and neural network models with different combinations of values of their hyper-parameters. We performed grid search and random search to find the most suitable hyper-parameters for each

Table 2. Precomputed similarity features.

Feature Name		
name_cos	short_description_cos	all_texts_cos
name_id	short_description_id	all_texts_id
name_brand	short_description_brand	all_texts_brand
name_numbers	short_description_numbers	all_texts_numbers
name_descriptives	short_description_descriptives	all_texts_descriptives
name_units	short_description_units	all_texts_units
name_words	short_description_words	all_texts_words
specification_text_numbers	long_description_cos	all_ids_list
specification_text_units	long_description_numbers	all_brands_list
specification_key	long_description_units	all_numbers_list
specification_key_value	long_description_descriptives	all_units_list
hash_similarity		

model. These hyper-parameter searches use 20 percent of the training set as a validation set. The possible parameter values for grid search are summarized in Table 3. The random search was performed with 100 samples and it uses the same parameter sets for categorical parameters and uses the full range between the minimum and maximum values for the numeric parameters. After the hyper-parameter search, the models are trained with the best hyper-parameters on the combined training and validation sets. We select the model that maximizes the F1 score. The threshold for classification is set such that the F1 score on the training set is maximized.

5.1   Amazon-Walmart and Amazon-Google Datasets

As the original Amazon-Walmart and Amazon-Google datasets are strongly unbalanced as to the number of matches and non-matches, we decided to shrink and sample these datasets to create datasets with similar sizes and ratios of matches and non-matches to ProMap datasets. This allows us to compare the experiments and results among all datasets more accurately.

Amazon-Walmart. The original Amazon-Walmart dataset contains 22,073 products from Amazon and 2,554 products from Walmart organized as pairs of 1,154 matches and 11,540 non-matches. We merged Amazon and Walmart products using the mappings provided in the dataset and we sampled 1,143 matches and 2,000 non-matches, so that the final dataset has similar distribution of matches and non-matches as the ProMap datasets. The data were split 80:20 into training and testing datasets. There are 1,640 unique Amazon products and 1,552 unique Walmart products. The Amazon-Walmart dataset contains only name1, short_description1, long_description1, price1, id1, name2, short_description2,

Table 3. Parameter settings for the grid search.

Model	Parameter	Possible Values
LogisticReg	penalty	l1, l2, elasticnet, none
	solver	lbfgs, newton-cg, liblinear, sag, saga
	max_iter	10, 20, 50, 100, 200, 500
SVM	kernel	linear, poly, rbf, sigmoid
	degree	2, 3, 4, 5
	max_iter	10, 20, 50, 100, 200, 500
DecisionTree	criterion	gini, entropy
	max_depth	5, 10, 15, 20
	min_samples_split	2, 5, 10, 15, 20
RandomForest	n_estimators	50, 100, 200, 500
	criterion	gini, entropy
	max_depth	5, 10, 20, 50
	min_samples_split	2, 5, 10, 20
NeuralNetwork	hidden_layer_sizes	(10, 10), (50, 50), (10, 50), (10, 10, 10)
		(50, 50, 50), (50, 10, 50), (10, 50, 10)
	activation	relu, logistic, tanh
	solver	adam, sgd, lbfgs
	learning_rate	constant, invscaling, adaptive
	learning_rate_init	0.01, 0.001, 0.0001
	max_iter	50, 100, 500

long_description2, price2, id2 and match columns. Image_hashes and specification columns are not present in this dataset.

Amazon-Google. The Amazon-Google dataset consists of 1,363 products from Amazon and 3,326 products from Google with 1,300 matches. We also merged the Amazon and Google products using the provided mappings to obtain a file of 1,300 pairs. We also created 1,935 non-matching pairs by random selection of Google products that do not match any of Amazon's products. We paired those with random products from the whole Amazon file in which all products were doubled to prevent too frequent occurrence of one Amazon product in the final dataset. The newly created Amazon-Google dataset has a total length of 3,235 pairs, of which 2,588 are in the training set and 647 in the testing set to preserve the 80:20 ratio as in ProMap datasets. The ratio of matches again approximately corresponds to the ratio of matches in the ProMap datasets. There are 1,341 unique Amazon products and 3,326 unique Google products. The created Amazon-Google dataset contains only name1, short_description1, price1, id1, name2, short_description2, price2, id2 and match columns. Image_hashes, long_description and specification columns are not present in this dataset.

Table 4. Comparison of several machine learning methods trained on ProMapCz, ProMapEn, Amazon-Walmart and Amazon-Google datasets evaluated on their test datasets. Results are from the models with the best parameters from random and grid searches. The model name contains the base model and the type of hyper-parameter search used.

Model	ProMapCz			ProMapEn			Am-Walmart			Am-Google		
	F1	Prec	Rec	F1	Prec	Rec	F1	Prec	Rec	F1	Prec	Rec
LR-Rand	0.754	0.774	0.735	0.692	0.683	0.703	0.921	0.917	0.925	0.990	0.989	0.992
LR-Grid	0.750	0.766	0.735	0.701	0.731	0.673	0.923	0.921	0.925	0.990	0.989	0.992
SVM-Rand	0.733	0.659	0.827	0.631	0.627	0.634	0.916	0.895	0.939	0.985	0.992	0.977
SVM-Grid	0.739	0.714	0.765	0.664	0.628	0.703	0.927	0.908	0.947	0.985	0.996	0.973
DT-Rand	0.737	0.761	0.714	0.626	0.535	0.752	0.896	0.902	0.890	0.983	0.981	0.985
DT-Grid	0.732	0.740	0.724	0.626	0.535	0.752	0.901	0.926	0.877	0.987	0.989	0.985
RF-Rand	0.760	0.890	0.663	0.647	0.573	0.743	0.918	0.924	0.912	0.982	0.996	0.969
RF-Grid	0.759	0.868	0.673	0.658	0.574	0.772	0.924	0.911	0.939	0.985	0.992	0.977
NN-Rand	0.777	0.790	0.765	0.690	0.697	0.683	0.912	0.872	0.956	0.990	0.996	0.985
NN-Grid	0.762	0.700	0.836	0.706	0.710	0.703	0.928	0.901	0.956	0.990	1.000	0.981

Preprocessing and Similarity Computations. We preprocessed columns in both Amazon-Walmart and Amazon-Google datasets and computed the similarities by the same technique as in ProMap datasets, which gave us the same columns of similarities except for specification and image-related ones in the case of Amazon-Walmart and except for specification, long description, and image-related ones in Amazon-Google. We publish newly created versions of Amazon-Walmart and Amazon-Google datasets including their train-test split and precomputed similarities along with our ProMap datasets.

Training and Evaluation. We have performed the same training and evaluation techniques on Amazon-Walmart and Amazon-Google as on the ProMap dataset including the parameters search. Moreover, we tested the transfer learning capabilities of the models by selecting the best model for every dataset and evaluating it in all the other datasets. As each of the datasets contain different attributes, we filled the values of the missing attributes with zeroes for models that were trained on datasets containing them and we removed the additional attributes for models trained on datasets not containing them.

5.2   Results

The best results were obtained by neural network-based models closely followed by logistic regression and random forests. The order of the models is the same across all datasets confirming the huge potential in the neural network-based models, see Table 4. It seems that neural networks tend to have the most balanced precision and recall, while the random forests can have larger differences between these values. Support vector machines seem to be inappropriate for

Table 5. Comparison of the best neural network models. One model was trained on each dataset and evaluated on all datasets to show the difficulty of ProMap datasets.

	Train data											
	ProMapCz			ProMapEn			AmWa			AmGo		
Test data	F1	Prec	Rec	F1	Prec	Rec	F1	Prec	Rec	F1	Prec	Rec
ProMapCz	0.762	0.700	0.836	0.710	0.647	0.786	0.678	0.569	0.837	0.582	0.438	0.867
ProMapEn	0.631	0.619	0.644	0.706	0.710	0.703	0.618	0.612	0.624	0.589	0.523	0.673
AmWa	0.859	0.863	0.855	0.915	0.909	0.921	0.928	0.901	0.956	0.990	1.000	0.981
AmGo	0.968	0.962	0.973	0.983	0.988	0.977	0.899	0.868	0.931	0.990	1.000	0.981

such predictions in general. Models trained on ProMapCz have 6–10 percentage points better results than models trained on ProMapEn which can be caused by several reasons such as language differences, differences in the complexities of the English and Czech dataset, selection of products, or sparsity of the data.

Complexity of the Datasets. Overall results of models for Amazon-Walmart and Amazon-Google datasets are much higher than for ProMap datasets confirming that ProMap datasets are more challenging. This hypothesis is also confirmed by evaluating the best model trained on every dataset on all other datasets (i.e. in a transfer learning setting), see Table 5. The difference between the complexity of Amazon-Walmart & Amazon-Google and ProMapCZ & ProMapEn is significant and we can observe it on every model independent of the training datasets.

Grid Search vs Random Search. There does not seem to be much difference between the results found by the grid search and the random search.

Close Non-matches vs Medium Non-matches. In order to evaluate the effect of the close and medium non-matches, we have selected the two best neural network models (NN-Grid): one trained on ProMapCz and the other trained on ProMapEn. We have evaluated them on four test datasets, two created by removing medium non-matches and two by removing close non-matches from the original ProMapCz and ProMapEn test datasets respectively. We used them to compare the differences in complexity between both types of non-matches. The results prove that close non-matches are more difficult for both models to distinguish from matches than medium non-matches – confirming the increased complexity of both the ProMap datasets, see Table 6.

Czech vs English. On the Czech dataset, the models consistently achieve better results than on the English datasets indicating that the English data are more difficult than the Czech data. There could be several reasons for that, such as fewer IDs in the product descriptions in the English e-shops, different selection of products and categories, or the need for further finetuning of preprocessing. More detailed analysis of these differences is left as a future work.

Table 6. Evaluation of the selected best neural network models on modified test dataset to compare differences in difficulties of close and medium non-matches in ProMapCz and ProMapen.

Test dataset version	Dataset length	F1	Prec	Rec
ProMapCz (matches+close nonmatches)	187	0.806	0.774	0.837
ProMapCz (matches+medium nonmatches)	210	0.859	0.881	0.837
ProMapEn (matches+close nonmatches)	195	0.743	0.789	0.703
ProMapEn (matches+medium nonmatches)	217	0.780	0.877	0.703

6   Conclusion

We created two datasets for the product mapping task by manual matching and automated extraction of all possible attributes characterizing products from different categories from two different e-shops in the Czech and English languages. These datasets categorize the non-matches into two categories of close and medium non-matches, making a more detailed analysis of product matching models viable. Additionally, even the medium non-matches are pairs of much more similar products than non-matches in existing product mapping datasets, making the new datasets more challenging for product mapping models and better suited for their evaluation.

We performed text and image preprocessing and we trained several machine learning models to predict matching and non-matching pairs of products between two e-shops. We also preprocessed two existing datasets for the same problem and trained several machine learning models on those to prove the increased complexity of our datasets. The result prove that the ProMap datasets are more complex and the presence of close and medium non-matches makes more realistic evaluation of machine learning models possible.

In this work, we focused on the description of the datasets and demonstration, how the close and medium non-matches help to better evaluate the product matching models. Evaluation of the datasets with more complex, state-of-the-art models is left as a future work.

7   Future Work

In future work, we would like to extend the ProMap dataset collection by-products from different e-shops and include more languages. We would like to extend ProMapCz and ProMapEn datasets with distant non-matches to compare them more accurately with Amazon-Walmart and Amazon-Google Datasets. Moreover, shrinking and sampling of Amazon-Walmart and Amazon-Google Datasets was performed only once and in future work, it would be interesting to repeat this process several times and create several versions of these datasets to verify the experiments on a wider set. We also intend to test other dataset pre-processing methods such as replacing tf.idf with other embedding creation techniques and involve advanced deep learning techniques such as recurrent neural networks or Transformer models to develop stronger predictive models. Finally, as product mapping is a subarea of more general entity matching, we would like to include more general entity-matching techniques and compare those with our neural network approaches.

Acknowledgements. This research was partially supported by SVV project number 260 698 and by TAILOR, a project funded by EU Horizon 2020 research and innovation programme under GA No 952215.

References

1. Bañón, M., et al.: Web-scale acquisition of parallel corpora. In: Proceedings of the 58th Annual Meeting of the Association for Computational Linguistics, pp. 4555–4567. Association for Computational Linguistics, Online (Jul 2020). 10.18653/v1/2020.acl-main.417, https://aclanthology.org/2020.acl-main.417
2. Köpcke, H., Thor, A., Rahm, E.: Evaluation of entity resolution approaches on real-world match problems. Proc. VLDB Endow. 3(1–2), 484–493 (Sep 2010). 10.14778/1920841.1920904. https://doi.org/10.14778/1920841.1920904
3. Naumann, F.: Amazon-walmart dataset. https://hpi.de/naumann/projects/repeatability/datasets/amazon-walmart-dataset.html
4. Primpeli, A., Peeters, R., Bizer, C.: The WDC training dataset and gold standard for large-scale product matching. In: Companion Proceedings of The 2019 World Wide Web Conference, pp. 381–386. WWW '19, Association for Computing Machinery, New York, NY, USA (2019). https://doi.org/10.1145/3308560.3316609,https://doi.org/10.1145/3308560.3316609
5. Rahm, E., Peukert, D.E., Saeedi, A., Nentwig, M.: Benchmark datasets for entity resolution. https://dbs.uni-leipzig.de/research/projects/object_matching/benchmark_datasets_for_entity_resolution
6. Sedláček, R., Smrž, P.: A new czech morphological analyser ajka. In: Matoušek, V., Mautner, P., Mouček, R., Taušer, K. (eds.) Text, Speech and Dialogue, pp. 100–107. Springer, Berlin Heidelberg, (2001)
7. Yang, B., Gu, F., Niu, X.: Block mean value based image perceptual hashing. In: 2006 International Conference on Intelligent Information Hiding and Multimedia, pp. 167–172 (2006). https://doi.org/10.1109/IIH-MSP.2006.265125

Context-Driven Interactive Query Simulations Based on Generative Large Language Models

Björn Engelmann[1]([✉])(ID), Timo Breuer[1](ID), Jana Isabelle Friese[2](ID),
Philipp Schaer[1](ID), and Norbert Fuhr[2](ID)

[1] TH Köln, University of Applied Sciences, Cologne, Germany
{bjoern.engelmann,timo.breuer,philipp.schaer}@th-koeln.de
[2] University of Duisburg-Essen, Duisburg, Germany
{jana.friese,norbert.fuhr}@uni-due.de

Abstract. Simulating user interactions enables a more user-oriented evaluation of information retrieval (IR) systems. While user simulations are cost-efficient and reproducible, many approaches often lack fidelity regarding real user behavior. Most notably, current user models neglect the user's context, which is the primary driver of perceived relevance and the interactions with the search results. To this end, this work introduces the simulation of context-driven query reformulations. The proposed query generation methods build upon recent Large Language Model (LLM) approaches and consider the user's context throughout the simulation of a search session. Compared to simple context-free query generation approaches, these methods show better effectiveness and allow the simulation of more efficient IR sessions. Similarly, our evaluations consider more interaction context than current session-based measures and reveal interesting complementary insights in addition to the established evaluation protocols. We conclude with directions for future work and provide an entirely open experimental setup.

Keywords: User Simulation · Interactive Retrieval · Query Generation

1 Introduction

The Cranfield paradigm is the de facto standard approach for evaluating IR methods, allowing a fair and reproducible comparison of different retrieval systems. However, these merits come at the cost of a strong abstraction of the user behavior. The underlying user model of Cranfield-style experiments assumes that the user formulates a single query, scans the result list in its entirety until a fixed rank, and judges the relevance irrespective of earlier seen search results. In the real world, users behave differently.

Simulating user interactions offers a more cost-efficient and reproducible alternative to real-world user experiments. It allows us to conclude the generalizability of experimental results beyond the boundaries of Cranfield-style

N. Goharian et al. (Eds.): ECIR 2024, LNCS 14609, pp. 173–188, 2024.
https://doi.org/10.1007/978-3-031-56060-6_12

experiments. However, current endeavors lack the inclusion of the user's context in the simulations. To this end, our work analyzes the contextual influence on simulated users, focusing on query generation.

To address the changing knowledge state of the user, we propose **two new query generation methods** based on generative LLMs that consider different types of context information. The methods leverage this information by incorporating it into the reformulations, thereby enhancing the fidelity of the generated queries to real-world user interactions and improving the quality of the user simulation in total.

In addition, we provide **a first-of-its-kind comparison of lexical (sparse) and Transformer-based (dense) retrieval methods** for simulated interactive retrieval. While dense retrieval methods typically show higher retrieval effectiveness, they come at a significantly higher cost in terms of time and resources compared to sparse methods. The simulated environment in this work allows for a direct comparison of both approaches in a user-oriented setting.

To assess the results of the simulations, we conduct **an in-depth evaluation of the simulations**. For a holistic understanding of the considered factors and their effects on retrieval effectiveness, the evaluation includes different perspectives of retrieval effectiveness in simulated interactive retrieval sessions. While prevalent evaluation measures report the information gain during a search session, this work also considers the effort required to acquire that knowledge beyond the reformulation of queries.

Besides the reported findings, we contribute additions for the TREC test collections in the form of new query datasets. Moreover, we provide an **entirely open and reusable experimental setup.**[1]

2   Related Work

Early experiments with user simulations date back to the 1980s [31,32], but recently, the topic got more attention from the IR community [3,4]. Like earlier works, we implement the simulations with a user model covering 1) query formulations, 2) scanning of the retrieved lists, 3) selecting and clicking appealing items, 4) reading and judging documents for relevance, and 5) inspecting other items in the result list and making stop decisions either leading to query reformulations or abandoning the search session (cf. Fig. 1). The literature provides different simulation frameworks [9,24,28,39]. However, this work mainly aligns with the *Complex Searcher Model* implemented by the SimIIR toolkit [23,38].

Generating high-fidelity queries is an important aspect of interactive user simulations [6,7,9,11]. Earlier works propose different query generation approaches that either rely on principled rules or language models. In recent times, there has been a notable focus on leveraging LLMs for query generation [21,35,36]. Doc2Query is a method that produces a set of questions that a given document may answer. The first presented use of these questions was to

---

[1] https://github.com/irgroup/SUIR.

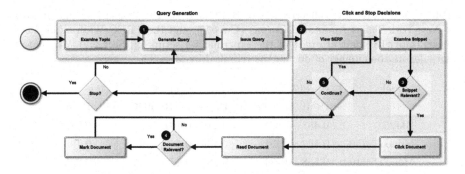

Fig. 1. Focus of this work in the context of the Complex Searcher Model [24].

enrich documents during indexing [27]. This enrichment significantly increased the retrieval effectiveness for term-based ranking methods like BM25. [10] have shown that this method is also suitable for varying queries in a simulation environment. Recently, instruction-tuned LLMs were used to generate queries based on a given prompt providing context information about the information need [1]. This approach has not been used for simulating interactive IR sessions, which is a contribution of this work.

A commonly used measure to simulate sessions with multiple query reformulations is the Session-based DCG (sDCG) [16], which adds an additional discount to the discounted cumulated gain (DCG) [15] by the query position in the session. More recently, a session-based variant of rank-biased precision (RBP) [25] that also builds upon a query-based discount was introduced as Session RBP (sRBP) [18]. The experiments by Lipani et al. [18] suggest that sDCG and sRBP lead to different outcomes and let the authors conclude that both measures provide different evaluation perspectives. For a more in-depth review of user simulations, we refer the reader to Balog and Zhai [5].

3   Methodology

Similar to earlier works [9,24,28,39], we ground our simulations on a defined sequence of interactions with the retrieval system and its search results, visualized in Fig. 1. In this work, the particular focus is the **query generation** and the **interaction with search engine result pages (SERPs)** in different *contexts*. We evaluate simulated sessions under two different retrieval paradigms: sparse, i.e., lexical-based ranking methods, and dense, i.e., Transformer-based methods (cf. Sect. 3.1). Regarding the interaction with SERPs, we analyze user-specific click decisions and context-dependent stopping strategies (cf. Sect. 3.2). Additionally, we analyze the user's *scope of the information need* and the implications of *interactive feedback* for query reformulations (cf. Sect. 3.3). The simulation runs we examine are analyzed with three different measures (cf. Sect. 3.4).

Table 1. Adhoc retrieval effectiveness of BM25 and BM25+MonoT5 evaluated with the New York Times Annotated Corpus (**Core17**) and the TREC Washington Post Corpus (**Core18**) that are part of the experimental setup (cf. Sect. 3.5). The cutoff values denote the number of documents reranked by MonoT5.

Model	Core17			Core18		
	P@10	nDCG@10	Bpref	P@10	nDCG@10	Bpref
BM25	0.458	0.372	0.216	0.404	0.371	0.222
MonoT5@50	0.562	0.451	0.139	0.480	0.454	0.195
MonoT5@100	0.558	0.454	0.185	0.482	0.449	0.232
MonoT5@200	0.574	0.468	0.232	0.476	0.447	0.261
MonoT5@500	0.566	0.460	0.283	0.460	0.435	0.292
MonoT5@1000	0.550	0.447	0.312	0.466	0.439	0.311

3.1   Retrieval Models

The retrieval model is a key element in our simulations. In the experiments, we evaluate the benefits of using a more effective model than a baseline ranking. To this end, we compare BM25 [29] against MonoT5 [26] that is known to outperform the former method in a zero-shot setting. To better understand the quality of search results shown to the simulated users, we evaluate the retrieval effectiveness in a preliminary test collection-based adhoc retrieval experiment.

Table 1 compares BM25 to MonoT5 at different cutoffs, i.e., the cutoff level determines how many documents are retrieved by the first-stage ranker and will be re-ranked by MonoT5. It is important to emphasize that MonoT5 works according to the retrieve and rerank paradigm, which means that in our case, a set of documents is retrieved in the first phase with a sparse model (BM25 in our case) and reranked in a second phase by a dense model. In principle, the re-ranker can rely on a larger set of potentially relevant documents and place them at higher ranks. However, an increasing cutoff level does not substantially improve P@10 and nDCG@10, which can be explained by unjudged documents placed at higher ranks, as also shown by the increasing Bpref scores of MonoT5.

In our simulations, we want to compare BM25 against a strong competitor that is also computationally efficient, as the user simulations imply several other computations as well. For this reason, we consider MonoT5@100 to be sufficient for our experiments. A lower cutoff level substantially reduces the computation time and also helps to reduce the environmental impact of our research [30]. Higher cutoff levels do not lead to remarkable improvements in precision, which is of primary interest for our simulation experiments.

From a more general perspective, these evaluations further reveal the limitations of using ad-hoc test collections for simulated interactive retrieval experiments. When there are many unjudged documents in the rankings, the simulated users are indecisive about clicks, and in this case, the true potential of Transformer-based models cannot be fully analyzed in simulations when the test

Table 2. Overview of our user simulation configurations based on query generation methods and click models.

(a) Query generation methods

Strategy	Topic	Feedback	Generation
GPT	–	–	Probabilistic
GPT+	+	–	Probabilistic
GPT*	+	–	Rule-based
GPT**	–	–	Rule-based
D2Q	–	–	Rule-based
D2Q+	–	+	Rule-based
D2Q++	+	+	Rule-based

(b) Click models

Model	rel	nrel
Perfect	1.0	0.0
Navigational	0.9	0.1
Informational	0.8	0.4
Almost random	0.6	0.4

collection is biased towards a particular kind of retrieval method. We consider it a part of the future work to provide adequate resources, data, and tools in this regard.

3.2 Click and Stop Decisions

To be in line with earlier work [14], we compare four different types of click behavior in our experiments. All of the methods are based on probabilistic decisions biased towards the level of relevance and are implemented as shown in Table 2b. We note that other ways exist to simulate clicks [8], when there are historical interaction logs, which unfortunately were not available for our experimental setup. The *perfect* and *almost random* browsing behaviors serve as sky- and baselines, respectively. The *perfect* user always clicks on relevant results and never clicks non-relevant or unjudged documents, i.e., the user *scents* the relevant information in a strongly idealistic way. In contrast, the *almost random* behavior mainly neglects the relevance labels. In between, there are the *informational* and *navigational* models with different degrees of exploratory behavior. We model the stop decisions in two different ways.

Static. The first stopping criterion is based on the naive assumption that the user browses a fixed number of *results per page (rpp)* and then continues the session by reformulating another query about the topic.

Dynamic. In addition, we implement a time-based stopping criterion, first applied by Maxwell [22] in simulated IR sessions, that is based on the *give-up* heuristic by Krebs et al. [17] stemming from the foraging theory. More precisely, the simulated user has a fixed time budget (*tnr*) that starts to deplete after the last seen relevant search result. If another relevant document is seen, the time budget resets. If the budget is depleted, the user reformulates the query.

3.3   Query Generation

In the following, we describe our query generation methods, for which an overview is given in Table 2a. Generally, we employ two different *strategies*: prompting GPT to output query strings and summarizing document contents with the help of Doc2Query (D2Q). Depending on the simulated context, we provide the simulated user with the contents of the *topic* file, i.e., the description and narrative, and let the user consider the *feedback* of earlier seen search results. On the one side, the generation methods can be classified as fully *probabilistic* since they rely entirely on the language models' outputs. Conversely, there are *rule-based* methods that expand the query stem based on the topic's title with newly generated query terms in a principled way.

Prompting Query Reformulations. Given the information need described by the topic file, we construct a topic-specific prompt for the LLM. More specifically, we let the LLM generate outputs based on the prompt template given below.

> **Prompt template for the query strategies "GPT" and "GPT+"**
>
> ```
> Please generate one-hundred keyword queries about <title>.
> <description> <narrative>
> ```

`<title>`, `<description>`, and `<narrative>` are taken from the respective fields of the particular topic. To better understand the effects of different prompting strategies, we consider the first generation method **GPT** (■) that omits the additional topic fields `<description>` and `<narrative>` in the prompt to have context-free query variants. The second method **GPT+** (■■) makes use of the entire prompt to generate query reformulations.

We implement two additional methods to better understand the benefits of having different query variants with semantics. **GPT*** uses the topic's title as the seed query for each query (re)formulation and expands it with a single term out of the vocabulary generated for the topic by the **GPT+** (■■) strategy. Similarly, **GPT**** is also a rule-based variant based on the context-free vocabulary of **GPT** (■). For each topic in the test collections (cf. Sect. 3.5), we adapt the prompting strategy to query OpenAI's API and parse the outputs of GPT-3.5 (more specifically, gpt-3.5-turbo-0301[2]). In total, we generated 400 queries for all 50 topics in two test collections, which resulted in a total of 40,000 generated queries that stem from the outputs of OpenAI's LLM.

Doc2Query. Doc2Query relies on a set of documents to generate queries. To obtain the first set of documents, a seed query is required, which equals the `<title>` of the corresponding topic in this case. Our Doc2Query generation approach is based on [10] and extends the methodology by integrating background information. Unlike the GPT variants, this method allows to generate

---

[2] https://platform.openai.com/docs/model-index-for-researchers.

queries dynamically at runtime, taking seen results and, therefore, the context of the current session into account. To factor in the user's changing knowledge state, terms are extracted from seen documents and added to the current knowledge state. This approach assumes that terms that come from documents actually seen are more likely to satisfy the information need than random terms from the corpus. For each query reformulation, a term is taken from the knowledge state and added to the seed query. The knowledge state at the time of the i-th query KS_i is a set of terms defined according to:

$$KS_i = \bigcup_{j \in \{1,...,i\}} \phi(\theta(Q_j)). \tag{1}$$

$\theta(Q_j)$ is the set of seen documents for the j-th query. The terms extracted for a given set of documents are defined by:

$$\phi(D) = \bigcup_{d \in D} \{t \in d2q(d) \mid idf(t) < 0.5 \wedge t \notin S\}. \tag{2}$$

All terms identified by the Doc2Query function $d2q$ are joined, except for those with an inverse document frequency (IDF) below 0.5 and stopwords.

To determine the effect of background information and relevance feedback on cumulative retrieval effectiveness, we present three different Doc2Query variants. **D2Q** serves as the baseline of the Doc2Query variants, which generates queries without background information and feedback. Here, the knowledge state is based solely on terms taken from seen documents. **D2Q+** integrates the relevance feedback of the simulated user. In addition to KS_i, a knowledge state KS_i^{rel} is used, which only contains terms from seen documents that have been marked as relevant by the simulated user. As long as KS_i^{rel} is not empty, $D2Q+$ uses these terms to generate queries. Otherwise, terms from KS_i are used:

$$Q_{i+1} = \begin{cases} Q_0 \cup t^{rel} : t^{rel} \in KS_i^{rel} & KS_i^{rel} \neq \emptyset \\ Q_0 \cup t \quad : t \quad \in KS_i & else. \end{cases} \tag{3}$$

D2Q++ extends the $D2Q+$ mechanism with background information. For this purpose, a set of terms is generated from the `description` and `narrative` fields. Because these terms are topic-specific, they add another form of context to the query generation. The topic-specific terms are used until no more new queries can be generated, and then the $D2Q+$ method takes over.

3.4   Evaluation Measures

Effort vs. Effect. At the most granular level, we evaluate the effect of a search session by determining the cumulated information gain (IG) over the costs of all logged interactions, i.e., the effort, by:

$$\text{Effect} = \sum_{s \in S} \text{IG}(s), \quad \text{IG}(s) = \begin{cases} \text{rel}_d, \text{ if } s = s_{rel} \\ 0, \quad else. \end{cases} \tag{4}$$

rel_d is the relevance level of the document d, $\mathcal{S}$ is the set of all logged interactions, and s_{rel} denotes the particular session log considering the read document as relevant. We note that in this case, the costs are modeled by several additional factors besides the query formulations, including the click decision based on the snippet, the reading of the corresponding document, and making a judgment of the document's relevance for a given topic, and there is no discount over the progress of the session. For the costs of each action, we use the default values of SimIIR.

Session-Based DCG. As proposed in earlier work, we evaluate the simulated sessions by the sDCG [16], which shows high correlations with real user effectiveness [12]. sDCG discounts the cumulated gain document- and also query-wise by the logarithm and the corresponding base bq as well as by the query position i in a session:

$$\text{sDCG} = \sum_{i \in \{1,\dots,n_j\}} \frac{\text{DCG}_{q_i}}{1 + \log_{bq}(i)}, \qquad \text{DCG}_{q_i} = \sum_{r \in \{1,\dots,n_d\}} \frac{2^{\text{rel}_r} - 1}{\log_2(r + 1)}, \qquad (5)$$

where n_j denotes the number of available queries for a topic j and bq is a free parameter set to 4. The DCG [15] is defined by the log-harmonic discounted sum of relevance rel_r of n_d documents in the ranking corresponding to query q_i. Obviously, sDCG exclusively considers query formulations as *costs* and models the user's stopping behavior by a log-harmonic probability distribution over the number of queries and documents.

Session RBP. Similar to sDCG, there is a session-based variant of RBP [25], which was introduced by Lipani et al. [18] and is defined as follows:

$$\text{sRBP} = (1 - p) \sum_{i \in \{1,\dots,n_j\}} \left(\frac{p - bp}{1 - bp} \right)^{i-1} \sum_{r \in \{1,\dots,n_d\}} (bp)^{r-1} \cdot \text{rel}_r, \qquad (6)$$

where p models the user's persistence similar to RBP and b is introduced as a parameter that balances reformulating a query and continuing to browse the result list. sRBP discounts later results more than sDCG. Because of the considerable length of the simulated search sessions, we set p to 0.99 to ensure that results found later in the sessions would still contribute to the outcome. For a good trade-off between results later in the queries and results from later queries, we set b to 0.9. Like the comparison between DCG and RBP, evaluating sDCG and sRBP provides two different perspectives of retrieval effectiveness that can be explained by the underlying user models based on different probability distributions.

3.5   Implementation Details and Datasets

We implement the experimental setup with a rich set of state-of-the-art software libraries, including `Pyterrier` [20], `ir_datasets` [19], `SIMIIR v2.0` [38], and HuggingFace's `transformers` [37]. We ground the experiments on two TREC newswire test collections, i.e., the New York Times Annotated Corpus[3] used as part of TREC Common Core 2017 (Core17) [2], and the TREC Washington Post Corpus[4] used as part of TREC Common Core 2018 (Core18) [34]. Here it is important to note that Core18 includes ~26k (~4k positive) relevance labels, while Core17 includes ~30k (~9k positive) labels. For generating queries with Doc2Query, we use the available pre-trained model without finetuning[5]. All of the experiments are run on a Dell workstation with an Intel i9-12900K CPU, 64 GB of RAM, and an NVIDIA GeForce RTX 3070 GPU on Ubuntu 22.04 LTS.

4   Experimental Results

The following experiments investigate how the information gain develops for increasing costs. In our case, the cost is either the number of queries or the time spent, expressed in time units. The evaluations of the simulation runs are based on different user behaviors regarding click and stop decisions (cf. Sect. 3.2), two retrieval systems (cf. Sect. 3.1), and seven different methods for varying queries (cf. Sect. 3.3). Three measures (cf. Sect. 3.4) are evaluated, providing different perspectives on cost-benefit trade-offs and using two different test collections. To determine the effect of specific methods on the information gain, a default configuration is set for each method type and varied only over the aspect that is being evaluated. This default configuration is: *BM25 - GPT - informational - 10rpp*. In our study, we evaluate the differences between the various configurations in an exploratory manner. To make statements about the effectiveness of the different configurations, we compare both their order and the trend of the information gain pairwise over the simulated runs.

4.1   Retrieval Models and Users

This section investigates the influence of the retrieval system and different types of simulated users. Both user characteristics, i.e., the click behavior (cf. Table 2b) and the stop decisions, are evaluated with BM25 and MonoT5. In Fig. 2, we see a clear order if we look at the sDCG values for the different click profiles. The closer the user is to the perfect click behavior, the higher their information gain. Comparing all pairs of retrieval models over the click profiles (e.g., BM25-perfect vs. MonoT5-perfect) shows a slight dominance of MonoT5. This effect corresponds to the results of the ad hoc retrieval from Table 1. Compared to the sDCG plots, the effort-based evaluation shows a clear difference in user behavior, while the retrieval model plays a lesser role.

---

[3] https://catalog.ldc.upenn.edu/LDC2008T19.

[4] https://trec.nist.gov/data/wapost/.

[5] https://github.com/terrierteam/pyterrier_doc2query.

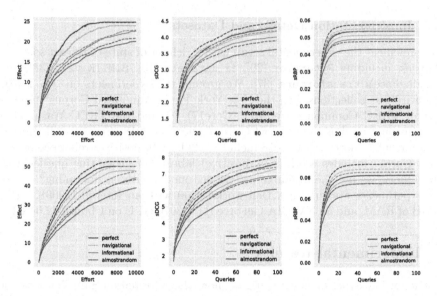

Fig. 2. Results of the simulated sessions for different click behaviors. Core18 (top) and Core17 (bottom). BM25 in solid lines, MonoT5 dashed, respectively.

We distinguish two types of stop decisions (cf. Sect. 3.2). Either a new query is formulated when a fixed number of snippets has been considered (*rpp*) or when a defined time has passed since the last relevant document was examined (*tnr*). Each type is evaluated with two parameters: $tnr \in \{50, 110\}$ and $rpp \in \{10, 20\}$. In Fig. 3, it can be seen that the dynamic click behavior positively affects the information gain. In addition, both an increase in the number of documents examined and an increase in the dynamic time budget have a positive effect. Again, the effort-based measure weights the impact of user behavior more heavily than the effectiveness of the retrieval model. Furthermore, sRBP is shown to punish dynamic stop behavior.

To put the different stop profiles in an insightful context, Fig. 4 shows the distribution over the average number of snippets considered across the issued queries. First, it is noticeable that both runs of *rpp* vary much less than the *tnr* variants. This effect was expected since the dynamic criterion's stop decision depends on the relevant documents in the result list. In contrast, the upper limit is strictly fixed for the static criterion, and downward outliers only occur when the result list is smaller than the fixed value or the global time limit is reached. Furthermore, the dynamic variants examine more snippets on average at the beginning of a session and fewer snippets during the session than the corresponding static variants. For *110nr*, a termination can be seen at about 80 queries. This is because the global time budget is reached here.

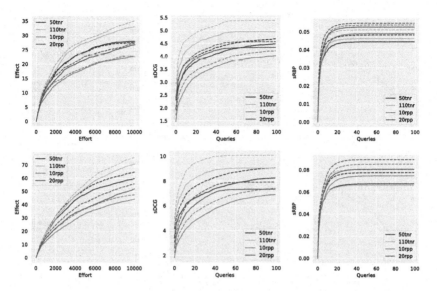

Fig. 3. Results of the simulated sessions for different stop decisions. Core18 (top) and Core17 (bottom). BM25 in solid lines, MonoT5 dashed, respectively.

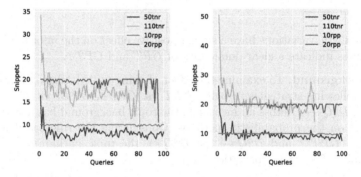

Fig. 4. Distributions for the average number of snippets examined for i-th query. Core18 (left) and Core17 (right).

4.2 Query Variation

Since it has been shown in Sect. 4.1 that MonoT5 does not cause any substantial differences in the trends of the curves, we continue with only BM25 due to the enormous computational effort. We examine the following aspects: query generation type, topic background, and feedback (cf. Table 2a). The plots of the seven different variants are shown in Fig. 5.

Query Generation Type. To determine what effect the query variation procedure has, we distinguish between probabilistic and rule-based query variations as described in Sect. 3.3. In our case, *GPT* and *GPT+* stand for probabilistic variation, while the rest works ruled-based. In Fig. 5, it can be clearly seen that

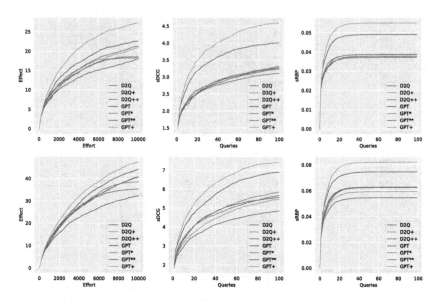

Fig. 5. Results of the simulated retrieval sessions for different query generation methods. Core18 (top) and Core17 (bottom).

the probabilistic variations have a strong positive effect on the information gain. All measures indicate a clear dominance of *GPT* and *GPT+*.

Topic Background. To examine the effect of using topic background, we compare the following strategies in pairs. *GPT* vs. *GPT+*, *GPT** vs. *GPT*** and *D2Q+* vs. *D2Q++*. Across all measures, the topic background has a strong positive effect on the probabilistic methods. Considering the rule-based methods (*GPT** vs. *GPT*** and *D2Q++* vs. *D2Q+*), the topic background does not substantially affect the information gain.

Feedback. The effect of feedback on information gain is evaluated in our setting by comparing the mechanisms *D2Q+* and *D2Q*. As described in Sect. 3.3, *D2Q+* integrates the relevance feedback of the simulated user and accordingly favors terms that come from relevant documents. Comparing the plots displays that feedback leads to a clear increase in information gain.

5   Discussion

The results in Sect. 4.1 show that although the dense retrieval model gives slightly better results, these differences stay the same throughout the simulation. Minor differences in trends align with other studies [13,33], and in our perspective, user behavior leads to a much greater variance in effectiveness. In particular, taking into account the context within the search session— modeled by dynamic click behavior, background knowledge, and the integration of feedback—substantially impacts information gain. From our point of view, it is crucial for future simulation experiments to integrate the user context instead of only investigating cost-intensive retrieval models. For both test collections, the trends were similar in all evaluations, even though the absolute information gain differed. We suspect this is related to the substantial difference in the number of documents rated as relevant (cf. Sect. 3.5).

In future evaluations, the costs for specific user actions used for the effort-based measure should be specified and justified. Still, the order and trends of the effort-based evaluations resemble the sDCG evaluations closely. Furthermore, the effort-based measure can reveal effects that would remain hidden if only the number of queries were considered. For example, Fig. 3 shows that the difference between 50tnr and 20rpp is more pronounced for the effort-based measure since the larger number of examined documents comes with high costs for the user that sDCG does not consider.

Regarding sRBP, sDCG also mainly supports the produced orders of the different configurations. However, the information gain stagnates more quickly, and a difference in the orders can be seen in Fig. 3, so dynamic click behavior is penalized by sRBP. Both findings can be explained by the substantially stronger discounting of late results by sRBP. Interestingly, dynamic stop behavior is notably rewarded by the other measures. As Fig. 4 indicates, the larger number of examined documents can likely further be attributed to this opposing evaluation behavior. Determining not only realistic durations for user actions but also estimating suitable parameters for sRBP is vital for future work to enable more accurate evaluations and could be done by analyzing real user behavior.

The use of probabilistic query generation methods shows promising results. Not only do they dominate rule-based approaches, but they are also more capable of integrating background knowledge of the topic. Integrating feedback into probabilistic approaches would be an exciting direction for future work, as our rule-based approach has clearly shown that feedback is beneficial.

Limitations of this work are, on the one hand, a need for estimates of how similar the simulated behavior (e.g., query generation) corresponds to real users. On the other hand, the results obtained cannot be generalized arbitrarily since we only examined a limited variation of simulated users for a specific search task, and both data sets correspond to the same domain.

6   Conclusion

In this work, we propose novel ways to simulate context-dependent query formulations that are evaluated with a state-of-the-art experimental setup, including LLMs and dense retrieval methods. To better understand the effects of context information, we compare context-sensitive query simulation methods against context-free variants. Our experimental results suggest that both the inclusion of comprehensive descriptions of the information need as well as the feedback inferred from earlier seen search results impact the progress of a simulation session, leading to better search effectiveness. In this regard, we envision experiments with user simulations of higher fidelity by considering contextual information.

Acknowledgements. This work was supported by Klaus Tschira Stiftung (JoIE - 00.003.2020) and Deutsche Forschungsgemeinschaft (RESIRE - 509543643).

References

1. Alaofi, M., Gallagher, L., Sanderson, M., Scholer, F., Thomas, P.: Can generative LLMs create query variants for test collections? An exploratory study. In: SIGIR, pp. 1869–1873. ACM (2023)
2. Allan, J., Harman, D., Kanoulas, E., Li, D., Gysel, C.V., Voorhees, E.M.: TREC 2017 common core track overview. In: TREC. NIST Special Publication 500-324. National Institute of Standards and Technology (NIST) (2017)
3. Azzopardi, L., Järvelin, K., Kamps, J., Smucker, M.D.: Report on the SIGIR 2010 workshop on the simulation of interaction. SIGIR Forum **44**(2), 35–47 (2010)
4. Balog, K., Maxwell, D., Thomas, P., Zhang, S.: Sim4IR: the SIGIR 2021 workshop on simulation for information retrieval evaluation. In: SIGIR, pp. 2697–2698. ACM (2021)
5. Balog, K., Zhai, C.: User simulation for evaluating information access systems. CoRR abs/2306.08550 (2023)
6. Baskaya, F., Keskustalo, H., Järvelin, K.: Modeling behavioral factors in interactive information retrieval. In: He, Q., Iyengar, A., Nejdl, W., Pei, J., Rastogi, R. (eds.) 22nd ACM International Conference on Information and Knowledge Management, CIKM 2013, San Francisco, CA, USA, 27 October–1 November 2013, pp. 2297–2302. ACM (2013). https://doi.org/10.1145/2505515.2505660
7. Breuer, T., Fuhr, N., Schaer, P.: Validating simulations of user query variants. In: Hagen, M., et al. (eds.) ECIR 2022. LNCS, vol. 13185, pp. 80–94. Springer, Cham (2022). https://doi.org/10.1007/978-3-030-99736-6_6
8. Breuer, T., Fuhr, N., Schaer, P.: Validating synthetic usage data in living lab environments. J. Data Inf. Qual. (2023, accepted). https://doi.org/10.1145/3623640
9. Carterette, B., Bah, A., Zengin, M.: Dynamic test collections for retrieval evaluation. In: Allan, J., Croft, W.B., de Vries, A.P., Zhai, C. (eds.) Proceedings of the 2015 International Conference on the Theory of Information Retrieval, ICTIR 2015, Northampton, Massachusetts, USA, 27–30 September 2015, pp. 91–100. ACM (2015). https://doi.org/10.1145/2808194.2809470
10. Engelmann, B., Breuer, T., Schaer, P.: Simulating users in interactive web table retrieval. In: Proceedings of the 32nd ACM International Conference on Information and Knowledge Management, CIKM 2023, pp. 3875–3879. Association for Computing Machinery, New York (2023). https://doi.org/10.1145/3583780.3615187
11. Günther, S., Hagen, M.: Assessing query suggestions for search session simulation. In: Sim4IR: The SIGIR 2021 Workshop on Simulation for Information Retrieval Evaluation (2021). https://ceur-ws.org/Vol-2911/paper6.pdf
12. Hagen, M., Michel, M., Stein, B.: Simulating ideal and average users. In: Ma, S., Wen, J.-R., Liu, Y., Dou, Z., Zhang, M., Chang, Y., Zhao, X. (eds.) AIRS 2016. LNCS, vol. 9994, pp. 138–154. Springer, Cham (2016). https://doi.org/10.1007/978-3-319-48051-0_11

13. Hersh, W.R., et al.: Do batch and user evaluation give the same results? In: Yannakoudakis, E.J., Belkin, N.J., Ingwersen, P., Leong, M.K. (eds.) Proceedings of the 23rd Annual International ACM SIGIR Conference on Research and Development in Information Retrieval, SIGIR 2000, 24–28 July 2000, Athens, Greece, pp. 17–24. ACM (2000). https://doi.org/10.1145/345508.345539

14. Hofmann, K., Schuth, A., Whiteson, S., de Rijke, M.: Reusing historical interaction data for faster online learning to rank for IR. In: Leonardi, S., Panconesi, A., Ferragina, P., Gionis, A. (eds.) Sixth ACM International Conference on Web Search and Data Mining, WSDM 2013, Rome, Italy, 4–8 February 2013, pp. 183–192. ACM (2013). https://doi.org/10.1145/2433396.2433419

15. Järvelin, K., Kekäläinen, J.: IR evaluation methods for retrieving highly relevant documents. In: SIGIR, pp. 41–48. ACM (2000)

16. Järvelin, K., Price, S.L., Delcambre, L.M.L., Nielsen, M.L.: Discounted cumulated gain based evaluation of multiple-query IR sessions. In: Macdonald, C., Ounis, I., Plachouras, V., Ruthven, I., White, R.W. (eds.) ECIR 2008. LNCS, vol. 4956, pp. 4–15. Springer, Heidelberg (2008). https://doi.org/10.1007/978-3-540-78646-7_4

17. Krebs, J.R., Ryan, J.C., Charnov, E.L.: Hunting by expectation or optimal foraging? A study of patch use by chickadees. Anim. Behav. 22, 953–964 (1974). https://doi.org/10.1016/0003-3472(74)90018-9

18. Lipani, A., Carterette, B., Yilmaz, E.: From a user model for query sessions to session rank biased precision (sRBP). In: Fang, Y., Zhang, Y., Allan, J., Balog, K., Carterette, B., Guo, J. (eds.) Proceedings of the 2019 ACM SIGIR International Conference on Theory of Information Retrieval, ICTIR 2019, Santa Clara, CA, USA, 2–5 October 2019, pp. 109–116. ACM (2019). https://doi.org/10.1145/3341981.3344216

19. MacAvaney, S., Yates, A., Feldman, S., Downey, D., Cohan, A., Goharian, N.: Simplified data wrangling with ir_datasets. In: Diaz, F., Shah, C., Suel, T., Castells, P., Jones, R., Sakai, T. (eds.) The 44th International ACM SIGIR Conference on Research and Development in Information Retrieval, SIGIR 2021, Virtual Event, Canada, 11–15 July 2021, pp. 2429–2436. ACM (2021). https://doi.org/10.1145/3404835.3463254

20. Macdonald, C., Tonellotto, N., MacAvaney, S., Ounis, I.: PyTerrier: declarative experimentation in python from BM25 to dense retrieval. In: CIKM, pp. 4526–4533. ACM (2021)

21. Mackie, I., Chatterjee, S., Dalton, J.: Generative relevance feedback with large language models. In: Proceedings of the 46th International ACM SIGIR Conference on Research and Development in Information Retrieval, SIGIR 2023, July 2023, pp. 2026–2031. ACM (2023). https://doi.org/10.1145/3539618.3591992

22. Maxwell, D.: Modelling search and stopping in interactive information retrieval. Ph.D. thesis, University of Glasgow, UK (2019)

23. Maxwell, D., Azzopardi, L.: Simulating interactive information retrieval: SimIIR: a framework for the simulation of interaction. In: SIGIR, pp. 1141–1144. ACM (2016)

24. Maxwell, D., Azzopardi, L., Järvelin, K., Keskustalo, H.: Searching and stopping: an analysis of stopping rules and strategies. In: CIKM, pp. 313–322. ACM (2015)

25. Moffat, A., Zobel, J.: Rank-biased precision for measurement of retrieval effectiveness. ACM Trans. Inf. Syst. 27(1), 2:1–2:27 (2008)

26. Nogueira, R.F., Jiang, Z., Pradeep, R., Lin, J.: Document ranking with a pretrained sequence-to-sequence model. In: EMNLP (Findings). Findings of ACL, EMNLP 2020, pp. 708–718. Association for Computational Linguistics (2020)

27. Nogueira, R.F., Yang, W., Lin, J., Cho, K.: Document expansion by query prediction. CoRR abs/1904.08375 (2019)
28. Pääkkönen, T., Kekäläinen, J., Keskustalo, H., Azzopardi, L., Maxwell, D., Järvelin, K.: Validating simulated interaction for retrieval evaluation. Inf. Retr. J. **20**(4), 338–362 (2017)
29. Robertson, S.E., Zaragoza, H.: The probabilistic relevance framework: BM25 and beyond. Found. Trends Inf. Retr. **3**(4), 333–389 (2009)
30. Scells, H., Zhuang, S., Zuccon, G.: Reduce, reuse, recycle: green information retrieval research. In: SIGIR, pp. 2825–2837. ACM (2022)
31. Tague, J., Nelson, M.J.: Simulation of user judgments in bibliographic retrieval systems. In: Crouch, C.J. (ed.) Theoretical Issues in Information Retrieval, Proceedings of the Fourth International Conference on Information Storage and Retrieval, Oakland, California, USA, 31 May–2 June 1981, pp. 66–71. ACM (1981). https://doi.org/10.1145/511754.511764
32. Tague, J., Nelson, M.J., Wu, H.: Problems in the simulation of bibliographic retrieval systems. In: Oddy, R.N., Robertson, S.E., van Rijsbergen, C.J., Williams, P.W. (eds.) Information Retrieval Research, Proceedings of the Joint ACM/BCS Symposium in Information Storage and Retrieval, Cambridge, UK, June 1980, pp. 236–255. Butterworths (1980). https://dl.acm.org/citation.cfm?id=636684
33. Turpin, A., Hersh, W.R.: Why batch and user evaluations do not give the same results. In: Croft, W.B., Harper, D.J., Kraft, D.H., Zobel, J. (eds.) Proceedings of the 24th Annual International ACM SIGIR Conference on Research and Development in Information Retrieval, SIGIR 2001, 9–13 September 2001, New Orleans, Louisiana, USA, pp. 225–231. ACM (2001). https://doi.org/10.1145/383952.383992
34. Voorhees, E.M., Ellis, A. (eds.): Proceedings of the Twenty-Seventh Text REtrieval Conference, TREC 2018, Gaithersburg, Maryland, USA, 14–16 November 2018, NIST Special Publication, 500-331. National Institute of Standards and Technology (NIST) (2018). https://trec.nist.gov/pubs/trec27/trec2018.html
35. Wang, L., Yang, N., Wei, F.: Query2doc: query expansion with large language models. In: Conference on Empirical Methods in Natural Language Processing, pp. 9414–9423. Association for Computational Linguistics (2023). https://api.semanticscholar.org/CorpusID:257505063
36. Wang, X., MacAvaney, S., Macdonald, C., Ounis, I.: Generative query reformulation for effective adhoc search (2023)
37. Wolf, T., et al.: Transformers: state-of-the-art natural language processing. In: Liu, Q., Schlangen, D. (eds.) Proceedings of the 2020 Conference on Empirical Methods in Natural Language Processing: System Demonstrations, EMNLP 2020, Demos, Online, 16–20 November 2020, pp. 38–45. Association for Computational Linguistics (2020). https://doi.org/10.18653/V1/2020.EMNLP-DEMOS.6
38. Zerhoudi, S., et al.: The SimIIR 2.0 framework: user types, Markov model-based interaction simulation, and advanced query generation. In: CIKM, pp. 4661–4666. ACM (2022)
39. Zhang, Y., Liu, X., Zhai, C.: Information retrieval evaluation as search simulation: a general formal framework for IR evaluation. In: ICTIR, pp. 193–200. ACM (2017)

A Deep Learning Approach for Selective Relevance Feedback

Suchana Datta[1]([envelope]) [ID], Debasis Ganguly[2] [ID], Sean MacAvaney[2] [ID],
and Derek Greene[1] [ID]

[1] University College Dublin, Dublin, Ireland
suchana.datta@ucdconnect.ie, derek.greene@ucd.ie
[2] University of Glasgow, Glasgow, UK
{debasis.ganguly,sean.macavaney}@glasgow.ac.uk

Abstract. Pseudo-relevance feedback (PRF) can enhance average retrieval effectiveness over a sufficiently large number of queries. However, PRF often introduces a drift into the original information need, thus hurting the retrieval effectiveness of several queries. While a selective application of PRF can potentially alleviate this issue, previous approaches have largely relied on unsupervised or feature-based learning to determine whether a query should be expanded. In contrast, we revisit the problem of selective PRF from a deep learning perspective, presenting a model that is entirely data-driven and trained in an end-to-end manner. The proposed model leverages a transformer-based bi-encoder architecture. Additionally, to further improve retrieval effectiveness with this selective PRF approach, we make use of the model's confidence estimates to combine the information from the original and expanded queries. In our experiments, we apply this selective feedback on a number of different combinations of ranking and feedback models, and show that our proposed approach consistently improves retrieval effectiveness for both sparse and dense ranking models, with the feedback models being either sparse, dense or generative.

1 Introduction

The keywords that a user enters as query to a search engine are often insufficient to express the user's information need, resulting in a *lexical gap* between the text in the query and the relevant documents [2]. Standard pseudo-relevance feedback (PRF) methods, such as the relevance model [19] and its variants [14,26,35, 36], can overcome this problem and ultimately yield improvements in retrieval effectiveness. Generally speaking, PRF methods are designed to enrich a user's initial query with distinctive terms from the top-ranked documents [27,34,43]. Despite the demonstrated success of PRF in improving retrieval effectiveness, a number of studies have identified certain limitations of this strategy [3,9,13,25]. For the most part, these limitations share a common theme: there is no consistent PRF setting that works well across a wide range of queries; to put in simple words, *one size does not fit all*. Figure 1 illustrates such a situation, where nearly

N. Goharian et al. (Eds.): ECIR 2024, LNCS 14609, pp. 189–204, 2024.
https://doi.org/10.1007/978-3-031-56060-6_13

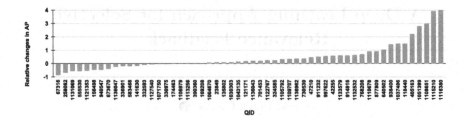

Fig. 1. Relative changes in AP, i.e., (AP(post-fdbk) - AP(pre-fdbk))/AP(pre-fdbk), for TREC DL'20 queries. We observe that many queries are negatively impacted by PRF (bars below the x-axis).

38.9% of queries from TREC DL'20 topic set are penalized as a result of PRF. It has been shown that not all documents contribute equally well to PRF, as certain documents may impair retrieval effectiveness when used to expand a query [1,20]. This can even be true when relevant documents are used to enrich a query's representation [39]. It has also been observed that some queries are amenable to more aggressive query expansion, while others work better with more conservative settings [32]. Moreover, not all terms might contribute equally well in terms of enriching the representation of a query [4,16], which suggests that a selective approach to PRF can potentially improve the overall IR effectiveness.

Rather than following the previous approaches on adapting the number of feedback terms [32] or attempting to choose a robust subset of documents for PRF [1,20], we rather focus on solving the more fundamental decision question of *"whether or not to apply PRF for a given query"* [9,25] with the help of a supervised data-driven approach. We hypothesise that selectively applying feedback to only those queries that are amenable to PRF can improve the overall retrieval effectiveness by avoiding query drift in cases where feedback would not be beneficial. Our idea is depicted in Fig. 2.

The main novelty of our proposed selective pseudo relevance feedback (SRF) approach is that in contrast to existing work on selective PRF, we propose a data-driven supervised neural model for predicting which queries are conducive to PRF. More specifically, during the training phase we make use of the relevance assessments to learn a decision function that, given the query and the top-retrieved set of documents both with and without feedback, predicts whether it is useful to apply PRF. During the inference phase, we make use of only a part of the shared parameter network which, given a query and its top-retrieved document set, predicts whether PRF is to be applied (schematically illustrated in Fig. 2). This way of inference reduces computational costs for queries where PRF should eventually be ignored.

A key advantage of our SRF approach is that it can be applied to the output ranked list obtained by **any retrieval model**, ranging from sparse models (e.g., BM25, LM-Dir etc.) to dense ones (e.g., MonoBERT [31]). Moreover, in the SRF workflow it also is possible to use **any PRF model** to enrich a query's

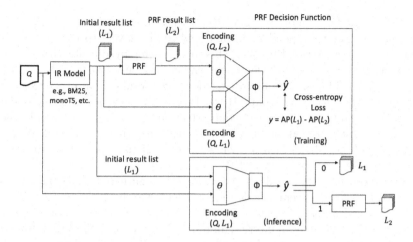

Fig. 2. A schematic diagram of selective feedback. The main contribution of this paper is a supervised data-driven approach towards realising the decision function.

representation, ranging from sparse models (e.g., RLM [19]) to dense ones (e.g., ColBERT-PRF [41]) to even generative ones (e.g., GRF [26]).

2   Related Work

The evolution of relevance feedback in IR spans from traditional query expansion models [4,32] to cluster-based feedback document selection [16,20]. While prior research has considered both unsupervised selective feedback [9] and feature-driven methods [25], we introduce a data-driven neural strategy for selective relevance feedback. Several existing methods, both supervised and unsupervised, hinge on *decision-based relevance feedback*. One unsupervised approach uses Query Performance Prediction (QPP) scores [10,15,37,38,47], which we include as a baseline. The higher the QPP score, the greater the chance of identifying relevant documents at the top rank positions with the initial query. However, high variances in retrieval status values, as seen in neural re-rankers like MonoBERT [31], can make QPP scores deceptive. To avoid such heuristics, our method focuses solely on query terms and the documents retrieved by that query in order to learn the selection function.

PRF on and for Dense Retrieval. Recently, the community has seen a significant interest in feedback for dense retrieval to boost performance. Precursors to dense feedback models made use of word embeddings for PRF, e.g., KDERLM [35] which proposed a generalised RLM with word embeddings, and PRF-NMF [45], which leveraged matrix factorisation to bridge the semantic gap between terms from a query and its top-retrieved documents.

The study by [44] explored relevance feedback principles within dense retrieval models. Li et al. [22] analyzed feedback signal quality, comparing traditional models like Rocchio [34] with dense retrievers like ANCE-based retrievers

[42], finding the latter more resilient. Representation models, such as ColBERT [18], can allow us to append additional embedding layers to the query representation, as demonstrated by [40]. This method employed contextualized PRF to cluster and rank feedback document embeddings in order to select suitable expansion embeddings, thus improving document ranking. In other work, [48] leveraged implicit feedback from historical clicks for relevance feedback in dense retrieval. The authors introduced counterfactual-based learning-to-rank, showing that historic clicks can be highly informative in terms of relevance feedback. Lastly, [23] proposed the idea of fusing feedback signals from both sparse and dense retrievers in the context of PRF.

More recently, PRF on dense IR models has garnered significant interest [21,29,41,46]. The concept of 'dense for PRF' was first emphasized in [28], which proposed a reinforcement-based learning algorithm designed to explore and exploit various retrieval metrics, aiming to learn an optimized PRF function. With the recent success of LLMs, [26] proposed a generative feedback method (GRF) that makes use of LLM generated long-form texts instead of first pass retrieved results to build a probabilistic feedback model. In contrast, our work aims to develop a generic PRF strategy that does not apply feedback blindly, but rather learns a selection function in a supervised manner to analyze the suitability of relevance feedback for each query irrespective of sparse or generative PRF.

Selective PRF. Prior work on selective PRF has considered either fully unsupervised approaches [9] or feature-based supervised approaches [25] for selective relevance feedback (SRF). The former makes use of query performance prediction (QPP) based measures to predict if a query should be expanded, where the decision depends on whether the QPP score exceeds a given threshold. On the other hand, existing supervised approaches first represent each query as a bag of characteristic features derived from its top-retrieved set of documents. A classifier is then trained on these features to predict whether or not a query should be expanded [25].

3   Model Description

3.1   A Generic Decision Framework for PRF

Given a set of queries $\mathcal{Q} = \{Q_1, \ldots, Q_n\}$, a standard relevance feedback model M uses the information from the top-retrieved documents of each query to enrich its representation, i.e., $M : Q \mapsto \phi_M(Q)$. Consequently, each query $Q \in \mathcal{Q}$ is transformed to an enriched representation $\phi_M(Q)$, which is then used either for re-ranking the initial list, or to execute a second-step retrieval.

Unlike the standard PRF setting, a decision-based selective PRF framework first applies a *decision function*, $\theta : Q \mapsto \{0, 1\}$, which outputs a Boolean to indicate if the retrieval results for Q is likely to be improved after application of PRF. As per our proposal, the overall PRF process on the set of queries $\mathcal{Q}$ does not simply replace each query Q with its enriched form $\phi_M(Q)$. Rather, it makes use of the function $\theta(Q)$ for each query Q to decide whether to output the initial ranked list or to make use of the enriched query representation $\phi_M(Q)$, as obtained by a PRF model M (leading to either re-ranking the initial list or re-retrieving a new list via a second stage retrieval). The top-k ranked list of documents, $L_k(Q) = \{D_1^Q, \ldots, D_k^Q\}$, retrieved for a query Q, in addition to being a function of the query Q itself, is thus also a function of i) the feedback model M, ii) the enriched query representation $\phi_M(Q)$, and iii) the decision function θ, i.e.,

$$L_k(Q) = \begin{cases} \sigma(Q), & \text{if } \theta(Q) = 0 \\ \sigma(\phi_M(Q)), & \text{if } \theta(Q) = 1, \end{cases} \tag{1}$$

where $\sigma(Q)$ denotes a retrieval model, e.g., BM25, that outputs an ordered set of k documents sorted by the similarity scores.

Previous approaches have explored the use of both unsupervised and supervised approaches for addressing this decision problem. We now briefly explain both strategies in our own context.

Unsupervised Decision Function. An unsupervised approach, such as [9], applies a threshold parameter on a QPP estimator function, $\theta_{\text{QPP}} : Q \mapsto [0,1]$. More concretely, if the predicted QPP score is lower than the threshold parameter, it is likely to indicate that the retrieval performance for the query has scopes for further improvement and subsequently PRF is applied for this query. Formally speaking, the decision function of an unsupervised approach takes the form

$$\theta(Q) \stackrel{\text{def}}{=} \mathbb{I}(\theta_{\text{QPP}} < \tau), \tag{2}$$

where $\tau \in [0,1]$ is the threshold parameter.

Supervised Decision Function. In the supervised approach, the decision function also depends on the enriched query representation and its top-retrieved documents. More precisely, a supervised PRF decision is a parameterized function of features of i) the query Q, ii) its top-retrieved documents $L_k(Q)$, iii) the enriched query $\phi_M(Q)$, and iv) its top-retrieved set $L_k(\phi_M(Q))$ [25]. The training process itself makes use of a set of queries $\mathcal{Q}_{\text{train}}$ for which ground-truth indicator labels are computed by evaluating the relative effectiveness obtained with the original query vs. the enriched query with the help of available relevance assessments. Formally,

$$y(Q) = \mathbb{I}(\text{AP}(\phi_M(Q)) > \text{AP}(Q)), \tag{3}$$

where $\text{AP}(Q)$ denotes the average precision of a query $Q \in \mathcal{Q}_{\text{train}}$. The indicator values of $y(Q)$ are used to learn the parameters of a classifier function to yield a supervised version of the decision function θ:

$$\theta(Q) \stackrel{\text{def}}{=} \zeta \cdot \mathbf{z}_{Q,\phi_M(Q)}, \text{ where } \theta(Q) \approx \underset{\zeta}{\text{argmin}} \sum_{Q' \in \mathcal{Q}'_{\text{train}}} (y(Q') - \zeta \cdot \mathbf{z}_{Q',\phi_M(Q')})^2. \tag{4}$$

In Eq. 4, ζ represents a set of learnable parameters, with $\mathbf{z}_{Q',\phi_M(Q')}$ denoting a set of features extracted from both the original query Q' and the enriched query

$\phi_M(Q')$ along with the features from their top-retrieved set of documents $L_k(Q')$ and $L_k(\phi_M(Q'))$. The variable $y(Q')$, as per the definition of $y(Q)$, denotes the ground-truth indicating if PRF should be applied for Q'. The optimal parameter vector ζ, as learned from a training set of queries $\mathcal{Q}_{\text{train}}$ (Eq. 4) is then used to predict the decision for any new query Q. The features we use are described later in the paper in Sect. 4.2.

3.2   Transformer-Based Encoding for PRF Decision

We now describe a data-driven approach for learning the decision function with deep neural networks. Instead of making use of a specific set of extracted features as used in the QPP model in [11], the learning objective of a transformer-based PRF model makes use of the terms present in the documents and the queries. As with Eq. 4, we make use of both the content of the original query Q and its enriched form $\phi_M(Q)$, along with their top-retrieved sets. More formally,

$$\theta(Q) \overset{\text{def}}{=} \zeta \cdot (\mathcal{E}(Q, D_1^Q, \ldots, D_k^Q) \oplus \mathcal{E}(\phi_M(Q), D_1^{\phi_M(Q)}, \ldots, D_k^{\phi_M(Q)})), \quad (5)$$

where $\theta(Q)$ is learned by computing

$$\underset{\zeta}{\text{argmin}} \sum_{Q' \in \mathcal{Q}'_{\text{train}}} (y(Q') - \zeta \cdot (\mathcal{E}(Q', L_k(Q')) \oplus \mathcal{E}(\phi_M(Q'), L_k(\phi_M(Q')))))^2. \quad (6)$$

In Eq. 6, $\mathcal{E}$ is a parameterized function for encoding the interaction between a query Q and its top-retrieved documents, $L_k(Q)$. This encoding function, generally speaking, maps a query (a sequence of query terms) and a sequence of documents (which are themselves sequences of their constituent terms) into a fixed length vector, i.e., $\mathcal{E} : Q, L_k \mapsto \mathbb{R}^p$ (p an integer, e.g., for BERT embeddings $p = 768$). Here $\oplus$ indicates an interaction operation (e.g., a merge layer in a neural network) between the query-document encodings corresponding to the original query and the enriched one.

The transformer-based encoding uses the BERT architecture which takes as input the contextual embeddings of the terms for each pair comprising a query Q and its top-retrieved document $D_i^Q \in L_k(Q)$. The 768 dimensional '[CLS]' representations of each *query-document* pair is then encoded with LSTMs as a realisation of the encoded representation of a query and its top-retrieved set, i.e., to define $\mathcal{E}(Q, L_k(Q))$ as per the notation of Eq. 6. We further obtain a BERT-based encoding of the expanded query $\phi_M(Q)$ and its top-retrieved set and then merge the two representations before passing them through a feed-forward network. More formally,

$$\mathcal{E}(\bar{Q}, L_k(\bar{Q})) = \text{LSTM}(\text{BERT}(\bar{Q}, D_1^{\bar{Q}})_{[\text{CLS}]}, \ldots, \text{BERT}(\bar{Q}, D_k^{\bar{Q}})_{[\text{CLS}]}). \quad (7)$$

The variable $\bar{Q} \in \{Q, \phi_M(Q)\}$, i.e., in one branch of the network it corresponds to the original query, whereas in the other it corresponds to the expanded one. Figure 3 shows the transformer-specific implementation of the encoding function. The set of learnable parameters ζ in this case comprises of the LSTM and

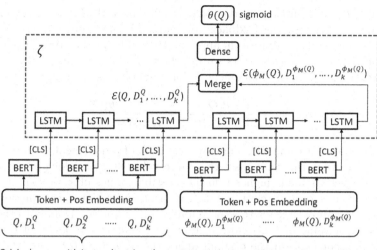

Fig. 3. Training of a transformer based query-document architecture with shared parameters for selective PRF. During inference, only the left part of the network is used to output whether to apply PRF or not for a given query.

the fully connected (dense) layer parameters, as shown in Fig. 3. We name this particular model **Deep-SRF-BERT** (Deep Selective Relevance Feedback with the use of BERT transformers).

Model Inference. During inference, only the part corresponding to the original query and its top-retrieved set of documents is used to predict the output variable (a sigmoid) which if higher than 0.5 indicates that PRF should be applied.

3.3   Model Confidence-Based PRF Calibration

Prior work has applied confidences of prediction models to improve retrieval effectiveness [5]. In our work, we use the uncertainties in the prediction of the decision function to further improve search results. Rather than only reporting either results with or without relevance feedback, we make use of the confidence of the decision function $\theta(Q)$ to combine the results from the two lists – one without feedback and the other with feedback. Specifically, if the supervised model outlined in Sect. 3.1 is decisive in its choice between $L_k(Q)$ (original query retrieved list) and $L_k(\phi_M(Q))$ (the list retrieved with the expanded query), then one of the rankings is expected to dominate over the other. However, when the model $\theta(Q)$ itself is not confident about the prediction, we can potentially achieve better results if we 'meet somewhere in the middle'.

Formally, we propose a rank-fusion based method, where the fusion weights are obtained from the predictions of the PRF decision model $\theta(Q)$. The predicted value $\theta(Q)$ being a sigmoid represents the probability of classifying the decision into one of the two outcomes - the closer $\theta(Q)$ is to 0, the higher is the model's confidence in not applying feedback, and similarly the closer $\theta(Q)$ is to 1, the higher is the model's confidence in applying PRF. The predicted value of $\theta(Q) \in [0, 1]$ can thus be used as weights to fuse the two different ranked lists, i.e., the fusion score $\sigma_F(Q, D)$ of a document D for a query Q is given by

$$\sigma_F(Q, D) = \frac{1 - \theta(Q)}{\text{Rank}(D, L_k(Q))} + \frac{\theta(Q)}{\text{Rank}(D, L_k(\phi_M(Q)))}, \tag{8}$$

where the notation $\text{Rank}(D, L)$ denotes the rank of a document D in a list L.

If $D \notin L$, then the rank is set to a large value $\aleph(> k)$, which in our experiments was set to 1000 (higher than all possible values of k we experimented with). For values of $\theta(Q)$ close to 0.5 (the most uncertainty in prediction), the fusion-based approach leads to a more uniform contribution from both the lists. In contrast, a value of $\theta(Q)$ close to 0 ensures that the majority of the score contribution comes from the original query (because $1 - \theta(Q) >> \theta(Q)$), and a similar argument applies for $\theta(Q) \rightarrow 1$, in which case the major contribution comes from the second term in the right-hand side of Eq. 8.

4   Evaluation

4.1   Research Questions

Since a primary contribution of this paper is the idea of applying a fully data-driven approach, the first research question that we investigate is whether a shift from the existing feature-based approach for selective PRF to a data-driven one does indeed result in improved retrieval effectiveness. Therefore, we formulate our first research question as follows.

- **RQ1:** Does SRF lead to overall improvements in IR effectiveness over non-selective and other baseline approaches?

Our second research question aims to investigate whether the model prediction uncertainty-based fusion of the two ranked lists – one retrieved with the original query and the other with the expanded one, can potentially improve the retrieval effectiveness further.

- **RQ2:** Can we use the confidence estimates from our selective PRF model for a *soft selection* of information from both pre-feedback and post-feedback sources to further improve IR effectiveness?

In our third research question, the aim is to investigate how effectively the selection strategy in PRF *transfers* across different feedback approaches, i.e., training the SRF approach only once on a PRF model (e.g., RLM), and then apply the decision model on other PRF models (e.g., ColBERT-PRF).

- **RQ3:** Does selection effectively transfers the learning across different PRF approaches?

4.2   Methods Investigated

We consider a range of unsupervised and supervised methods, described below. Some baselines refer to existing methods, while others are extensions of alternative approaches to allow us to provide a fair comparison, such as by using a more recent QPP method instead of the originally-proposed clarity score [9].

PRF is a standard non-selective relevance feedback model, namely the relevance model (RLM) [19]. We use the RM3 version of the relevance model as reported in [17], which is a linear combination of the weights of the original query model and new expansion terms. In fact, we use RLM as one of the base PRF model M which means that the standard RLM degenerates to a specific case of the generic selective PRF framework of Eq. 1 with $\theta(Q) = 1 \, \forall Q \in \mathcal{Q}$, i.e., when for each query we use its enriched form $\phi_M(Q)$.

R2F2 refers to an adaptation of the Reciprocal Rank-based Fusion (RRF) [6], a simple yet effective approach for combining the document rankings from multiple IR systems. For our task, instead of combining ranked lists from two different retrieval models, we merge the ranked lists of the original and the expanded queries, i.e., $L_k(Q)$ and $L_k(\phi_M(Q))$ as per our notations. We name the adapted method Reciprocal Rank Fusion-based Feedback (R2F2). Formally, the score for document D after fusion is given by

$$\sigma_F(Q, D) = \frac{1 - \alpha}{\text{Rank}(D, L_k(Q))} + \frac{\alpha}{\text{Rank}(D, L_k(\phi_M(Q)))}, \qquad (9)$$

where, similar to Eq. 8 $\text{Rank}(D, L)$ denotes the rank of a document in a list L (this being a large number $\aleph$ if $D \notin L$), and $\alpha \in [0, 1]$ is a linear combination hyper-parameter that we adjust with grid search on each training fold. A lower value of α puts more emphasis on the initial retrieval list, whereas a higher value ensures that the feedback rank of a document contributes more. Equation 9 is a special case of Eq. 8 with a constant value of $\theta(Q) = \alpha$ for each query Q.

QPP-SRF is an adaptation of the method proposed in [9], where the QPP score of a query is used as estimate to decide if PRF should be applied for that query (see $\theta(Q)$ in Sect. 3.1). The idea here is that a high QPP score is already indicative of an effective retrieval performance, in which case, the method avoids any further risk of potentially degrading the retrieval quality with query expansion. We refer to this method as QPP-based selective relevance feedback (QPP-SRF). The method requires a base QPP estimator to provide θ_{QPP} scores.

To choose the QPP estimator, we conducted a set of initial experiments using several standard unsupervised QPP approaches. The recently introduced supervised QPP method qppBERT-PL [12] demonstrated the best downstream retrieval effectiveness. Therefore, we report results of QPP-SRF combined with qppBERT-PL, where training is conducted using the settings as reported in [12]. A key parameter for QPP-SRF is the threshold value τ ($\tau \in [0, 1]$) which controls the decision around whether PRF is applied or not. In our experiments we tune

Table 1. Summary of the data used in our experiments. The columns '$|\bar{Q}|$' and '$\#\bar{Rel}$' denote average number of query terms and average number of relevant documents.

Collection	#Docs	Topic Set	#Topics	$|\bar{Q}|$	$\#\bar{Rel}$
MS MARCO Passage	8,841,823	MS MARCO Train	502,939	5.97	1.06
		TREC DL'19	43	5.40	58.16
		TREC DL'20	54	6.04	30.85

τ on the train folds. To ensure that the threshold can be applied for any QPP estimate, we normalize the QPP estimates in the range $[0, 1]$.

TD2F is an unsupervised selective feedback approach that is conceptually similar to QPP-SRF [9]. Rather than using a QPP method, it computes the difference of the term weight distributions across the sets of documents retrieved with the original and the expanded queries, i.e., the sets $L_k(Q)$ and $L_k(\phi_M(Q))$ as per our notations introduced in Sect. 3.1. Formally,

$$\theta(Q) = \frac{1}{|V|} \sum_{t \in V} \log P(t|L_k(Q)) - \log P(t|L_k(\phi_M(Q))), \tag{10}$$

where the set V denotes the vocabulary of the two lists, i.e., $V = V_{L_k} \cup V_{L_k(\phi_M(Q))}$. As per [9], we set the feedback decision threshold τ to a value such that over 95% of the queries satisfy the criterion that $\theta(Q) \leq \tau$. We name this method as Term Distribution Divergence based Feedback, or TD2F for short.

LR-SRF is the only existing supervised method that uses the query features, along with their top-retrieved documents, to predict the PRF decision [25]. The ground-truth labels for learning the decision function is obtained for a training set of queries with existing relevance assessments, i.e. $y(Q) = \mathbb{I}(\text{AP}(\phi_M(Q)) > \text{AP}(Q))$. The method then uses Eq. 4 to train a feature-based logistic regression classifier. In particular, the experiments reported in [25] used the following features for training the logistic regression model: i) the clarity [10] of top-retrieved documents, ii) the absolute divergence between the query model Q and the relevance model [19], iii) the Jensen-Shannon divergence [24] between the language model of the feedback documents, and iv) the clarity of the query language model. We name this method as Regression-based Selective relevance Feedback (LR-SRF).

Proposed Methods. In addition to conducting experiments with our proposed model Deep-SRF-BERT (Fig. 3), we also incorporate confidence-based calibration (as per objective **RQ2**) with rank fusion (Eq. 8 and 9), which we denote by adding the suffix R2F2[1].

---

[1] Implementation available at: https://github.com/suchanadatta/AdaptiveRLM.git.

4.3   Experimental Setup

Dataset and Train-Test Splits. Our retrieval experiments are conducted on a standard ad-hoc IR dataset, the MS MARCO passage collection [30]. The relevance of the passages in the MS MARCO collection are more of personalized in nature. A common practice is to use the TREC DL topic sets, which contains depth-pooled relevance assessments on the passages of the MS MARCO collection. For TREC DL, we conduct experiments on a total of 97 queries from the years 2019 and 2020 [7,8]. Table 1 provides an overview of the dataset.

We use a random sample of 5% of queries (constituting a total of about 40K queries) to train the supervised models in our experiments, whereas evaluation is conducted on the TREC DL (both '19 and '20) query sets. We use a small sample from the training set since the training process requires executing a feedback model (e.g., RLM) for all queries. Therefore, the model needs to learn a task-specific encoding for each query-document pair, both for the original and the expanded queries.

To investigate the generalisation of our selective feedback model, we employ RLM as the feedback approach to train the decision function (Fig. 3). During inference, we employ three different PRF approaches, namely RLM, ColBERT-PRF [41] and GRF [26] to test the effectiveness of selective feedback.

Parameter Settings. A common parameter for all the methods is the number of top-retrieved documents k used for the feedback process and also for training the supervised PRF decision models. For each method we tune the $k \in [5, 40]$ via grid search on the training folds, and use the optimal value on the test fold. We use the same approach to tune the parameter α in Eq. 8, which controls the importance of the feedback process for the rank-based fusion methods. For the R2F2-based methods, we conduct a grid search for α in the set $\{0, 0.1, \ldots, 1\}$. The number of terms used for relevance feedback was tuned for the collection and we use the optimal value across all the methods considered.

To obtain the initial retrieval list, we use both sparse and dense models. As a sparse model, we employ BM25 [33] to retrieve the top-1000 results from MS MARCO collection and a supervised neural model, namely, MonoT5 [31] which operates by reranking the top-1000 of BM25. MonoT5 model was trained on the MS MARCO training queries.

4.4   Results and Discussion

Main Observations. The key findings of our experiments are reported in Table 2. We observe that the accuracy of the decisions is quite satisfactory, even for the unsupervised threshold-based approaches. The results also indicate that more accurate PRF decisions usually lead to an increase in retrieval effectiveness.

For **RQ1**, we find that supervised selective PRF approaches yield improved results over their unsupervised counterparts. Of particular interest is the fact that a data-driven approach (as per our proposal in this paper) outperforms the feature-based approach LR-SRF [25], which answers RQ1 in the affirmative.

Table 2. Comparison of different SRF approaches on the TREC DL (2019 and 2020) topic sets with BM25 and MonoT5 set as the initial retrieval models. MAP values are computed for top-1000 documents. Paired t-test ($p < 0.05$) shows a significant improvement of Deep-SRF over the best performing baselines (comparing bold-faced results with the underlined ones).

		BM25 (ϕ: RLM)			BM25 (ϕ: GRF)			BM25 (ϕ: ColBERT-PRF)		
	Methods	Accuracy	MAP	nDCG@10	Accuracy	MAP	nDCG@10	Accuracy	MAP	nDCG@10
Baselines	No PRF	N/A	0.3766	0.5022	N/A	0.3766	0.5022	N/A	0.3766	0.5022
	PRF	N/A	0.4321	0.5134	N/A	0.4883	0.6226	N/A	0.4514	0.6067
	R2F2	N/A	0.4381	0.5140	N/A	0.5094	0.6332	N/A	0.4968	0.6184
	QPP-SRF	0.7835	0.4400	0.5152	<u>0.7844</u>	<u>0.5321</u>	<u>0.6667</u>	0.7742	0.5238	0.6400
	TD2F	0.7611	0.4392	0.5135	0.7580	0.4579	0.5900	0.7642	0.4910	0.6038
	LR-SRF	<u>0.7842</u>	<u>0.4411</u>	<u>0.5154</u>	0.7784	0.5107	0.6512	<u>0.7854</u>	<u>0.5254</u>	<u>0.6414</u>
Ours	Deep-SRF-BERT	**0.8081***	0.4705	0.5374	**0.8093**$_*$	0.5654	0.6821	**0.8165***	0.5631	0.6765
	Deep-SRF-BERT-R2F2		**0.4961**	**0.5486**		**0.5730**	**0.6839**		**0.5785**	**0.6873**
	Oracle	1.0000	0.5038	0.5528	1.0000	0.5876	0.6941	1.0000	0.5820	0.6936

		MonoT5 (ϕ: RLM)			MonoT5 (ϕ: GRF)			MonoT5 (ϕ: ColBERT-PRF)		
	Methods	Accuracy	MAP	nDCG@10	Accuracy	MAP	nDCG@10	Accuracy	MAP	nDCG@10
Baselines	No PRF	N/A	0.5062	0.6451	N/A	0.5062	0.6451	N/A	0.5062	0.6451
	PRF	N/A	0.5081	0.6463	N/A	0.5200	0.6487	N/A	0.5297	0.6491
	R2F2	N/A	0.5112	0.6484	N/A	0.5241	0.6494	N/A	0.5324	0.6502
	QPP-SRF	<u>0.7963</u>	<u>0.5189</u>	<u>0.6559</u>	0.7871	0.5313	0.6604	0.7900	0.5419	<u>0.6673</u>
	TD2F	0.7789	0.5071	0.6453	0.7670	0.4991	0.6403	0.7612	0.5179	0.5986
	LR-SRF	0.7958	0.5180	0.6543	<u>0.7980</u>	<u>0.5422</u>	<u>0.6628</u>	<u>0.7928</u>	<u>0.5500</u>	0.6654
Ours	Deep-SRF-BERT	**0.8152***	0.5306	0.6640	**0.8160**$_*$	0.5529	0.6694	**0.8067**$_*$	0.5624	0.6733
	Deep-SRF-BERT-R2F2		**0.5317**	**0.6659**		**0.5607**	**0.6719**		**0.5711**	**0.6746**
	Oracle	1.0000	0.5416	0.6786	1.0000	0.5722	0.6803	1.0000	0.5801	0.6821

In relation to **RQ2**, we see that a soft combination of the initial and the feedback lists via a confidence-based calibration (Deep-SRF-BERT-R2F2) improves results further.

An interesting finding is that the SRF decision function trained on RLM on a set of queries generalises well not only for a different set of queries (the test set), but also across different feedback models. This suggests that the queries which improve with RLM also improve with other feedback models, such as GRF or ColBERT-PRF. This can be seen from the GRF and the ColBERT-PRF group of results for both BM25 and MonoT5. This entails that the SRF based decision function does not need to be trained for specific PRF approaches, which makes it more suitable to use in a practical setup, affirming **RQ3**.

We observe that the best results obtained by our method are close to those achieved by an 'oracle'. In the ideal oracle scenario, PRF is applied *only if* the AP of a query is actually improved (i.e., the oracle uses the relevance assessments for the test queries). The fact that the results from Deep-SRF-BERT are close to the oracle suggests that further attempts to increase the accuracy of PRF decisions may have little impact on retrieval effectiveness, likely due to saturation effects.

Table 3. Contingency tables of the Deep-SRF-BERT model with sample queries from TREC DL. Here, $|Q|$ is the count of queries for each of the 4 possible cases of prediction (true/false positives and true/false negatives), and $\overline{\Delta AP}$ denotes the average ΔAP values of each cell, where $\Delta AP(Q) = \frac{AP(\psi_M(Q)) - AP(Q)}{AP(Q)}$.

		Actual			
		$\Delta AP > 0$		$\Delta AP \leq 0$	
Predicted	$\Delta AP > 0$	What is active margin? Exon definition Biology	$|Q| = 59$ $\overline{\Delta AP} = 0.1302$	Why is Pete Rose banned from hall of fame? What are best foods to lower cholesterol?	$|Q| = 8$ $\overline{\Delta AP} = 0.0525$
	$\Delta AP \leq 0$	Define BMT medical Who is Robert Gray?	$|Q| = 11$ $\overline{\Delta AP} = 0.0246$	Do Google docs auto save? How many sons Robert Kraft has?	$|Q| = 19$ $\overline{\Delta AP} = 0.0737$

Per-Query Analysis. Table 3 shows examples of queries from the TREC DL dataset. Firstly, we see that the average differences in the AP values before and after feedback are mostly higher for the green cells, which indicates that the penalty incurred due to queries for which the model (Deep-SRF-BERT) predicts incorrectly is not too high. This also conforms to the fact that at close to 80% accuracy, Deep-SRF-BERT achieves results close to the oracle. Secondly, a manual inspection of the examples reveals that the queries for which the Deep-SRF-BERT model correctly decides to apply PRF appear to be those with under-specified information needs, i.e., those queries which are likely to be benefited from enrichment, e.g., the query 'what is active margin' as seen from Table 3.

5 Conclusions and Future Work

In this paper, we proposed a selective relevance feedback framework that includes a data-driven supervised neural approach to optimize retrieval effectiveness by applying feedback on queries in a selective fashion. By testing this approach using multiple PRF models over sparse and dense architectures, we observed that it performs favorably compared to alternative strategies, approaching the performance of an oracle system.

This work opens the door for interesting future studies. Although our method is effective, it requires executing PRF to gauge result quality. Exploring techniques to determine the necessity of the PRF step could reduce computational costs for queries where PRF is ultimately ignored. Further work could also examine strategies for predicting the parameters of PRF itself, such as the number of relevant documents.

Acknowledgement. The first and the fourth authors were partially supported by Science Foundation Ireland (SFI) grant number SFI/12/RC/2289_P2.

References

1. Bashir, S., Rauber, A.: Improving retrievability of patents with cluster-based pseudo-relevance feedback documents selection. In: Proceedings of CIKM 2009, pp. 1863–1866. ACM, New York (2009)
2. Belkin, N.J., Oddy, R.N., Brooks, H.M.: Ask for information retrieval: Part I. Background and theory. J. Doc. **38**(2), 61–71 (1982)
3. Billerbeck, B., Zobel, J.: Questioning query expansion: an examination of behaviour and parameters. In: Proceedings of 15th Australasian Database Conference - Volume 27, ADC 2004, pp. 69–76. Australian Computer Society Inc, AUS (2004)
4. Cao, G., Nie, J.Y., Gao, J., Robertson, S.: Selecting good expansion terms for pseudo-relevance feedback. In: Proceedings of SIGIR 2008, pp. 243–250. ACM, New York (2008)
5. Cohen, D., Mitra, B., Lesota, O., Rekabsaz, N., Eickhoff, C.: Not all relevance scores are equal: efficient uncertainty and calibration modeling for deep retrieval models. In: Proceedings of SIGIR 2021, pp. 654–664. ACM, New York (2021)
6. Cormack, G.V., Clarke, C.L.A., Buettcher, S.: Reciprocal rank fusion outperforms condorcet and individual rank learning methods. In: Proceedings of SIGIR 2009, pp. 758–759. ACM, New York (2009)
7. Craswell, N., Mitra, B., Yilmaz, E., Campos, D.: Overview of the TREC 2020 deep learning track. In: Proceedings of TREC 2020, vol. 1266. NIST Special Publication (2020)
8. Craswell, N., Mitra, B., Yilmaz, E., Campos, D., Voorhees, E.M.: Overview of the TREC 2019 deep learning track (2019)
9. Cronen-Townsend, S., Zhou, Y., Croft, W.B.: A framework for selective query expansion. In: Proceedings of CIKM 2004, pp. 236–237. ACM, New York (2004)
10. Cronen-Townsend, S., Zhou, Y., Croft, W.B.: Predicting query performance. In: Proceedings of SIGIR 2002, pp. 299–306. ACM, New York (2002)
11. Datta, S., Ganguly, D., Greene, D., Mitra, M.: Deep-QPP: a pairwise interaction-based deep learning model for supervised query performance prediction. In: Proceedings of WSDM 2022, pp. 201–209. ACM, New York (2022)
12. Datta, S., MacAvaney, S., Ganguly, D., Greene, D.: A 'pointwise-query, listwise-document' based query performance prediction approach. In: Proceedings of SIGIR 2022, pp. 2148–2153. ACM, New York (2022)
13. Deveaud, R., Mothe, J., Ullah, M.Z., Nie, J.Y.: Learning to adaptively rank document retrieval system configurations. ACM Trans. Inf. Syst. **37**(1), 1–41 (2018)
14. Ganguly, D., Leveling, J., Jones, G.J.F.: Cross-lingual topical relevance models. In: COLING, pp. 927–942. Indian Institute of Technology Bombay, India (2012)
15. He, B., Ounis, I.: Combining fields for query expansion and adaptive query expansion. Inf. Process. Manage. **43**(5), 1294–1307 (2007)
16. He, B., Ounis, I.: Finding good feedback documents. In: Proceedings of CIKM 2009, pp. 2011–2014. ACM, New York (2009)
17. Jaleel, N.A., et al.: UMass at TREC 2004: novelty and HARD. In: TREC 2004 (2004)
18. Khattab, O., Zaharia, M.: ColBERT: efficient and effective passage search via contextualized late interaction over BERT, pp. 39–48. ACM, New York (2020)
19. Lavrenko, V., Croft, W.B.: Relevance based language models. In: Proceedings of SIGIR 2001, pp. 120–127. ACM, New York (2001)
20. Lee, K.S., Croft, W.B., Allan, J.: A cluster-based resampling method for pseudo-relevance feedback. In: Proceedings of SIGIR 2008, pp. 235–242. ACM, New York (2008)

21. Li, C., et al.: NPRF: a neural pseudo relevance feedback framework for ad-hoc information retrieval. In: Proceedings of EMNLP 2018, Brussels, Belgium, pp. 4482–4491. ACL (2018)
22. Li, H., Mourad, A., Koopman, B., Zuccon, G.: How does feedback signal quality impact effectiveness of pseudo relevance feedback for passage retrieval. In: Proceedings of SIGIR 2022, pp. 2154–2158. ACM, New York (2022)
23. Li, H., et al.: To interpolate or not to interpolate: PRF, dense and sparse retrievers. In: Proceedings of SIGIR 2022, pp. 2495–2500. ACM, New York (2022)
24. Lin, J.: Divergence measures based on the shannon entropy. IEEE Trans. Inf. Theor. **37**(1), 145–151 (2006)
25. Lv, Y., Zhai, C.: Adaptive relevance feedback in information retrieval. In: Proceedings of CIKM 2009, pp. 255–264. ACM, New York (2009)
26. Mackie, I., Chatterjee, S., Dalton, J.: Generative relevance feedback with large language models. In: Proceedings of SIGIR 2023, pp. 2026–2031. ACM, New York (2023)
27. Mitra, M., Singhal, A., Buckley, C.: Improving automatic query expansion. In: Proceedings of SIGIR 1998, pp. 206–214. ACM, New York (1998)
28. Montazeralghaem, A., Zamani, H., Allan, J.: A reinforcement learning framework for relevance feedback. In: Proceedings of SIGIR 2020, pp. 59–68. ACM, New York (2020)
29. Naseri, S., Dalton, J., Yates, A., Allan, J.: CEQE: contextualized embeddings for query expansion. In: Hiemstra, D., Moens, M.F., Mothe, J., Perego, R., Potthast, M., Sebastiani, F. (eds.) ECIR 2021. LNCS, vol. 12656, pp. 467–482. Springer, Cham (2021). https://doi.org/10.1007/978-3-030-72113-8_31
30. Nguyen, T., et al.: MS MARCO: a human generated machine reading comprehension dataset. In: CoCo@NIPS. CEUR Workshop Proceedings, vol. 1773 (2016)
31. Nogueira, R.F., Yang, W., Cho, K., Lin, J.: Multi-stage document ranking with BERT. CoRR abs/1910.14424 (2019)
32. Ogilvie, P., Voorhees, E., Callan, J.: On the number of terms used in automatic query expansion. Inf. Retrieval **12**(6), 666–679 (2009)
33. Robertson, S., Walker, S., Beaulieu, M., Gatford, M., Payne, A.: Okapi at TREC-4 (1996)
34. Rocchio, J.J.: Relevance Feedback in Information Retrieval. Prentice Hall, Englewood Cliffs (1971)
35. Roy, D., Ganguly, D., Mitra, M., Jones, G.J.: Word vector compositionality based relevance feedback using kernel density estimation. In: Proceedings of CIKM 2016, pp. 1281–1290. ACM, New York (2016)
36. Salakhutdinov, R., Mnih, A.: Bayesian probabilistic matrix factorization using Markov chain Monte Carlo. In: Proceedings of ICML 2008, pp. 880–887. ACM, New York (2008)
37. Shtok, A., Kurland, O., Carmel, D.: Using statistical decision theory and relevance models for query-performance prediction. In: Proceedings of SIGIR 2010, pp. 259–266. ACM, New York (2010)
38. Shtok, A., Kurland, O., Carmel, D., Raiber, F., Markovits, G.: Predicting query performance by query-drift estimation. ACM Trans. Inf. Syst. **30**(2), 1–35 (2012)
39. Terra, E., Warren, R.: Poison pills: harmful relevant documents in feedback. In: Proceedings of CIKM 2005, pp. 319–320. ACM, New York (2005)
40. Wang, X., Macdonald, C., Tonellotto, N., Ounis, I.: Pseudo-relevance feedback for multiple representation dense retrieval. In: ICTIR, pp. 297–306. ACM, New York (2021)

41. Wang, X., MacDonald, C., Tonellotto, N., Ounis, I.: ColBERT-PRF: semantic pseudo-relevance feedback for dense passage and document retrieval. ACM Trans. Web **17**(1), 1–39 (2023)
42. Xiong, L., et al.: Approximate nearest neighbor negative contrastive learning for dense text retrieval. In: ICLR (2021)
43. Xu, J., Croft, W.B.: Improving the effectiveness of information retrieval with local context analysis. ACM Trans. Inf. Syst. **18**(1), 79–112 (2000)
44. Yu, H., Xiong, C., Callan, J.: Improving query representations for dense retrieval with pseudo relevance feedback, pp. 3592–3596. ACM, New York (2021)
45. Zamani, H., Dadashkarimi, J., Shakery, A., Croft, W.B.: Pseudo-relevance feedback based on matrix factorization. In: Proceedings CIKM 2016, pp. 1483–1492. ACM, New York (2016)
46. Zheng, Z., Hui, K., He, B., Han, X., Sun, L., Yates, A.: BERT-QE: contextualized query expansion for document re-ranking. In: Findings of the ACL: EMNLP 2020, pp. 4718–4728. ACL (2020)
47. Zhou, Y., Croft, W.B.: Query performance prediction in web search environments. In: Proceedings of SIGIR 2007, pp. 543–550. ACM, New York (2007)
48. Zhuang, S., Li, H., Zuccon, G.: Implicit feedback for dense passage retrieval: a counterfactual approach. In: Proceedings of SIGIR 2022, pp. 18–28. ACM, New York (2022)

Investigating the Robustness of Sequential Recommender Systems Against Training Data Perturbations

Filippo Betello[1]([✉]) [iD], Federico Siciliano[1] [iD], Pushkar Mishra[2] [iD], and Fabrizio Silvestri[1] [iD]

[1] Sapienza University of Rome, Rome, Italy
{betello,siciliano,fsilvestri}@diag.uniroma1.it
[2] AI at Meta, London, UK
pushkarmishra@meta.com

Abstract. Sequential Recommender Systems (SRSs) are widely employed to model user behavior over time. However, their robustness in the face of perturbations in training data remains a largely understudied yet critical issue. A fundamental challenge emerges in previous studies aimed at assessing the robustness of SRSs: the Rank-Biased Overlap (RBO) similarity is not particularly suited for this task as it is designed for infinite rankings of items and thus shows limitations in real-world scenarios. For instance, it fails to achieve a perfect score of 1 for two identical finite-length rankings. To address this challenge, we introduce a novel contribution: Finite Rank-Biased Overlap (FRBO), an enhanced similarity tailored explicitly for finite rankings. This innovation facilitates a more intuitive evaluation in practical settings. In pursuit of our goal, we empirically investigate the impact of removing items at different positions within a temporally ordered sequence. We evaluate two distinct SRS models across multiple datasets, measuring their performance using metrics such as Normalized Discounted Cumulative Gain (NDCG) and Rank List Sensitivity. Our results demonstrate that removing items at the end of the sequence has a statistically significant impact on performance, with NDCG decreasing up to 60%. Conversely, removing items from the beginning or middle has no significant effect. These findings underscore the criticality of the position of perturbed items in the training data. As we spotlight the vulnerabilities inherent in current SRSs, we fervently advocate for intensified research efforts to fortify their robustness against adversarial perturbations. Code is available at https://github.com/siciliano-diag/finite_rank_biased_rbo.git.

Keywords: Recommender Systems · Evaluation of Recommender Systems · Model Stability · Input Data Perturbation

This work was partially supported by projects FAIR (PE0000013) and SERICS (PE00000014) under the MUR National Recovery and Resilience Plan funded by the European Union - NextGenerationEU. Supported also by the ERC Advanced Grant 788893 AMDROMA, EC H2020RIA project "SoBigData++" (871042), PNRR MUR project IR0000013-SoBigData.it. This work has been supported by the project NEREO (Neural Reasoning over Open Data) project funded by the Italian Ministry of Education and Research (PRIN) Grant no. 2022AEFHAZ.

N. Goharian et al. (Eds.): ECIR 2024, LNCS 14609, pp. 205–220, 2024.
https://doi.org/10.1007/978-3-031-56060-6_14

1 Introduction

Recommender systems have become ubiquitous in our daily lives [1], playing a key role in helping users navigate the vast amounts of information available online. Thanks to the global spread of e-commerce services, social media and streaming platforms, recommender systems have become increasingly important for personalized content delivery and user engagement [45]. In recent years, Sequential Recommender Systems (SRSs) have emerged as a popular approach to modeling user behavior over time [29], leveraging the temporal dependencies in users' interaction sequences to make more accurate predictions.

However, despite their success, the robustness of SRSs against perturbations in the training data remains an open research question [22]. In real-world scenarios, disruptions may occur when users employ different services for the same purpose. Data becomes fragmented and divided between a service provider and its competitors in such cases. Nevertheless, the provider must train a recommender system with such incomplete data while ensuring robustness to perturbations. This challenge is accentuated when we scrutinize previous attempts [24] to assess the robustness of SRSs: Rank-Biased Overlap (RBO) [41], designed explicitly for infinite lists, reveals limitations when applied to real-world scenarios.

Our experiments revolve around the following research questions:

- **RQ1:** Do changes in the training seed heavily impact rankings?
- **RQ2:** How does the type of removal influence the model's performance?
- **RQ3:** Does more item removed significantly decrease in performance?

The contribution of this study is two-fold. Firstly, we propose the novel Finite Rank-Biased Overlap (FRBO) measure. Unlike RBO, which is only suited for infinite rankings, FRBO is specifically designed to assess the robustness of SRSs within finite ranking scenarios, aligning seamlessly with real-world settings. Secondly, we empirically assess the impact of item removal from user interaction sequences on SRS performance. Our investigation shows that the most recent user interaction sequence items are critical for accurate recommendation performance. When these items are removed, there is a significant drop in all metrics.

2 Related Works

2.1 Sequential Recommender Systems

SRSs are algorithms that leverage a user's past interactions with items to make personalized recommendations over time and they have been widely used in various applications, including e-commerce [18,31], social media [3,14], and music streaming platforms [2,32,33]. Compared to traditional recommender systems, SRSs consider the order and timing of these interactions, allowing for more accurate predictions of a user's preferences and behaviors [39]. Various techniques have been proposed to implement SRSs. Initially, Markov Chain models were employed in Sequential Recommendation [12,13], but they struggle to capture complex dependencies in long-term sequences. In recent years, Recurrent Neural Networks (RNNs) have emerged as one of the most promising approaches in this field [9,16,30]. These methods encode the users' historical preferences into a vector that gets updated at each time step and is used to predict the next item in the sequence. Despite their success, RNNs may face challenges dealing with long-term dependencies and generating diverse recommendations. Another approach comes from the use of the attention mechanism [38]: two different examples are SASRec [20] and BERT4Rec [36] architectures. This method dynamically weighs the importance of different sequence parts to better capture the important features and improve the prediction accuracy. Recently, Graph Neural Networks have become popular in the field of recommendations system [28,43], especially in the sequential domain [7,11].

2.2 Robustness in Sequential Recommender Systems

Robustness is an important aspect of SRSs as they are vulnerable to noisy and incomplete data. Surveys on the robustness of recommender systems [6,17] discussed the challenges in developing robust recommender systems and presented various techniques for improving the robustness of recommender systems.

Many tools have been developed to test the robustness of different algorithms: [25], provided both a formalisation of the topic and a framework; the same was done more recently by [26], who developed a toolkit called RGRecSys which provides a unified framework for evaluating the robustness of SRSs.

Few recent work has focused on studying the problem and trying to increase robustness: [37] focuses on robustness in the sense of training stability while [24] investigated the robustness of SRSs to interaction perturbations. They showed that even small perturbations in user-item interactions can lead to significant changes in the recommendations. They proposed Rank List Sensitivity (RLS), a measure of the stability of rankings produced by recommender systems.

Our work expands these, making a more accurate investigation of the effect different types of perturbation can have on the models' performance. While [24] perturbs a single interaction in the whole dataset, we perturb the sequences of all users and analyze the performance as the number of perturbations changes. We also provide a theoretical contribution in a new sensitivity evaluation measure for finite rankings, presented in Sect. 3.2.

3 Methodology

3.1 Setting

In Sequential Recommendation, each user u is represented by a temporally ordered sequence of items $S_u = (I_1, I_2, ..., I_j, ..., I_{L_u-1}, I_{L_u})$ with which it has interacted, where L_u is the length of the sequence for user u.

User-object interactions in real-world scenarios are often fragmented across services, resulting in a lack of comprehensive data. For example, in the domains of movies and TV shows, a single user may interact with content on TV, in a movie theater, or across multiple streaming platforms. To mimic this real-world scenario in our training data perturbations, we considered three different cases, each removing n items at a specific position in the sequence:

– **Beginning:** $S_u = (I_{n+1}, \ldots, I_{L_u-1})$. This represents a user who signs up for a new service, so all his past interactions, i.e., those at the beginning of the complete sequence, were performed on other services.
– **Middle:** $S_u = (I_1, \ldots, I_{\lfloor \frac{L_u-1-n}{2} \rfloor}, I_{\lfloor \frac{L_u-1+n}{2} \rfloor} \ldots, I_{L_u-1})$. This represents a user who takes a break from using the service for a certain period and resumes using it. Still, any interactions they had during the considered period are not available to the service provider.
– **End:** $S_u = (I_1, \ldots, I_{L_u-1-N})$. This represents a user who has stopped using the service, so the service provider loses all the subsequent user interactions. The service provider still has an interest in winning the user back through their platform or other means, such as advertising. Thus, it is essential to have a robust model to continue providing relevant items to the user.

with $n \in \{1, 2, \ldots, 10\}$. In practice, the data is first separated into training, validation and test set (always composed by I_{L_u}). Subsequently, only the training data are perturbed, with a methodology dependent on the scenario considered and the model is then trained on these. The models, trained on data perturbed in a different manner, are therefore always tested on the same data.

3.2   Metrics

To evaluate the performance of the models, we employ traditional evaluation metrics used for Sequential Recommendation: Precision, Recall, MRR and NDCG.

Moreover, to investigate the stability of the recommendation models, we employ the Rank List Sensitivity (RLS) [24]: it compares two lists of rankings $\mathcal{X}$ and $\mathcal{Y}$, one derived from the model trained under standard conditions and the other derived from the model trained with perturbed data.

Therefore, having these two rankings, and a similarity function sim between them, we can formalize the RLS measure as:

$$\mathbf{RLS} = \frac{1}{|\mathcal{X}|} \sum_{k=1}^{|\mathcal{X}|} \mathrm{sim}(R^{X_k}, R^{Y_k}) \tag{1}$$

where X_k and Y_k represent the k-th ranking inside $\mathcal{X}$ and $\mathcal{Y}$ respectively.

RLS's similarity measure can be chosen from two possible options:

– **Jaccard Similarity (JAC)** [19] is a normalized measure of the similarity of the contents of two sets. A model is stable if its Jaccard score is close to 1.

$$\mathbf{JAC(X, Y)} = \frac{|X \cap Y|}{|X \cup Y|} \tag{2}$$

- **Rank-Biased Overlap (RBO)** [41] measures the similarity of orderings between two rank lists. Higher values indicate that the items in the two lists are arranged similarly:

$$\textbf{RBO}(\textbf{X}, \textbf{Y}) = (1 - p) \sum_{d=1}^{+\infty} p^{d-1} \frac{|X[1 : d] \cap Y[1 : d]|}{d} \tag{3}$$

In the domain of recommendation systems, it is customary to compute metrics using finite-length rankings, typically denoted by appending "@k" to the metric's name, such as NDCG@k. While traditional metrics (e.g. NDCG, MRR, etc.) readily adapt to finite-length rankings, maintaining their core meaning, the same behaviour does not extend to RLS when employing RBO. The reason lies in Eq. 3, which exhibits a notable limitation: it fails to converge to one, even when applied to identical finite-length lists. To overcome this limitation, we introduce the Finite Rank-Biased Overlap (FRBO) similarity, denoted as FRBO@k, which represents a novel formulation engineered to ensure convergence to a value of 1 for identical lists and a value of 0 for entirely dissimilar lists.

Theorem 1. *Given a set of items* $I = \{I_1, ..., I_{N_I}\}$, *two rankings* $X = (x_1, ..., x_k)$ *and* $Y = (y_1, ..., y_k)$, *such that* $x_i, y_i \in I$, *and* $k \in \mathbb{N}^+$

$$\textbf{FRBO}(\textbf{X}, \textbf{Y})@\textbf{k} = \frac{\text{RBO}(X, Y)@k - min_{X,Y} \text{RBO}@k}{max_{X,Y} \text{RBO}@k - min_{X,Y} \text{RBO}@k} \tag{4}$$

$$\min_{X,Y} \text{FRBO}(X, Y)@k = 0, \ \max_{X,Y} \text{FRBO}(X, Y)@k = 1$$

Proof. This follows from the fact that given a function $f : A \to [a, b]$ and another function $g = \frac{f-a}{b-a}$, then $g : A \to [0, 1]$, where A is any set. □

To normalize RBO, we need to identify its minimum and maximum values when the summation is carried out up to the top-k elements of the ranking, simultaneously proving that these values are not naturally constrained to be 0 and 1.

Lemma 1. *Given a set of items* $I = \{I_1, ..., I_{N_I}\}$, *two rankings* $X = (x_1, ..., x_k)$ *and* $Y = (y_1, ..., y_k)$, *such that* $x_i, y_i \in I$, *and* $k \in \mathbb{N}^+$, *the following holds:*

$$\min_{X,Y} \text{RBO}@k = \begin{cases} 0, & \text{if } k \leq \lfloor \frac{N_I}{2} \rfloor \\ (1 - p) \left(2 \frac{p^{\lfloor \frac{N_I}{2} \rfloor} - p^{N_I}}{1-p} - N_I \ell \right) & \text{otherwise} \end{cases} \tag{5}$$

where $\text{RBO}(X, Y)@k = (1 - p) \sum_{d=1}^{k} p^{d-1} \frac{|X[1 : d] \cap Y[1 : d]|}{d}$

and $\ell = p^{\lfloor \frac{N_I}{2} \rfloor} \Phi(p, 1, \lfloor \frac{N_I}{2} \rfloor + 1) - p^{N_I} \Phi(p, 1, N_I + 1)$

Proof. For the first part, it suffices to consider two rankings X, Y that share no common elements, so that $|X[1:d] \cap Y[1:d]| = 0$. This leads to:

$$\min_{X,Y} \text{RBO@k} = (1-p) \sum_{d=1}^{k} p^{d-1} \frac{0}{d} = 0$$

However, since the number of items I is not infinite, if $k > \lfloor \frac{N_I}{2} \rfloor$, at least one element must necessarily be in common between the two rankings:

$$|X[1:k] \cap Y[1:k]| = 0 \iff |X[1:k] \cup Y[1:k]| = 2k \le N_I \iff k \le \lfloor \frac{N_I}{2} \rfloor$$

Consequently, the similarity can't assume a value of 0 if $k > \lfloor \frac{N_I}{2} \rfloor$.

Given that $\text{RBO}(X,Y)@k+1 \ge \text{RBO}(X,Y)@k \,\forall k \in \mathbb{N}^+$, similarity it's minimized when intersections between the two rankings occur as far down the list as possible. When $d > \lfloor \frac{N_I}{2} \rfloor$, there are at least $2d - N_I$ intersections, because items in d-th position in one ranking are necessarily contained in the other.

$$\min_{X,Y} \text{RBO@k} = (1-p) \left(0 + \sum_{d=\lfloor \frac{N_I}{2} \rfloor+1}^{N_I} p^{d-1} \frac{2d - N_I}{d} \right)$$

$$= (1-p) \left(2 \sum_{d=\lfloor \frac{N_I}{2} \rfloor+1}^{N_I} p^{d-1} - N_I \sum_{d=\lfloor \frac{N_I}{2} \rfloor+1}^{N_I} \frac{p^{d-1}}{d} \right)$$

The first series can be regarded as a finite geometric series, for which we can apply the formula for the sum of a geometric series: $\sum_{n=1}^{k} ar^{n-1} = \frac{a(1-r^k)}{1-r}$:

$$\sum_{d=\lfloor \frac{N_I}{2} \rfloor+1}^{N_I} p^{d-1} = p^{\lfloor \frac{N_I}{2} \rfloor} \sum_{d=1}^{N_I - \lfloor \frac{N_I}{2} \rfloor} p^{d-1} = \frac{p^{\lfloor \frac{N_I}{2} \rfloor} - p^{N_I}}{1-p}$$

Using Lerch transcendent function $\Phi(z,s,\alpha) = \sum_{n=0}^{+\infty} \frac{z^n}{(n+\alpha)^s}$ in second series:

$$\sum_{d=\lfloor \frac{N_I}{2} \rfloor+1}^{N_I} \frac{p^{d-1}}{d} = \sum_{d=0}^{N_I - \lfloor \frac{N_I}{2} \rfloor - 1} \frac{p^{d+\lfloor \frac{N_I}{2} \rfloor}}{d + \lfloor \frac{N_I}{2} \rfloor + 1}$$

$$= \sum_{d=0}^{+\infty} \frac{p^{d+\lfloor \frac{N_I}{2} \rfloor}}{d + \lfloor \frac{N_I}{2} \rfloor + 1} - \sum_{d=N_I - \lfloor \frac{N_I}{2} \rfloor}^{+\infty} \frac{p^{d+\lfloor \frac{N_I}{2} \rfloor}}{d + \lfloor \frac{N_I}{2} \rfloor + 1}$$

$$= p^{\lfloor \frac{N_I}{2} \rfloor} \Phi(p, 1, \lfloor \frac{N_I}{2} \rfloor + 1) - p^{N_I} \Phi(p, 1, N_I + 1)$$

$\square$

Lemma 2. *Given a set of items* $I = \{I_1, ..., I_{N_I}\}$, *two rankings* $X = (x_1, ..., x_k)$ *and* $Y = (y_1, ..., y_k)$, *such that* $x_i, y_i \in I$, *and* $k \in \mathbb{N}^+$, *the following holds:*

$$\max_{X,Y} \text{RBO@k} = 1 - p^k \tag{6}$$

Proof. RBO@k reaches its maximum value when the two rankings are identical:

$$|X[1:d] \cap Y[1:d]| = d \ \forall d \in \{1, ..., k\}$$

Referring to the result for the geometric series, we can compute:

$$max_{X,Y} \text{RBO@k} = (1-p) \sum_{d=1}^{k} p^{d-1} \frac{d}{d} = (1-p) \sum_{d=1}^{k} p^{d-1} = 1 - p^k$$

$\square$

Corollary 1. *Given a set of items* $I = \{I_1, ..., I_{N_I}\}$ *and* $k \in \mathbb{N}^+$:

$$\min_{X,Y} \text{RBO}(X,Y)@k \neq 0, \ \max_{X,Y} \text{RBO}(X,Y)@k \neq 1$$

$$\exists X, Y \in \{(x_1, ..., x_k) | x_i \in I \wedge x_i \neq x_j \forall i \neq j\}$$

$$s.t. \ \text{RBO}(X,Y)@k \neq \text{FRBO}(X,Y)@k$$

Proof. This follows from Theorems 1 and 2. $\square$

It's worth considering that in real-world scenarios, the number of possible items N_I is significantly larger compared to the length k of the rankings used to compute the metrics, i.e., $N_I >> k$. In this context, we can safely omit the minimum value from Eq. 4, resulting in:

$$\textbf{FRBO(X,Y)@k} = \frac{\text{RBO}(X,Y)@k}{\max_{X,Y} \text{RBO@k}} = \frac{1-p}{1-p^k} \sum_{d=1}^{k} p^{d-1} \frac{|X[1:d] \cap Y[1:d]|}{d}$$

In this section, we have shown that RBO is not an adequate similarity score when dealing with finite-length rankings. So, we have derived expressions that quantify the minimum and maximum values of RBO, allowing us to compute a normalized version of RBO.

Kendall's Tau [21] assumes that two rankings contain precisely the same items. However, this assumption may not hold for finite top-k ranked lists. In addition, average overlap [10,42] has a peculiar property of monotonicity in depth, where greater agreement with a deeper ranking does not necessarily lead to a higher score, and less agreement does not necessarily lead to a lower score [41].

4   Experiments

4.1   Datasets

We use four different datasets:

MovieLens [15] → This benchmark dataset is often used to test recommender systems. In this work, we use the 100K version and 1M version.

Foursquare [44] → This dataset contains check-ins from New York City and Tokyo collected over a period of approximately ten months.

The statistics for all the datasets are shown in Table 1.

We select datasets widely used in the literature and with a high number of interactions per user. The limitation in dataset selection arises from our intention to assess the robustness against the removal of up to 10 elements. Therefore, the dataset must satisfy the following constraint: $L_u > 10 \quad \forall u \in U$, where L_u is the number of interactions of user u, i.e. the length of the sequence S_u of interactions, and U is the set of all users in the dataset. If the condition is not met, we delete all the items for a user with less than ten interactions. In this case, we cannot train the model on this particular user. We have, thus, decided to exclude datasets such as Amazon [23] for they cannot meet the previous criteria.

Table 1. Dataset statistics after preprocessing

Dataset	Users	Items	Interactions	Average $\frac{\text{Actions}}{\text{User}}$	Median $\frac{\text{Actions}}{\text{User}}$
MovieLens 1M	6040	3952	1M	165	96
MovieLens 100K	943	1682	100K	106	65
Foursquare Tokyo	2293	61858	537703	250	153
Foursquare New York	1083	38333	227428	210	173

4.2   Architectures

In our study, we use two different architectures to validate the results:

- **SASRec** [20] uses self-attention processes to determine the importance of each interaction between the user and the item.
- **GRU4Rec** [16] is a recurrent neural network architecture that uses gated recurrent units (GRUs) [8] to improve the accuracy of the prediction.

We choose to use these two models because both have demonstrated excellent performance in several benchmarks and have been widely cited in the literature. Furthermore, as one model employs attention while the other utilizes RNN, their network functioning differs, which makes evaluating their behavior under training perturbations particularly interesting. We use the models' implementation provided by the RecBole Python library [46], with their default hyperparameters.

Table 2. Variation of metrics between two seeds. Metrics for the 4 datasets considered for GRU4Rec and SASRec. For Precision (Prec.), Recall, MRR and NDCG, it is shown the percentage variation between the obtained performance using two different initialization seeds for the models. For each metric, the value corresponding to the dataset where a model is less robust is highlighted in **bold**. For each dataset, the value corresponding to the model that is the least robust of the two given a metric is underlined.

	SASRec						GRU4Rec					
	Prec	Recall	MRR	NDCG	FRBO	JAC	Prec	Recall	MRR	NDCG	FRBO	JAC
ML 100k	0.5%	0.5%	0.5%	<u>0.3%</u>	.466	.489	**<u>1.7%</u>**	**<u>1.7%</u>**	<u>1.2%</u>	0.1%	<u>.337</u>	<u>.413</u>
ML 1M	<u>0.3%</u>	<u>0.4%</u>	**0.5%**	**0.5%**	.549	.569	0.2%	0.2%	<u>3.5%</u>	<u>2.5%</u>	<u>.311</u>	<u>.347</u>
FS NYC	<u>0.9%</u>	<u>0.9%</u>	0.0%	<u>0.3%</u>	**.398**	**.273**	0.1%	0.1%	<u>0.6%</u>	0.1%	**<u>.110</u>**	**<u>.083</u>**
FS TKY	**<u>1.4%</u>**	**<u>1.4%</u>**	0.4%	0.02%	.418	.267	0.8%	0.8%	**<u>4.7%</u>**	**<u>3.4%</u>**	<u>.210</u>	<u>.165</u>

4.3 Experimental Setup

All the experiments are performed on a single NVIDIA RTX 3090. The batch size is fixed to 4096. Adam optimizer is used with a fixed learning rate of $5 * 10^{-4}$. The number of epochs is set to 300, but in order to avoid overfitting, we stop the training when the NDCG@20 does not improve for 50 epochs. The average duration of each run is 1.5 h. The RecBole library was utilized for conducting all the experiments, encompassing data preprocessing, model configuration, training, and testing. This comprehensive library ensures the reproducibility of the entire process. All the evaluation metrics are calculated with a cut-off K of 20. To validate the effective degradation in performance and rankings similarity, we employed the paired Student's t-test [35], after testing normality of distributions with Shapiro-Wilk test [34], and a significance level of 10^{-3}. The code required to replicate the experiments is accessible in our GitHub repository[1].

5 Results

5.1 Intrinsic Models Instability (RQ1)

To measure the inherent robustness of the models, i.e. in the baseline case (without removal of items), we train the model twice using different initialization seeds and compute the percentage discrepancy between the Precision, Recall, MRR and NDCG obtained by the two rankings. The results are shown in Table 2: it can be seen that in general the discrepancy is negligible, almost always less than 1%. On the other hand, the two RLS, calculated using FRBO and Jaccard respectively, show us the similarity between the two rankings produced with different initialization seeds. The results deviate significantly from the ideal value of 1, indicating considerably different rankings. These combined results indicate

---

[1] https://github.com/siciliano-diag/finite_rank_biased_rbo.git.

to us that the models converge to an adequate performance beyond the initialization seed, but that the actual rankings produced are heavily influenced by it. The **bold** represent the dataset with the least robust result for each metric. No datasets stands out significantly, yet Foursquare Tokyo seems to give more problems regarding standard evaluation metrics, while for Foursquare New York City it seems more difficult to produce stable rankings. The underlined values instead compare, for each metric and dataset, which of the two models is the least robust. If we consider metrics that do not consider the position of the positive item in the ranking, i.e. Precision and Recall, GRU4Rec seems more robust than SASRec. If, on the other hand, we look at metrics that penalize relevant items in positions too low in the ranking, we see that the opposite happens. This suggests to us that GRU4Rec can return a better set of results, but in a less relevant order than SASRec does. As proof of this, if we check the RLS values, we see that GRU4Rec is always the least robust model as the initialization seed changes.

5.2   Comparison of the Position of Removal (RQ2)

Table 3 compares performance and stability when ten items are removed from the sequence versus retaining all items (reference value) in the training set with a consistent initialization seed. It's observed that discarding items from the beginning or middle of the sequence does not significantly impair the model's performance; only a minor decline is noted, potentially attributable to the marginally reduced volume of total training data. Instead, it can be observed how removing items from the end of the sequence leads to a drastic reduction in metrics: in the case of SASRec applied to the MovieLens 1M dataset, the NDCG more than halves. Finally, we can see how the difference between the three settings, although maintaining the same trend, is less marked for GRU4Rec applied to the Foursquare NYC dataset. This may be due to the generally higher performance of GRU4Rec [20]. A more in-depth analysis is presented in Sect. 5.4.

Table 3 also shows the RLS values, computed using FRBO and Jaccard similarity, on the same model-dataset pairs: removing items at the end of the sequence leads to considerable variation in the rankings produced by the models. The values approach 0, meaning that the produced rankings share almost no items. Our results are in contrast to those of [24], which instead claim that an initial perturbation of a user sequence leads to a higher impact on the RLS. However, their experimental setting is different than ours, as explained in Sect. 2.

5.3   Effect of the Number of Elements Removed (RQ3)

As we discussed in the previous section, removing elements that are at the beginning or in the middle of the temporally ordered sequence has no effect on performance. This is also confirmed by Fig. 1, where we can also see, however, that for the above-mentioned cases there is no variation as the number of removed elements increases. The deviation of the RLS displayed in Fig. 1d will be analyzed in more detail in Sect. 5.4. For the remaining setting, the one where we remove

Table 3. Variations in metrics for ten-item removal. Metrics for the 3 scenarios considered for SASRec on ML-1M and GRU4Rec on FS-NYC. For Precision, Recall, MRR, and NDCG, it is shown the percentage variation between removing ten items and the reference value. For each metric, in **bold** it is highlighted the value representing the less robust model. † indicates a statistically significant result.

Model	Removal	Prec.	Recall	MRR	NDCG	FRBO	JAC
SASRec ML-1M	Beginning	−0.23%	−0.23%	−0.15%	−0.07%	.399†	.368†
	Middle	−0.35%	−0.29%	−1.09%	−0.73%	.385†	.356†
	End	**−15.5%**†	**−15.3%**†	**−45.9%**†	**−56.0%**†	.080†	.106†
GRU4Rec FS-NYC	Beginning	−0.23%	−0.23%	−1.47%	−0.94%	.105†	.075†
	Middle	−0.93%	−0.93%	−0.18%	−0.44%	.110†	.074†
	End	**−4.92%**†	**−4.86%**†	**−8.42%**†	**−7.39%**†	.089†	.062†

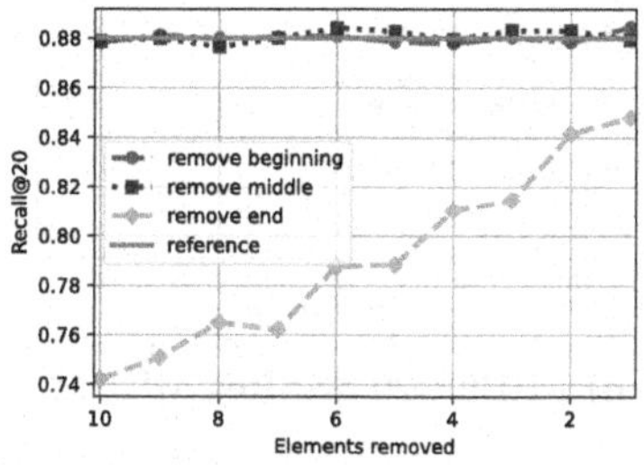

(a) HR@20 SASRec ML-100k

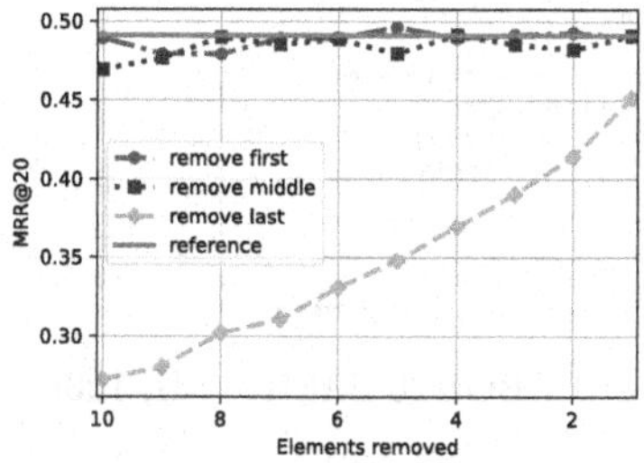

(b) MRR GRU4Rec ML-1M

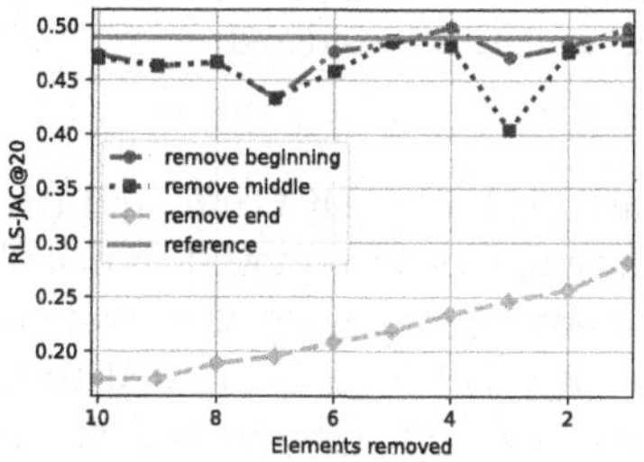

(c) RLS-JAC@20 SASRec ML-100k

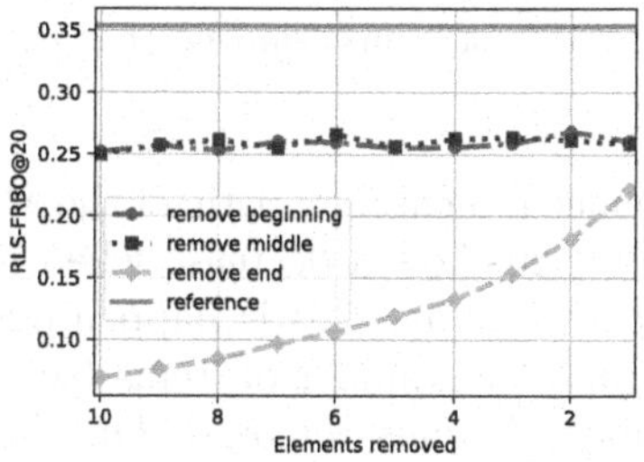

(d) RLS-FRBO@20 GRU4Rec ML-1M

Fig. 1. Plots of various metrics for the ML-100k and ML-1M datasets as the number of removed elements increases. The baseline is shown as a horizontal solid line, while dashed lines show the metrics as the number of items removed changes for the three scenarios considered.

items at the end of the sequence, the effect of the number of items removed is evident: the metrics drop drastically as the number of items removed increases. This result holds true for both models considered and for all four datasets tested.

5.4 Differences Between the Datasets

Figures 2 and 3 show the performance for the three different settings for the SASRec model applied to all datasets. From Figs. 2a, 2b, 2c, 2d we see that the downward trend of the metric when removing items at the end of the

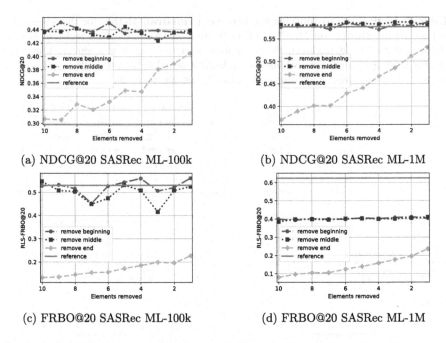

(a) NDCG@20 SASRec ML-100k

(b) NDCG@20 SASRec ML-1M

(c) FRBO@20 SASRec ML-100k

(d) FRBO@20 SASRec ML-1M

Fig. 2. Plots of NDCG and FRBO for SASRec on the ML-100K and ML-1M datasets. The baseline is shown as a horizontal solid line, while dashed lines show the metrics as the number of items removed changes for the three scenarios considered.

sequence is a characteristic of the MovieLens dataset: both NDCG@20 and RLS-FRBO@20 show a decrease when increasing the number of removed items. We hypothesize that this is happening because the average number of actions per user and the number of items (see Table 1) are not that large compared to the number of items removed. On the other hand, the Foursquare datasets (3a, 3b, 3c, 3d) do not suffer major performance degradation, probably due to the higher average number of actions per user and the number of items (see Table 1) than MovieLens. In addition to this, and probably for the same motivation, the degradation of the RLS is lower with respect to that displayed in the MovieLens datasets. Finally, it is interesting to note that on MovieLens 1M, there is a consistent performance degradation even when elements at the beginning and in the middle of the temporally ordered sequence are removed. This can be observed as a small decrease in the NDCG@20 (Fig. 2b), but a sharp decrease in the value of the RLS-FRBO (Figs. 1d, 2d). This means that even if the model performs approximately the same, the rankings produced vary greatly. The cause may be the fact that the MovieLens 1M dataset, among those considered, has the largest number of users and interactions.

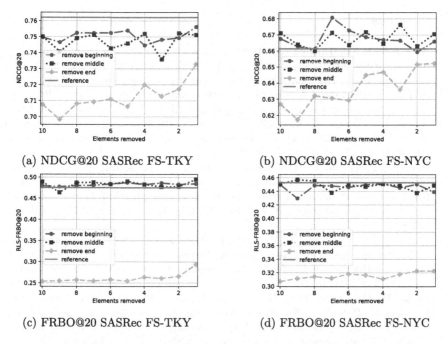

(a) NDCG@20 SASRec FS-TKY   (b) NDCG@20 SASRec FS-NYC

(c) FRBO@20 SASRec FS-TKY   (d) FRBO@20 SASRec FS-NYC

Fig. 3. Plots of NDCG and FRBO for SASRec on FS-TKY and FS-NYC datasets. The baseline is shown as a horizontal solid line, while dashed lines show the metrics as the number of items removed changes for the three scenarios considered.

6 Conclusion

In this work, we have analyzed the importance of the position of items in a temporally ordered sequence for training SRSs. For this purpose, we introduced Finite RBO, a version of RBO for finite-length ranking lists and proved its normalization in [0,1]. Our results demonstrate the importance of the most recent elements in users' interaction sequence: when these items are removed from the training data, there is a significant drop in all evaluation metrics for all case studies investigated and this reduction is proportional to the number of elements removed. Conversely, this reduction is not as pronounced when elements at the beginning and middle of the sequence are removed. We validated our hypothesis using four different datasets and two different models, using traditional evaluation metrics such as NDCG, Recall, but also RLS, a measure specifically designed to measure Sensitivity. Future work in this direction could first extend our results to more models and more datasets, and then investigate a way to make the models robust to the removal of training data. We hypothesize that the solution may lie in using different training strategies [27], robust loss functions [5,40], or different optimization objectives [4].

References

1. Adomavicius, G., Tuzhilin, A.: Toward the next generation of recommender systems: a survey of the state-of-the-art and possible extensions. IEEE Trans. Knowl. Data Eng. **17**(6), 734–749 (2005). https://doi.org/10.1109/TKDE.2005.99
2. Afchar, D., Melchiorre, A., Schedl, M., Hennequin, R., Epure, E., Moussallam, M.: Explainability in music recommender systems. AI Mag. **43**(2), 190–208 (2022)
3. Amato, F., Moscato, V., Picariello, A., Sperlí, G.: Recommendation in social media networks. In: 2017 IEEE Third International Conference on Multimedia Big Data (BigMM), pp. 213–216. IEEE (2017)
4. Bacciu, A., Siciliano, F., Tonellotto, N., Silvestri, F.: Integrating item relevance in training loss for sequential recommender systems. In: Proceedings of the 17th ACM Conference on Recommender Systems, RecSys 2023, pp. 1114–1119. Association for Computing Machinery, New York (2023). https://doi.org/10.1145/3604915.3610643
5. Bucarelli, M.S., Cassano, L., Siciliano, F., Mantrach, A., Silvestri, F.: Leveraging inter-rater agreement for classification in the presence of noisy labels. In: Proceedings of the IEEE/CVF Conference on Computer Vision and Pattern Recognition, pp. 3439–3448 (2023)
6. Burke, R., O'Mahony, M.P., Hurley, N.J.: Robust collaborative recommendation. In: Ricci, F., Rokach, L., Shapira, B. (eds.) Recommender Systems Handbook, pp. 961–995. Springer, Boston (2015). https://doi.org/10.1007/978-1-4899-7637-6_28
7. Chang, J., et al.: Sequential recommendation with graph neural networks. In: Proceedings of the 44th International ACM SIGIR Conference on Research and Development in Information Retrieval, SIGIR 2021, pp. 378–387. Association for Computing Machinery, New York (2021). https://doi.org/10.1145/3404835.3462968
8. Cho, K., et al.: Learning phrase representations using RNN encoder-decoder for statistical machine translation (2014)
9. Donkers, T., Loepp, B., Ziegler, J.: Sequential user-based recurrent neural network recommendations. In: Proceedings of the Eleventh ACM Conference on Recommender Systems, RecSys 2017, pp. 152–160. Association for Computing Machinery, New York (2017). https://doi.org/10.1145/3109859.3109877
10. Fagin, R., Kumar, R., Sivakumar, D.: Comparing top k lists. SIAM J. Discret. Math. **17**(1), 134–160 (2003)
11. Fan, Z., Liu, Z., Zhang, J., Xiong, Y., Zheng, L., Yu, P.S.: Continuous-time sequential recommendation with temporal graph collaborative transformer. In: Proceedings of the 30th ACM International Conference on Information & Knowledge Management, CIKM 2021, pp. 433–442. Association for Computing Machinery, New York (2021). https://doi.org/10.1145/3459637.3482242
12. Fouss, F., Faulkner, S., Kolp, M., Pirotte, A., Saerens, M., et al.: Web recommendation system based on a Markov-chainmodel. In: ICEIS (4), pp. 56–63 (2005)
13. Fouss, F., Pirotte, A., Saerens, M.: A novel way of computing similarities between nodes of a graph, with application to collaborative recommendation. In: The 2005 IEEE/WIC/ACM International Conference on Web Intelligence (WI 2005), pp. 550–556. IEEE (2005)
14. Guy, I., Zwerdling, N., Ronen, I., Carmel, D., Uziel, E.: Social media recommendation based on people and tags. In: Proceedings of the 33rd International ACM SIGIR Conference on Research and Development in Information Retrieval, SIGIR 2010, pp. 194–201. Association for Computing Machinery, New York (2010). https://doi.org/10.1145/1835449.1835484

15. Harper, F.M., Konstan, J.A.: The movielens datasets: history and context. ACM Trans. Interact. Intell. Syst. **5**(4) (2015). https://doi.org/10.1145/2827872
16. Hidasi, B., Karatzoglou, A., Baltrunas, L., Tikk, D.: Session-based recommendations with recurrent neural networks (2016)
17. Hurley, N.J.: Robustness of recommender systems. In: Proceedings of the Fifth ACM Conference on Recommender Systems, RecSys 2011, pp. 9–10. Association for Computing Machinery, New York (2011). https://doi.org/10.1145/2043932. 2043937
18. Hwangbo, H., Kim, Y.S., Cha, K.J.: Recommendation system development for fashion retail e-commerce. Electron. Commer. Res. Appl. **28**, 94–101 (2018)
19. Jaccard, P.: The distribution of the flora in the alpine zone. 1. New Phytol. **11**(2), 37–50 (1912)
20. Kang, W.C., McAuley, J.: Self-attentive sequential recommendation. In: 2018 IEEE International Conference on Data Mining (ICDM), pp. 197–206. IEEE (2018)
21. Kendall, M.G.: Rank correlation methods (1948)
22. Li, Y., Chen, H., Fu, Z., Ge, Y., Zhang, Y.: User-oriented fairness in recommendation. In: Proceedings of the Web Conference 2021, WWW 2021, pp. 624–632. Association for Computing Machinery, New York (2021). https://doi.org/10.1145/ 3442381.3449866
23. Ni, J., Li, J., McAuley, J.: Justifying recommendations using distantly-labeled reviews and fine-grained aspects. In: Proceedings of the 2019 Conference on Empirical Methods in Natural Language Processing and the 9th International Joint Conference on Natural Language Processing (EMNLP-IJCNLP), pp. 188–197 (2019)
24. Oh, S., Ustun, B., McAuley, J., Kumar, S.: Rank list sensitivity of recommender systems to interaction perturbations. In: Proceedings of the 31st ACM International Conference on Information & Knowledge Management, CIKM 2022, pp. 1584–1594. Association for Computing Machinery, New York (2022). https://doi. org/10.1145/3511808.3557425
25. O'Mahony, M., Hurley, N., Kushmerick, N., Silvestre, G.: Collaborative recommendation: a robustness analysis. ACM Trans. Internet Technol. (TOIT) **4**(4), 344–377 (2004)
26. Ovaisi, Z., Heinecke, S., Li, J., Zhang, Y., Zheleva, E., Xiong, C.: RGRecSys: a toolkit for robustness evaluation of recommender systems. In: Proceedings of the Fifteenth ACM International Conference on Web Search and Data Mining, pp. 1597–1600 (2022)
27. Petrov, A., Macdonald, C.: Effective and efficient training for sequential recommendation using recency sampling. In: Proceedings of the 16th ACM Conference on Recommender Systems, pp. 81–91 (2022)
28. Purificato, A., Cassarà, G., Liò, P., Silvestri, F.: Sheaf neural networks for graph-based recommender systems (2023)
29. Quadrana, M., Cremonesi, P., Jannach, D.: Sequence-aware recommender systems. ACM Comput. Surv. (CSUR) **51**(4), 1–36 (2018)
30. Quadrana, M., Karatzoglou, A., Hidasi, B., Cremonesi, P.: Personalizing session-based recommendations with hierarchical recurrent neural networks. In: Proceedings of the Eleventh ACM Conference on Recommender Systems, RecSys 2017, pp. 130–137. Association for Computing Machinery, New York (2017). https:// doi.org/10.1145/3109859.3109896
31. Schafer, J.B., Konstan, J.A., Riedl, J.: E-commerce recommendation applications. Data Min. Knowl. Disc. **5**, 115–153 (2001)

32. Schedl, M., Knees, P., McFee, B., Bogdanov, D., Kaminskas, M.: Music recommender systems. In: Ricci, F., Rokach, L., Shapira, B. (eds.) Recommender Systems Handbook, pp. 453–492. Springer, Boston (2015). https://doi.org/10.1007/978-1-4899-7637-6_13

33. Schedl, M., Zamani, H., Chen, C.W., Deldjoo, Y., Elahi, M.: Current challenges and visions in music recommender systems research. Int. J. Multimed. Inf. Retrieval **7**, 95–116 (2018)

34. Shapiro, S.S., Wilk, M.B.: An analysis of variance test for normality (complete samples). Biometrika **52**(3/4), 591–611 (1965)

35. Student: The probable error of a mean. Biometrika **6**(1), 1–25 (1908)

36. Sun, F., et al.: BERT4Rec: sequential recommendation with bidirectional encoder representations from transformer. In: Proceedings of the 28th ACM International Conference on Information and Knowledge Management, CIKM 2019, pp. 1441–1450. Association for Computing Machinery, New York (2019). https://doi.org/10.1145/3357384.3357895

37. Tang, J., et al.: Improving training stability for multitask ranking models in recommender systems. arXiv preprint arXiv:2302.09178 (2023)

38. Vaswani, A., et al.: Attention is all you need. In: Proceedings of the 31st International Conference on Neural Information Processing Systems, NIPS 2017, pp. 6000–6010. Curran Associates Inc., Red Hook (2017)

39. Wang, S., Hu, L., Wang, Y., Cao, L., Sheng, Q.Z., Orgun, M.: Sequential recommender systems: challenges, progress and prospects. arXiv preprint arXiv:2001.04830 (2019)

40. Wani, F.A., Bucarelli, M.S., Silvestri, F.: Combining distance to class centroids and outlier discounting for improved learning with noisy labels (2023)

41. Webber, W., Moffat, A., Zobel, J.: A similarity measure for indefinite rankings. ACM Trans. Inf. Syst. **28**(4) (2010). https://doi.org/10.1145/1852102.1852106

42. Wu, S., Crestani, F.: Methods for ranking information retrieval systems without relevance judgments. In: Proceedings of the 2003 ACM Symposium on Applied Computing, pp. 811–816 (2003)

43. Wu, S., Sun, F., Zhang, W., Xie, X., Cui, B.: Graph neural networks in recommender systems: a survey. ACM Comput. Surv. **55**(5), 1–37 (2022)

44. Yang, D., Zhang, D., Zheng, V.W., Yu, Z.: Modeling user activity preference by leveraging user spatial temporal characteristics in LBSNs. IEEE Trans. Syst. Man Cybern.: Syst. **45**(1), 129–142 (2014)

45. Zhang, S., Yao, L., Sun, A., Tay, Y.: Deep learning based recommender system: a survey and new perspectives. ACM Comput. Surv. **52**(1) (2019). https://doi.org/10.1145/3285029

46. Zhao, W.X., et al.: RecBole: towards a unified, comprehensive and efficient framework for recommendation algorithms (2021)

Exploring Large Language Models and Hierarchical Frameworks for Classification of Large Unstructured Legal Documents

Nishchal Prasad[✉] , Mohand Boughanem[✉] , and Taoufiq Dkaki[✉]

Institut de Recherche en Informatique de Toulouse (IRIT), Toulouse, France
{Nishchal.Prasad,Mohand.Boughanem,Taoufiq.Dkaki}@irit.fr

Abstract. Legal judgment prediction suffers from the problem of long case documents exceeding tens of thousands of words, in general, and having a non-uniform structure. Predicting judgments from such documents becomes a challenging task, more so on documents with no structural annotation. We explore the classification of these large legal documents and their lack of structural information with a deep-learning-based hierarchical framework which we call MESc; "Multi-stage Encoder-based Supervised with-clustering"; for judgment prediction. Specifically, we divide a document into parts to extract their embeddings from the last four layers of a custom fine-tuned Large Language Model, and try to approximate their structure through unsupervised clustering. Which we use in another set of transformer encoder layers to learn the inter-chunk representations. We analyze the adaptability of Large Language Models (LLMs) with multi-billion parameters (GPT-Neo, and GPT-J) with the hierarchical framework of MESc and compare them with their standalone performance on legal texts. We also study their intra-domain(legal) transfer learning capability and the impact of combining embeddings from their last layers in MESc. We test these methods and their effectiveness with extensive experiments and ablation studies on legal documents from India, the European Union, and the United States with the ILDC dataset and a subset of the LexGLUE dataset. Our approach achieves a minimum total performance gain of approximately 2 points over previous state-of-the-art methods.

Keywords: Legal judgment prediction · Long document classification · Multi-stage hierarchical classification framework

1 Introduction

A legal case proceeding cycle[1] involves analyzing vast amounts of data and legal precedents, which can be a time-consuming process given the complexity and

---

[1] https://www.law.cornell.edu/wex/civil_procedure.

© The Author(s), under exclusive license to Springer Nature Switzerland AG 2024
N. Goharian et al. (Eds.): ECIR 2024, LNCS 14609, pp. 221–237, 2024.
https://doi.org/10.1007/978-3-031-56060-6_15

length of the case. The number of legal cases in a country is also proportionally related to its population. This leads to a backlog of cases, especially in highly populated countries, ultimately setting back the progress of its legal system[2] [17]. Automating such legal case procedures can help speed up and strengthen the decision-making process, saving time and benefiting both the legal authorities and the people involved. One of the fundamental problems that deal with this larger component is the prediction of the outcome based just on the case's raw texts (which can include facts, arguments, appeals, etc. except the final decision), as in a typical real-life (raw) setting.

Several machine learning techniques have been applied to legal texts to predict judgments as a text classification problem [12,14]. While it seems like a general text classification task, legal texts differ from general texts and are rather more complex, broadly in two ways, i.e. structure and syntax and, lexicon and grammar [7,23,43]. The structure of legal case documents is not uniform in most settings and their complex syntax and lexicon make it more difficult and expensive to annotate, requiring only legal professionals. This adds to another challenge of the long lengths of these documents, reaching more than 10000 words (Table 2). This lack of structure information and the long lengths of these legal documents pose a challenge in predicting judgments. In our work we explore this problem on four fronts, by (a) developing a hierarchical framework for the classification of large unstructured legal documents, (b) exploring the adaptability of billion-parameter large language models (LLMs) to this framework, (c) analyzing the performance of these LLMs without this framework and (d) checking the intra-domain(legal) transfer learning capability of domain-specific pre-trained LLMs. This is summarized below:

- We explore the problem of judgment prediction from large unstructured legal documents and propose a hierarchical multi-stage neural classification framework named "Multi-stage Encoder-based Supervised with-clustering" (MESc). This works by extracting embeddings from the last four layers of a fine-tuned encoder of a large language model (LLM) and using an unsupervised clustering mechanism to approximate the structure. Alongside the embeddings, these approximated structure labels are processed through another set of transformer encoder layers for final classification.
- We show the effect of combining features from the last layers of transformer-based LLMs (BERT [13], GPT-Neo [3], GPT-J [32]), along with the impact on classification upon using the approximated structure.
- We study the adaptability of domain-specific pre-trained multi-billion parameter LLMs to such documents and study their intra-domain(legal) transfer learning capability (both with fine-tuning and in MESc).
- We performed extensive experiments and analysis on four different datasets (ILDC [20] and LexGLUE's [9] ECtHR(A), ECtHR(B), and SCOTUS) and achieved a total gain of ≈ 2 points in classification on these datasets.

---

[2] https://www.globaltimes.cn/page/202204/1260044.shtml.

2   Related Works

Several strategies have been investigated to predict the result of legal cases in specific categories (criminal, civil, etc.) with rich annotations (Xiao et al. [34], Xu et al. [35], Zhong et al. [42], Chen et al. [10]). These studies on well-structured and annotated legal documents show the effect and importance of having good structural information. While creating such a dataset is both time and resource (highly skilled) demanding, researchers have worked on legal documents in a more general and raw setting. Chalkidis et al. [5] presented a dataset of European Court of Human Rights case proceedings in English, with each case assigned a score indicating its importance. They described a Legal Judgment Prediction (LJP) task, which seeks to predict the outcome of a court case using the case facts and law violations. They also create another version of this dataset [8] to give a rational explanation for the predictions. In the US legal case setting, Kaufman et al. [18] used AdaBoost decision tree to predict the U.S. Supreme Court rulings. Tuggener et al. [30] proposed LEDGAR, a multilabel dataset of legal provisions in US contracts. Malik et al. [20] curated the Indian Legal Document Corpus (ILDC) of unannotated and unstructured documents, and used it to build baseline models for their Case Judgment Prediction and Explanation (CJPE) task upon which Prasad et al. [25] showed the possibility of intra-domain(legal) transfer learning using LEGAL-BERT on Indian legal texts.

Pretrained large language models (LLMs) based on transformers (Devlin et al. [13], Vaswani et al. [31]) have shown widespread success in all fields of natural language processing (NLP) but only for short texts spanning a few hundred tokens. There have been several approaches to handle longer sequences with long sequence transformer-based LLMs (Beltagy et al. [2], Kitaev et al. [19], Zaheer et al. [39], Ainslie et al. [1]). These architectures display similar performance as the hierarchical adaptation of their vanilla counterparts [6,9,11], and since we try to learn and approximate the structure information of the document, we choose to process the document in short sequences rather than as a whole. So, we take a different approach to handle large documents with LLMs (such as BERT [13]) based on the hierarchical idea of "divide, learn and combine" (Chalkidis et al. [6], Zhang et al. [40], Yang et al. [37]), where the document is split (into parts then sentences and words, etc.) and features of each component are learned and combined hierarchically from bottom-up to get the whole document's representation. Also with the unavailability of the domain-specific pre-trained checkpoints of these long sequence LLMs and considering their expensive pretraining, we choose to use the vanilla models and develop a hierarchical adaptation.

Moreover, the domain-specific pre-training of transformer encoders has accelerated the development of NLP in legal systems with better performance as compared to the general pre-trained variants (Chalkidis et al. [7]'s LEGAL-BERT trained on court cases of the US, UK, and EU, Zheng et al. [41]'s BERT trained on US court cases dataset CaseHOLD, Shounak et al. [24]'s InLegal-BERT and InCaseLawBERT trained on the Indian legal cases). Recently, with the emergence of multi-billion parameter LLMs such as GPT-3 [4], LLaMA [28], LaMDA [27], and their superior performance in natural language understanding, researchers have tried to adapt their variants (with few-shot learning) to legal texts (Trautmann et al. [29], Yu et al. [38]). In this paper, with full-fine tuning, we check the adaptability of these billion parameter LLMs with the hierarchical framework and also their intra-domain(legal) transfer-learning compared to the intra-domain pre-training (as done in LEGAL-BERT, InLegal-BERT). To do so we use three such variants of GPT (GPT-Neo (1.3 and 2.7) [3], GPT-J [32]) pre-trained on Pile [15], which has a subset (FreeLaw) of court opinions of US legal cases.

3 Method: Classification Framework (MESc)

To handle large documents MESc architecture shares the general hierarchical idea of divide, learn, and combine [6, 37, 40] but differs from the previous works in the following: (a) It uses the last four layers of the fine-tuned transformer-based LLM for extracting global representations for parts(chunks) of the document. (b) Approximating the document structure by applying unsupervised learning (clustering) on these representations' embeddings and using this information alongside, for classification. (c) Instead of only RNNs, different combinations of transformer encoder layers are tested to get a global document representation. (d) Divide the process into four stages, fine-tuning, extracting embeddings, processing the embeddings (supervised + unsupervised learning), and classification.

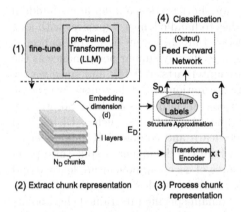

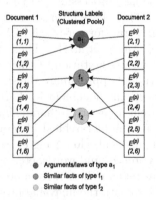

Fig. 1. Multi-stage Encoder-based Supervised with-clustering (MESc) framework.

Fig. 2. An example of clustering of chunk representations of two documents to generate structure labels.

An overview of MESc can be seen in Fig. 1. An input document D is tokenized into a sequence of tokens, $D = \{t_{1,D}, t_{2,D}, \cdots, t_{L_D,D}\}$ via a tokenizer specific to a chosen pre-trained transformer-based language model (BERT, GPT etc.), where $t \in \mathbb{N}$ and $\mathbb{N}$ is the vocabulary of the tokenizer. This token sequence is split into a set of blocks (chunks) $\{C_{1,D}, C_{2,D}, \cdots, C_{N_D,D}\}$ with overlaps(o). Where each chunk, $C_{i,D} = \{t_{((i-1)\times(c-o))}, \cdots, t_{((i-1)\times(c-o)+c)}\}$ with c being the maximum number of tokens in the chunk, which is a predefined parameter for MESc (e.g. 512). $N_D = \lceil \frac{L_D}{c-o} \rceil$ is the total number of chunks for a document having L tokens in total, with $o \ll c$. N_D varies with the length of the document.

Stage 1 - Fine-Tuning: To each chunk of a document, we associate the gold class label of the document l_D, and combine them to form a token matrix:

$$I_D \in \mathbb{R}^{N_D \times c \times 1} \leftarrow [\{C_{1,D}, l_D\}, \{C_{2,D}, l_D\}, \cdots, \{C_{N_D,D}, l_D\}] \tag{1}$$

This is used as input for the document for fine-tuning the pre-trained LLM, where N_D is the batch size for one pass through the encoder. This allows the encoder to adapt to the domain-specific legal texts, which helps get richer features for the next stage.

Stage 2 - Extracting Chunk Embeddings: Different layers of transformer models learn varied representations of the input sentence [16,26,36]. When simultaneously used alongside each other these representations can be used to give varied features for further learning. Since the last pre-trained LLM layer captures the final representation of a chunk, we use it alongside its immediate lower layers to extract the chunk representation. We use the immediate lower layers with the intuition that the representations learned are not heavily but enough varied from the final layer.

For a document, we pass its chunks C_i through the fine-tuned encoder and extract its representation embeddings $(E_{i,D})$ from the last l layers. $E_{i,D} \in \mathbb{R}^{l \times d}$, where d is the dimension of the features (we use $l = 4$). The representation embeddings can be either the first token (as in BERT) or the last token for causal language models (as in GPT). We accumulate all $E_{i,D}$ of a document to form an embedding matrix:

$$E_D \in \mathbb{R}^{N_D \times l \times d} \leftarrow [E_{1,D}, E_{2,D}, \cdots, E_{N_D,D}] \tag{2}$$

The $E_{i,D}$ acts as a representation of the chunk in this context, and combining them yields an approximate representation of the entire document. Doing this for all the documents gives us generated training data.

Stage 3 - Processing the Extracted Representations: Since the features extracted from the last layers of a fine-tuned encoder have different embedding spaces, they can contribute to give varied features. So for this stage, we choose to combine the last $p < l$ layers in E_D for further training. We experiment with different p before fixing one value as discussed in Sect. 4. This gives $E_D^{(p)} \in \mathbb{R}^{N_D \times p \times d}, p \in \{1, 2, 3, 4\}$. We used $p = 1, 2$, and 4 in our experiments to compare their effects. We concatenate together the representations from these p layers to get,

$$E_{i,D}^{(p)} \in \mathbb{R}^{pd \times 1} \leftarrow [E_{i,D}^{(l)} | E_{i,D}^{(l-1)} | \cdots | E_{i,D}^{(l-p-1)}] \tag{3}$$

This gives,

$$\widehat{E}_D^{(p)} \in \mathbb{R}^{N_D \times pd} \leftarrow \{E_{1,D}^{(p)} | E_{2,D}^{(p)} | \cdots | E_{N,D}^{(p)}\} \tag{4}$$

We also experimented with the element-wise addition of representations in $E_D^{(p)}$ and found their performance to be lower in most of the experiments of Sect. 4, hence we exclude it here.

a. **Approximating the structure labels** (S_D) (Unsupervised learning): To get the information on the document's structure i.e. its parts (facts, arguments, concerned laws, etc.), we use a clustering algorithm (HDBSCAN [21]). We cluster the p chosen extracted chunk embeddings, $\widehat{E}_D^{(p)}$ to map similar parts of different documents together where the labels of one part of a document are learned by its similarity with another part of another document. The idea is that the embeddings of similar parts from different documents will group forming a pool of labeled clusters that can help identify its part in the document. A synthetic example can be seen in Fig. 2, where the $E_{i,D}^{(p)}$ of documents 1 and 2 learn their cluster (label) pool for, arguments of type $a_1 = \{E_{1,1}^{(p)}, E_{1,2}^{(p)} E_{2,1}^{(p)}\}$, facts of type $f_1 = \{E_{1,3}^{(p)}, E_{1,5}^{(p)}, E_{2,2}^{(p)}, E_{2,3}^{(p)}, E_{2,5}^{(p)}\}$, facts of type $f_2 = \{E_{1,4}^{(p)}, E_{1,6}^{(p)}, E_{2,4}^{(p)}, E_{2,6}^{(p)}\}$. So for document 1 the approximated structure then becomes $S_1 = \{a_1, a_1, f_1, f_2, f_1, f_2\}$ and for document 2 it is $S_2 = \{a_1, f_1, f_1, f_2, f_1, f_2\}$. It is to be noticed that this distinction if it's a fact or an argument etc. is done here for representation. In an actual setting, this is unknown and the labels don't carry any specific name or meaning except for the model to give an approximation of its structure.

Since the performance of the HDBSCAN clustering mechanism decreases significantly with an increase in data dimension, we use a dimensionality reduction algorithm (pUMAP [22]), before clustering. For all the chunks of a document, their approximated structure labels are combined with the output of stage 3(b), before processing through the final classification stage (4).

b. **Global document representation** (Supervised learning): For intra-chunk attention, we use transformer encoder layers (Vaswani et al. [31]), for a chunk to attend to another through its multi-head attention and feed-forward neural network (FFN) layer. This helps the chunk representations to attend to one another in parallel. For a chunk's position in the document, we add its positional embeddings ([13]) in $E_D^{(p)}$ and process it through t transformer layers $T_{\{h,d_f\}}^{(t)}$, with h attention heads and $d_f = pd$ as the dimension of the FFN. t and h are both hyperparameters whose choice depends upon the input feature lengths. Section 4 evaluates different values of these parameters, but $t \geq 3$ sometimes overfits the model in our experiments, hence we fix $t = 2$ for MESc. The output is max-pooled and passed through a feed-forward neural network FFN_T of 128 nodes to get:

$$G\left(\widehat{E}_D^{(p)}\right) = FFN_T\left(maxpool\left(T_{\{h,d_f\}}^{(t)}\left(\widehat{E}_D^{(p)}\right)\right)\right) \in \mathbb{R}^{128} \qquad (5)$$

Stage 4 - Classification: The structure labels along with the output of the feed-forward network of stage 3(b) are concatenated together and passed through an internal feed-forward network FFN_i (32 nodes, with softmax activation) and an external feed-forward network FFN_e (u label/class nodes with task-specific activation function sigmoid or softmax) giving the output $O(D)$ for a document D (Eq. 6). O and G are learnt together while S_D is learnt independently.

$$O(D) = FFN_e\left(FFN_i\left(\left(\left[G\left(\widehat{E}_D^{(p)}\right) | S_D\right]\right)\right)\right) \in \mathbb{R}^u \qquad (6)$$

The code and trained models for the above implementation can be found at GitHub[3] and our finetuned LLMs at HuggingFace[4].

3.1 Experimental Setup

Table 1 lists the major details for the experimental setup. For our backbone transformer-based language model, we used domain-specific models LEGAL-BERT [7], InLegalBERT [24], and for multi-billion parameter LLMs we chose GPT-Neo [3], GPT-J [32]. The tokenizers used are from the respective backbone transformer encoders. We abbreviate the encoders fine-tuned on 512 input length as (α) and, for ones fine-tuned with 2048 input length as (γ).

These hyperparameters (Table 1) were used based on the guidelines of the respective language models and several of our previous experiments and dataset analyses. We list out some of them further in the paper and in discussions while referring to Table 4, Table 5, Table 2, and Fig. 3.

3.2 Dataset

We chose the legal datasets having large documents with a nonuniform structure throughout and without any structural annotations. The ILDC dataset [20] includes highly unstructured 39898 English-language case transcripts from the Supreme Court of India (SCI), where the final decisions have been removed from the document. Upon analyzing the documents from their sources and the dataset we found that they are highly unstructured and noisy. The initial decision between "rejected" and "accepted" made by the SCI judge(s) is used to classify each document and serves as their decision label. The LexGLUE dataset [9] comprises a set of seven datasets from the European Union and US court case setting, for uniformly assessing model performance across a range of legal NLP tasks, from which we choose ECtHR (Task A), ECtHR (Task B), and SCO-TUS as they are classification tasks involving long unstructured legal documents. ECtHR (A and B) are court cases from the European Convention on Human Rights (ECHR) for articles that were violated or allegedly violated. The dataset contains factual paragraphs from the description of the cases. SCOTUS consists of court cases from the highest federal court in the United States of America, with metadata from SCDB[5]. The details of the number of labels, the document lengths (in tokens), task description, and class distribution can be found in Table 2 and Table 3. The tokenization Table 2 is done using the tokenizer of GPT-J.

---

[3] https://github.com/NishchalPrasad/MESc.

[4] https://huggingface.co/nishchalprasad.

[5] http://scdb.wustl.edu/.

Table 1. Experimental setup for different stages of MESc architecture.

Stage 1
BERT-based LLM: chunk-size = 512 tokens (90 token overlaps), [CLS] token to test.
GPT-based LLM: max input length = 2048, chunk-size $\in \{512, 2048\}$, last token to test.
For all (α) GPT we compare with (α) BERT-based LLM on 512 input length.
Finetuned for e=4 epochs, chose best e for Stage 2 and evaluation.

Stage 2 (Embedding Extraction)	Stage 3 & 4
	Optimizer = Adam (learning rate = $3.5e^{-6}$)
BERT-based LLM:	*Loss func.:* multi-label: categorical cross-entropy
[CLS] token for each chunk	binary & multi-class: binary cross-entropy
	$t = \{1, 2, 3\}$, $h = 8$, e = 5 epochs (best e for evaluation)
GPT-based LLM:	*Structure approximation:* pUMAP[a] (64 dimensions)
last token for each chunk	HDBSCAN (15 min clusters)

GPU used: Nvidia V100 & A100, with ZERO-3 in Deepspeed[b] with Accelerate[c]
Maximum fine-tune time (hours/epoch) for GPTs (6 Nvidia A100):
GPT-Neo-1.3B = 2.1, GPT-Neo-2.7B = 4, GPT-J = 8

[a] https://umap-learn.readthedocs.io/en/latest/parametric_umap.html
[b] https://www.deepspeed.ai/
[c] https://huggingface.co/docs/accelerate/index

Table 2. Dataset statistics

Name		ILDC	ECtHR(A)	ECtHR(B)	SCOTUS
No. of Docs.	Train	37387	9000	9000	5000
	Val.	994	1000	1000	1400
	Test	1517	1000	1000	1400
Average tokens	Train	4120 501275	2011 46500	2011 46500	8291 126377
	Val.	8048 51045	2210 18352	2210 18352	12639 56310
Max tokens	Test	5238 55703	2401 20835	2401 20835	12597 124955
No. of labels		2	10	10	13
Problem Type		Binary	Multi-Label	Multi-Label	Multi-Class

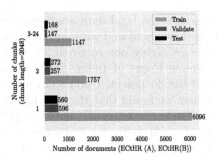

Fig. 3. Number of documents vs. the number of chunks for ECtHR.

Table 3. Class distribution of the datasets.

(Problem type)		class : # documents											
ILDC (Binary)	Train	0: 22067	1: 15320										
	Val.	0: 497	1: 497										
	Test	0: 755	1: 762										
ECtHR (A) (Multi-label)	Train	0: 505	1: 1349	2: 1368	3: 4704	4: 710	5: 41	6: 291	7: 110	8: 141	9: 1421		
	Val.	0: 57	1: 193	2: 187	3: 300	4: 87	5: 4	6: 42	7: 33	8: 18	9: 139		
	Test	0: 56	1: 189	2: 166	3: 299	4: 123	5: 5	6: 77	7: 37	8: 16	9: 122		
ECtHR (B) (Multi-label)	Train	0: 623	1: 1740	2: 1623	3: 5437	4: 1056	5: 81	6: 441	7: 162	8: 444	9: 1558		
	Val.	0: 75	1: 236	2: 219	3: 394	4: 153	5: 9	6: 64	7: 39	8: 34	9: 168		
	Test	0: 76	1: 234	2: 196	3: 394	4: 188	5: 11	6: 106	7: 43	8: 32	9: 155		
SCOTUS (Multi-class)	Train	0: 1011	1: 811	2: 423	3: 193	4: 45	5: 35	6: 255	7: 1043	8: 717	9: 191	10: 53	11: 220	12: 3
	Val.	0: 360	1: 218	2: 108	3: 70	4: 22	5: 35	6: 51	7: 226	8: 165	9: 83	10: 14	11: 38	12: 10
	Test	0: 372	1: 222	2: 88	3: 51	4: 28	5: 17	6: 24	7: 260	8: 200	9: 83	10: 15	11: 37	12: 3

For performance comparison on LexGLUE, we used the SOTA benchmark of Chalkidis et al. [9], Condevaux et al.'s LSG [11], Chalkidis et al.'s HAT [6] and for ILDC we used its benchmark from [20] and of Shounak et al. [24]'s experiments.

4   Results and Discussion

μ-F1 (micro) and m-F1 (macro) are used to measure the performance for the LexGLUE dataset, and accuracy(%) and macro-F1 for the ILDC dataset. These metrics were chosen partly to compare with the previous benchmark models (stated in Table 4) conforming to their original results and metrics. We list out the detailed experimental results for best configurations of MESc in Table 5 and the fine-tuned performance of the LLMs used in Table 4.

Intra-domain(Legal) Transfer Learning: Based on the analysis of ILDC by Malik et al. [20] we use the last chunk of the documents for evaluation. As can be seen from Table 4, for LexGLUE's subset, all the GPTs used here adapt better than the BERT-based models with a minimum of ≈ 3 points gain on μ-F1 and a minimum of ≈ 6 points on m-F1 score. On the other hand in the ILDC dataset, for the α variants with 512 input lengths for evaluation, the performance dropped or remained similar to the InLegalBERT, while upon increasing the evaluation input length to 2048 we can see an increase of more than 1 point in the performance. When fine-tuned with 2048 input length, the performance of GPT-J (γ) compared to its α and β variant is at least ≈ 2 points higher for all the datasets. We can see that an increase in the input length for fine-tuning helps to capture more feature information for such documents. Also going from GPT-Neo-1.3B to GPT-Neo-2.7B to GPT-J-6B, the performance increases by a margin of 2 points at minimum, here we see the parameter count playing an important role in adapting and understanding these documents. Even though GPT-Neo and GPT-J are pre-trained on US legal cases (Pile [15]) they adapt better to the European and Indian legal documents, with a minimum gain of $\approx$ 7 points (γ) on the ECtHR(A & B) and the ILDC dataset over their domain-specific pre-trained counterparts LEGAL-BERT and InLegalBERT respectively. These results show the transfer learning capacity of LLMs between different legal domains with different settings, which can be a better alternative with limited resources compared to expensive domain-specific pre-training.

Performance with MESc: Looking at Table 5 we interpret the results in two directions.

(a) Encoders fine-tuned on 512 input length (α): For LEGAL-BERT and InLegalBERT in all datasets, MESc achieves a significant increase in performance by at least 4 points in all metrics than their fine-tuned LLM counterparts with just the last layer ($p = 1$). Combining the last four layers with $t = 1$ encoder layer yields a performance boost of 4 points or more in ECtHR datasets while there is not much improvement in ILDC and SCOTUS.

Table 4. Fine-tuned results on the last chunk for the chosen LLMs (Sect. 3.1)

α: fine-tuned and evaluated with 512 input length, β: evaluating α on its maximum input length, γ: fine-tuned and evaluated with its maximum input length. All measures are in (%). e = epoch.

Dataset		LEGAL-BERT (μ-F1/m-F1)	GPT-Neo 1.3B (μ-F1/m-F1)	GPT-Neo 2.7B (μ-F1/m-F1)	GPT-J 6B (μ-F1/m-F1)
LexGLUE's subset	ECtHR (A)	(α) 62.85/48.66 (e = 4)	(α)66.19/56.59 (β)66.20/57.16 (e = 2)	(α)68.49/54.45 (β)68.11/56.49 (e = 2)	(α)71.42/59.27 (β)73.30/62.45 (γ)**74.51/64.67** (e = 3)
	ECtHR (B)	(α) 70.89/64.05 (e = 3)	(α)75.42/70.91 (β)75.74/70.09 (e = 2)	(α)74.48/68.26 (β)75.13/70.72 (e = 2)	(α)77.15/73.26 (β)80.49/76.31 (γ)**83.16/79.27** (e = 3)
	SCOTUS	(α) 68.76/53.57 (e = 6)	(α)71.14/60.35 (β)73.71/63.10 (γ)75.02/64.38 (e = 2)	(α)70.57/60.25 (β)73.64/65.64 (γ)76.36/66.19 (e = 1)	(α)72.00/62.76 (β)75.71/66.25 (γ)**78.50/71.96** (e = 3)
		InLegalBERT (Acc./m-F1)		Accuracy (Acc.) / m-F1	
ILDC		(α) **76.00/76.10** (e = 4)	(α)72.91/72.91 (β)77.26/77.25 (e=1)	(α)74.29/74.24 (β)81.21/81.18 (e=1)	(α)73.96/73.96 (β)81.93/81.92 (γ)**83.72/83.66** (e=1)

Table 5. Test results for different configurations of MESc. We show the maximum scores attained in all the runs. The bold-faced values also signify statistically significant findings in 5 different runs. (The baseline results are from their original papers.)

* is the fine-tuned LLM used for embedding extraction (Table 4).
p = last p layers of the * model; t transformer encoder layers; S_D = approximated structure.

			ECtHR (A)	ECtHR (B)	SCOTUS		ILDC
			(%) μ-F1/m-F1				(%) Accuracy/m-F1
Chalkidis et al.[9]			71.2/64.7	80.4/74.7	76.6/66.5		
LSG [11]			71.7/63.9	81.0/75.1	74.5/62.6	Malik et al.[20]	77.78/77.79
HAT [6]			-	80.8/79.8	-	Shounak et al.[24]	-/83.09
MESc							
	p, t	S_D					
	$p=1, t=1$	✗	68.25/58.06	74.18/68.90	71.36/59.16		83.72/83.73
		✓	-	-	-		83.65/83.65
	$p=1, t=2$	✗	69.23/59.35	73.86/67.42	71.52/58.17		83.45/83.47
		✓	-	-	-		83.78/83.78
LEGAL-BERT* (α)	$p=4, t=1$	✗	75.46/62.26	81.02/75.73	73.96/58.65	InLegalBERT* (α)	83.41/83.41
		✓	75.82/63.78	81.22/77.25	75.25/61.94		**84.15/84.15**
	$p=4, t=2$	✗	75.43/63.37	81.18/75.64	74.31/60.54		83.72/83.68
		✓	**76.18/65.08**	**81.57/76.70**	**75.50/62.08**		84.11/84.13
	$p=4, t=3$	✗	75.23/63.11	81.32/76.99	73.99/56.35		-
		✓	75.10/63.09	81.00/76.21	73.92/57.83		-
GPT-Neo 1.3B* (α)	$p=2, t=2$	✗	71.15/63.59	80.30/77.02	75.36/64.79		-
		✓	**72.73/64.48**	**80.40/78.08**	**76.46/65.92**		-
	$p=4, t=2$	✗	71.46/62.77	80.86/76.64	74.29/63.52		-
		✓	70.68/64.10	80.60/77.57	74.18/63.77		-
GPT-Neo 2.7B* (α)	$p=2, t=2$	✗	74.57/62.24	79.49/76.20	76.76/65.70	GPT-Neo 2.7B* (α)	82.97/82.79
		✓	**75.67/66.44**	**80.72/76.96**	**76.27/66.30**		83.65/83.64
	$p=4, t=2$	✗	75.24/63.55	79.40/75.03	75.77/65.54		83.01/83.00
		✓	75.87/65.61	79.35/76.35	76.41/67.75		83.22/83.21
GPT-J 6B* (α)	$p=2, t=2$	✗	72.22/62.63	79.31/76.92	75.05/66.58	GPT-J 6B* (α)	82.84/82.78
		✓	**71.63/64.06**	**79.77/77.60**	**75.98/67.15**		83.21/83.19
	$p=4, t=2$	✗	71.56/61.18	78.00/76.05	74.90/63.33		82.73/82.73
		✓	72.19/64.37	77.95/76.25	74.85/65.93		83.37/83.36
GPT-J 6B* (γ)	$p=2, t=2$	✗	73.84/64.34	80.94/76.75	76.88/67.73		
		✓	**74.70/65.71**	**81.69/78.01**	78.14/68.53		
	$p=4, t=2$	✗	72.96/63.33	81.13/77.63	77.28/67.86		
		✓	74.84/65.48	81.34/78.02	**78.67/69.66**		

With S_D, the approximated structure labels, there is a slight performance increase in the ILDC. The same goes for SCOTUS with ≈ 1 point increase. With the same configuration and $t = 2$ encoder layers, we can see a much bigger performance with the structure labels achieving new baseline scores in ECtHR (A), ECtHR (B), and ILDC datasets. For SCOTUS, this improvement from the baseline is not much. This is because of the high skew of class labels in the test dataset (for example label 5 has only 5 samples). With these results, we fixed certain parameters in MESc for further experiments with the extracted embeddings from GPT-Neo and GPT-J. For them, we ran experiments with $t = 2$ encoders and the last layer ($p = 1$) and gained lesser performance than $p = 2$ (or 4) layers and $t = 2$ encoders, which we exclude in this paper. For ECtHR(A&B) and SCOTUS, concatenating the embeddings from the last two layers of GPT-Neo or GPT-J had a significant impact above their vanilla fine-tuned variants by a minimum margin of 3 points for GPT-Neo-1.3B, and 1 point for GPT-Neo-2.7B and GPT-J. This increases further by a minimum of 1 point when including the approximated structure labels, showing the impact of having structural information. For ILDC, concatenating the last four layers didn't have much improvement in the performance while including the generated structure labels increased the performance.

(b) Encoders fine-tuned on 2048 input length (γ): Referring to Table 4 and Table 5 for the documents we did a comparative study of MESc(on GPT-J 6B* (γ))'s performance with its backbone fine-tuned LLM (GPT-J 6B (γ)) to see the effect of increasing the number of parameters and the input length. GPT-J 6B (γ) fine-tuned on its maximum input length (2048) achieves better (or similar) performance than its MESc overhead trained on its extracted embeddings. For SCOTUS, MESc achieves better performance (2 points, m-F1) in the test set. Almost similar performance (m-F1) in ECtHR(B), 1 point higher (m-F1) in ECtHR(A)'s test set, and lesser in ILDC. To check if this is the case with GPT-Neo-1.3B and GPT-Neo-2.7B we fine-tuned them with their maximum input length (2048) on SCOTUS (which through our experiments can be seen are more difficult to classify). We found that fine-tuning GPT-Neo (1.3B and 2.7B) on its maximum input length didn't show the same results as with the GPT-J. We find that for both GPT-Neo-1.3B(γ) and GPT-Neo-2.7B(γ) even the MESc (GPT-Neo-1.3B*(α)) and MESc (GPT-Neo-2.7B*(α)) performs better (> 1 point m-F1) respectively. To analyze this, we plot the distribution of the number of documents with respect to their chunk counts (chunk length = 2048) in the datasets, one such example of ECtHR can be found in Fig. 3 (we accumulate the document counts for chunks 3 to 24 for clarity). As observed, most of the documents can fit in 1 or 2 chunks (median = 1), which means that with the longer input of 2048, most of the important information is not fragmented during the fine-tuning process (stage 1) and prediction. Along with this, the higher number of parameters in GPT-J helps it adapt better to most of the documents. We observe that most ($>90\%$) of the documents can fit in very few chunks, deepening the models with extra layers (stages 3 & 4) does not have any added value.

The results obtained are statistically significant[6] [33].

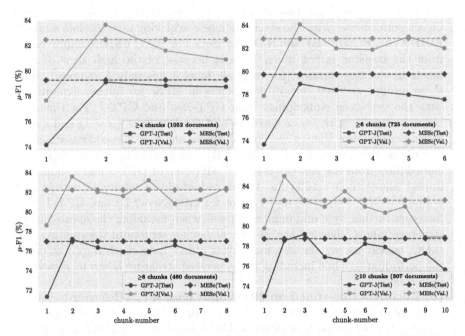

Fig. 4. μ-F1 for chunk-number for GPT-J (γ) vs MESc (GPT-J* (γ)) in SCOTUS on both Validation(Val.) and Test set.

Analysis on Long Documents: To analyze the performance of MESc and its standalone LLM with the document length we ran experiments with GPT-J (γ) on documents with minimum lengths of 4, 6, 8, and 10 chunks of the documents. The predictions are made using the nth chunk of all the documents. We show these results from the SCOTUS dataset in Fig. 4, where, for the input chunk of the documents, we plot its corresponding μ-F1 score. Since MESc has no input length limit and takes all the chunks at once we plot its prediction for all chunks considered as constant lines. The performance with GPT-J (γ) fluctuates with the input chunk, with the worst performance when using the first chunk and the best on the second/third chunk. This shows that for these documents in the test set the second/third chunk has a higher probability of containing the important information for a more robust prediction. Choosing which chunk to use for an unseen test set becomes more difficult as the document length increases and there is no prior information on its important parts. The fluctuations become worse for documents with a minimum of 10 chunks. While for MESc the performance is overall better than GPT-J (γ) in all the lengths considered. This shows that the hierarchical framework (such as MESc) is more

---

[6] We performed student's t-test (p-value < 0.05).

reliable than its LLM counterpart on longer documents and when the important parts of the document are unknown.

With these results on MESc, we find that:

1. Concatenating embeddings from the last two layers in GPT-Neo (1.3B, 2.7B) or GPT-J or, the last 4 layers in BERT-based models, provides the optimum number of feature variances. Globally, concatenating the embeddings helped to get a better approximation of the structure labels and improved performance.
2. MESc works better than its counterpart LLM under the condition that the length of most of the documents in the dataset is much greater than the maximum input length of the LLM.
3. For long documents when their important parts are unknown MESc performs better than its counterpart LLMs.

5   Conclusion

We explore the problem of classification of large and unstructured legal documents and develop a multi-stage hierarchical classification framework (MESc). We test the effect of including our approximated structure and the impact of combining the embeddings from the last layers of a fine-tuned transformer-based LLM in MESc. Along with BERT-based LLMs, we explored the adaptability of LLMs with billion parameters (GPT-Neo and GPT-J) to MESc and analyzed its limits (Sect. 4) with these LLMs suggesting the optimal condition for its performance. The benchmark performance of GPT-Neo and GPT-J on the legal cases from India and Europe shows the intra-domain(legal) transfer learning capability of these billion-parameter language models. Most of all, our experiments achieve a new baseline in the classification of the ILDC and the LexGLUE subset (ECtHR (A), ECtHR (B), and SCOTUS). In our future work, we aim to analyze the clusters and how they contribute to the prediction. We aim to develop an explanation algorithm to explain the predictions while also leveraging this work in-domain, on the French and European legal cases to further explore the problem of length and the non-uniform structure.

6   Ethical Considerations

This work conforms to the ethical consideration of the datasets (ILDC [20] and LexGLUE [9])) used here. The framework developed here is in no way to create a "robotic" judge or replace one in real life, but rather to help analyze how LLMs and our hierarchical framework can be applied to long legal documents to predict judgments. These methods are not foolproof to predict judgments, and should not be used for the same in real-life settings (courts) or used to guide people unfamiliar with legal proceedings. The results from our framework are not reliable enough to be used by a non-professional to make high-stakes decisions in one's life concerning legal cases.

Acknowledgements. This work is supported by the LAWBOT project (ANR-20-CE38-0013) and was performed using HPC resources from GENCI-IDRIS (Grant 2022-AD011013937).

References

1. Ainslie, J., et al.: ETC: encoding long and structured inputs in transformers. In: Proceedings of the 2020 Conference on Empirical Methods in Natural Language Processing (EMNLP), pp. 268–284. Association for Computational Linguistics (2020). https://doi.org/10.18653/v1/2020.emnlp-main.19, https://aclanthology.org/2020.emnlp-main.19

2. Beltagy, I., Peters, M.E., Cohan, A.: Longformer: the long-document transformer (2020). https://doi.org/10.48550/ARXIV.2004.05150, https://arxiv.org/abs/2004.05150

3. Black, S., Leo, G., Wang, P., Leahy, C., Biderman, S.: GPT-neo: large scale autoregressive language modeling with mesh-tensorflow (2021). https://doi.org/10.5281/zenodo.5297715

4. Brown, T., et al.: Language models are few-shot learners. In: Larochelle, H., Ranzato, M., Hadsell, R., Balcan, M., Lin, H. (eds.) Advances in Neural Information Processing Systems, vol. 33, pp. 1877–1901. Curran Associates, Inc. (2020)

5. Chalkidis, I., Androutsopoulos, I., Aletras, N.: Neural legal judgment prediction in English. In: Proceedings of the 57th Annual Meeting of the Association for Computational Linguistics, pp. 4317–4323. Association for Computational Linguistics, Florence (2019). https://doi.org/10.18653/v1/P19-1424, https://aclanthology.org/P19-1424

6. Chalkidis, I., Dai, X., Fergadiotis, M., Malakasiotis, P., Elliott, D.: An exploration of hierarchical attention transformers for efficient long document classification (2022). https://arxiv.org/abs/2210.05529

7. Chalkidis, I., Fergadiotis, M., Malakasiotis, P., Aletras, N., Androutsopoulos, I.: LEGAL-BERT: the muppets straight out of law school. In: Findings of the Association for Computational Linguistics: EMNLP 2020, pp. 2898–2904. Association for Computational Linguistics (2020). https://aclanthology.org/2020.findings-emnlp.261

8. Chalkidis, I., Fergadiotis, M., Tsarapatsanis, D., Aletras, N., Androutsopoulos, I., Malakasiotis, P.: Paragraph-level rationale extraction through regularization: a case study on European court of human rights cases. In: Proceedings of the 2021 Conference of the North American Chapter of the Association for Computational Linguistics: Human Language Technologies, pp. 226–241. Association for Computational Linguistics (2021). https://doi.org/10.18653/v1/2021.naacl-main.22, https://aclanthology.org/2021.naacl-main.22

9. Chalkidis, I., et al.: LexGLUE: a benchmark dataset for legal language understanding in English. In: Proceedings of the 60th Annual Meeting of the Association for Computational Linguistics (Volume 1: Long Papers), pp. 4310–4330. Association for Computational Linguistics, Dublin (2022). https://aclanthology.org/2022.acl-long.297

10. Chen, H., Cai, D., Dai, W., Dai, Z., Ding, Y.: Charge-based prison term prediction with deep gating network. In: Proceedings of the 2019 Conference on Empirical Methods in Natural Language Processing and the 9th International Joint Conference on Natural Language Processing (EMNLP-IJCNLP), pp. 6362–6367. Association for Computational Linguistics, Hong Kong (2019). https://doi.org/10.18653/v1/D19-1667, https://aclanthology.org/D19-1667

11. Condevaux, C., Harispe, S.: LSG attention: extrapolation of pretrained transformers to long sequences. In: Kashima, H., Ide, T., Peng, W.C. (eds.) PAKDD 2023. LNCS, vol. 13935, pp. 443–454. Springer, Cham (2023). https://doi.org/10.1007/978-3-031-33374-3_35

12. Cui, J., Shen, X., Nie, F., Wang, Z., Wang, J., Chen, Y.: A survey on legal judgment prediction: datasets, metrics, models and challenges. ArXiv abs/2204.04859 (2022)

13. Devlin, J., Chang, M., Lee, K., Toutanova, K.: BERT: pre-training of deep bidirectional transformers for language understanding. In: Burstein, J., Doran, C., Solorio, T. (eds.) Proceedings of the 2019 Conference of the North American Chapter of the Association for Computational Linguistics: Human Language Technologies, NAACL-HLT 2019, Minneapolis, MN, USA, 2–7 June 2019, Volume 1 (Long and Short Papers), pp. 4171–4186. Association for Computational Linguistics (2019). https://doi.org/10.18653/v1/n19-1423

14. Feng, Y., Li, C., Ng, V.: Legal judgment prediction: a survey of the state of the art. In: Raedt, L.D. (ed.) Proceedings of the Thirty-First International Joint Conference on Artificial Intelligence, IJCAI-22, pp. 5461–5469. International Joint Conferences on Artificial Intelligence Organization (2022). https://doi.org/10.24963/ijcai.2022/765

15. Gao, L., et al.: The pile: an 800gb dataset of diverse text for language modeling. CoRR **abs/2101.00027** (2021). https://arxiv.org/abs/2101.00027

16. Jawahar, G., Sagot, B., Seddah, D.: What does BERT learn about the structure of language? In: Korhonen, A., Traum, D., Màrquez, L. (eds.) Proceedings of the 57th Annual Meeting of the Association for Computational Linguistics, pp. 3651–3657. Association for Computational Linguistics, Florence (2019). https://doi.org/10.18653/v1/P19-1356, https://aclanthology.org/P19-1356

17. Katju, J.M.: Backlog of cases crippling judiciary (2019). https://www.tribuneindia.com/news/archive/comment/backlog-of-cases-crippling-judiciary-776503

18. Kaufman, A.R., Kraft, P., Sen, M.: Improving supreme court forecasting using boosted decision trees. Polit. Anal. **27**(3), 381–387 (2019). https://doi.org/10.1017/pan.2018.59

19. Kitaev, N., Kaiser, L., Levskaya, A.: Reformer: the efficient transformer. In: 8th International Conference on Learning Representations, ICLR 2020, Addis Ababa, Ethiopia, 26–30 April 2020. OpenReview.net (2020). https://openreview.net/forum?id=rkgNKkHtvB

20. Malik, V., et al.: ILDC for CJPE: Indian legal documents corpus for court judgment prediction and explanation. CoRR **abs/2105.13562** (2021). https://arxiv.org/abs/2105.13562

21. McInnes, L., Healy, J., Astels, S.: HDBSCAN: hierarchical density based clustering. J. Open Sour. Softw. **2**(11), 205 (2017). https://doi.org/10.21105/joss.00205 https://doi.org/10.21105/joss.00205 https://doi.org/10.21105/joss.00205

22. McInnes, L., Healy, J., Melville, J.: UMAP: uniform manifold approximation and projection for dimension reduction (2018). https://doi.org/10.48550/ARXIV.1802.03426

23. Nallapati, R., Manning, C.D.: Legal docket classification: where machine learning stumbles. In: Proceedings of the 2008 Conference on Empirical Methods in Natural Language Processing, pp. 438–446. Association for Computational Linguistics, Honolulu (2008). https://aclanthology.org/D08-1046

24. Paul, S., Mandal, A., Goyal, P., Ghosh, S.: Pre-training transformers on Indian legal text (2022). https://doi.org/10.48550/ARXIV.2209.06049

25. Prasad, N., Boughanem, M., Dkaki, T.: Effect of hierarchical domain-specific language models and attention in the classification of decisions for legal cases. In: Proceedings of the 2nd Joint Conference of the Information Retrieval Communities in Europe (CIRCLE 2022), Samatan, Gers, France, 4–7 July 2022. CEUR Workshop Proceedings, vol. 3178. CEUR-WS.org (2022). http://ceur-ws.org/Vol-3178/CIRCLE_2022_paper_21.pdf

26. Song, Y., Wang, J., Liang, Z., Liu, Z., Jiang, T.: Utilizing BERT intermediate layers for aspect based sentiment analysis and natural language inference. CoRR **abs/2002.04815** (2020). https://arxiv.org/abs/2002.04815

27. Thoppilan, R., et al.: LaMDA: language models for dialog applications (2022)

28. Touvron, H., et al.: LLaMA: open and efficient foundation language models (2023)

29. Trautmann, D., Petrova, A., Schilder, F.: Legal prompt engineering for multilingual legal judgement prediction. CoRR abs/2212.02199 (2022). https://doi.org/10.48550/arXiv.2212.02199

30. Tuggener, D., von Däniken, P., Peetz, T., Cieliebak, M.: LEDGAR: a large-scale multi-label corpus for text classification of legal provisions in contracts. In: Proceedings of the Twelfth Language Resources and Evaluation Conference, pp. 1235–1241. European Language Resources Association, Marseille (2020). https://aclanthology.org/2020.lrec-1.155

31. Vaswani, A., et al.: Attention is all you need. CoRR abs/1706.03762 (2017). http://arxiv.org/abs/1706.03762

32. Wang, B., Komatsuzaki, A.: GPT-J-6B: a 6 billion parameter autoregressive language model (2021). https://github.com/kingoflolz/mesh-transformer-jax

33. Welch, B.L.: The generalization of 'student's' problem when several different population variances are involved. Biometrika **34**(1/2), 28–35 (1947). http://www.jstor.org/stable/2332510

34. Xiao, C., et al.: CAIL2018: a large-scale legal dataset for judgment prediction. CoRR abs/1807.02478 (2018). http://arxiv.org/abs/1807.02478

35. Xu, N., Wang, P., Chen, L., Pan, L., Wang, X., Zhao, J.: Distinguish confusing law articles for legal judgment prediction. In: Proceedings of the 58th Annual Meeting of the Association for Computational Linguistics, pp. 3086–3095. Association for Computational Linguistics (2020). https://doi.org/10.18653/v1/2020.acl-main.280, https://aclanthology.org/2020.acl-main.280

36. Yang, J., Zhao, H.: Deepening hidden representations from pre-trained language models for natural language understanding. CoRR abs/1911.01940 (2019). http://arxiv.org/abs/1911.01940

37. Yang, Z., Yang, D., Dyer, C., He, X., Smola, A., Hovy, E.: Hierarchical attention networks for document classification. In: Proceedings of the 2016 Conference of the North American Chapter of the Association for Computational Linguistics: Human Language Technologies, pp. 1480–1489. Association for Computational Linguistics, San Diego (2016). https://doi.org/10.18653/v1/N16-1174, https://aclanthology.org/N16-1174

38. Yu, F., Quartey, L., Schilder, F.: Legal prompting: teaching a language model to think like a lawyer (2022)

39. Zaheer, M., et al.: Big bird: transformers for longer sequences. In: Larochelle, H., Ranzato, M., Hadsell, R., Balcan, M., Lin, H. (eds.) Advances in Neural Information Processing Systems, vol. 33, pp. 17283–17297. Curran Associates, Inc. (2020)
40. Zhang, X., Wei, F., Zhou, M.: HIBERT: document level pre-training of hierarchical bidirectional transformers for document summarization. In: Proceedings of the 57th Annual Meeting of the Association for Computational Linguistics, pp. 5059–5069. Association for Computational Linguistics, Florence (2019). https://doi.org/10.18653/v1/P19-1499, https://aclanthology.org/P19-1499
41. Zheng, L., Guha, N., Anderson, B.R., Henderson, P., Ho, D.E.: When does pre-training help? Assessing self-supervised learning for law and the casehold dataset of 53,000+ legal holdings. In: Proceedings of the Eighteenth International Conference on Artificial Intelligence and Law, ICAIL 2021, pp. 159–168. Association for Computing Machinery, New York (2021). https://doi.org/10.1145/3462757.3466088
42. Zhong, H., Guo, Z., Tu, C., Xiao, C., Liu, Z., Sun, M.: Legal judgment prediction via topological learning. In: Proceedings of the 2018 Conference on Empirical Methods in Natural Language Processing, pp. 3540–3549. Association for Computational Linguistics, Brussels (2018). https://doi.org/10.18653/v1/D18-1390, https://aclanthology.org/D18-1390
43. Zhong, H., Xiao, C., Tu, C., Zhang, T., Liu, Z., Sun, M.: How does NLP benefit legal system: a summary of legal artificial intelligence. In: Proceedings of the 58th Annual Meeting of the Association for Computational Linguistics, pp. 5218–5230. Association for Computational Linguistics (2020). https://aclanthology.org/2020.acl-main.466

Emotional Insights for Food Recommendations

Mehrdad Rostami[1]([✉]) [iD], Ali Vardasbi[2] [iD], Mohammad Aliannejadi[2] [iD],
and Mourad Oussalah[1] [iD]

[1] University of Oulu, Center for Machine Vision and Signal Analysis, Oulu, Finland
`mehrdad.rostami@oulu.fi`
[2] University of Amsterdam, Information Retrieval Lab, Amsterdam, Netherlands

Abstract. Food recommendation systems have become pivotal in offering personalized suggestions, enabling users to discover recipes in line with their tastes. However, despite the existence of numerous such systems, there are still unresolved challenges. Much of the previous research predominantly lies on users' past preferences, neglecting the significant aspect of discerning users' emotional insights. Our framework aims to bridge this gap by pioneering emotion-aware food recommendation. The study strives for enhanced accuracy by delivering recommendations tailored to a broad spectrum of emotional and dietary behaviors. Uniquely, we introduce five novel scores for *Influencer-Followers*, *Visual Motivation*, *Adventurous*, *Health* and *Niche* to gauge a user's inclination toward specific emotional insights. Subsequently, these indices are used to re-rank the preliminary recommendation, placing a heightened focus on the user's emotional disposition. Experimental results on a real-world food social network dataset reveal that our system outperforms alternative emotion-unaware recommender systems, yielding an average performance boost of roughly 6%. Furthermore, the results reveal a rise of over 30% in accuracy metrics for some users exhibiting particular emotional insights.

Keywords: Recommender system · Food recommendation · Cognitive science · Emotional insights

1 Introduction

In today's digital world, an immense amount of information is continuously uploaded to the Internet by various online services and applications. This led to an information overload challenge where users struggled to sort through massive data to find relevant content [29]. To address this challenge, researchers have turned to machine learning and deep learning to develop Recommender Systems (RSs) as key tools. These RSs not only filter out information overload efficiently but also use principles from information retrieval theory to ensure content is accurate and relevant, helping users easily find items that match their preferences [19]. Food recommender systems aim to offer food suggestions to users by assisting them in discovering their preferred food choices. They became

N. Goharian et al. (Eds.): ECIR 2024, LNCS 14609, pp. 238–253, 2024.
https://doi.org/10.1007/978-3-031-56060-6_16

a vital component in a variety of lifestyle applications and significantly influenced various dietary practices [23,34]. In today's fast-paced world, where food choices abound, food RSs emerge as valuable tools in shaping our dietary habits. While numerous food RSs exist in the literature [10,34], many challenges remain unaddressed. Notably, most prior research has focused mainly on users' historical preferences [25], often overlooking the crucial aspect of identifying emotional users' insights. The impact of informational stimuli on consumer decision-making is moderated by emotional responses [3]. Yang et al. [37] suggest that when users browse fashion brand websites, feelings of dominance and arousal are triggered. These emotions enhance pleasure and subsequently lead to a positive attitude toward the brands. In the realm of dietary choices, the behavioral, emotional, and cognitive profiles of users provide valuable insights into the underlying motivations and triggers that dictate their food preferences and diet style [20]. Based on a survey by the National Restaurant Association regarding ethnic food consumption, 20% of participants were identified as adventurous eaters, expressing a keen interest in trying dishes they have never tasted before [30]. Therefore, classic recommendation systems that emphasize only on the past user's interactions, and neglect their behavioral and emotional insights may struggle to provide efficient recommendations in this regard. Moreover, previous studies examining behavioral patterns of users on food reveal that some users place a higher emphasis on the visual attraction of foods compared to other attributes when making their food choices [4,9]. As a result, traditional food recommender systems, which predominantly rely on user ratings and comments as their primary data resource, fall short of offering appropriate recommendations for users with specific emotional insight. This underscores the need for a more nuanced approach to designing systems that take into account user emotions.

Drawing from prior literature in food science, we can identify diverse emotional insights in interactions with online food and recipes. Each pattern represents a distinct combination of dietary styles and emotions. First, *Social Influencer-followers* (in short, *followers*) align with digital trends, gravitating towards popular items and culinary movements. These trends are often dictated by the narratives promoted by social influencers [24,28]. *Visual Motivated* users in food social media are driven by the aesthetic appeal of dishes, prioritizing the photogenic quality of food presentations in their choices and interactions [33]. In contrast, *Adventurous Eaters* in food-focused social media are the culinary explorers, continuously delving into varied foods and always on the hunt for a diverse array of recipes. *Health Motivated* group prioritizes nutritional value and wellness, often seeking recipes and foods that align with a holistic and health-conscious lifestyle. Lastly, the *Niche Users* group is passionate about discovering lesser-known foods and exotic flavors, often engaging with less-explored gastronomic delights. Together, these user emotional insights underscore the diverse motivations and inclinations driving user engagement with food on social media.

In this study, our primary objective is to address the oversight of emotional patterns in food recommendations on social media. To capture the diversity in users' emotional patterns and attitudes, we innovate by introducing a re-ranking method that integrates the identified emotional attitudes of users. Using these five scores, we adjust the initial recommendation list. Experiments conducted on a renowned food recommendation dataset showed that our model significantly improves the accuracy for both users with specific emotional patterns and the average community group. Importantly, our emotional-aware food recommendation approach enhances the performance of food recommendation systems by approximately 6%. Our contributions can be summarized as follows:

- To the best of our knowledge, this study is the first emotional user pattern-aware food recommendation system tackled in the information retrieval community. Although the features used to identify these patterns are straightforward and simple, this research lays the groundwork for future advancements in incorporating emotional patterns into a wide variety of RSs.
- We get inspirations from cognitive science and bring them into RS, contributing to a growing body of research aimed at improving personalization.
- We investigate the impact of our model on users with varied patterns. Results showed that users with pronounced health and visual motivation patterns experienced improvements of 30% and 18% in NDCG@10, respectively.
- We examine the significance of the emotion-aware model and its influence on aspects beyond accuracy, such as diversity and health.

2   Related Work

In contrast to conventional recommendation systems that only make use of content or collaborative filtering signals to model user preferences, context-aware recommendation systems take different contextual attributes into consideration and try to capture user preferences more correctly [16]. One of the approaches in context-aware recommendation is contextual post-filtering which involves two steps: generation of predicted ratings similar to conventional RSs, and then, use of contextual information to adjust those ratings for every user [22]. A special case of context-aware RSs that has gained attention in recent years is using personality [1,6–8,31], and other cognitive aspects such as behavior, and attitude [2,5,18,35] to enhance recommendation accuracy. Personality-aware RSs are based on measurable scales of personality traits, defined by personality theories. The Five Factors Model (FFM) [11] is one of the most popular in RS literature. The human personality is defined by five factors in FFM, namely Openness to experience, Conscientiousness, Extraversion, Agreeableness, and Neuroticism. For instance, the impact of personality modeling using FFM was analyzed in [14] on music recommendation error. In [36], the impact of popularity bias on different personality traits was investigated. In [6], both users and items were grouped according to the influence of the recommendation on users' behavior, and the extreme scenarios of recommending influential items to sensitive users were studied. In [8] four different personality models were combined to represent the user's personality in a more precise way that fits various recommendation scenarios.

Many previous traditional behavior and personality-aware methods that categorize users based solely on behavioral or personality traits often oversimplify the vast emotional spectrum by categorizing them into *positive* or *negative* groups.

This reductionist approach misses the depth and intricacies of human emotions, limiting the potential for truly personalized recommendations. Embracing a multidimensional emotional representation and creating an emotional user-item profile, as we propose in this study, recognizes the intricate nature of emotions. This approach sets the stage for a more efficient emotion-aware model.

3 Proposed Method

The diversity of emotional insights reflects the vast array of individual experiences, backgrounds, and perspectives in human interactions and decision-making [12, 26]. Diversity in emotional insights greatly influences decision-making, both on social media platforms and in the real world. Diversity plays a pivotal role when considering the emotional insights of individuals in the context of food choices. These patterns can vary significantly among individuals and can be even more pronounced due to the unique criteria and decision-making patterns each user employs when deciding what to eat. For instance, while one person might prioritize the visual appeal, shape, and imagery of a recipe, another might be more inclined to consider the healthiness of the food. This vast array of preferences suggests that a one-size-fits-all system may not cater adequately to the diverse needs of users. Therefore, it becomes imperative to first identify and analyze these emotional insights to subsequently personalize the system to better cater to individual preferences.

In this section, we introduce a framework that incorporates identified emotional user insights in food recommendations. The primary objective of this framework is to establish an emotion-aware model to improve overall performance by providing more tailored recommendations based on diverse emotional dietary preferences. It is important to note that behavioral patterns, such as Adventurous and Healthy motivation, do not inherently conflict with one another, and it is an oversimplification to categorize users solely based on a behavioral aspect. For instance, a user can prioritize Adventurous while also being Health-motivated. This nuanced understanding sets our RS apart from previous community detection-based models that pigeonhole users based on specific measures.

3.1 Problem Formulating

Consider a food recommendation system with n users and m foods. Assume $U = \{u_1, u_2, u_3, ..., u_n\}$ and $F = \{f_1, f_2, f_3, ..., f_m\}$ represent the sets of users and foods, respectively. Let R denote the user–food matrix containing the user ratings assigned to individual food items. In the scenario of not considering emotional insights, a list of top-N items, $L(u) = \{f_1, f_2, f_3, ..., f_N\}$, is suggested to the user $u \in U$ employing a base recommendation model. Given that basic food recommendations often overlook the intricacies of emotional insights, they tend to suggest items that might not align with individuals whose behaviors diverge from societal norms. To address this, we employ a re-ranking binary matrix, $A = [A_{ij}]_{n \times m}$, to determine if a food f_j should be included in the emotion-aware list for the user u_i or not.

3.2   Emotional Insights Identification

This method aims to provide more personalized emotion-aware food recommendations by adjusting the initial lists produced by traditional models. As previously pointed out, users interacting with online food and recipes typically exhibit five predominant emotional insights: Followers, Visual Motivated, Adventurous, Health Motivated, and Niche. In the subsequent phase, using these identified attitudes, we ascertain an individual's inclination weight towards each of them. To achieve this, we introduce for the first time, five novel criteria to score user's tendency towards these emotional attitudes and re-rank the initial recommendation list accordingly. In this step, as opposed to other behavioral-based food recommendation models that categorize users into predefined categories, we assign weight values to different emotional aspects. In fact, we aim to personalize the recommendation list by placing a greater emphasis on the emotional attitude of users. We describe these emotion scores in detail in the following subsections.

Followers: In the expansive realm of food-centric digital networks, the phenomenon of Social Influencer-followers stands out distinctly. These individuals not only demonstrate a keen sensitivity to the rapidly changing digital landscape but also exhibit a marked preference for following and engaging with popular foods and the latest culinary trends. Their behaviors and choices consistently align with these prevailing trends, which are predominantly influenced and shaped by the compelling narratives crafted and propagated by leading social influencers. In this study, we quantify the Followers' score by calculating the Social Influence value of the recipe items rated by them. As mentioned in [21], this value can be determined by considering three factors of average rating, the number of recorded ratings, and the number of comments as below:

$$Influence(f_i) = 2 \times \frac{AveRating(f_i) + Rating(f_i) + Comment(f_i)}{3} - 1 \quad (1)$$

where, $AveRating(f_i)$, $Rating(f_i)$ and $Comment(f_i)$ represent the normalized values of the average rating, the number of recorded ratings, and the number of comments, respectively, each ranging from 0 to 1. Given this normalization, the final computed $Influence(f_i)$ value will fall within the range $[-1, 1]$, where -1 and 1 respectively indicate low and high tendencies to follow influencers and trends. Next, the Social Influencer-followers value of user u_i calculated as:

$$InfluencerValue(u_i) = \sum_{f_j \in F_{u_i}} Rate(u_i, f_j) \times (Influence(f_j)) \quad (2)$$

where F_{u_i} is set of rated recipes by user u_i, $Rate(u_i, f_j)$ is our new defined rating scale denoting the rating of user u_i to recipe f_j. The new scale transforms the original ratings, ranging from $[1, 5]$, into a new range of -1 to $+1$, indicating levels of strong disagreement to strong agreement. When a user consistently rates food items with high Influence Values highly or gives lower ratings to food items with low Influence Values, their Social Influencer-followers value is increased. As a result, Social Influencer-followers values will also range from -1 to $+1$.

Visual Attraction: Users influenced by the Visual Attraction emotional attitude are profoundly driven by the aesthetic allure of recipes, prioritizing the photogenic quality of food presentations in their interactions. These users, captivated by visual appeal, often base their choices and dining decisions on the visual spectacle of the food. This underscores the significant power of a food's visual presentation in the digital age. In this study, we determine the Visual Attraction value for users by computing the Visual Attraction value of a recipe item they rated. Drawing inspiration from [33], this value is determined by evaluating four factors: Sharpness, Contrast, Saturation, and Brightness:

$$Attraction(f_i) = \frac{Sharpness(f_i)+Contrast(f_i)+Saturation(f_i)+Brightness(f_i)}{2} - 1 \quad (3)$$

where, $Sharpness(f_i)$, $Contrast(f_i)$, $Saturation(f_i)$, and $Brightness(f_i)$ represent the normalized values of the Sharpness, Contrast, Saturation, and Brightness of the food f_i, respectively, each ranging from 0 to 1. Given this normalization, the final computed $Attraction(f_i)$ value will fall within the range $[-1, 1]$, where -1 and 1 respectively indicate low and high tendencies to follow Visual Attraction of the recipes. Next, the Visual Attraction value of user u_i can be calculated as below:

$$VisualValue(u_i) = \sum_{f_j \in F_{u_i}} Rate(u_i, f_j) \times (Attraction(f_j)) \quad (4)$$

Adventurous: Adventurous explorers are characterized not just by their curiosity, but by their continuous pursuit of diverse recipes, their openness to unfamiliar flavors, and their relentless quest for a broad spectrum of foods and recipes. The Adventurous value for a given user u_i can be defined as:

$$AdventurousValue(u_i) = \sum_{f_j, f_k \in F_{u_i}} Rate(u_i, f_j) \times Rate(u_i, f_k) \times (1 - Sim(f_j, f_k))$$

where $sim(f_i, f_j)$ is the similarity of f_i and f_j which as calculated as below:

$$Sim(f_i, f_j) = 2 \times \frac{\widetilde{G}_i . \widetilde{G}_j}{\left\|\widetilde{G}_i\right\|_2 \times \left\|\widetilde{G}_j\right\|_2} - 1 \quad (5)$$

where $\widetilde{G}_i$ is deep embedding vector of food f_i based on its ingredients.

Health Motivation: This emotional insight places paramount importance on nutritional value and overall wellness. These users are consistently on the lookout for recipes and foods that resonate with a holistic and health-conscious approach. To determine the Health Motivation value of a user, we must first compute the health factor of recipe items as follows:

$$Health(f_i) = \frac{12 - FSA(f_i)}{4} - 1 \quad (6)$$

where, $FSA(f_i)$ is the health level for food f_i. In this study, to calculate the healthiness of a specific recipe, we utilize the "traffic light" system [32]. This method, a universally acknowledged standard set by the UK Food Standard Agency (FSA) for food labeling and health assessment as referenced by Starke et al. [27], evaluates the health score of a food item based on four macronutrients: fats, saturated fats, salt, and sugar content. The FSA score assigned to each recipe falls between 4 and 12. Consequently, the ultimate $Health(f_i)$ value will span the range $[0, 1]$. Following the determination of a recipe item's health level, the Health Motivation for user u_i can be computed as follows:

$$HealthValue\,(u_i) = \sum_{f_j \in F_{u_i}} Rate\,(u_i, f_j) \times (Health(f_j)) \tag{7}$$

Niche Motivation: This emotional insight is mainly devoted to unearthing lesser-known recipes, and users who are mainly into exotic flavors. The Niche Motivation of user u_i can be defined as:

$$NicheValue\,(u_i) = \sum_{f_j \in F_{u_i}} Rate\,(u_i, f_j) \times (Novelty(f_j)) \tag{8}$$

where $Novelty(f_j)$ is the level of novelty of f_j can be calculated as:

$$Novelty(f_j) = 2 \times \frac{\sum_{f_k \in F, f_k \neq f_j} (1 - Sim(f_k, f_j))}{m - 1} - 1 \tag{9}$$

where m is the number of all foods in the dataset.

3.3 Re-ranking Recommendation

In this phase, we present a framework for harnessing the determined emotional tendencies of users. The objective is to generate a recommendation list that prioritizes user-specific emotional states and preferences. By doing so, this framework aims to produce a balanced and highly personalized food recommendation system that respects the unique emotional attributes of each user.

$$Maximize \quad \sum_{i=1}^{n} \left[\sum_{j=1}^{m} A_{ij} \times S_{ij} \right. \tag{10}$$

$$+ \sum_{j=1}^{m} A_{ij} \times Influence(f_j) \times InfluencerValue(u_i)$$

$$+ \sum_{j=1}^{m} A_{ij} \times Attraction(f_j) \times VisualValue(u_i)$$

$$+ \sum_{j=1}^{m} A_{ij} \times \sum_{f_k \in F_{u_i}} (1 - Sim(f_k, f_j)) \times AdventurousValue(u_i)$$

$$+ \sum_{j=1}^{m} A_{ij} \times Health(f_j) \times HealthValue(u_i)$$

$$+ \sum_{j=1}^{m} A_{ij} \times Novelty(f_j) \times NicheValue(u_i) \Bigg],$$

$$\text{subject to} \quad \sum_{j=1}^{m} A_{ij} = K$$

4   Experimental Results and Discussion

In this section, we assess the performance of the developed method through a set of comprehensive experiments using a real-world dataset.

Baselines: We discuss the effectiveness of our model in comparison to other emotion-unaware models. Below, we provide a brief overview of the baseline recommendation methods in the experiments.

- **Collaborative Filtering (CF)** predicts user favorite items by examining both user interactions and similarities among users.
- **Matrix Factorization (MF)** [15] takes into account users, items, and global bias terms for generating recommendations.
- **Neural Collaborative Filtering model (NeuMF)** [13] leverages a deep neural network to learn a function that effectively pairs users with items.
- **Variational Autoencoder Collaborative Filtering (VAECF)** [17] employs variational autoencoders to implement a generative approach, supported by a multinomial likelihood function.

Dataset: To assess the efficacy of our developed approach, we use a large-scale dataset featuring millions of ratings sourced from AllRecipes.com [10]. In this dataset, a total of 52,821 recipes were collected. The dataset includes user ratings, user comments, ingredients, and nutritional information about the recipes. After a pre-processing stage, the dataset contains 68,768 users, 45,630 food items with 33,147 ingredients, and 1,093,845 user interactions.

Evaluation metrics: In this study, we employ four accuracy evaluation metrics, including Precision@K (P@K), Recall@K (R@K), F1@K, and NDCG@K (N@K), to assess the performance of the models being compared. In addition to accuracy metrics, we introduce four alternative metrics to evaluate models from different perspectives: the Health, Popularity, Novelty, and Diversity Indexes. The first three metrics are defined as follows:

$$MetricIndex\,(A) = \frac{\sum_{i=1}^{n} \sum_{j=1}^{m} A_{ij} \times \text{Metric}\,(f_j)}{n \times K}$$

where Metric $\in \{Health, Popularity, Novelty\}$, while the Diversity Index is defined as follows:

$$DiversityIndex\,(A) = \frac{\sum_{i=1}^{n} \sum_{j=1}^{m} \sum_{k=j+1}^{m} A_{ij} \times A_{ik} \times (1 - Sim\,(f_j, f_k))}{\left(\frac{K(K-1)}{2}\right)}$$

Table 1. Recommendation accuracy performance of different models. * indicates statistically significant improvements of each Emotion-Aware model, compared to their Emotion-Unaware model (Friedman test, p-value < 0.05).

Method	Emotion-Unaware				Emotion-Aware (ours)			
	P@10	R@10	F1@10	N@10	P@10	R@10	F1@10	N@10
CF	0.0601	0.0604	0.0645	0.0385	0.0643*	0.0652*	0.0647*	0.0411*
MF	0.0684	0.0681	0.0682	0.0465	0.0728*	0.0723*	0.0725*	0.0484*
NeuMF	0.0707	0.0685	0.0696	0.0467	0.0751*	0.0724*	0.0737*	0.0492*
VAECF	0.0712	0.0691	0.0701	0.0490	0.0761*	0.0741*	0.0751*	0.0518*

Table 2. Recommendation beyond-accuracy performance of different models

Method	Emotion-Unaware				Emotion-Aware (ours)			
	Health	Popularity	Novelty	Diversity	Health	Popularity	Novelty	Diversity
CF	-0.03	0.24	0.12	0.16	-0.04	0.28	0.14	0.17
MF	0.04	0.20	0.17	0.24	0.03	0.23	0.18	0.21
NeuMF	-0.05	0.23	0.14	0.21	-0.07	0.27	0.12	0.22
VAECF	0.06	0.21	0.15	0.22	0.04	0.26	0.14	0.20

It is important to note that all of these metrics range from -1 to 1. For the Health, Novelty, and Diversity Indexes, higher values denote better recommendation lists. However, for the Popularity index, a higher value could be interpreted as an indication of system bias.

Results: In Table. 1, we report the accuracy performance metrics including P@10, R@10, F1@10, and N@10. These are compared between two recommendation scenarios: Emotion-Unaware and Emotion-Aware. Our analysis suggests a consistent trend: using emotional insights to inform food recommendations enhances performance across all metrics. For example, in the case of VAECF, improvements of 6.88%, 7.24%, 7.13%, and 5.71% are observed for P@10, R@10, F1@10, and N10, respectively, indicating the advantage of understanding user's emotional tendencies. This enables more tailored personalized recommendations. Furthermore, we note that all the improvements between the emotion-aware models are statistically significant, when compared to their emotion-aware models, indicating the significance of incorporating emotion-inspired scores in the recommendation algorithm.

Additionally, in Table. 2, we detail the metrics focused on beyond-accuracy metrics, specifically the Health, Popularity, Novelty, and Diversity Indexes. An intriguing observation is that when emotional insights are incorporated into the recommendation system, there is a potential decline in the Health Index. A plausible explanation for this trend lies in individual preferences; many people are naturally drawn to unhealthy food options. Consequently, these identified emotional insights might inadvertently amplify the recommendation of less healthy food options. In other words, the preference of many users towards

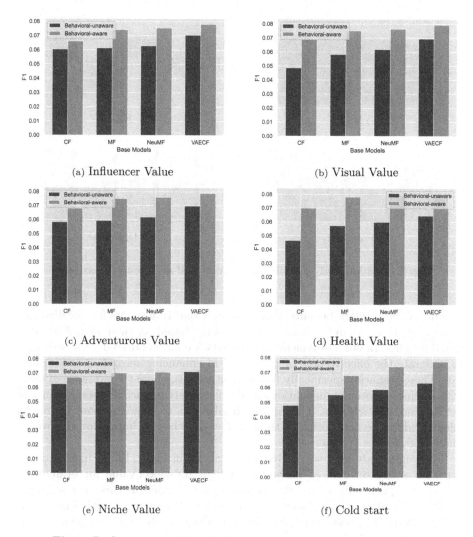

(a) Influencer Value

(b) Visual Value

(c) Adventurous Value

(d) Health Value

(e) Niche Value

(f) Cold start

Fig. 1. Performance improvement across various user emotion insights

unhealthy food choices prompts the RS to echo this behavior, recommending more unhealthy foods. A similar trend is observed in the Popularity index. This indicates that the system identifies a predominant emotional insight aligning with popular preferences, leading to recommendations of more popular food items. It is worth noting that a rise in the Popularity Index for recommended foods could also point to an inherent bias in the system recommending items with higher Influence value (i.e., average rating, number of ratings, and comments).

However, the situation differs for Novelty and Diversity. In certain cases, our model performs better on these criteria. This inconsistency might arise because the system contains an equal number of users who are hesitant to try novel food recipes as there are users who display a penchant for trying unknown foods.

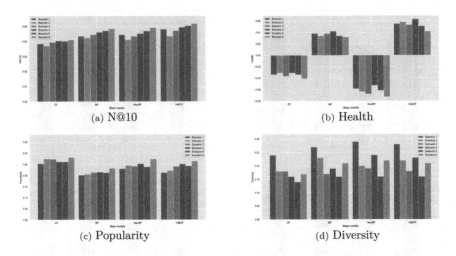

(a) N@10   (b) Health

(c) Popularity   (d) Diversity

Fig. 2. Ablation study on different accuracy and non-accuracy measures

One of the primary benefits of our model is that it accounts for the diversity in user emotions, thereby enhancing re-ranking based on their identified emotional insights. In our subsequent experiments, we have crafted diverse scenarios to illustrate the effectiveness of our model in enhancing recommendation metrics. These scenarios specifically focus on situations where the presence of users with distinct emotional inclinations might negatively impact the performance of recommendation systems. This approach allows us to showcase how our model adeptly addresses and mitigates these challenges, thereby improving the overall quality of RSs.

To demonstrate this, we outline five scenarios in Fig. 1. In the first scenario, we focus solely on 10% of users with a high InfluencerValue for evaluation. This is to specifically gauge the impact of our model on this user segment. For the second to fifth scenarios, we consider 10% of users with high levels of VisualValue, AdventurousValue, HealthValue, and NicheValue, respectively. The reported results indicate significant performance improvement for these user groups. For instance, for the top 10% of users with a high InfluencerValue, our model exhibits an average improvement of 13% in terms of F1@10. Furthermore, we see that users with a high HealthValue experience a maximum improvement of 36% in terms of F1@10. Conversely, we observe the minimum improvement for users with a high NicheValue, i.e., 8% in terms of F1@10. In addition to analyzing five user segments, we also assess the performance of various RSs for cold start users. The results demonstrate that our emotional-aware model significantly improves recommendation performance for these users.

Our model's primary contribution lies in its ability to identify user inclinations toward the aforementioned five distinct emotional insights. This part of our analysis assesses the significance of each user insight in improving the quality of food recommendations generated by our model. To illustrate this, we conduct an ablation study, examining how the performance of our model is affected when each of these user insights is excluded from the optimization equation (i.e., Eq (10)). Specifically, we outline five scenarios labeled as Scenario 1 through 5. In each, we exclude the Influence, Visual Motivation, Adventurous, Health Motivation, and Niche emotional insights, respectively. Moreover in Scenario 6, we consider all of these emotional user insights. Based on the described scenarios, we conduct experiments using various evaluation metrics. Figure 2 shows the results of the ablation study concerning accuracy and beyond-accuracy measures. Examining the accuracy performance (i.e., N@10) reveals maximum drops in Scenarios 2 and 3, indicating the importance of Visual Motivation and Adventurous. Upon examining various beyond-accuracy measures, it becomes evident that each term uniquely influences these measures. In terms of the Health measures, the findings are more interesting. The figure shows that omitting the Health term from the optimization equation enhances the final Health measure. This result arises because including this term often results in recommending unhealthier food options to those who already have unhealthy eating emotional insights, thereby directing them towards even more unhealthy recipes. As for the Popularity measure, the results are anticipated. The Influence and Niche terms emerge as the most pivotal. Additionally, our analysis shows that excluding the Influence term positively impacts the Diversity measure.

5   Discussion and Conclusion

Food RSs help users find their preferred food choices and play a pivotal role in many lifestyle apps. While many of these systems exist, they often focus on users' past preferences including user ratings and comments, neglecting their emotional insights. This makes recommendations less effective for users with specific emotional insights. In our study, we specifically addressed this gap and tested our approach on the social media food dataset. Specifically, our approach introduces a re-ranking mechanism that integrates users' behavioral and emotional insights. We assessed each user's emotional inclination toward each of the five factors: Social Influence Follower, Visual Motivation, Adventure, Health Motivation, and Niche. These factors are then used to refine the recommendation list. Testing on a well-known food recommendation dataset revealed that our approach enhances the accuracy of recommendations by boosting the match with users' preferred recipes by around 6%. The results also indicated that incorporating emotional insights into a recommendation system can lead to a decrease in the Health Index due to users' emotional inclination towards unhealthy food options, resulting in the system recommending such foods more frequently. A similar trend is observed in the Popularity Index, recommending more popular food items. Additionally, the results showed that specific users with high Health-Value and VisualValue yield improvements of 36% in F1@10, 30% in N@10, 27% in F1@10, and 18% in N@10. This underscores the adaptability of our model to cater to a broad spectrum of users with varying emotional aspects, rather than merely focusing on the community's average.

Our initial findings are promising, hinting at potential applications of emotional analysis and cognitive science in enhancing personalized recommendations. Our approach to Emotional Insights for Food Recommendations can be extended to other areas of recommender systems. By identifying specific emotional insights for users and items in various domains, such as E-Commerce, Fashion and Lifestyle, Travel and Tourism, our method can be adapted for emotional-aware recommendations in these fields. This generalization allows for a more nuanced understanding of user preferences across different sectors, leveraging the emotional aspect to enhance the recommendation experience. However, several areas require further investigation to enhance the performance of our model:

- *Concept Drift:* The current model of the RS relies on static data and does not account for potential shifts in user's emotional attitudes towards food, diet, or lifestyle. For a more accurate recommendation, it is essential to detect concept drift, which refers to changes in the relationship between users and their emotional responses.

- *Controllable Recommendation:* Controllable RSs allow users to influence recommendations, enhancing transparency and trust. Our emotion-aware model does not consider controllability. This omission could lead users to perceive the system as overly deterministic, possibly causing dissatisfaction if they feel their emotions are misinterpreted. Providing user control, especially regarding emotional states, is vital for user acceptance.

- *Intensification of Negative Attitude*: The emotion-aware system could inadvertently intensify negative emotional attitudes. If users display emotions related to an unhealthy diet, the system might predominantly recommend unhealthy foods. This could unintentionally promote detrimental dietary habits, with the system potentially acting as an enabler of such behaviors. Therefore, in future work, we should emphasize the role of AI in promoting a healthy lifestyle, rather than solely focusing on increasing user satisfaction.

- *Not Considering Seasonality and Social Media Trends*: Key elements such as seasonality, which plays a crucial role in the availability and popularity of certain foods and recipes, are not integrated into our system. This omission means that the recommendations may not always align with seasonal products or dishes, potentially leading to less relevant or appealing suggestions for users. Additionally, our system does not factor in the dynamic and influential social media trends. Given that social media significantly shapes culinary trends and public interest in specific types of cuisine or dishes, this exclusion could result in the system being out of sync with current food fads or popular eating habits.

Acknowledgements. The project is supported by the Research Council of Finland (former Academy of Finland) and Profi5 DigiHealth Research Program (project number 326291).

References

1. Beheshti, A., Moraveji-Hashemi, V., Yakhchi, S., Motahari-Nezhad, H.R., Ghafari, S.M., Yang, J.: personality2vec: Enabling the analysis of behavioral disorders in social networks. In: Proceedings of the 13th International Conference on Web Search and Data Mining, pp. 825–828 (2020)
2. Beheshti, A., Yakhchi, S., Mousaeirad, S., Ghafari, S.M., Goluguri, S.R., Edrisi, M.A.: Towards Cogn. Recommender Syst. Algorithms 13(8), 176 (2020)
3. Bigne, E., Chatzipanagiotou, K., Ruiz, C.: Pictorial content, sequence of conflicting online reviews and consumer decision-making: the stimulus-organism-response model revisited. J. Bus. Res. 115, 403–416 (2020)
4. Carvalho, M., Cadène, R., Picard, D., Soulier, L., Thome, N., Cord, M.: Cross-modal retrieval in the cooking context: Learning semantic text-image embeddings. In: The 41st International ACM SIGIR Conference on Research and Development in Information Retrieval, pp. 35–44 (2018)
5. Contreras, D., Salamó, M.: A cognitively inspired clustering approach for critique-based recommenders. Cogn. Comput. 12(2), 428–441 (2020)
6. De Biasio, A., Monaro, M., Oneto, L., Ballan, L., Navarin, N.: On the problem of recommendation for sensitive users and influential items: simultaneously maintaining interest and diversity. Knowl.-Based Syst. 275, 110699 (2023)
7. Dhelim, S., Aung, N., Bouras, M.A., Ning, H., Cambria, E.: A survey on personality-aware recommendation systems. Artif. Intell. Rev., pp. 1–46 (2022)
8. Dhelim, S., Chen, L., Aung, N., Zhang, W., Ning, H.: A hybrid personality-aware recommendation system based on personality traits and types models. J. Ambient. Intell. Humaniz. Comput. 14(9), 12775–12788 (2023)
9. Elsweiler, D., Trattner, C., Harvey, M.: Exploiting food choice biases for healthier recipe recommendation. In: Proceedings of the 40th International ACM SIGIR Conference on Research and Development in Information Retrieval, pp. 575–584 (2017)
10. Gao, X., et al.: Hierarchical attention network for visually-aware food recommendation. IEEE Trans. Multimedia 22(6), 1647–1659 (2019)
11. Goldberg, L.R.: An alternative" description of personality": the big-five factor structure. J. Pers. Soc. Psychol. 59(6), 1216 (1990)
12. Gutnik, L.A., Hakimzada, A.F., Yoskowitz, N.A., Patel, V.L.: The role of emotion in decision-making: a cognitive neuroeconomic approach towards understanding sexual risk behavior. J. Biomed. Inform. 39(6), 720–736 (2006)
13. He, X., Liao, L., Zhang, H., Nie, L., Hu, X., Chua, T.S.: Neural collaborative filtering. In: Proceedings of the 26th International Conference on World Wide Web, pp. 173–182 (2017)
14. Kleć, M., Wieczorkowska, A., Szklanny, K., Strus, W.: Beyond the big five personality traits for music recommendation systems. EURASIP J. Audio, Speech, Music Process. 2023(1), 4 (2023)
15. Koren, Y., Bell, R., Volinsky, C.: Matrix factorization techniques for recommender systems. Computer 42(8), 30–37 (2009)
16. Kulkarni, S., Rodd, S.F.: Context aware recommendation systems: a review of the state of the art techniques. Comput. Sci. Rev. 37, 100255 (2020)
17. Liang, D., Krishnan, R.G., Hoffman, M.D., Jebara, T.: Variational autoencoders for collaborative filtering. In: Proceedings of the 2018 World Wide Web Conference, pp. 689–698 (2018)

18. Moscato, V., Picariello, A., Sperlí, G.: An emotional recommender system for music. IEEE Intell. Syst. **36**(5), 57–68 (2021)

19. Neophytou, N., Mitra, B., Stinson, C.: Revisiting Popularity and Demographic Biases in Recommender Evaluation and Effectiveness. In: Hagen, M., Verberne, S., Macdonald, C., Seifert, C., Balog, K., Nørvåg, K., Setty, V. (eds.) ECIR 2022. LNCS, vol. 13185, pp. 641–654. Springer, Cham (2022). https://doi.org/10.1007/978-3-030-99736-6_43

20. Paans, N.P., Bot, M., Brouwer, I.A., Visser, M., Gili, M., Roca, M., Hegerl, U., Kohls, E., Owens, M., Watkins, E., et al.: Effects of food-related behavioral activation therapy on eating styles, diet quality and body weight change: results from the moodfood randomized clinical trial. J. Psychosom. Res. **137**, 110206 (2020)

21. Park, D.H., Lee, J., Han, I.: The effect of on-line consumer reviews on consumer purchasing intention: the moderating role of involvement. Int. J. Electron. Commer. **11**(4), 125–148 (2007)

22. Ramirez-Garcia, X., García-Valdez, M.: Post-Filtering for a Restaurant Context-Aware Recommender System. In: Castillo, O., Melin, P., Pedrycz, W., Kacprzyk, J. (eds.) Recent Advances on Hybrid Approaches for Designing Intelligent Systems. SCI, vol. 547, pp. 695–707. Springer, Cham (2014). https://doi.org/10.1007/978-3-319-05170-3_49

23. Rehman, F., Khalid, O., Bilal, K., Madani, S.A., et al.: Diet-right: a smart food recommendation system. KSII Trans. Internet Inf. Syst. (TIIS) **11**(6), 2910–2925 (2017)

24. Rogers, A., Wilkinson, S., Downie, O., Truby, H.: Communication of nutrition information by influencers on social media: a scoping review. Health Promot. J. Austr. **33**(3), 657–676 (2022)

25. Rostami, M., Farrahi, V., Ahmadian, S., Jalali, S.M.J., Oussalah, M.: A novel healthy and time-aware food recommender system using attributed community detection. Expert Syst. Appl. **221**, 119719 (2023)

26. Shah, A.M., Abbasi, A.Z., Yan, X.: Do online peer reviews stimulate diners' continued log-in behavior: Investigating the role of emotions in the O2O meal delivery apps context. J. Retail. Consum. Serv. **72**, 103234 (2023)

27. Starke, A.D., Willemsen, M.C., Trattner, C.: Nudging healthy choices in food search through visual attractiveness. Frontiers Artif. Intell. **4**, 621743 (2021)

28. Taillon, B.J., Mueller, S.M., Kowalczyk, C.M., Jones, D.N.: Understanding the relationships between social media influencers and their followers: the moderating role of closeness. J. Prod. Brand Manage. **29**(6), 767–782 (2020)

29. Thonet, T., Renders, J.-M., Choi, M., Kim, J.: Joint Personalized Search and Recommendation with Hypergraph Convolutional Networks. In: Hagen, M., Verberne, S., Macdonald, C., Seifert, C., Balog, K., Nørvåg, K., Setty, V. (eds.) ECIR 2022. LNCS, vol. 13185, pp. 443–456. Springer, Cham (2022). https://doi.org/10.1007/978-3-030-99736-6_30

30. Thorn, B.: Beyond fuel: modern eating linked to identity, community. Nation's Restaurant News **49**, 12 (2015)

31. Tkalčič, M.: Emotions and personality in recommender systems: Tutorial. In: Proceedings of the 12th ACM Conference on Recommender Systems, pp. 535–536 (2018)

32. Trattner, C., Elsweiler, D.: Investigating the healthiness of internet-sourced recipes: implications for meal planning and recommender systems. In: Proceedings of the 26th International Conference on World Wide Web, pp. 489–498 (2017)

33. Trattner, C., Moesslang, D., Elsweiler, D.: On the predictability of the popularity of online recipes. EPJ Data Sci. **7**(1), 1–39 (2018)

34. Wang, W., Duan, L.Y., Jiang, H., Jing, P., Song, X., Nie, L.: Market2dish: health-aware food recommendation. ACM Trans. Multimedia Comput. Communi. Appl. (TOMM) 17(1), 1–19 (2021)
35. Yakhchi, S., Beheshti, A., Ghafari, S.M., Orgun, M.: Enabling the analysis of personality aspects in recommender systems. arXiv preprint. arXiv:2001.04825 (2020)
36. Yalcin, E., Bilge, A.: Popularity bias in personality perspective: an analysis of how personality traits expose individuals to the unfair recommendation. Concurrency Comput.: Pract. Experience 35(9), e7647 (2023)
37. Yang, K., Kim, H.M., Zimmerman, J.: Emotional branding on fashion brand websites: harnessing the pleasure-arousal-dominance (PAD) model. J. Fashion Mark. Manage.: Int. J. 24(4), 555–570 (2020)

HyperPIE: Hyperparameter Information Extraction from Scientific Publications

Tarek Saier[1]([✉]) [iD], Mayumi Ohta[2] [iD], Takuto Asakura[3] [iD],
and Michael Färber[1] [iD]

[1] Karlsruhe Institute of Technology, Karlsruhe, Germany
{tarek.saier,michael.faerber}@kit.edu
[2] Fraunhofer Institute for Systems and Innovation Research, Karlsruhe, Germany
mayumi.ohta@isi.fraunhofer.de
[3] The University of Tokyo, Tokyo, Japan
takuto@is.s.u-tokyo.ac.jp

Abstract. Automatic extraction of information from publications is key to making scientific knowledge machine-readable at a large scale. The extracted information can, for example, facilitate academic search, decision making, and knowledge graph construction. An important type of information not covered by existing approaches is hyperparameters. In this paper, we formalize and tackle hyperparameter information extraction (HyperPIE) as an entity recognition and relation extraction task. We create a labeled data set covering publications from a variety of computer science disciplines. Using this data set, we train and evaluate BERT-based fine-tuned models as well as five large language models: GPT-3.5, GALACTICA, Falcon, Vicuna, and WizardLM. For fine-tuned models, we develop a relation extraction approach that achieves an improvement of 29% F_1 over a state-of-the-art baseline. For large language models, we develop an approach leveraging YAML output for structured data extraction, which achieves an average improvement of 5.5% F_1 in entity recognition over using JSON. With our best performing model we extract hyperparameter information from a large number of unannotated papers, and analyze patterns across disciplines. All our data and source code is publicly available at https://github.com/IllDepence/hyperpie.

Keywords: Information Extraction · Scientific Text · Hyperparameter

1 Introduction

Models capable of extracting fine-grained information from publications can make scientific knowledge machine-readable at a large scale. Aggregated, such information can fuel platforms like Papers with Code[1] and the Open Research Knowledge Graph [3,26], and thereby facilitate academic search, recommendation, and reproducibility. Accordingly, a variety of approaches for information extraction (IE) from scientific text have been proposed [11–13,16,19].

---

[1] See https://paperswithcode.com/.

© The Author(s), under exclusive license to Springer Nature Switzerland AG 2024
N. Goharian et al. (Eds.): ECIR 2024, LNCS 14609, pp. 254–269, 2024.
https://doi.org/10.1007/978-3-031-56060-6_17

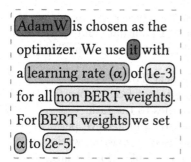 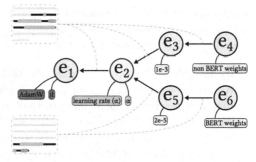

text example   relation   entities ◯, relations ◄, and
based on arXiv:2005.00512   evidence   entity mentions ◌ extracted

Fig. 1. Illustration of hyperparameter information in a text example alongside the extracted entities and relations. Entity types are research artifact, parameter, value, and context. Relations are indicated by arrows.

However, to the best of our knowledge, no approaches exist for the extraction of structured information on hyperparameter use from publications. That is, information on *with which parameters* researchers use methods and data. We refer to this information as "hyperparameter information" (see Fig. 1). Hyperparameter information is important for several reasons. (1) First, its existence in a paper is an indicator for reproducibility [22] and, when extracted automatically, can improve automated reproduction of results [25]. (2) Second, in aggregate it can inform on both conventions in a field as well as trends over time. (3) Lastly, it enables more fine-grained paper representations benefiting downstream applications based on document similarity, such as recommendation and search. Hyperparameter information is challenging to extract, because (1) it is usually reported in a dense format, (2) often includes special notation, and (3) operates on domain specific text (e.g. "For Adam we set α and β to 1e-3 and 0.9 respectively.").

To address the lack of approaches for extracting this type of information, we define the task of "hyperparameter information extraction" (HyperPIE) and develop several approaches to it. Specifically, we formalize HyperPIE as an entity recognition (ER) and relation extraction (RE) task. We create a labeled data set spanning a variety of computer science disciplines from machine learning (ML) and related areas. The data set is created by manual annotation of paper full-texts, which is accelerated by a pre-annotation mechanism based on an external knowledge base. Using our data set, we train and evaluate both BERT-based [10] fine-tuned models as well as large language models (LLMs). For the former, we develop a dedicated relation extraction model that achieves an improvement of 29% F_1 compared to a state-of-the-art baseline. For LLMs, we develop an approach leveraging YAML output for structured data extraction, which achieves a consistent improvement in entity recognition across all tested models, averaging at 5.5% F_1. Using our best performing model, we extract hyperparameter information from 15,000 unannotated papers, and analyze patterns across ML disciplines of how authors report hyperparameters. All our data and source code is made publicly available.[2] In summary, we make the following contributions.

1. We formalize a novel and relevant IE task (HyperPIE).
2. We create a high quality, manually labeled data set from paper full-texts, enabling the development and study of approaches to the task.
3. We develop two lines of approaches to HyperPIE and achieve performance improvements in both of them over solutions based on existing work.
4. We demonstrate the utility of our approaches by application on large-scale, unannotated data, and analyze the extracted hyperparameter information.

2   Related Work

Fine-Tuned Models. Named entity recognition (NER) and RE from publications in ML and related fields have been tackled by SciERC [19] and subsequently SciREX [13]. The entity types considered are methods, tasks, data sets, and evaluation metrics. Proposed methods for the task utilize BiLSTMs, BERT and SciBERT [5]. With both approaches, there is a partial overlap in entity types to our task, as we also extract methods and data sets. The key difference arises though the parameter and value entities we cover, which are a challenge in part due to their varied forms of notation (e.g. α / alpha, or 0.001 / 1×10^{-3} / 1e-3).

IE models aiming to relate natural language to numerical values and mathematical symbols have been introduced at SemEval 2021 Task 8 [12] and SemEval 2022 Task 12 [16] respectively. Most of the proposed models base their processing of natural language on BERT or SciBERT. To handle numbers and symbols rendered in LaTeX, as well as to accomplish RE between entity types with highly regular writing conventions (e.g. numbers and units such as "5 ms"), rule-based approaches or dedicated smaller neural networks are commonly used.

Similarly, we find a level of regularity in how authors report parameters and values, and make use of that in our approach accordingly. In line with related work using fine-tuned models, we also use BERT and SciBERT for contextualized token embeddings.

LLMs. With the recent advances in LLMs, there has been a surge in efforts to utilize them for IE from scientific text. Nevertheless, their performance is not on par with dedicated models for NER and RE yet [32].

An improtant concept for IE with LLMs is introduced by Agrawal et al. [1]: a "resolver" is a function that maps the potentially ambiguous output of an LLM to a defined, task specific output space. In their work, the authors extract singular values and lists from clinical notes using GPT-3. They use a variety of resolvers that perform steps like tokenization, removal of specific symbols or words, and pattern matching using regular expressions.

---

[2] See https://github.com/IllDepence/hyperpie.

Work with similar output data complexity (values and lists) has also been done in the area of material science. Xie et al. [30] use GPT-3.5 to extract information on solar cells from paper full-text. Similarly, Polak et al. [20] use Chat-GPT to extract material, value, and unit information from sentences of material science papers. They define a conversational progression, in which they prompt the model generate tables, which are processed using simple string parsing rules.

An approach for IE of more complex information is proposed by Dunn et al. [11]. They use GPT-3.5 to extract material information from materials chemistry papers. Given the hierarchical nature of the information to be extracted, the authors find simple output formats insufficient. To overcome this, they prompt the model to output the data in JSON format.

Given hyperparameter information also is hierarchical (see Fig. 1), we adopt prompting LLMs to output data in a text based data serialization format. Different from the related work introduced above, we do not limit our experiments to API access based closed source LLMs, but also evaluate various open LLMs, because we recognize the importance of contributing efforts to the advancement of the more transparent, accountable, and reproducibility friendly side of this new and rapidly evolving area of research [17].

Besides IE from scientific publications, there have been efforts to extract hyperparameter schemata and constraints from Python docstrings [4] using CNL grammars [15], and from Python code [23] using static analysis. Compared to our task setting, these rely on a known context (e.g. a `fit` method) and operate on constrained input (generated docstrings and source code instead).

3   Hyperparameter Information Extraction

3.1   Task Definition

We define HyperPIE as an ER+RE task with four entity classes "research artifact", "parameter", "value", and "context", and a single relation type. Briefly illustrated by a minimal example, in the sentence *"During fine-tuning, we use the Adam optimizer with $\alpha = 10^{-4}$."*, the research artifact *Adam* has the parameter α which is set to the value 10^{-4} in the context *During fine-tuning*.

The entity classes are characterized as follows. A "research artifact", within the scope of our task, is an entity used for a specific purpose with a set of variable aspects that can be chosen by the user. These include methods, models, and data sets.[3] A "parameter" is a variable aspect of an artifact. This includes model parameters, but also, for example, the size of a sub-sample of a data set. A "value" expresses a numerical quantity and in our task is treated like an entity rather than a literal. Lastly, a "context" can be attached to a value if the value is only valid in that specific context. The single relation type relates entities as follows: parameter → research artifact, value → parameter, and context → value.

---

[3] Broader definitions in other contexts also include software in general, empirical laws, and ideas [18]. For our purposes, however, above specific definition is more useful.

Co-reference relations implicitly exist between the mentions of a common entity (e.g. "AdamW" and "it" in Fig. 1). That is, if an entity has multiple mentions within the text, they are considered co-references to each other.

The scope of the IE task comprises the extraction of entities, their relations, and the identification of all their mentions in the text (and thereby implicitly co-references). Furthermore, we specifically consider IE from text, and not from tables, graphs, or source code.[4]

3.2   Data Set Construction

Because HyperPIE is a novel task, we cannot rely on existing data sets for training and evaluating our approaches. We therefore create a new data set by manually annotating papers. As our data source we chose unarXive [24], because it includes paper full-texts and, most importantly, retains mathematical notation as LaTeX. This is crucial because parsing such notation from PDFs is prone to noise, which would be problematic for our parameter and value entities.

To ensure we cover a wide variety of artifacts and discipline specific writing conventions, we use papers from multiple ML related fields. Specifically, these are Machine Learning (ML), Computation and Language (CL), Computer Vision (CV), and Digital Libraries (DL), which make up 143,203 papers in unarXive.[5]

We base our annotation guidelines on the widely used ACL RD-TEC guideline[6] [21]. To make sure our resulting annotations are able to properly capture how authors report hyperparameters in text, we perform two annotation rounds: (1) an initial exploratory round, the results of which are used to refine the annotation guidelines and inform later model development, and (2) the main annotation round, the results of which constitute our data set used for model training and evaluation. In the following, both steps are described in more detail.

Initial Annotation Round. We heuristically pre-filter our ML paper corpus for sections reporting on hyperparameters.[7] Annotators then inspect these sections, select a continuous segment of text that contain hyperparameter information, and make their annotations. This task is performed independently by two annotators and results in a total of 151 text segments (131 unique, 2×10 annotated by both). The annotated text segments contain 1,345 entities and 1,110 relations.

As shown in Fig. 2a, we observe text segments reporting on hyperparameters to generally have a length below 600 characters. We furthermore see that most text segments contain between 3 and 15 entities. Lastly, in Fig. 2b, show distances between artifacts and their parameters, as well as parameters and their values. We see that artifacts usually are mentioned before their parameters (78%), and

---

[4] We leave investigating multi-modal IE pipelines (text/code/graphs) for future work.

[5] The respective arXiv categories are cs.LG, cs.CL, cs.AI, and cs.DL. See https://arxiv.org/category_taxonomy for a more detailed description.

[6] See http://pars.ie/publications/papers/pre-prints/acl-rd-tec-guidelines-ver2.pdf.

[7] We filter based on key phrases ("use", "set", etc.), numbers, and LaTeX math content.

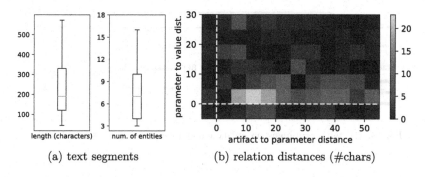

(a) text segments   (b) relation distances (#chars)

Fig. 2. Observations of initial annotation round

parameters before their values (93%). The reverse cases also exists, but are less common. Additionally, we can see that values are most commonly reported right after their parameter, while there is a higher variability in distances between parameters and artifacts. Based on above observations we determine the unit of annotation for the final round to be one paragraph (on average 563.4 characters long in our corpus), as it is sufficient to capture hyperparameters being reported.

The inter annotator agreement (IAA, reported as Cohen's kappa) of the text segments annotated by both annotators is 0.867 for entities and 0.737 for relations[8] (strong to almost perfect agreement) which is compares favorably to SciERC [19] with an IAA of 0.769 for entities and 0.678 for relations.

Main Annotation Round. In our main annotation round we annotate whole papers (paragraph by paragraph) instead of pre-filtered text-segments. This is done to ensure that the final annotation result reflects data as it will be encountered by a model during inference—that is, containing a realistic amount of paragraphs that have no information on hyperparameters, or, for example, only mention research artifacts but no parameters.

Similar to related work [13], we use Papers with Code as an external knowledge base to pre-annotate entity candidates to make the annotation process more efficient. In a similar fashion, we use annotator's previously annotated entity mentions for pre-annotation. Pre-annotated text spans are, as the name suggests, set automatically, but need to be checked by annotators manually.

Through this process we annotate 444 paragraphs, which contain 1,971 entities and 614 relations. The entity class distribution is 1,134 research artifacts, 131 parameters, 662 values, and 44 contexts. The annotation data is provided in a JSON structure as shown in Fig. 1, as well as in the W3C Web Annotation Data Model[9] to facilitate easy re-use and compatibility with existing systems.

---

[8] Measured by the character level entity class and character level relation target span agreement respectively.

[9] See https://www.w3.org/TR/annotation-model/.

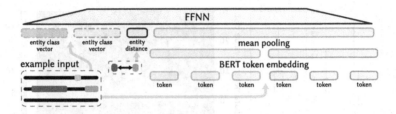

Fig. 3. RE with emphasis on entity candidate pair types and distance.

4   Methods

We approach hyperparameter information extraction in two ways. First, we build upon established ER+RE methods and develop an approach using a fine-tuned model in a supervised learning setting. Second, given the recent promising advances with LLMs, we develop an approach utilizing LLMs in a zero-shot and few-shot setting.

4.1   Fine-Tuned Models

We base our fine-tuned model approach on PL-Marker [33], the currently best performing model on SciERC. Specifically, we use the ER component of PL-Marker. Our reason is that (1) the text our model will be applied on is of the same type as in SciERC (ML publications), and (2) there is some correspondence between the entities to be identified—namely our entity class "research artifact" including methods and datasets, which are both entity classes in SciERC.

For RE we develop an approach that utilizes token embeddings as well as relative entity distance and entity class pairings. This is motivated by the fact that (a) we observed a high level or regularity in the relative distance of research artifact, parameter, and value mentions[10] (see Fig. 2), and (b) relations only exist between specific pairs of entity types.

In Fig. 3 we show a schematic depiction of our new relation extraction component. Entity candidate pair classes as well as the relative distance between the entities in the text are used as a dedicated model input, BERT token embeddings of the entity mentions are combined using mean pooling. These inputs are fed into a feed-forward neural network FFNN for prediction. Formally, the model performs pairwise binary classification as $\text{FFNN}(E_0^c, E_1^c, E^d, E^T)$, where E_i^c are class vectors, E^d encodes candidate distance, and E^T is the token pair embedding calculated as $E^T = \frac{1}{|T|} \sum_{i=0}^{|T|} \text{BERT}(t_i)$, the mean of the pair's tokens $t_i \in |T|$.

During the development of our model we also experiment with concatenation in favor of mean pooling to preserve information on the order of the entities, but find that mean pooling results in better performance. Furthermore, we

---

[10] We note that these observations where made during the initial exploratory annotation round (Sec. 3.2) and not during annotation of the evaluation data.

investigated the use of SciBERT instead of BERT, but find that regular BERT embeddings give us better results, despite our model handling scientific text.

4.2 LLM

We develop our LLM approach for a zero-shot and a few-shot setting. This means the models perform the IE task based on either instructions only (zero-shot), or instructions and a small number of examples (few-shot).

Performing IE using LLMs in zero-shot or few-shot settings requires the desired structure of the output data to be specified within the model input. In simple cases (e.g. numbers or yes/no decisions) this can be achieved by an in-line specification of the format in natural language (e.g. "The answer (arabic numerals) is") [14]. IE from scientific publications, however, often seeks to extract more complex information. To achieve this, the model can be tasked to produce output in a text based data serialization format such as JSON, as done in previous work [11]. Especially for complex structured predictions, few-shot prompting has been shown to further boost in-context learning (ICL) accuracy and consistency at inference time [6].

Drawing from techniques used in previous work approaching other IE tasks, we investigate several prompting strategies to build our approach.

1. *Multi-stage prompting* [20]: first determine the presence of hyperparameters information; if present, extract the list of entities; lastly, determine relations.
2. *In-text annotation* [29]: let the input text be repeated with entity annotations, e.g. repeat "We use BERT for ..." as "We use [a1|BERT] for ...".
3. *Data serialization format* [11]: specify a serialization format in the promt that is parsed afterwards; then match in-text mentions in the input.
4. *(3)+(2)*: prompt as in (3); then match in-text mentions using (2).

We find (1) to lead to problems with errors propagation along steps. With (2) and (4) we frequently see alterations in the reproduced text. Accordingly, we use prompt type (3) for our approach—specifying a data serialization format in the prompt. While existing work uses the JSON format for this [11], we use YAML, as it is less prone to "delimiter collision" problems due to its minimal requirements for structural characters.[11] In doing so, we expect to avoid problems with LLM output not being parsable. Our overall LLM approach looks as follows.

Zero-Shot. We build our zero-shot prompts from the following consecutive components: [task][input text][format][completion prefix]. In [task] we specify the information to extract, i.e. research artifacts, their parameters, etc. [input text] is the paragraph from which to extract the information. [format] defines the output YAML schema.[12] [completion prefix] is a piece of text that directly precedes the LLM's output, such as "ASSISTANT: ". To generate

---

[11] See https://yaml.org/spec/1.2.2/.

[12] Examples can be found at https://github.com/IllDepence/hyperpie.

predictions based on LLM output, we pass it to a standard YAML parser after cleansing (e.g. removing text around the YAML block). For each used LLM model, we individually perform prompt tuning. Here we determine, for example, if a model gives better results when the [completion prefix] includes the beginning of the serialized output (e.g., "---\ntext_contains_entities:") or if this leads to a deterioration in output quality.

Few-Shot. Our few-shot approach makes the following adjustments to the method described above. Prompts additionally include a component [examples], which are valid input output pairs sampled by their cosine similarity to the input text. Specifically, for an input text from a document X, we sample the five most similar paragraphs from all ground truth documents excluding X. As these examples can be confused with the input text, we re-position the input text to appear *after* the examples. The resulting prompt structure we use for our few-shot approach is as follows: [task][format][examples][input text] [completion prefix].

LLMs reaching a sufficient context size for a few-shot approach to our task are a recent development. We can therefore additionally make use of other recently added capabilities. Specifically, we make use of generation constrains via a gBNF grammar[13] to enforce LLM output according to our data scheme, allowing us to mitigate parsing errors.

5   Experiments

We evaluate the fine-tuned models and LLM approach against baselines from existing work. Both evaluations are performed on our data set described in Sect. 3.2. Metrics used to measure prediction performance are precision, recall and F_1 score, abbreviated as P, R and F_1 respectively.

5.1   Fine-Tuned Models

We use PL-Marker, the currently best performing model on SciERC, as our baseline. Models are trained and evaluated using 5-fold cross validation (3 folds training, 1 dev, 1 test). We train the ER component of PL-Marker as done in [33], using *scibert-scivocab-uncased* as the encoder, Adam as the optimizer, a learning rate of 2e-5, and 50 training epochs. Regarding the two RE components we compare, the PL-Marker RE component is trained using *bert-base-uncased*, Adam, a learning rate of 2e-5, and 10 training epochs. Our own RE component also uses *bert-base-uncased*, Adam as the optimizer, and is trained with a learning rate of 1e-3 for 90 epochs.[14] The models are trained and evaluated on a server with a single GeForce RTX 3090 (24 GB).

---

[13] See https://github.com/ggerganov/llama.cpp/pull/1773.

[14] The two RE models we compare require different learning rates and number of training epochs, because their architecture varies significantly.

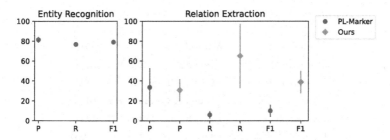

Fig. 4. Fine-tuned model evaluation (5-fold cross validation).

Results. In Fig. 4 we show the results of PL-Marker ER (used for both models) as well as the PL-Marker RE component and our RE model. For ER we evaluate exact matches (no partial token overlap). In the case of RE, each entity pair is predicted as having a relation or not—as there is just one relation type.

Mean ER performance is 81.5, 76.8, and 79.0 (P, R, F_1). For RE, the precision of PL-Marker and our model are similar at 33.5 and 30.7 respectively, but our model performs more consistent. PL-Marker only achieves a very low recall of 5.9, whereas our model, while showing large variability, achieves a mean of 65.0. The resulting F_1 scores are 9.9 for PL-Marker and 38.8 for our model.

Analysis. Token level ER performance across entity classes (none, artifact, parameter, value, context) is at 98.5%, 77.8%, 47.9%, 84.4%, 0% F_1. That is, the model does not predict contexts and struggles with parameters, but artifacts and values are predicted reliably. For our RE model, we observe that value-parameter relations are more reliably predicted than parameter-artifact relations.

Table 1. Ablation study

Used	P [%]	R [%]	F_1 [%]
_CD	15.5	8.8	11.1
T_D	16.6	29.8	19.6
TC_	26.5	65.0	35.5
TCD	**30.7**	**65.0**	**38.8**

To assess the impact of the different components in our RE model, we perform an ablation study with the same 5-fold cross-validation setup as above. In Table 1, showing its results, we can see that removing the BERT token embeddings (T) results in the largest performance loss, followed by entity class embeddings (C) and entity distance (D). Removing any of the inputs results in worse predictions.

Finally, we apply our full model to a random sample of 15,000 papers. Analyzing the results, we find hyperparameters (artifact, parameter, value triples) are reported in 36% of ML papers, 42% of CV papers, 36% of CL papers, and 7% of DL papers. In Fig. 5 we further look at the distribution of the information

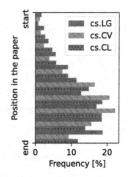

Fig. 5. Mentioning position in papers.

across the length of papers (excluding DL as not being representative). We can see a clear tendency towards the latter half of papers.

Table 2. LLM selection (size in number of parameters).

Model	Variant	Size
WizardLM [31]	`WizardLM-13B-V1.1`	13 B
Vicuna$_{4k}$ [9]	`vicuna-13b-v1.3`	13 B
Vicuna$_{16k}$ [9]	`vicuna-13b-v1.5-16k`	13 B
Falcon [2]	`falcon-40b-instruct`	40 B
GALACTICA [28]	`galactica-120b`	120 B
GPT-3.5 [7]	`text-davinci-003`	175 B

5.2 LLMs

For our LLM experiments we chose a variety of models, with sizes ranging from 13 B to 175 B parameters, as shown in Table 2. We chose WizardLM [31] as it is meant to handle complex instructions, Vicuna [9] due to its performance relative to its size, Falcon [2] because of its alleged performance, and GALACTICA [28] because it was trained on scientific text. Vicuna$_{16k}$ is a model extended using Position Interpolation [8] based on Rotary Positional Embeddings [27], which makes it the only model in our experiments with a sufficient context size for a few-shot evaluation.

The models are run as follows. GPT-3.5 is accessed through its official API. All open models are run on a high performance compute cluster. Vicuna$_{4k}$ and WizardLM are run on nodes with 4×NVIDIA Tesla V100 (32 GB). GALACTICA, Falcon, and Vicuna$_{16k}$ are run on nodes with 4×NVIDIA A100 (80 GB).

As a baseline, we use a JSON variant for each model, where the [format] and [examples] compontents of prompts use JSON, and compare it to the respective YAML version. All models are used with greedy decoding (temperature = 0) for the sake of reproducibility.

Results. In Table 3, show the prediction performance of all models and prompt variants. Overall, LLM performance does not reach the level of our pre-trained models. For zero-shot, we observe the best performance with both GPT-3.5 variants, where YAML outperforms JSON (+3.6% ER and +0.6% RE in F_1 score). The second highest ER F_1 score by model is achieved by Vicuna$_{4k}$ (22.3), despite its size being less than a 10th that of GPT-3.5. For RE, however, even the best model only reaches 7.8%. With our few-shot approach, we are able to considerably improve performance between Vicuna models (+27% ER and +6% RE in F_1), surpassing the zero-shot performance of GPT-3.5 in ER. Lastly, we see that using YAML leads to better ER results accross all six models, with ER performance being comparable or improved as well.

Analysis. In Fig. 6 we show an analysis of the steps leading up to model prediction. Focussing first on the zero-shot models (upper five) we observe the following

Table 3. Prediction performance of LLM models. Subscripts ($_{\Delta\pm n}$) show the delta in F_1 from JSON to YAML output of each model. Format: **best**, <u>second</u>.

Zero-shot		Entity Recognition			Relation Extraction		
Model	Output	P [%]	R [%]	F_1 [%]	P [%]	R [%]	F_1 [%]
WizardLM	JSON	6.9	11.3	8.6	0.1	0.8	0.1
	YAML	9.7	35.6	15.3$_{\Delta+6.7}$	0.1	1.5	0.1$_{\Delta+0.0}$
Vicuna$_{4k}$	JSON	15.1	9.3	11.5	0.7	3.8	1.2
	YAML	17.3	31.5	22.3$_{\Delta+10.8}$	0.0	0.8	0.1$_{\Delta-1.1}$
Falcon	JSON	**37.1**	5.9	10.2	0.0	0.0	0.0
	YAML	32.7	14.2	19.8$_{\Delta+9.6}$	0.0	0.0	0.0$_{\Delta+0.0}$
GALACTICA	JSON	25.9	15.7	19.5	0.1	2.3	0.3
	YAML	23.1	19.5	21.1$_{\Delta+1.6}$	0.0	0.8	0.1$_{\Delta-0.2}$
GPT-3.5	JSON	27.9	**42.8**	<u>33.8</u>	<u>5.4</u>	<u>10.7</u>	<u>7.2</u>
	YAML	<u>34.0</u>	<u>41.7</u>	**37.4**$_{\Delta+3.6}$	**5.8**	**12.2**	**7.8**$_{\Delta+0.6}$
5-shot		Entity Recognition			Relation Extraction		
Vicuna$_{16k}$	JSON	<u>34.4</u>	<u>46.7</u>	<u>39.6</u>	<u>0.8</u>	<u>4.6</u>	<u>1.3</u>
	YAML	**43.9**	**44.1**	**44.0**$_{\Delta+0.4}$	**4.5**	**9.9**	**6.1**$_{\Delta+4.8}$

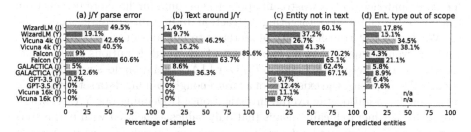

Fig. 6. Parsing success, format adherence, hallucinations, and scope adherence of LLM generated JSON (J) and YAML (Y).

across the four plots from left to right. (a) For three of five models, prompting for YAML leads to fewer parsing errors. (b) Unwanted text around the extracted data is generated more/less by two models each. (c) Hallucinated entities and (d) out of scope entities appear overall slightly more often for in YAML compared to JSON. For our few-shot approach (bottom model), we see that the use of a grammar (a, b) prevents all output format issues. Furthermore (c) hallucinated entities are reduced. (d) Out of scope entities can not be evaluated, because our in-context examples lead to frequent omission of type information in the output.

Through manual analysis we identify a common cause for parsing errors in JSON output to be boolean values (e.g. for "`text_contains_entities:`") being copied by the LLM as "true/false" from the promt. We furthermore find that "entities not in the text" can arise from unsolicited LATEX parsing by the LLM (e.g. "`\lambda`" in text → "λ" in YAML). Prompting for *verbatim* parameter-/value strings does not mitigate this.

6   Discussion

Our overall results, with a top performance of 79% F_1 for entity recognition and 39% F_1 for relation extraction, show that extraction of hyperparameter information from scientific text can be accomplished to a degree that yields sound results. There are, however, challenges that remain, such as more reliable entity recognition of parameters and contexts, as well as more reliable relation extraction in general. Our novel data set enables further development of approaches from hereon. Our IE results on large-scale unannotated data give an indication of possible downstream analyses and applications. Here we see large potential for reproducibility research, faceted search, and recommendation.

Our LLM evaluation shows that for IE tasks dealing with complex information, the choice of text based data serialization format can have a considerable impact on performance, even when using grammar based generation constrains. Additionally, we can see that in-context learning enabled by larger context sizes, as well as grammars, are an effective method to improve IE performance.

Limitations. (1) Our work considers HyperPIE from text. This is sensible for a focussed approach, but downstream applications could furthermore benefit from composite pipelines also targeting extraction from tables, source code, etc. (2) We do not test transferability of methods to domains outside of ML related fields. It would require domain expertise to find useful definitions for hyperparameters in each respective domain. (3) Our LLM evaluation does not cover fine-tuning. Presupposing the existence of a large enough training data set, this would be a valuable addition the overall investigation. (4) Defining our YAML/JSON output format hierarchically means that only values associated with parameters and parameters associated with artifacts can be extracted. (5) Lastly, our data and experiments unfortunately are limited to English text only and do not cover other languages.

7   Conclusion

We formalize the novel ER+RE task HyperPIE and develop approaches for it, thereby expanding IE from scientific text to hyperparameter information. To this end, we create a manually labeled data set spanning various ML fields. In a supervised learning setting, we propose a BERT-based model that achieves an improvement of 29% F_1 in RE compared to a state-of-the-art baseline. Using the model, we perform IE on a large amount of unannotated papers, and analyze patterns of hyperparameter reporting across ML disciplines. In a zero-/few-shot setting, we propose an LLM based approach using YAML for complex IE, achieving an average improvement of 5.5% F_1 in ER over using JSON. We furthermore achieve large performance gains for LLMs using grammar based generation constrains and in-context learning. In future work, we plan to investigate fine-tuning LLMs, as well as additional practical use cases for data extracted from large publication corpora, such as knowledge graph construction.

Acknowledgements. This work was partially supported by the German Federal Ministry of Education and Research (BMBF) via [KOM,BI], a Software Campus project (01IS17042). The authors acknowledge support by the state of Baden-Württemberg through bwHPC. We thank Nicholas Popovic for extensive feedback on the experiment design and prompt engineering. We thank Tarek Gaddour for feedback during the annotation scheme development, and Xiao Ning for input during early model development.

Author contributions. Tarek Saier: Conceptualization, Data curation, Formal analysis, Methodology, Software, Visualization, Writing – original draft, Writing – review & editing. Mayumi Ohta: Conceptualization (LLM few-shot), Formal analysis (LLM few-shot), Methodology (LLM few-shot), Software (LLM few-shot), Writing – original draft (support). Takuto Asakura: Conceptualization, Writing – review & editing. Michael Färber: Writing – review & editing.

References

1. Agrawal, M., Hegselmann, S., Lang, H., Kim, Y., Sontag, D.: Large language models are few-shot clinical information extractors. In: Proceedings of the 2022 Conference on Empirical Methods in Natural Language Processing, pp. 1998–2022 (Dec 2022)
2. Almazrouei, E., et al.: Falcon-40B: an open large language model with state-of-the-art performance (2023)
3. Auer, S., Oelen, A., Haris, M., Stocker, M., D'Souza, J., Farfar, K.E., Vogt, L., Prinz, M., Wiens, V., Jaradeh, M.Y.: Improving access to scientific literature with knowledge graphs. Bibliothek Forschung und Praxis **44**(3), 516–529 (2020). https://doi.org/10.1515/bfp-2020-2042
4. Baudart, G., Kirchner, P.D., Hirzel, M., Kate, K.: Mining documentation to extract hyperparameter schemas. In: Proceedings of the 7th ICML Workshop on Automated Machine Learning (AutoML 2020) (2020)
5. Beltagy, I., Lo, K., Cohan, A.: SciBERT: A pretrained language model for scientific text. In: Proceedings of the 2019 Conference on Empirical Methods in Natural Language Processing and the 9th International Joint Conference on Natural Language Processing (EMNLP-IJCNLP), pp. 3615–3620. Association for Computational Linguistics (Nov 2019). https://doi.org/10.18653/v1/D19-1371
6. Brown, T., Mann, B., Ryder, N., Subbiah, M., Kaplan, J.D., Dhariwal, P., Neelakantan, A., Shyam, P., Sastry, G., Askell, A., et al.: Language models are few-shot learners. Adv. Neural. Inf. Process. Syst. **33**, 1877–1901 (2020)
7. Brown, T.B., et al.: Language models are few-shot learners. In: Proceedings of the 34th International Conference on Neural Information Processing Systems. NIPS'20 (2020)
8. Chen, S., Wong, S., Chen, L., Tian, Y.: Extending context window of large language models via positional interpolation. arXiv preprint. arXiv:2306.15595 (2023)
9. Chiang, W.L., et al.: Vicuna: An open-source chatbot impressing GPT-4 with 90%* chatgpt quality (March 2023), https://lmsys.org/blog/2023-03-30-vicuna/
10. Devlin, J., Chang, M.W., Lee, K., Toutanova, K.: BERT: Pre-training of deep bidirectional transformers for language understanding. In: Proceedings of the 2019 Conference of the North American Chapter of the Association for Computational Linguistics: Human Language Technologies, Volume 1 (Long and Short Papers), pp. 4171–4186. Association for Computational Linguistics (Jun 2019). https://doi.org/10.18653/v1/N19-1423

11. Dunn, A., Dagdelen, J., Walker, N., Lee, S., Rosen, A.S., Ceder, G., Persson, K., Jain, A.: Structured information extraction from complex scientific text with fine-tuned large language models (Dec 2022). https://doi.org/10.48550/arXiv.2212.05238

12. Harper, C., Cox, J., Kohler, C., Scerri, A., Daniel Jr., R., Groth, P.: SemEval-2021 task 8: MeasEval - extracting counts and measurements and their related contexts. In: Proceedings of the 15th International Workshop on Semantic Evaluation (SemEval-2021), pp. 306–316 (Aug 2021). https://doi.org/10.18653/v1/2021.semeval-1.38

13. Jain, S., van Zuylen, M., Hajishirzi, H., Beltagy, I.: SciREX: A Challenge dataset for Document-level information extraction. In: Proceedings of the 58th Annual Meeting of the Association for Computational Linguistics, pp. 7506–7516. Association for Computational Linguistics (Jul 2020). https://doi.org/10.18653/v1/2020.acl-main.670

14. Kojima, T., Gu, S.S., Reid, M., Matsuo, Y., Iwasawa, Y.: Large Language Models are Zero-Shot Reasoners. Adv. Neural. Inf. Process. Syst. **35**, 22199–22213 (2022)

15. Kuhn, T.: A survey and classification of controlled natural languages. Comput. Linguist. **40**(1), 121–170 (mar 2014). https://doi.org/10.1162/COLI_a_00168

16. Lai, V., Pouran Ben Veyseh, A., Dernoncourt, F., Nguyen, T.: SemEval 2022 task 12: Symlink - linking mathematical symbols to their descriptions. In: Proceedings of the 16th International Workshop on Semantic Evaluation (SemEval-2022), pp. 1671–1678 (Jul 2022). https://doi.org/10.18653/v1/2022.semeval-1.230

17. Liesenfeld, A., Lopez, A., Dingemanse, M.: Opening up chatgpt: Tracking openness, transparency, and accountability in instruction-tuned text generators. In: Proceedings of the 5th International Conference on Conversational User Interfaces. CUI '23, New York, NY, USA (2023). https://doi.org/10.1145/3571884.3604316

18. Lin, J., Yu, Y., Song, J., Shi, X.: Detecting and analyzing missing citations to published scientific entities. Scientometrics **127**(5), 2395–2412 (2022). https://doi.org/10.1007/s11192-022-04334-5

19. Luan, Y., He, L., Ostendorf, M., Hajishirzi, H.: Multi-task identification of entities, relations, and coreferencefor scientific knowledge graph construction. In: Proc. Conf. Empirical Methods Natural Language Process. (EMNLP) (2018)

20. Polak, M.P., Morgan, D.: Extracting accurate materials data from research papers with conversational language models and prompt engineering - Example of ChatGPT (Mar 2023). https://doi.org/10.48550/arXiv.2303.05352

21. QasemiZadeh, B., Schumann, A.K.: The ACL RD-TEC 2.0: A language resource for evaluating term extraction and entity recognition methods. In: Proceedings of the Tenth International Conference on Language Resources and Evaluation (LREC'16), pp. 1862–1868. European Language Resources Association (ELRA) (May 2016)

22. Raff, E.: A step toward quantifying independently reproducible machine learning research. In: Adv. Neural Info. Process. Syst. vol. 32. Curran Associates, Inc. (2019)

23. Rak-Amnouykit, I., Milanova, A., Baudart, G., Hirzel, M., Dolby, J.: Extracting Hyperparameter Constraints from Code. In: ICLR Workshop on Secur. Saf. Mach. Learn. Syst. (May 2021). https://hal.science/hal-03401683

24. Saier, T., Krause, J., Färber, M.: unarXive 2022: All arXiv Publications Pre-Processed for NLP, Including Structured Full-Text and Citation Network. In: Proceedings of the 23rd ACM/IEEE Joint Conference on Digital Libraries. JCDL '23 (2023)

25. Sethi, A., Sankaran, A., Panwar, N., Khare, S., Mani, S.: Dlpaper2code: Auto-generation of code from deep learning research papers. Proc. AAAI Conf. Artif. Intell. 32(1) (Apr 2018). https://doi.org/10.1609/aaai.v32i1.12326

26. Stocker, M., Oelen, A., Jaradeh, M.Y., Haris, M., Oghli, O.A., Heidari, G., Hussein, H., Lorenz, A.L., Kabenamualu, S., Farfar, K.E., Prinz, M., Karras, O., D'Souza, J., Vogt, L., Auer, S.: Fair scientific information with the open research knowledge graph. FAIR Conn. 1(1), 19–21 (2023). https://doi.org/10.3233/FC-221513

27. Su, J., Lu, Y., Pan, S., Murtadha, A., Wen, B., Liu, Y.: Roformer: enhanced transformer with rotary position embedding. arXiv preprint. arXiv:2104.09864 (2021)

28. Taylor, R., et al.: GALACTICA: A large language model for science (2022)

29. Wang, S., et al.: GPT-NER: named entity recognition via large language models (May 2023). 10.48550/arXiv. 2304.10428

30. Xie, T., et al.: Large language models as master key: unlocking the secrets of materials science with GPT (Apr 2023). https://doi.org/10.48550/arXiv.2304.02213

31. Xu, C., et al.: Wizardlm: Empowering large language models to follow complex instructions (2023)

32. Yang, J., et al.: Harnessing the power of LLMs in Practice: a survey on ChatGPT and Beyond (Apr 2023). https://doi.org/10.48550/arXiv.2304.13712

33. Ye, D., Lin, Y., Li, P., Sun, M.: Packed levitated marker for entity and relation extraction. In: Proceedings of the 60th Annual Meeting of the Association for Computational Linguistics (Volume 1: Long Papers), pp. 4904–4917. Association for Computational Linguistics (May 2022). https://doi.org/10.18653/v1/2022.acl-long.337

LaQuE: Enabling Entity Search at Scale

Negar Arabzadeh[1(✉)], Amin Bigdeli[1], and Ebrahim Bagheri[2]

[1] University of Waterloo, Waterloo, Canada
{narabzad,abigdeli}@uwaterloo.ca
[2] Toronto Metropolitan University, Toronto, Canada
bagheri@torontomu.ca

Abstract. Entity search plays a crucial role in various information
access domains, where users seek information about specific entities.
Despite significant research efforts to improve entity search methods,
the availability of large-scale resources and extensible frameworks has
been limiting progress. In this work, we present LaQuE (Large-scale
Queries for Entity search), a curated framework for entity search, which
includes a reproducible and extensible code base as well as a large relevance judgment collection consisting of real-user queries based on the
ORCAS collection. LaQuE is industry-scale and suitable for training
complex neural models for entity search. We develop methods for curat-
ing and judging entity collections, as well as training entity search methods based on LaQuE. We additionally establish strong baselines within
LaQuE based on various retrievers, including traditional bag-of-words-
based methods and neural-based models. We show that training neural
entity search models on LaQuE enhances retrieval effectiveness compared
to the state-of-the-art. Additionally, we categorize the released queries in
LaQuE based on their popularity and difficulty, encouraging research on
more challenging queries for the entity search task. We publicly release
LaQuE at https://github.com/Narabzad/LaQuE.

1 Introduction

The importance of entity search has grown significantly in various information
access domains, where users seek to find information on specific entities such as
individuals, organizations, and places and their associated attributes [16,24,25].
Research suggests that more than 40% of web search queries revolve around entities, prompting search engines to rely on knowledge graphs to provide relevant
and reliable responses [5,15,36,44]. Entity search finds applications in diverse
areas, including vertical search, which may only display a limited number of
entities due to space constraints on the screen; enterprise search, focusing on
entities within a specific organization; and social networks, emphasizing on the
relationships between people [6,26], among others. The task of entity search is
defined as retrieving a ranked list of entities from a knowledge graph, such as
Wikipedia, in response to an input keyword query. The entity search task dif-
ferentiates itself from the more traditional ad hoc retrieval task by capitalizing
on additional knowledge graph semantics such as relations, types, and attributes
[22,35,51,52,57].

© The Author(s), under exclusive license to Springer Nature Switzerland AG 2024
N. Goharian et al. (Eds.): ECIR 2024, LNCS 14609, pp. 270–285, 2024.
https://doi.org/10.1007/978-3-031-56060-6_18

The entity search task has witnessed a growing range of methods including traditional bag-of-word-based approaches to more complex methods that incorporate neural embeddings, anchor texts, structural components of Wikipedia, and associated categories, to name a few [8,12,16,24,26,56,57]. While there has been significant research focused on improving method performance on this task [20,28,29], the development of large-scale frameworks consisting of query collections and resources has not kept up the pace. One of the main reasons for this relates to the time and resource-intensive nature of identifying and maintaining a comprehensive set of user queries and their relevant entities. As recent advances in ad hoc retrieval and the experience with the MS MARCO dataset show, neural methods are data-hungry and are often effective when trained on large relevance judgement collections [42]. However, the scarcity of such resources makes it challenging to develop and benchmark strong and generalizable entity search models [3,34,42].

From among the available resources curated specifically for the entity search task, TREC Complex Answer Retrieval (CAR)[1] stands out in terms of its size and coverage [17,18]. TREC CAR was designed initially for the complex question-answering task that involves retrieving a ranked list of relevant entities and their supporting passages for each section of a given complex topic query. To create the TREC CAR dataset, topics, outlines, and paragraphs were extracted from the English version of Wikipedia. In addition to the manual ground truth, automatic ground truths were also curated in the CAR collection. The automatic ground truth is released for all training sets and is determined by whether a paragraph is contained within the page/section, making it relevant, or if it is not contained, making it non-relevant. The ground truth is provided at three levels of granularity: paragraph contained in the section (hierarchical), paragraph contained in section hierarchy below the top-level section (top-level), and paragraph contained anywhere in the page (article). While the TREC CAR collection constitutes a valuable resource for the community, it *is not designed to include real-world user queries*. Given the nature of the CAR task, the queries in this collection are section titles from Wikipedia pages; therefore, these queries are not considered to be real-world user-generated queries and differ substantially in characteristics from real user queries.

There are however other available entity search datasets that have user-generated queries and manually-labelled ground truths such as the DBpedia-Entity (v1 and v2)[2] dataset, which is a widely recognized and standard test collection for evaluating entity search methods [7,27]. The purpose of this test collection is to assess the performance of retrieval systems in generating ranked lists of entities in response to user queries expressed in free text. *The limitation of the two versions of DBpedia-Entity is that they only include 485 and 467 queries* from INEX, QALD, SemSearch and TREC Entity benchmarks, respectively. The low number of queries prevents researchers from training deep learning models for this task.

---

[1] https://trec-car.cs.unh.edu/.

[2] https://github.com/iai-group/DBpedia-Entity.

In this paper, our main focus is to propose a large-scale publicly accessible framework, called LaQuE (pronounced as léjk - layk), along with supporting code and dataset for entity search. LaQuE is based on an intuitive idea that helps with the automated generation of a large-scale dataset for entity search with two important characteristics: (1) LaQuE includes real-user queries; and, (2) LaQuE is large-scale such that neural methods can be trained and tested on it.

In order to enable these two main characteristics, we propose and implement an intuitive idea to leverage the Open Resource for Click Analysis in Search (ORCAS) dataset. ORCAS is a large-scale click-based resource curated for the TREC Deep Learning Track. The extensive set of queries within ORCAS has already facilitated research in various areas of information retrieval (IR) and natural language processing (NLP), including query autocompletion and web mining [1,13,21,40,41]. We intuitively propose that the clickthrough data between user queries and their corresponding relevant web documents in the ORCAS dataset can be considered to be a form of *pseudo-relevance feedback*. On this basis, we establish specific criteria for filtering queries from the ORCAS dataset by only considering those queries whose relevant clicked web pages refer to links pointing to Wikipedia entities. This way, we identify the subset of queries from the ORCAS dataset where the users have determined the relevant document to be a Wikipedia entity. This subset of queries are those that require the retrieval of an entity to be satisfied.

On the basis of this idea, we incorporate such queries and their relevant clicked entities within LaQuE and offer supporting methods and code to work with the query collection. LaQuE delivers over 2 million pairs of queries and clicked Wikipedia entities. To facilitate training neural models, LaQuE offers separate standard train, test and development sets. Furthermore, for benchmarking purposes, LaQuE offers implementation for state-of-the-art first-stage retrievers, ranging from traditional bag-of-words-based retrievers to more complex pre-trained neural models in order to offer out-of-the-box strong baselines.

In this paper, we empirically illustrate that training on datasets intended for other tasks, such as ad hoc retrieval, is not as effective as training neural models on the collection offered through the LaQue framework when it comes to the task of entity retrieval. In addition, we report on our detailed investigative studies on different cuts of queries offered through LaQuE. We categorize queries based on *popularity*, and *difficulty*. LaQuE offers information on entity popularity by collecting page views of the relevant entities on Wikipedia (number of times the relevant entity page was viewed on Wikipedia) and categorizes them based on the number of views they received. Based on LaQuE, and in this paper, we investigate whether popular entities are easier to retrieve and whether language models have an inherent bias towards more popular entities rather than rare ones. Finally, inspired by previous work [2,4,10,19], LaQuE offers an additional categorization of entities based on their difficulty. This categorization encourages the research community to not only focus on improving overall performance but also specifically tackle more challenging queries [2] (Fig. 1).

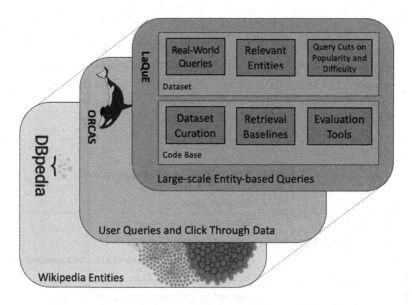

Fig. 1. Overview of the LaQuE Framework.

Licensing: In compliance with open data principles and to facilitate transparent and accessible research, we have made the LaQuE framework openly available on Zenodo. It has been assigned a Digital Object Identifier (DOI) for easy identification and citation, ensuring its long-term accessibility and proper attribution. We have chosen to license it under the Creative Commons license, which allows for broad use and redistribution, provided proper attribution is given. The dataset is released under anonymity given this stage of peer-review. As such, the authors' identities are not disclosed at this time. Researchers interested in accessing and utilizing LaQuE at this time can access it through the anonymized link: https:// anonymous.4open.science/r/LaQuE-0CDD/.

2   The LaQuE Framework

In this section, we will discuss the intuitive idea behind the resources curated and made available through LaQuE as well as provide some of their statistical characteristics.

2.1   Dataset Curation in LaQuE

The data curated by LaQuE are derived from the ORCAS dataset, which consists of a vast collection of 18 million relations between 10 million distinct queries and relevant document URLs. There are at least two main advantages to our approach for using the ORCAS queries: First, the query set is diverse as it was curated from millions of users, encompassing different topics. Second, the abundance of data in the ORCAS dataset allows for different cuts of the dataset for

Table 1. The statistics of the dataset from LaQuE in terms of queries and their relevant entities in train, dev, and test sets.

Split	#Queries	Avg $|q|$	Std $|q|$	#Related Entities	Avg #Entity
Train	2,019,183	2.793	1.344	2,176,400	1.08
Dev	112,176	2.792	1.345	120,081	1.07
Test	112,176	2.796	1.349	119,200	1.06
Total	2,243,535	2.793	1.345	2,415,681	1.08

Table 2. Sample queries and their relevant entities.

Query	Related Entity
what is phylogeny	`<dbpedia:Phylogenetics >`
who was melchizedek parents	`<dbpedia:Melchizedek >`
nashville actors	`<dbpedia:List_of_Nashville_cast_members >`
first iphone released	`<dbpedia:IPhone_(1st_generation) >`
fisher river	`<dbpedia:Fisher_River_Cree_Nation >`

various purposes, and even after applying these cuts, there remain a significant number of data points in each cut for model training and evaluation purposes. The LaQuE framework offers the possibility to extract queries and their relevant clicked documents from ORCAS based on an intuitive idea: From the URIs of the clicked documents in ORCAS, LaQuE applies a filtering strategy to retain only those URLs connected to the English version of Wikipedia. This allows LaQuE to identify entities from Wikipedia that were able to satisfy the information need behind a specific user query. This idea is inspired by previous work [11,14,37,50], where user clicks in search log files are considered implicit feedback for relevance.

Similar to prior works [23,27,30,43], LaQuE leverages the English subset of DBpedia version 3.7[3] as its main collection of entities. LaQuE ensures that all selected entities must have a title and abstract, specifically the *rdfs:label* and *rdfs:comment* predicates, and have excluded any category, redirect, and disambiguation pages. This provides LaQuE with access to a set of 4.6 million entities, each uniquely identifiable through their URI. By intersecting the filtered ORCAS dataset that is linked with documents that have a valid Wikipedia URI with the filtered DBpedia dataset consisting of 4.6 million entities, LaQuE curates and offers a dataset, which includes user-generated queries that are related to Entities in DBpedia.

In summary, the ORCAS dataset serves as a foundational resource for LaQuE. This is a strong advantage for LaQuE as ORCAS boasts a rich collection of user-generated queries, sourced from millions of users across various topics and domains. This diversity ensures that LaQuE also encompasses a wide spectrum of information needs, making it a valuable resource for training and

---

[3] http://downloads.dbpedia.org/wiki-archive/Downloads2015-10.html.

Table 3. Sample queries with more than one relevant entity.

Query	Relevant Entity
the temptations	`<dbpedia:Paul_Williams_(The_Temptations)>`
	`<dbpedia:The_Temptations>`
the texas rangers	`<dbpedia:Texas_Rangers_(baseball)>`
	`<dbpedia:Texas_Ranger_Division>`
the tracey ullman show	`<dbpedia:The_Tracey_Ullman_Show>`
	`<dbpedia:The_Simpsons_shorts>`
project management	`<dbpedia:Project_management_triangle>`
	`<dbpedia:Project_management>`
	`<dbpedia:Project_manager>`
	`<dbpedia:Project_management_software>`
progressivism definition	`<dbpedia:Progressivism>`
	`<dbpedia:Progressive_education>`
	`<dbpedia:Progressivism_in_the_United_States>`
prague	`<dbpedia:Prague,_Oklahoma>`
	`<dbpedia:Czech_Republic>`
	`<dbpedia:Prague>`
	`<dbpedia:Prague_astronomical_clock>`

evaluation purposes. We emphasize that LaQuE includes a careful selection of queries from ORCAS that explicitly exhibit a clear intent to retrieve Wikipedia entities. This intent-filtering approach is designed to ensure that the queries offered through LaQuE are relevant to the entity retrieval task. The process supported by LaQuE involves intersecting ORCAS queries with relevant entities in DBpedia. LaQuE ensures that relevant entities are available on DBpedia as this will allow entity retrieval methods to benefit from additional external sources of information such as content on DBpedia and knowledge graph embeddings. In summary, LaQuE is designed to benefit from the effective integration and intersection of ORCAS and DBpedia for entity retrieval. We believe that this approach strengthens the foundations of our work and ensures that LaQuE is a valuable framework for the research community.

2.2 LaQuE Statistics

The statistics of the data provided through the LaQuE framework in terms of the number of queries as well as their relevant entities are shown in Table 1. LaQuE offers over 2.2 million queries and their relevant entities. It randomly splits the queries into training, development (dev), and test sets, with a distribution of 90%, 5%, and 5%, respectively. As a result, the training set contains 2 million queries, while both the development and test sets consist of over 100,000 queries each. Furthermore, Table 1 presents the average number of query terms and

Table 4. Sample entities with more than one related query. Individual queries are separated by a semicolon;

Relevant Entity	Submitted Queries
`<dbpedia:Aaron>`	Aaron; aaron and moses; aaron bible; aaron brother of moses; aaron budjen; aaron from the bible; aaron high priest; aaron in bible; aaron in the bible; aaron in the bible facts; aaron meaning; aaron moses; aaron moses brother; aaron of the bible; aaron old testament
`<dbpedia:Belly_dance>`	arab belly dance; arabian dance; bally dance
`<dbpedia:Ballston,_New_York>`	Ballston; ballston lake; ballston lake new york; ballston lake ny
`<dbpedia:Lumbricus_terrestris>`	Anecic; canadian nightcrawlers; classification of earthworm; common earthworm
`<dbpedia:Common_krait>`	blue krait; blue krait snake; bungarus caeruleus; common krait; common krait snake

the standard deviation of the number of query terms. These statistics show LaQuE ensures that the data is well-distributed among the three splits. On average, each query is associated with 1.08 relevant entities, indicating that the majority of queries have only one relevant entity. Sparse labels, where queries have few relevant entities, are also observed in other well-known and widely-used benchmarks such as the MS MARCO dataset [3, 9, 39, 42, 45]. However, this does not undermine the reliability of the evaluation process as appropriate strategies can be employed to handle sparse labels [3, 9, 39]. In Table 2, we show a few sample queries from LaQuE accompanied with their relevant entities, generated from real users and adopted from the ORCAS dataset.

We have also conducted a detailed analysis of the distribution of entities on a per-query basis. While, on average, there are only 1.08 entities per query, LaQuE offers a substantial number of queries with multiple entities, owing to its large size. In Fig. 2(a), we present a logarithmic representation of the number of entities per query. We note that to avoid noise, we filter a number of queries that have more than 20 entities. Figure 2(a) shows that there are over 145,000 queries with 2 entities, over 10,000 queries with 3 entities, over 1,600 queries with more than 4 entities and so on. This abundance of queries with multiple relevant entities in LaQuE enables us to achieve better and more diverse training for entity search purposes. In Table 3, we present a few examples of queries that have more than one entity. We additionally investigate the mapping between individual entities and variations of queries with which they are associated. To do so, we demonstrate the histogram of the number of unique queries per entity in Fig. 2(b). LaQuE filters any noisy entities that have more than 100 different queries (such as 'www'). As shown in this Figure, there are an abundant number of entities with different queries mapped to them. This will allow such entities to be used in methods such as query transformation, query refinement, and query expansion. In Table 4, we provide a few examples of a single entity that have been considered relevant for different queries.

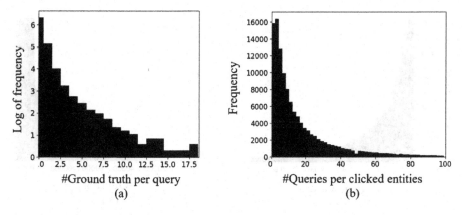

Fig. 2. (a) Distribution of the number of relevant entities per query. (b) Distribution of the number of unique queries per entity.

2.3   State of the Art Baselines in LaQuE

LaQuE offers a comprehensive set of retrievers, including both high-dimensional bag-of-word-based sparse retrievers and neural-based dense retrievers, for benchmarking purposes. In this paper, we only report the results of various retrievers based on the LaQuE dev set due to limited space. However, all complete results are available on our GitHub repository. For the sparse retrievers, LaQuE offers two methods, namely BM25 [49] and QL [54]. To enhance these sparse retrievers and investigate the impact of pseudo-relevance feedback and query expansion on the entity retrieval task, LaQuE incorporates the RM3 framework to create BM25-PRF and QL-PRF variants with pseudo-relevance feedback. For the dense retrievers, LaQuE provides a bi-encoder-based siamese network, as used in numerous previous studies, including [32,38,46,58]. The model consists of two separate encoder towers, with one encoding the query and the other encoding the candidate content. LaQuE provides the means to evaluate the performance using various transformer models, such as BERT, DistilBERT, DistilRoBERTa, and MiniLM, all of which exhibit promising results across different downstream tasks, including passage retrieval, entity retrieval, and question answering [32,38,46–48,55,58]. During the training phase, LaQuE optimizes multiple negatives ranking loss function, encouraging higher similarity scores for relevant query-candidate pairs and lower scores for irrelevant pairs. In the training set, LaQuE considers the query and entity pairs as positive (relevant) data points. It adopts a training strategy from the Sentence Transformer library [46], where negative pairs are randomly sampled from the top-1000 retrieved entities using BM25, similar to [32,47,58]. For the purposes of the experiments reported in this paper, we customize LaQuE to consider only one negative sample per pair and train the models for one epoch. During each epoch, a batch size of 64 is employed, and the process is initiated with a warm-up phase spanning 1,000 steps. Upon completing the training of the bi-encoder model, LaQuE proceeds

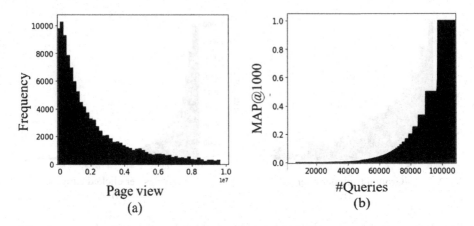

Fig. 3. Distribution of queries based on (a) number of page views (b) performance of BM25 in terms of MAP@1000.

Table 5. Performance of sparse and dense retrievers on the development set from LaQuE in terms of MAP@10, nDCG@10 and Recall@10 as well as MAP@1000, nDCG@1000 and Recall@1000.

	Retriever	Training	Cut-off@10			Cut-off@1000		
			MAP	nDCG	Recall	MAP	nDCG	Recall
Sparse Retrievers	BM25	–	0.2234	0.2662	0.3953	0.2338	0.3369	0.7965
	BM25 + RM3	–	0.2050	0.2624	0.3869	0.2157	0.3222	0.7965
	QL	–	0.2156	0.2581	0.3869	0.2260	0.3291	0.7891
	QL+RM3	–	0.1952	0.2414	0.3833	0.2058	0.3135	0.7918
Dense Retrievers	BERT	MS MARCO	0.3742	0.4218	0.5600	0.3820	0.4748	0.8607
		LaQuE	0.6018	0.6489	0.7801	0.6069	0.6781	0.6069
	DistilBERT	MS MARCO	0.4078	0.4553	0.5915	0.4155	0.5025	0.8505
		LaQuE	0.6179	0.6636	0.7900	0.6229	0.6920	0.9417
	DistilRoBERTa	MS MARCO	0.3335	0.3778	0.5068	0.3412	0.4272	0.7828
		LaQuE	0.5569	0.6056	0.7418	0.5629	0.6404	0.9289
	MiniLM	MS MARCO	0.4226	0.4664	0.5902	0.4294	0.5081	0.8184
		LaQuE	0.5731	0.6195	0.7481	0.5785	0.6501	0.9110

to build an index for the collection, which involves storing the embedding vectors of entities within the collection. To accomplish this, LaQuE leverages the capabilities of FAISS [31]. The choice of FAISS is motivated by its efficiency in conducting approximate nearest neighbor retrieval, a feature that significantly enhances the speed and effectiveness of retrieving relevant entities in response to queries during the inference phase. When a query is received, the trained model initially encodes it into a vector representation. This vector is then employed to locate relevant entities within the constructed index, utilizing the L^2 distance function as recommended in [33].

Table 6. Performance of established baselines on popularity-based query subsets in terms of MAP@1000

	Retriever	Train Set	Unpopular	Somewhat Popular	Popular	Highly Popular
Sparse Retrievers	BM25	–	0.3464	0.2545	0.1821	0.1217
	BM25 + RM3	–	0.3283	0.2447	0.1789	0.1229
	QL	–	0.5101	0.4292	0.3746	0.3193
	QL+RM3	–	0.5738	0.4927	0.4383	0.3781
Dense Retrievers	BERT	MS MARCO	0.5109	0.4146	0.3346	0.2377
		LaQuE	0.6102	0.6076	0.6030	0.5825
	DistilBERT	MS MARCO	0.5101	0.4293	0.3746	0.3193
		LaQuE	0.6104	0.6170	0.6228	0.6179
	DistilRoBERTa	MS MARCO	0.4231	0.3604	0.3130	0.2425
		LaQuE	0.5559	0.5612	0.5590	0.5488
	MiniLM	MS MARCO	0.5218	0.4381	0.3867	0.3397
		LaQuE	0.5840	0.5790	0.5704	0.5593

Table 5 demonstrates the performance of the various retrievers in terms of Mean Average Precision (MAP), normalized Discounted Cumulative Gain (nDCG), and Recall on the top-10 and top-1000 retrieved entities, specifically on the development set offered by LaQuE. As shown in this table and aligned with the performance of sparse versus dense retrievers on other downstream tasks, dense retrievers outperform sparse retrievers by a large margin. We also note that pseudo-relevance feedback (RM3) would not help addressing the queries since it did not lead to any consistent significant improvement on the results. Among the dense retrievers, we conducted experiments using a pre-trained model on the MS MARCO passage collection dataset and compared it with the model trained on LaQuE. While both models perform better than the set of sparse retrievers, the dense retrievers fine-tuned on the language model using LaQuE outperform the MS MARCO model. For example, taking DistilBERT as an example, it achieves a MAP@10 of 0.4078 and a recall@10 of 0.8505, whereas the same model trained on LaQuE for one epoch obtained a MAP@10 of 0.6179 and a recall@10 of 0.9417. This observation confirms how training on a large-scale entity retrieval task can significantly boost the performance of entity retrievers (Table 6).

3   Query Subsets in LaQuE

We delve deeper into entity retrieval task by examining queries based on their characteristics and the attributes of the related entities. LaQuE categorizes queries based on 1) popularity of the related entities; and 2) performance of the queries. This categorization approach encourages the research community to not only focus on the query set as a whole but also tackle more challenging and diverse queries.

Table 7. Performance of established baselines on difficulty-based query subsets in terms of MAP@1000

	Retriever	Train Set	Easy	Medium	Hard	Very Hard
Sparse Retrievers	BM25	–	0.8171	0.1585	0.0155	0.0004
	BM25 + RM3	–	0.7602	0.1716	0.0200	0.0007
	QL	–	0.7015	0.5264	0.3511	0.1597
	QL+RM3	–	0.7851	0.6035	0.4099	0.1921
Dense Retrievers	BERT	MS MARCO	0.6611	0.4711	0.3094	0.1579
		LaQuE	0.7599	0.7213	0.6251	0.3899
	DistilBERT	MS MARCO	0.7016	0.5265	0.3512	0.1597
		LaQuE	0.7632	0.7331	0.6471	0.4188
	DistilRoBERTa	MS MARCO	0.5604	0.4189	0.2861	0.1526
		LaQuE	0.6973	0.6553	0.5632	0.3769
	MiniLM	MS MARCO	0.7137	0.5472	0.3655	0.1639
		LaQuE	0.7415	0.6940	0.5959	0.3593

3.1 Popularity-Based Query Subsets

To determine the popularity of entities, LaQuE collects the total page views for each entity on Wikipedia from January 1, 2018, to December 31, 2022, spanning a period of five years. Analyzing the number of views received by these entities provides insights into whether popular entities are more likely to be retrieved by the retrieval systems. Figure 3(a) illustrates the histogram of page views for related entities in LaQuE. As depicted in the figure, the distribution of entity view counts follows a long-tailed pattern. Based on the range of page views, LaQuE divides the queries into four equally sized buckets, creating four query subsets: "Unpopular", "Somewhat Popular", "Popular", and "Highly Popular". In Table 7, we present the performance results of retrievers on these query subsets. As observed in the table, we consistently notice that the less popular a related entity is to a query, the higher the performance achieved by both sparse and pre-trained dense retrievers. However, this observation does not hold on models that were trained on LaQuE. We hypothesize that this trend occurs because, in cases where the information need of a user is less well-known, users tend to provide more detailed and elaborate queries. For instance, queries related to unpopular queries include examples such as 'apostrophe figure of speech' and 'which president moved thanksgiving up a week'. Such elaborate queries result in higher retrieval performance. Conversely, for popular queries, users often enter shorter queries, examples of which include 'alphabet' and 'lightning 2'. For these short queries, industry-scale search engines can find relevant entities by utilizing various user personalization, and trending information. However, without access to such information, retrieval methods will find it challenging to retrieve the relevant information for those queries.

3.2 Difficulty-Based Query Subsets

LaQuE also provides the means to evaluate query subsets based on their difficulty. Following the approach of previous studies [2,10,19], LaQuE classifies queries based on their level of difficulty. By evaluating query performance across different difficulty levels, LaQuE offers insights into the strengths and limitations of existing entity retrieval systems and identifies areas for improvement [53].

We report the performance of the widely used BM25 model on the queries from the LaQuE development set, using MAP@1000 as shown in Fig. 3(b). The figure reveals a long-tail distribution of retriever performance. This means that while the retrievers are effective for a subset of queries, their performance is poor for others, resulting in an imbalanced distribution of performance across all queries. For instance, more than 20,000 queries in the LaQuE development set have a MAP@1000 value of zero when being retrieved with BM25. This underscores the need for the research community to focus on addressing more challenging queries. Building upon this observation, LaQuE categorizes queries into four equal-sized buckets based on their performance in terms of MAP@1000 with the BM25 model. These categories are labeled as "Very Hard", "Hard", "Medium", and "Easy". The results for these query subsets are reported in the right side of Table 7. It is shown that query difficulty remains consistent across all the retrievers. In other words, the "very hard" subset of queries, which exhibits the lowest performance by BM25, also demonstrates the lowest performance even with dense retrievers. The poor performance on this subset compared to the other query subsets emphasizes the importance for the research community to focus on addressing more challenging queries.

4 Concluding Remarks

We have introduced the LaQuE framework, which offers an extensible code base as well as real-user queries and large-scale training data for the entity search task. LaQuE offers access to more than 2.2 million query-entity pairs divided into train, development, and test sets, facilitating the training and evaluation of neural models for entity search. Additionally, LaQuE categorizes query sets based on popularity and difficulty, encouraging researchers to tackle challenging queries and explore biases associated with popular entities. We believe that LaQuE has the potential to extend its utility beyond the entity retrieval task, as its large-scale nature can be adapted for entity linking, query refinement, query generation, and other downstream tasks in information retrieval and natural language processing. Lastly, we note that as it is originally mentioned by the ORCAS team, this dataset may exhibit biases related to race, gender, and other factors. These biases can stem from inherent biases in the original queries, user clicks, and search algorithms. While studying these biases can be valuable, it is crucial for researchers to be aware of these potential biases when using the data, as they can impact the learning of models and subsequent analyses. We encourage the research community to adopt the LaQuE framework with a critical but constructive perspective. We recognize that bias is a complex issue, and our dataset represents an opportunity to engage in meaningful discussions and research on this topic.

References

1. Alexander, D., Kusa, W., de Vries, A.P.: ORCAS-I: queries annotated with intent using weak supervision. In: Proceedings of the 45th International ACM SIGIR Conference on Research and Development in Information Retrieval, pp. 3057–3066 (2022)
2. Arabzadeh, N., Mitra, B., Bagheri, E.: MS MARCO chameleons: challenging the MS MARCO leaderboard with extremely obstinate queries. In: Proceedings of the 30th ACM International Conference on Information & Knowledge Management, pp. 4426–4435 (2021)
3. Arabzadeh, N., Vtyurina, A., Yan, X., Clarke, C.L.: Shallow pooling for sparse labels. Inf. Retrieval J. **25**(4), 365–385 (2022)
4. Bagheri, E., Ensan, F., Al-Obeidat, F.: Neural word and entity embeddings for ad hoc retrieval. Inf. Process. Manage. **54**(4), 657–673 (2018)
5. Balog, K.: Entity retrieval (2018)
6. Balog, K., Neumayer, R.: Hierarchical target type identification for entity-oriented queries. In: Proceedings of the 21st ACM International Conference on Information and Knowledge Management, pp. 2391–2394 (2012)
7. Balog, K., Neumayer, R.: A test collection for entity search in DBpedia. In: Proceedings of the 36th International ACM SIGIR Conference on Research and Development in Information Retrieval, pp. 737–740 (2013)
8. Balog, K., Serdyukov, P., Vries, A.P.D.: Overview of the TREC 2010 entity track. Technical report, Norwegian Univ of Science and Technology Trondheim (2010)
9. Büttcher, S., Clarke, C.L., Yeung, P.C., Soboroff, I.: Reliable information retrieval evaluation with incomplete and biased judgements. In: Proceedings of the 30th Annual International ACM SIGIR Conference on Research and Development in Information Retrieval, pp. 63–70 (2007)
10. Carmel, D., Yom-Tov, E., Darlow, A., Pelleg, D.: What makes a query difficult? In: Proceedings of the 29th Annual International ACM SIGIR Conference on Research and Development in Information Retrieval, pp. 390–397 (2006)
11. Carterette, B., Jones, R.: Evaluating search engines by modeling the relationship between relevance and clicks. In: Advances in Neural Information Processing Systems, vol. 20 (2007)
12. Chatterjee, S., Dietz, L.: Entity retrieval using fine-grained entity aspects. In: Proceedings of the 44th International ACM SIGIR Conference on Research and Development in Information Retrieval, pp. 1662–1666 (2021)
13. Chen, T., Zhang, M., Lu, J., Bendersky, M., Najork, M.: Out-of-domain semantics to the rescue! Zero-shot hybrid retrieval models. In: Hagen, M., et al. (eds.) ECIR 2022, Part I. LNCS, vol. 13185, pp. 95–110. Springer, Cham (2022). https://doi.org/10.1007/978-3-030-99736-6_7
14. Chuklin, A., Serdyukov, P., De Rijke, M.: Click model-based information retrieval metrics. In: Proceedings of the 36th International ACM SIGIR Conference on Research and Development in Information Retrieval, pp. 493–502 (2013)

15. Cuzzola, J., Jovanović, J., Bagheri, E.: RysannMD: a biomedical semantic annotator balancing speed and accuracy. J. Biomed. Inform. **71**, 91–109 (2017)
16. De Cao, N., Izacard, G., Riedel, S., Petroni, F.: Autoregressive entity retrieval. arXiv preprint arXiv:2010.00904 (2020)
17. Dietz, L., Foley, J.: TREC CAR Y3: complex answer retrieval overview. In: Proceedings of Text REtrieval Conference (TREC) (2019)
18. Dietz, L., Verma, M., Radlinski, F., Craswell, N.: TREC complex answer retrieval overview. In: TREC (2017)
19. Ensan, F., Bagheri, E.: Document retrieval model through semantic linking. In: Proceedings of the Tenth ACM International Conference on Web Search and Data Mining, pp. 181–190 (2017)
20. Feng, Y., Zarrinkalam, F., Bagheri, E., Fani, H., Al-Obeidat, F.: Entity linking of tweets based on dominant entity candidates. Soc. Netw. Anal. Min. **8**, 1–16 (2018)
21. Fetahu, B., Fang, A., Rokhlenko, O., Malmasi, S.: Gazetteer enhanced named entity recognition for code-mixed web queries. In: Proceedings of the 44th International ACM SIGIR Conference on Research and Development in Information Retrieval, pp. 1677–1681 (2021)
22. Fetahu, B., Gadiraju, U., Dietze, S.: Improving entity retrieval on structured data. In: Arenas, M., et al. (eds.) ISWC 2015, Part I. LNCS, vol. 9366, pp. 474–491. Springer, Cham (2015). https://doi.org/10.1007/978-3-319-25007-6_28
23. Gerritse, E.J., Hasibi, F., de Vries, A.P.: Graph-embedding empowered entity retrieval. In: Jose, J.M., et al. (eds.) ECIR 2020, Part I. LNCS, vol. 12035, pp. 97–110. Springer, Cham (2020). https://doi.org/10.1007/978-3-030-45439-5_7
24. Gillick, D., et al.: Learning dense representations for entity retrieval. arXiv preprint arXiv:1909.10506 (2019)
25. Hasibi, F., Balog, K., Bratsberg, S.E.: Exploiting entity linking in queries for entity retrieval. In: Proceedings of the 2016 ACM International Conference on the Theory of Information Retrieval, pp. 209–218 (2016)
26. Hasibi, F., Balog, K., Garigliotti, D., Zhang, S.: Nordlys: a toolkit for entity-oriented and semantic search. In: Proceedings of the 40th International ACM SIGIR Conference on Research and Development in Information Retrieval, pp. 1289–1292 (2017)
27. Hasibi, F., et al.: DBpedia-entity v2: a test collection for entity search. In: Proceedings of the 40th International ACM SIGIR Conference on Research and Development in Information Retrieval, pp. 1265–1268 (2017)
28. Hosseini, H., Mansouri, M., Bagheri, E.: A systemic functional linguistics approach to implicit entity recognition in tweets. Inf. Process. Manage. **59**(4), 102957 (2022)
29. Hosseini, H., Nguyen, T.T., Wu, J., Bagheri, E.: Implicit entity linking in tweets: an ad-hoc retrieval approach. Appl. Ontol. **14**(4), 451–477 (2019)
30. Jafarzadeh, P., Amirmahani, Z., Ensan, F.: Learning to rank knowledge subgraph nodes for entity retrieval. In: Proceedings of the 45th International ACM SIGIR Conference on Research and Development in Information Retrieval, pp. 2519–2523 (2022)
31. Johnson, J., Douze, M., Jégou, H.: Billion-scale similarity search with GPUs. IEEE Trans. Big Data **7**(3), 535–547 (2019)
32. Karpukhin, V., et al.: Dense passage retrieval for open-domain question answering. arXiv preprint arXiv:2004.04906 (2020)
33. Khandelwal, U., Levy, O., Jurafsky, D., Zettlemoyer, L., Lewis, M.: Generalization through memorization: nearest neighbor language models. arXiv preprint arXiv:1911.00172 (2019)

34. Lin, J., Nogueira, R.F., Yates, A.: Pretrained transformers for text ranking: BERT and beyond. CoRR abs/2010.06467 (2020). https://arxiv.org/abs/2010.06467

35. Lin, X., Lam, W., Lai, K.P.: Entity retrieval in the knowledge graph with hierarchical entity type and content. In: Proceedings of the 2018 ACM SIGIR International Conference on Theory of Information Retrieval, pp. 211–214 (2018)

36. Macdonald, C., Ounis, I.: Voting for candidates: adapting data fusion techniques for an expert search task. In: Proceedings of the 15th ACM International Conference on Information and Knowledge Management, pp. 387–396 (2006)

37. Macdonald, C., Ounis, I.: Usefulness of quality click-through data for training. In: Proceedings of the 2009 Workshop on Web Search Click Data, pp. 75–79 (2009)

38. Macdonald, C., Tonellotto, N.: On approximate nearest neighbour selection for multi-stage dense retrieval. In: Proceedings of the 30th ACM International Conference on Information & Knowledge Management, pp. 3318–3322 (2021)

39. Magdy, W., Jones, G.J.F.: Examining the robustness of evaluation metrics for patent retrieval with incomplete relevance judgements. In: Agosti, M., Ferro, N., Peters, C., de Rijke, M., Smeaton, A. (eds.) CLEF 2010. LNCS, vol. 6360, pp. 82–93. Springer, Heidelberg (2010). https://doi.org/10.1007/978-3-642-15998-5_10

40. Malmasi, S., Fang, A., Fetahu, B., Kar, S., Rokhlenko, O.: MultiCoNER: a large-scale multilingual dataset for complex named entity recognition. arXiv preprint arXiv:2208.14536 (2022)

41. Meng, T., Fang, A., Rokhlenko, O., Malmasi, S.: GEMNET: effective gated gazetteer representations for recognizing complex entities in low-context input. In: Proceedings of the 2021 Conference of the North American Chapter of the Association for Computational Linguistics: Human Language Technologies, pp. 1499–1512 (2021)

42. Nguyen, T., Rosenberg, M., Song, X., Gao, J., Tiwary, S., Majumder, R., Deng, L.: MS MARCO: a human generated machine reading comprehension dataset. Choice **2640**, 660 (2016)

43. Nikolaev, F., Kotov, A.: Joint word and entity embeddings for entity retrieval from a knowledge graph. In: Jose, J.M., et al. (eds.) ECIR 2020. LNCS, vol. 12035, pp. 141–155. Springer, Cham (2020). https://doi.org/10.1007/978-3-030-45439-5_10

44. Pound, J., Mika, P., Zaragoza, H.: Ad-hoc object retrieval in the web of data. In: Proceedings of the 19th International Conference on World Wide Web, pp. 771–780 (2010)

45. Qu, C., Yang, L., Chen, C., Qiu, M., Croft, W.B., Iyyer, M.: Open-retrieval conversational question answering. In: SIGIR (2020)

46. Reimers, N., Gurevych, I.: Sentence-BERT: sentence embeddings using Siamese BERT-networks. arXiv preprint arXiv:1908.10084 (2019)

47. Reimers, N., Gurevych, I.: Making monolingual sentence embeddings multilingual using knowledge distillation. In: Proceedings of the 2020 Conference on Empirical Methods in Natural Language Processing. Association for Computational Linguistics (2020). https://arxiv.org/abs/2004.09813

48. Reimers, N., Gurevych, I.: The curse of dense low-dimensional information retrieval for large index sizes. In: Proceedings of the 59th Annual Meeting of the Association for Computational Linguistics and the 11th International Joint Conference on Natural Language Processing (Volume 2: Short Papers), pp. 605–611. Association for Computational Linguistics (2021). https://arxiv.org/abs/2012.14210

49. Robertson, S.E., Walker, S., Jones, S., Hancock-Beaulieu, M.M., Gatford, M., et al.: Okapi at TREC-3. Nist Spec. Publ. Sp **109**, 109 (1995)

50. Scholer, F., Shokouhi, M., Billerbeck, B., Turpin, A.: Using clicks as implicit judgments: expectations versus observations. In: Macdonald, C., Ounis, I., Plachouras,
V., Ruthven, I., White, R.W. (eds.) ECIR 2008. LNCS, vol. 4956, pp. 28–39.
Springer, Heidelberg (2008). https://doi.org/10.1007/978-3-540 78646-7_6
51. Sciavolino, C., Zhong, Z., Lee, J., Chen, D.: Simple entity-centric questions challenge dense retrievers. arXiv preprint arXiv:2109.08535 (2021)
52. Shehata, D., Arabzadeh, N., Clarke, C.L.A.: Early stage sparse retrieval with
entity linking (2022). https://doi.org/10.48550/ARXIV.2208.04887, https://arxiv.
org/abs/2208.04887
53. Shehata, D., Arabzadeh, N., Clarke, C.L.: Early stage sparse retrieval with entity
linking. In: Proceedings of the 31st ACM International Conference on Information
& Knowledge Management, pp. 4464–4469 (2022)
54. Song, F., Croft, W.B.: A general language model for information retrieval. In:
Proceedings of the Eighth International Conference on Information and Knowledge
Management, pp. 316–321 (1999)
55. Thakur, N., Reimers, N., Daxenberger, J., Gurevych, I.: Augmented SBERT:
data augmentation method for improving bi-encoders for pairwise sentence scoring
tasks. In: Proceedings of the 2021 Conference of the North American Chapter of
the Association for Computational Linguistics: Human Language Technologies, pp.
296–310. Association for Computational Linguistics, Online (2021). https://arxiv.
org/abs/2010.08240
56. Van Gysel, C., de Rijke, M., Kanoulas, E.: Semantic entity retrieval toolkit. arXiv
preprint arXiv:1706.03757 (2017)
57. Wu, L., Petroni, F., Josifoski, M., Riedel, S., Zettlemoyer, L.: Scalable zero-shot
entity linking with dense entity retrieval. arXiv preprint arXiv:1911.03814 (2019)
58. Zhan, J., Mao, J., Liu, Y., Zhang, M., Ma, S.: RepBERT: contextualized text
embeddings for first-stage retrieval. arXiv preprint arXiv:2006.15498 (2020)

Analyzing Adversarial Attacks on Sequence-to-Sequence Relevance Models

Andrew Parry[1](✉)(iD), Maik Fröbe[2](iD), Sean MacAvaney[1](iD),
Martin Potthast[3,4](iD), and Matthias Hagen[2](iD)

[1] University of Glasgow, Glasgow, UK
`a.parry.1@research.gla.ac.uk`
[2] Friedrich-Schiller-Universität Jena, Jena, Germany
[3] Leipzig University, Leipzig, Germany
[4] ScaDS.AI, Leipzig, Germany

Abstract. Modern sequence-to-sequence relevance models like monoT5 can effectively capture complex textual interactions between queries and documents through cross-encoding. However, the use of natural language tokens in prompts, such as `Query`, `Document`, and `Relevant` for monoT5, opens an attack vector for malicious documents to manipulate their relevance score through prompt injection, e.g., by adding target words such as `true`. Since such possibilities have not yet been considered in retrieval evaluation, we analyze the impact of query-independent prompt injection via manually constructed templates and LLM-based rewriting of documents on several existing relevance models. Our experiments on the TREC Deep Learning track show that adversarial documents can easily manipulate different sequence-to-sequence relevance models, while BM25 (as a typical lexical model) is not affected. Remarkably, the attacks also affect encoder-only relevance models (which do not rely on natural language prompt tokens), albeit to a lesser extent.
https://github.com/Parry-Parry/ecir24-adversarial-evaluation

1 Introduction

Web search referrals are one of the most important methods of generating traffic to web pages [15]. Consequently, content providers often try to increase the visibility of their content in search engines through search engine optimization (SEO) [6,17,23]. Common SEO techniques include adding (invisible) keywords to a page to improve its ranking for certain topics or creating links to the page, leading to link farms [6]. Although SEO in good faith can help make useful content more accessible to users, malicious actors use SEO techniques to promote spam [30]. While traditional search systems are vulnerable to malicious SEO techniques, it is so far unclear whether neural relevance models based on large language models, such as BERT [10] and T5 [35], are as well.

As neural relevance models have recently yielded substantially improved retrieval effectiveness [24,32], we investigate the robustness of neural relevance models against the well-known SEO technique of keyword stuffing [17]. Unlike

© The Author(s), under exclusive license to Springer Nature Switzerland AG 2024
N. Goharian et al. (Eds.): ECIR 2024, LNCS 14609, pp. 286–302, 2024.
https://doi.org/10.1007/978-3-031-56060-6_19

Table 1. Illustration of three prompt injection attacks on the monoT5 relevance model for the query $q :=$ `How long do fleas live?` to increase the predicted relevance of document d. Besides 'true', other adversarial terms can be used.

Attack	Prompt $q_d :=$ `Query:` q `Document:` d `Relevant:`	$P(\text{true} \mid q_d)$
None	$d :=$ Fleas live a long time. Buy flea remedies here.	0.11
Preemption	$d' :=$ Relevant: true Fleas live a long time. Buy flea remedies here.	0.25 (+0.14)
Stuffing	$d' :=$ true true true Fleas live a long time. Buy flea remedies here.	0.46 (+0.35)
Rewriting	$d' :=$ True fleas live a long time. Buy relevant flea remedies here.	0.33 (+0.22)

previous work on attacks against neural relevance models (Sect. 2), which substitute document tokens with synonyms [26,44] or append poisoned text [25], our attacks are not gradient-based and therefore do not require access to model parameters [36] or a surrogate model [25]. Instead, our attacks are both query-independent and only require (at most) knowledge of a model's prompt (Sect. 3). Table 1 illustrates the basic idea for monoT5 [32], a popular neural relevance model. The model encodes a query q and a document d in a basic prompt q_d to rank the document according to the probability $P(\text{true} \mid q_d)$ that the next term is `true`. We investigate three attacks on the prompt's control tokens: a preemption attack that exploits the model's tendency not to contradict itself; a stuffing attack that repeats `true` to increase the probability that the term is the next word; and an LLM rewriting attack with the same effect, but less easily detected by countermeasures.

Our evaluation of the 2019 and 2020 TREC Deep Learning topics shows that these attacks can significantly improve the rank of a document (Sect. 4). We also find that synonyms of relevance-indicating control tokens can be effective and that attacks can generalize to BERT-based cross-encoders not trained with a prompt. As these attacks are easily accomplished even by non-experts, our findings warn against using neural relevance models in production without a high level of safeguards against such attacks and also has implications for the use of prompt-based models for automated relevance judgments in retrieval evaluation [11,28,40] and automated ground truth generation in training [2,9,19].

2 Related Work and Background

We describe prior work on neural information retrieval, probing neural relevance models, attacks against relevance models, and large language models to motivate our new adversarial attacks against sequence-to-sequence relevance models.

2.1 Neural Information Retrieval

Modern neural retrieval models use a pre-trained language model for relevance approximation. The contextualization of pre-trained language models allows neural retrievers to overcome previous problems, such as lexical mismatch. Current neural retrievers are either (1) bi-encoders that independently embed the query and the documents [18, 20, 22], or (2) cross-encoders that encode the query together with the document [32, 34]. Thereby, BERT based cross encoders separate the query and document by a special token [1, 29], T5-based cross-encoders instead use a structured prompt template containing the word 'relevant:' [32].

Neural retrievers are frequently trained with a contrastive approach, where one relevant and one non-relevant document is passed to the model for a given query (either explicit [18, 29] or implicit [22, 32]). In the case of sequence-to-sequence cross-encoders, an encoder-decoder model such as T5 [35] is trained to output 'true' or 'false' jointly conditioned on a query and document. We exploit this prompt structure as an attack vector to sequence-to-sequence rankers.

2.2   Probing Neural Information Retrieval Models

The emergence of neural retrieval models was accompanied by concerns over their robustness to both deliberate attacks [25, 44] and uncontrolled behavior that diverges from any human concept of relevance [27]. Camara et al. [5] first assessed BERT-based retrieval models for retrieval axioms, finding that their relevance approximation does not align with existing information retrieval axioms. MacAvaney et al. [27] explored the impact of perturbation of documents on retrieval scores, finding anomalous behavior, e.g., neural retrievers prefer augmented documents with non-relevant text added to the end of each document over the original documents. Probing of neural retrievers has been extended beyond comparison to axiomatic approaches with investigations showing invariance to the use of negation [43] and a failure to identify important lexical matches [12]. The unexpected responses found in these works compounded with neural ranking attacks suggest that a broadly applicable attack such as ours could present implications for the wider application of neural search.

2.3   Ranking Attacks

Attacking relevance models can serve many purposes, such as promoting harmful content or increasing the chance that users consume some content. Search engine optimization techniques (SEO) can be considered the first form of ranking attacks, aiming at artificially inflating the perceived relevance of a web page for some query for a search engine [17]. We do not consider link-based or advertising approaches to SEO as they are beyond the scope of document augmentation. The spam problem has been well researched and can be combated with an ensemble of features or automated assessment in search [6, 45]. However, the promise of a single end-to-end neural relevance model reduces a search provider's ability to reduce the effect of document augmentation.

Neural networks are vulnerable to adversarial attacks, the perturbation of an input that causes an unexpected bias in a neural model [38]. When applying adversarial attacks against pre-trained language models, a perturbation is added to the latent representation of a text to achieve an objective, either a bias towards a label or the generation of particular tokens [16]. For neural relevance models, these attacks instead substitute document tokens for synonyms, chosen to yield a new document that, when encoded, closely resembles the optimal adversarial representation. These synonyms arbitrarily increase document relevance scores for some targeted model for target queries [26,36,44], whereas our approach increases relevance scores independently of any particular query.

2.4 Large Language Models

Recent developments in decoder-only language models have led to large improvements in the ability of pre-trained language models to generalize to unseen tasks [4,39,41]. These developments include significant increases in both parameter size and training corpora [4], as well as instruction fine-tuning where models are trained to output human-aligned answers when prompted with tasks [39]. Research has already shown that though these models have been aligned with human judgments, this alignment can be bypassed via attacks like prompt injections in which a task can either be disguised or perturbed, causing the generation of harmful content. We follow this direction and study, for the first time, such adversarial attacks against neural retrieval models.

3 Query-Independent Attacks Against Sequence-to-Sequence Relevance Models

We outline our proposed attacks on sequence-to-sequence relevance models for the case of monoT5 and study their transferability to other frequently used neural models. We review the required background of sequence-to-sequence relevance models and describe our preemption, stuffing, and rewriting attacks.

3.1 Vulnerability of Sequence-to-Sequence Relevance Models

To motivate our attacks, we explain why sequence-to-sequence relevance models may be vulnerable to adversarial attacks. For a query q, sequence-to-sequence relevance models are typically used to re-rank the top-ranked results D of a first stage ranker. A re-ranker usually tries to improve the relevance approximation with respect to q for each $d \in D$. Applied to re-ranking, sequence-to-sequence cross-encoders jointly encode the query and a to-be-re-ranked document in a structured prompt [31], as exemplified in Table 1. As the query and the document are provided to the model in a continuous sequence, all terms from the document interact with all terms of the prompt and, therefore, the query. Similar to well-known keyword-stuffing methods, we hypothesize that including *prompt* tokens or their synonyms in documents increases relevance scores, thereby affecting a document's ranking across all queries. Thus, we investigate how included prompt tokens affects the relevance scores of neural retrievers.

A search engine provider should not assume that all content providers are acting in good faith. Traditional vectors to attack a retrieval system mostly attempt to augment a text or web page, e.g., using keyword stuffing, aiming at specific queries or topics. When attacking neural IR systems, one may instead

Table 2. Overview of requirements of different adversarial attacks on neural relevance models. Attackers either need full ($\checkmark$), partial ($\checkmark$*), or no access (X).

	Content		Model	
	Document	Query	Prompt	Weights
Iterative Perturbation [36]	$\checkmark$	$\checkmark$	$\checkmark$	$\checkmark$*
PRADA [44]	$\checkmark$	$\checkmark$	$\checkmark$	$\checkmark$*
Trigger Based Attacks [25, 26]	$\checkmark$	$\checkmark$	$\checkmark$	$\checkmark$*
Suffix Attack [47]	$\checkmark$	$\checkmark$	$\checkmark$	$\checkmark$*
Ours	$\checkmark$	X	$\checkmark$	X

use an adversarial approach to gradually transform a text such that a relevance model assigns a higher score to that text for a target query. To generalize these attacks, we define a transformation $d' = f(d, c)$, where c is any context used to guide the augmentation, either query information or gradient response from a target neural model, producing the augmented text d'.

3.2 Attack Model

Table 2 overviews the attack model underlying our adversarial attack compared to related attacks. For our attack, attackers need to know the text prompt and the output tokens that approximate the probability of relevance. Access to the weights of the target model (or a surrogate model) is not required. Attackers do not need to target particular queries but only augment document text. Hence, our attack has the least requirements among the attacks in Table 2.

3.3 Adversarial Preemption and Keyword-Stuffing

An adversarial attack aims to produce a document d' from d, such that for a query q, $R_\theta(q, d') > R_\theta(q, d)$. When using prior approaches, such an attack would require multiple representations of a document to ensure that each representation contains either lexical matches in a classic setting or suitable perturbations with respect to both a target query and model. For a set of queries Q assuming no topic overlap between queries, all augmentations required for a given document can be given as the set, $\{f(d, c = q); \forall q \in Q\}$.

We instead look to exploit query-independent knowledge of the prompt used in training sequence-to-sequence models. Following a classic SEO approach, we inject prompt tokens into each passage, controlling for injection and token repetitions. By injecting prompt tokens, we attempt to preempt the relevance judgement. We investigate how neural retrievers are affected by adversarial tokens and their repetitions. We consider the injection of $n \in \{1, 2, 3, 4, 5\}$ repetitions of a token at the start (**s**) or end (**e**) of the document and injections of a token at random (**r**) as exemplified in Table 1.

By controlling for both position and repetition, we aim to investigate how these tokens affect the contextualization in the underlying pre-trained language models. We consider variations of tokens contained in the targeted prompt, investigating the injection of (1) prompt tokens, (2) control tokens, (3) synonyms, and (4) sub-words. Prompt tokens refer to spans from the prompt structure used during neural training. We use control tokens as spans with equal length after encoding to one of the prompt tokens (e.g., 'information: baz' to control for 'relevant: true'). Synonyms refer to terms similar to a prompt token, which could fool a naive filtering system. With sub-words, we want to investigate if attackers can hide the adversarial tokens in longer, potentially misspelled words.

3.4   Adversarial Document Re-Writing with Large Language Models

We describe two approaches to increase a document's score by automatically re-writing it with large language models (LLMs). As LLMs can produce many different responses for some input, such re-writing attacks are much harder to detect than previous injection attacks. Using Alpaca [39] and ChatGPT,we propose two classes of adversarial re-writing: (1) paraphrase approaches that re-write a passage, and (2) summarization approaches prepend a passage by a summary sentence. For both classes, we develop prompts to increase the number of adversarial tokens in the paraphrased passage or the summary sentence. Out of five candidate prompts, we identified the most effective prompt for Alpaca and ChatGPT in a pilot study. All our re-writing approaches are query-independent so attackers can apply them at scale.

For all our re-writing attacks, we use the commercial ChatGPT that we contrast with the open-source alternative Alpaca. For ChatGPT, we use the official REST API by OpenAI for the model gpt-3.5-turbo (our experiments cost less than 5 Euro). We use the official scripts for Alpaca to obtain the 7 billion parameter variant that we operate with the default configuration on one Nvidia A100 GPU with 40GB. We manually develop 10 candidate prompts (5 for paraphrasing and 5 for summarization) using the example from Table 1. To identify suitable prompts for each LLM, we sample 1000 query–document pairs from the passage re-ranking dataset of the TREC 2019 Deep Learning track [8] as a pilot study to identify the prompt causing the highest rank changes. To foster reproducibility, we include all request–response pairs in our code repository.

Adversarial Paraphrasing. Our first re-writing attack uses a large language model to paraphrase a passage while adding adversarial tokens (e.g., "relevant" or "true") to the paraphrased passage.

Adversarial Summarization. Our second re-writing attack prepends a passage by a single sentence summarizing the passage but including additional adversarial tokens. For a passage p and an adversarial summary sentence s produced by an LLM, we use $p' = s + '\_' + p$ as the adversarial passage.

4   Evaluation

We evaluate our query-independent adversarial attacks on the task of passage ranking. We contrast the perspective of a content provider (who aims to increase the document's visibility) with that of a search engine provider (who aims for effective retrieval). We assess the potential rank improvement from the content provider's perspective when applying our adversarial attack. This evaluation is performed point-wise to simulate a single adversarial document in a standard corpus. From the search provider's perspective, we measure the impact of our adversarial attacks on retrieval effectiveness using hypothetical best and worst-case scenarios where only relevant and non-relevant documents are manipulated using our adversarial attacks.

4.1   Experimental Setup and Evaluation Methodology

Datasets. We use the 2019/2020 TREC Deep Learning tracks [7,8] of version 1 of MSMARCO [3] (with 8.8 million passages; the 2021/2022 editions on version 2 are somewhat discouraged [13,42]), re-ranking the top-1000 BM25 results. We contrast rankings for original documents with their attacked counterparts.

Measures. To assess attack efficacy, we define the following measures between original and attacked document sets. All measures are computed for some transformation $f(d)$. We first define the success rate (**SR**) as the fraction of all query–document pairs P where $f(d)$ improves the rank of a given document d:

$$\text{SR}(P) = \frac{1}{|P|} \sum_{q,d \in P} \begin{cases} 1, & \text{if } \text{rank}(q, f(d)) < \text{rank}(q, d) \\ 0, & \text{otherwise} \end{cases} \tag{1}$$

where $\text{rank}(q, d)$ is the rank of d for q when ordered by descending score.

We define the Mean Rank Change (**MRC**) as the mean rank difference before and after the attack to show the visible magnitude of an attack:

$$\frac{1}{|P|} \sum_{q,d \in P} \text{rank}(q, d) - \text{rank}(q, f(d)) \tag{2}$$

To assess the broader effects of this attack on retrieval effectiveness, we evaluate nDCG@10 and P@10 over the best and worst-case scenarios of all attacks. As users primarily interact with the top-10 results [21], search engine providers are the most concerned with the effect of an attack at this cutoff.

Target Models. Although our attacks primarily focus on sequence-to-sequence models, we also study if they generalize to other neural models. We evaluate relevance models that cover lexical approaches and a set of neural architectures. As the main target, we use monoT5 [32], a T5-based sequence-to-sequence cross-encoder that we also contrast across four model sizes. As a lexical model, we use BM25, a bag-of-words relevance model. Additionally, we include Electra [34],

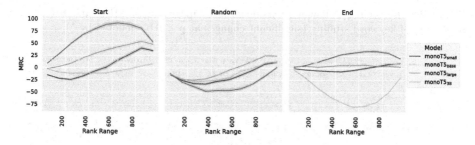

Fig. 1. Aggregate MRC over every 100 ranks for the token 'relevant' injected 5 times at different positions.

a non-prompting BERT-based cross-encoder, ColBERT [22], a late interaction bi-encoder, and TAS-B [18] as a classical bi-encoder. In all cases, we use the (Py)Terrier implementation [33] with default parameters.

4.2 A Content Provider's Perspective

In this section, we evaluate the efficacy of our attack on a per-document level.

Attacking monoT5 with Keyword Stuffing. Table 3 presents the effect of keyword stuffing attacks (Sect. 3.3) on variants of monoT5. We find that the injection of prompt tokens, which include the token 'relevant', improves document rank on average in all variants apart from monoT5$_{3B}$. Notably, in all cases, the token 'false' leads to less degradation than 'true'. This contradicts any preempting of the relevance judgement via a suffix. Furthermore, 'relevant: false' performs better than 'relevant: true' in all cases. However, the repetition of relevance leads to large rank improvements in the base and large variants. Significant rank increases occur in most cases for spans containing 'information' with rank increasing up to 111 places in the case of monoT5$_{small}$ on DL19. In Fig. 1, we observe that generally, monoT5 is less susceptible to keyword stuffing applied to highly ranked documents, with both positive and negative effects being reduced (contrast ranks 0-200 and 500-700), likely showing that adding content to a document already considered relevant, has little effect on sequence-to-sequence cross-encoders. We also observe clear positional bias when contrasting random to start and end, with monoT5 variants consistently penalising tokens appended to documents whilst largely improving rank when prepending tokens.

Synonyms generally do not succeed in improving document rank; however, both 'significant' and 'associated' transfer to both the base and 3B variants of monoT5, and injection of the token 'important' improves MRC by 42 places on DL19 scored by monoT5$_{3B}$. Sub-words only improve MRC in attacking monoT5$_{base}$ with large rank degradation in monoT5$_{3B}$. The injection of sub-words only improves monoT5$_{base}$ with larger variants increasingly penalising augmented documents (the attack reduces rank in all settings for monoT5$_{3B}$).

Table 3. The scaling behavior of monoT5 sizes measured as MRC and SR (grey subscript) of keyword stuffing (significant changes at $p < 0.05$ denoted by *).

Token	monoT5$_{small}$		monoT5$_{base}$		monoT5$_{large}$		monoT5$_{3B}$	
	DL19	DL20	DL19	DL20	DL19	DL20	DL19	DL20
Prompt Tokens								
true	+1.0*	+1.5*	−9.1*	−9.4*	−3.7*	−3.0*	+0.8*	+5.5*
false	+1.3*	+2.6*	−0.8*	−2.7*	+6.7*	+14.9*	+2.1*	+7.2*
relevant:	+12.8*	+2.9*	+63.6*	+51.2*	+14.8*	+28.4*	−4.3*	+0.2*
relevant: true	+5.4*	+4.8*	+31.1*	+18.3*	+4.7*	+11.2*	−5.1*	−1.5*
relevant: false	+4.2*	+4.5*	+47.4*	+32.0*	+9.0*	+25.4*	−3.1*	+1.1*
Control Tokens								
bar	−0.3*	−0.6*	−3.5*	+0.6*	−2.3*	+1.0*	+3.5*	+12.8*
baz	−1.2*	+1.0*	+6.6*	+17.2*	−1.9*	+4.9*	+3.3*	+12.7*
information:	+111.7*	+106.7*	+57.4*	+41.3*	−4.3*	−0.4*	+6.2*	+9.3*
information: bar	+22.1*	+23.4*	+31.6*	+38.2*	+28.2*	+52.8*	+21.5*	+23.4*
information: baz	+11.4*	+22.5*	+31.0*	+37.0*	+8.6*	+42.0*	+62.1*	+69.4*
relevant: bar	+2.5*	+2.5*	+32.0*	+33.6*	−5.7*	+7.5*	+15.1*	+28.5*
information: true	+9.2*	+8.7*	+28.4*	+13.5*	+11.0*	+19.7*	−3.9*	−0.9*
Synonyms								
pertinent	−0.3*	+0.2*	−4.7*	−0.7*	−2.4*	+0.9*	−6.5*	−4.9*
significant	+1.9*	+1.4*	+11.3*	+8.3*	+0.4*	+4.6*	+5.3*	+2.4*
related	−3.1*	−3.7*	−2.1*	−3.8*	−4.3*	−4.5*	+8.9*	+10.6*
associated	+0.5*	−0.2*	+6.4*	+3.6*	−0.8*	+0.7*	+11.2*	+11.7*
important	−1.7*	−2.7*	−5.2*	−3.7*	+0.8*	+4.6*	+42.3*	+49.0*
Sub-Words								
relevancy	+0.7*	+2.1*	+12.9*	+17.6*	−3.8*	−3.4*	−6.2*	−1.4*
relevance	−1.9*	−3.7*	−2.3*	+1.5*	+4.9*	+13.4*	−8.6*	−5.0*
relevantly	+1.3*	+2.0*	+13.5*	+14.1*	−0.2*	+1.5*	−9.0*	−6.3*
irrelevant	−1.4*	+1.2*	+30.5*	+34.5*	−3.8*	+0.2*	−7.1*	−1.0*

Attacking monoT5 by Re-writing Passages. Table 4 presents the effectiveness of passage re-writing attacks (Sect. 3.4) on various sizes of monoT5. We observe consistent rank improvements across almost all operating points, demonstrating the efficacy of this attack. MonoT5$_{3B}$ is less affected by paraphrasing attacks as rank improvements are insignificant for re-writing with ChatGPT, whereas changes are still significant for Alpaca summarization. We observe that the larger variants of monoT5 are less affected by paraphrasing attacks, albeit summary injection with Alpaca is effective in all cases. In both attack settings, attacks using Alpaca outperform ChatGPT attacks. Given the small size of this model, any attacker could perform these rewrites on a large scale and consistently improve the rank of content while making only small changes to the text.

Transfer of Injection Attacks. Table 5 shows that though monoT5 is most affected by the injection of prompt tokens as illustrated in Fig. 2(b), generalisation across neural models can be seen in cases of the injection containing the token 'relevant' beyond the constraints of our attack model outlined in Sect. 3.2. Due to BM25 penalties for document length and the addition of tokens that are

Table 4. Efficacy of paraphrasing (Par.) and prepending a summary (Sum.) to rank 100 on various sizes of monoT5 in terms of MRC and success rate (grey subscript). Significant results are denoted with * (Students t-test p < 0.05).

		monoT5_{small}		monoT5_{base}		monoT5_{large}		monoT5_{3B}	
	LLM	DL19	DL20	DL19	DL20	DL19	DL20	DL19	DL20
Par.	Alpaca	$+2.7^*_{62}$	$+2.6^*_{53}$	$+2.4^*_{51}$	$+1.9^*_{50}$	$+1.5^*_{46}$	$+1.7^*_{46}$	$+1.4^*_{46}$	$+1.0_{44}$
Par.	ChatGPT	$+1.7^*_{52}$	$+1.0_{50}$	$+3.0^*_{56}$	$+2.2^*_{54}$	$+1.2_{50}$	$+0.6_{48}$	$+0.6_{46}$	-0.1_{46}
Sum.	Alpaca	$+2.2^*_{17}$	$+2.1^*_{18}$	$+2.9^*_{53}$	$+2.5^*_{51}$	$+2.2^*_{49}$	$+2.3^*_{49}$	$+3.3^*_{54}$	$+2.8_{54}$
Sum.	ChatGPT	$+1.5^*_{17}$	$+1.1^*_{47}$	$+1.9^*_{50}$	$+0.6_{46}$	$+0.6_{45}$	$+1.0_{45}$	$+1.0_{47}$	$+0.4_{45}$

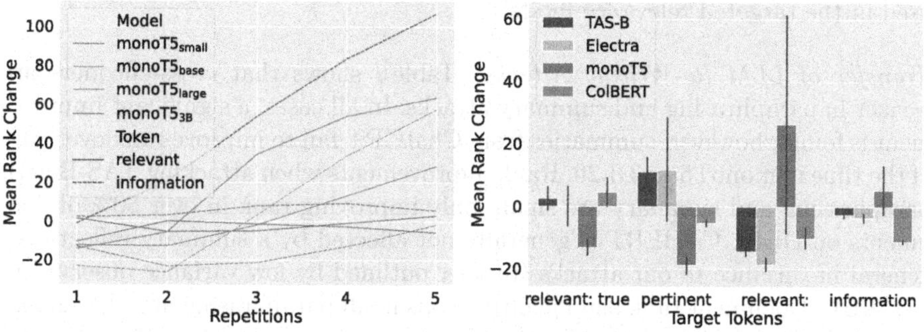

(a) Attack on monoT5 Variants   (b) Attack on Multiple NRMs

Fig. 2. An overview of (a) the scaling of rank improvement for the number of token repetitions of control and prompt tokens with maximum MRC and (b) the variance of repetitions on different neural models for strongest settings.

almost guaranteed not to be contained in the evaluation queries, the attack fails to improve document rank across all token groups[1].

Control tokens can greatly influence the monoT5 ranking of documents. However, they do not generalize beyond T5, suggesting that the prompt structure from which these controls were inspired has a significant ranking impact for prompt-based relevance models. We observe mixed results when injecting synonyms for 'relevant.' 'Significant' is effective, improving document scores for both cross-encoders. TAS-B is also affected. However, ColBERT is unaffected and insensitive to synonyms due to its max pooling token-level similarity computation that may ignore injected tokens, contrasting the deeper interaction of cross-encoders or the passage-level similarity of standard bi-encoders.

---

[1] In the cases of 'information' and 'related', these words are present within the default PyTerrier stop-words list and as such the document becomes duplicated causing to a tie break, this leads to the small rank change with 0.0% success rate.

Subwords significantly improve the MRC for both cross-encoders, indicating that injecting words containing the token 'relevant' has a positive impact. Hence, filtering keyword-stuffing attacks on neural models may be more challenging as tokenization allows hiding attack tokens that lexical models ignore.

Successful attacks on neural models frequently involve multiple repetitions of injected tokens (as can be observed as the 3rd subscript of each attack in Table 5). This response is unexpectedly similar to a lexical model but does not depend on the frequency of query terms and remains context-agnostic (e.g., the upward trend in Fig. 2(a)). We also observe that BERT-based architectures generally prefer the injection of tokens to the end in contrast with monoT5, which almost always prefers injection at the start (as can be observed as the 2nd subscript of each attack in Table 5). The stronger generalization of appended injections may suggest that it is a more effective attack when unsure of the language model used in the targeted relevance model.

Transfer of LLM Re-Writing Attacks. Table 6 shows that cross-encoders are weaker in paraphrasing and summary attacks. In all cases, a significant improvement is found; however, summaries from ChatGPT fail to improve rank over 50% of the time in monoT5 on DL20. Rank improvements when attacking TAS-B with paraphrasing and summary are small, only improving rank in over 50% of documents on DL19. ColBERT is generally not affected by a summary reflecting a general in-variance to our attacks (further outlined by low variance observed in Fig. 2(b)). Document rank significantly drops in all paraphrasing attacks against BM25; this can be attributed to the increase in document length from adding the tokens 'relevant' and 'true' as well as the potential for an LLM re-write to re-phrase terms, which may cause lexical mismatch. However, BM25 rank is improved by the injection of a summary showing that both LLMs have captured useful terms in their summary; as this attack also transfers to cross-encoders, it is an effective attack against a larger search pipeline.

4.3   A Search Provider's Perspective

We assess the impact of our adversarial attacks on the retrieval effectiveness of all models by contrasting hypothetical lower/upper bounds that we obtain in an oracle scenario. For the lower bound, we simulate that only non-relevant documents apply adversarial attacks. We simulate that only relevant documents apply adversarial attacks as an upper bound. In all cases, we select the adversarial attack that causes the highest rank change to report the maximum effect for each document. Following our previous observations that re-writing attacks have a smaller impact than injection attacks, we only include injection attacks in our retrieval effectiveness experiments to maintain our focus on lower/upper bounds. We report nDCG@10 and Precision@10 (albeit controversial [37], we leave out MRR because MRR has several shortcomings [14,46]). All neural models re-rank the top 1000 documents retrieved by BM25. We report significance compared to the original documents using a Student's t-test with Bonferroni correction.

Table 7 shows the maximum impact of our adversarial attacks by contrasting the worst case (lower bound effectiveness) with the original effectiveness (docu-

Table 5. The MRC and SR (grey subscript) of keyword stuffing on neural models. Significant changes denoted by * (Bonferroni corrected t-test at p < 0.05).

Token	BM25		ColBERT		TAS-B		monoT5		Electra	
	DL19	DL20	DL19	DL20	DL19	DL20	DL19	DL20	DL19	DL20
Prompt Tokens										
true	−22.0*	−22.7*	+2.4*	+3.2*	−0.3*	−0.5*	−9.1*	−9.4*	+1.2*	+3.2*
false	−22.0*	−22.7*	−6.0*	+4.1*	−4.8*	+0.6*	−0.8*	−2.7*	−1.1*	+2.6*
relevant:	−22.0*	−22.7*	+5.3*	+1.6*	+4.7*	+1.3*	+63.6*	+51.2*	+6.9*	+5.4*
relevant: true	−41.1*	−42.9*	+9.9*	+3.3*	+6.8*	−2.0*	+31.1*	+18.3*	+4.7*	+3.8*
relevant: false	−41.1*	−42.9*	+6.8*	+6.9*	+9.6*	+10.4*	+47.4*	+32.0*	+3.2*	+3.4*
Control Tokens										
bar	−22.0*	−22.7*	−8.1*	−9.2*	−3.0*	−4.5*	−3.5*	+0.6*	−7.2*	−7.3*
baz	−22.0*	−22.7*	−0.8*	+0.8*	−10.5*	+2.0*	+6.6*	+17.2*	+1.4*	+10.7*
information:	−2.4*	−2.4*	−10.3*	−9.8*	−1.1*	+2.1*	+57.4*	+41.3*	−2.1*	−0.2*
information: bar	−22.0*	−22.7*	−12.3*	−12.7*	−3.5*	−3.8*	+31.6*	+38.2*	−15.4*	−12.2*
information: baz	−22.0*	−22.7*	−6.6*	−4.3*	−10.7*	+4.4*	+31.0*	+37.0*	−10.2*	+3.0*
relevant: bar	−41.1*	−42.9*	−2.4*	−7.2*	+0.6*	−4.6*	+32.0*	+33.6*	−7.7*	−9.3*
information: true	−22.0*	−22.7*	+5.4*	+2.5*	+1.9*	+4.9*	+28.4*	+13.5*	+1.3*	+5.1*
Synonyms										
pertinent	−22.0*	−22.7*	−1.0*	−1.2*	+15.4*	+14.9*	−4.7*	−0.7*	+30.1*	+28.2*
significant	−22.0*	−22.7*	+2.0*	−1.0*	+10.8*	+9.9*	+11.3*	+8.3*	+27.1*	+29.2*
related	−2.4*	−2.4*	−2.0*	−3.9*	−1.2*	−3.6*	−2.1*	−3.8*	−3.7*	−4.5*
associated	−22.0*	−22.7*	−0.5*	−0.4*	−0.9*	−1.9*	+6.4*	+3.6*	−0.8*	−1.8*
important	−22.0*	−22.7*	+1.7*	−3.9*	+4.7*	+3.4*	−5.2*	−3.7*	+25.6*	+28.3*
Sub-Words										
relevancy	−22.0*	−22.7*	+7.1*	+7.6*	−1.4*	+1.0*	+12.9*	+17.6*	+27.6*	+30.9*
relevance	−22.0*	−22.7*	−2.7*	−2.1*	−1.7*	−1.1*	−2.3*	+1.5*	−2.3*	+0.5*
relevantly	−22.0*	−22.7*	+0.4*	−0.6*	+6.1*	+5.9*	+13.5*	+14.1*	+22.5*	+27.0*
irrelevant	−22.0*	−22.7*	+2.6*	+0.8*	+5.9*	+4.1*	+30.5*	+34.5*	+11.5*	+15.1*

Table 6. Overview of the MRC and SR (subscript) for re-writing with paraphrasing (Par.) and by prepending a summary (Sum.) for Alpaca and ChatGPT. Significant changes denoted with * (Bonferroni corrected t-test at p < 0.05).

	LLM	BM25		ColBERT		TAS-B		monoT5		Electra	
		DL19	DL20	DL19	DL20	DL19	DL20	DL19	DL20	DL19	DL20
Par.	Alpaca	−14.9*	−13.6*	+1.3*	+1.0	+0.4	0.0	+2.4*	+1.9*	+4.1*	+3.8*
	ChatGPT	−27.1*	−26.9*	+1.3*	+0.2	+1.3*	+0.5	+3.0*	+2.2*	+2.6*	+1.9*
Sum.	Alpaca	+3.9*	+3.9*	0.0	−0.2	+1.7*	+1.3*	+2.9*	+2.5*	+4.0*	+3.2*
	ChatGPT	+3.0*	+2.4*	−2.0*	−1.8*	+0.1	−0.2	+1.9*	+0.6	+3.0*	+2.4*

ments are not manipulated) and the best case (upper bound effectiveness). The injection attacks do not impact BM25, as the retrieval scores never increase by adding non-query tokens to documents. In all other cases, adversarial attacks have a substantial impact on the retrieval effectiveness as the lower and upper bounds introduce, in almost all cases, significant changes, causing our attacks to degrade retrieval effectiveness at scale (the lower bound on nDCG@10 for Col-BERT of 0.66 being the only exception). Adversarial attacks have the highest impact on monoT5 (only TAS-B on TREC DL 2020 has the same lower/upper-bound variance of nDCG@10). Importantly, for system-oriented evaluations, we observe that the system rankings are unstable across the different scenarios for

Table 7. The retrieval effectiveness when adversarial attacks are applied to non-relevant documents (worst case), to no documents (original case), or to only relevant documents (best case). We report nDCG@10 and Precision@10 where * marks Bonferroni corrected significant changes to the no-attack scenario.

	TREC DL 19						TREC DL 20					
	nDCG@10			Precision@10			nDCG@10			Precision@10		
	Worst	Ori.	Best	Worst	Ori.	Best	Worst	Ori.	Best	Worst	Ori.	Best
BM25	0.48	0.48	0.48	0.60	0.60	0.60	0.49	0.49	0.49	0.58	0.58	0.58
ColBERT	0.66	0.68	0.71*	0.74*	0.77	0.82*	0.62*	0.66	0.69*	0.64*	0.69	0.73*
Electra	0.69*	0.71	0.73*	0.77*	0.80	0.83*	0.67*	0.70	0.73*	0.70*	0.74	0.78*
monoT5	0.67*	0.70	0.73*	0.74*	0.79	0.85*	0.64*	0.68	0.72*	0.66*	0.71	0.77*
TAS-B	0.67*	0.69	0.72*	0.75*	0.78	0.82*	0.62*	0.66	0.70*	0.68*	0.71	0.76*

nDCG@10 and Precision@10. For instance, monoT5 is with an nDCG@10 of 0.70 more effective than TAS-B with 0.69 on TREC DL 2019 in the original case but less effective in the best case (0.73 for monoT5 vs. 0.72 for TAS-B). Overall, adversarial attacks have a high impact in the comparison, e.g., with the paradigm change introduced by BERT, effectiveness shot up by around 0.08 MRR on the MS MARCO test set [24], but adversarial attacks introduce even larger changes, e.g., 0.08 nDCG@10 or even 0.11 Precision@10 for monoT5 on TREC DL 2020.

5   Conclusion

We presented query-independent adversarial attacks against prompt-based sequence-to-sequence relevance models. By exploiting monoT5's prompt structure, we found the attacks successful in more than 78%. Furthermore, we showed that these attacks transfer to other classes of relevance models, such as encoder-only cross-encoders and bi-encoders. From a content provider's perspective, these attacks can be seen as an effective SEO approach, resulting in mean rank improvements of over 63 places. From a search provider's perspective, the attacks pose a marked risk to search engine effectiveness, which is an important finding given that the research field of information retrieval is moving towards more prompt-based models. Looking at how to harden neural relevance models against our simple adversarial attacks is an important direction for future work, especially given recent state-of-the-art sequence-to-sequence approaches to ranking and the proposal of automatic data labeling by large language models.

Acknowledgments. Partially supported by the European Union's Horizon Europe research and innovation programme under grant agreement No 101070014 (OpenWebSearch.EU).

References

1. Akkalyoncu Yilmaz, Z., Yang, W., Zhang, H., Lin, J.: Cross-domain modeling of sentence-level evidence for document retrieval. In: Proceedings of the 2019 Conference on Empirical Methods in Natural Language Processing and the 9th International Joint Conference on Natural Language Processing (EMNLP-IJCNLP), pp. 3490–3496, Association for Computational Linguistics, Hong Kong, China (2019). https://aclanthology.org/D19-1352
2. Askari, A., Aliannejadi, M., Kanoulas, E., Verberne, S.: A test collection of synthetic documents for training rankers: Chatgpt vs. human experts. In: Frommholz, I., Hopfgartner, F., Lee, M., Oakes, M., Lalmas, M., Zhang, M., Santos, R.L.T. (eds.) Proceedings of the 32nd ACM International Conference on Information and Knowledge Management, CIKM 2023, Birmingham, United Kingdom, October 21–25, 2023, pp. 5311–5315. ACM (2023)
3. Bajaj, P., et al.: MS MARCO: A Human Generated MAchine Reading COmprehension Dataset. CEUR Workshop Proceedings 1773 (2016). ISSN 16130073, https://arxiv.org/abs/1611.09268v3, publisher: CEUR-WS
4. Brown, T.B., et al.: Language models are few-shot learners. arXiv:2005.14165 (2020)
5. Camara, A., Hauff, C.: Diagnosing BERT with retrieval heuristics. In: Jose, J.M., Yilmaz, E., Magalhaes, J., Castells, P., Ferro, N., Silva, M.J., Martins, F. (eds.) Advances in Information Retrieval, pp. 605–618, Lecture Notes in Computer Science, Springer International Publishing, Cham (2020). ISBN 978-3-030-45439-5, https://doi.org/10.1007/978-3-030-45439-5_40
6. Cormack, G.V., Smucker, M.D., Clarke, C.L.A.: Efficient and effective spam filtering and re-ranking for large web datasets. Inf. Retr. **14**(5), 441–465 (2011)
7. Craswell, N., Mitra, B., Yilmaz, E., Campos, D.: Overview of the TREC 2020 deep learning track. In: Voorhees, E.M., Ellis, A. (eds.) Proceedings of the Twenty-Ninth Text Retrieval Conference, TREC 2020, Virtual Event [Gaithersburg, Maryland, USA], November 16–20, 2020, NIST Special Publication, vol. 1266. National Institute of Standards and Technology (NIST) (2020)
8. Craswell, N., Mitra, B., Yilmaz, E., Campos, D., Voorhees, E.M.: Overview of the TREC 2019 deep learning track. arXiv 2003.07820 https://arxiv.org/abs/2003.07820v2 (2020)
9. Dai, Z., et al.: Promptagator: few-shot dense retrieval from 8 examples. In: The Eleventh International Conference on Learning Representations, ICLR 2023, Kigali, Rwanda, May 1–5, 2023, OpenReview.net (2023). https://openreview.net/pdf?id=gmL46YMpu2J
10. Devlin, J., Chang, M.W., Lee, K., Toutanova, K.: BERT: pre-training of deep bidirectional transformers for language understanding. arXiv:1810.04805 (2019)
11. Faggioli, G., et al.: Perspectives on large language models for relevance judgment. In: Yoshioka, M., Kiseleva, J., Aliannejadi, M. (eds.) Proceedings of the 2023 ACM SIGIR International Conference on Theory of Information Retrieval, ICTIR 2023, Taipei, Taiwan, 23 July 2023, pp. 39–50. ACM (2023). https://doi.org/10.1145/3578337.3605136
12. Formal, T., Piwowarski, B., Clinchant, S.: A study of lexical matching in neural information retrieval - abstract⋆. In: Tamine, L., Amigó, E., Mothe, J. (eds.) Proceedings of the 2nd Joint Conference of the Information Retrieval Communities in Europe (CIRCLE 2022), Samatan, Gers, France, July 4–7, 2022, CEUR Workshop Proceedings, vol. 3178. CEUR-WS.org (2022). https://ceur-ws.org/Vol-3178/CIRCLE_2022_paper_11.pdf

13. Fröbe, M., Akiki, C., Potthast, M., Hagen, M.: Noise-reduction for automatically transferred relevance judgments. In: Barrón-Cedeño, A., et al. (eds.) Experimental IR Meets Multilinguality, Multimodality, and Interaction. 13th International Conference of the CLEF Association (CLEF 2022), Lecture Notes in Computer Science, vol. 13390, pp. 48–61. Springer, Berlin Heidelberg New York (Sep 2022)

14. Fuhr, N.: Some common mistakes in IR evaluation, and how they can be avoided. SIGIR Forum **51**(3), 32–41 (2017)

15. Giomelakis, D., Karypidou, C., Veglis, A.A.: SEO inside newsrooms: reports from the field. Future Internet **11**(12), 261 (2019)

16. Goodfellow, I.J., Shlens, J., Szegedy, C.: Explaining and harnessing adversarial examples. arXiv:1412.6572 (2015)

17. Gyöngyi, Z., Garcia-Molina, H.: Spam: it's not just for inboxes anymore. Computer **38**(10), 28–34 (2005)

18. Hofstatter, S., Althammer, S., Schroder, M., Sertkan, M., Hanbury, A.: Improving Efficient Neural Ranking Models with Cross-Architecture Knowledge Distillation. arXiv:2010.02666 (2021)

19. Jeronymo, V., et al.: Inpars-v2: large language models as efficient dataset generators for information retrieval. CoRR abs/2301.01820, https://doi.org/10.48550/arXiv.2301.01820 (2023)

20. Karpukhin, V., et al.: Dense passage retrieval for open-domain question answering. In: Proceedings of the 2020 Conference on Empirical Methods in Natural Language Processing (EMNLP), pp. 6769–6781 (2020). https://aclanthology.org/2020.emnlp-main.550

21. Kelly, D., Azzopardi, L.: How many results per page?: a study of SERP size, search behavior and user experience. In: SIGIR, pp. 183–192. ACM (2015)

22. Khattab, O., Zaharia, M.: ColBERT: efficient and effective passage search via contextualized late interaction over BERT. In: SIGIR 2020 - Proceedings of the 43rd International ACM SIGIR Conference on Research and Development in Information Retrieval, pp. 39–48 (2020). https://arxiv.org/abs/2004.12832v2, ISBN: 9781450380164 Publisher: Association for Computing Machinery Inc

23. Lewandowski, D., Sünkler, S., Yagci, N.: The influence of search engine optimization on google's results: a multi-dimensional approach for detecting SEO. In: Hooper, C., Weber, M., Weller, K., Hall, W., Contractor, N., Tang, J. (eds.) WebSci 2021: 13th ACM Web Science Conference 2021, Virtual Event, United Kingdom, June 21–25, 2021, pp. 12–20. ACM (2021)

24. Lin, J., Nogueira, R.F., Yates, A.: Pretrained Transformers for Text Ranking: BERT and Beyond. Morgan & Claypool Publishers, Synthesis Lectures on Human Language Technologies (2021)

25. Liu, J., et al.: Order-disorder: imitation adversarial attacks for black-box neural ranking models. In: Proceedings of the 2022 ACM SIGSAC Conference on Computer and Communications Security, pp. 2025–2039, CCS 2022, Association for Computing Machinery, New York (2022), ISBN 978-1-4503-9450-5, https://dl.acm.org/doi/10.1145/3548606.3560683

26. Liu, Y.A., et al.: Topic-oriented adversarial attacks against black-box neural ranking models. arXiv:2304.14867 (2023)

27. MacAvaney, S., Feldman, S., Goharian, N., Downey, D., Cohan, A.: ABNIRML: analyzing the behavior of neural IR models. Trans. Assoc. Comput. Linguist. **10**, 224–239 (2022). https://aclanthology.org/2022.tacl-1.13

28. MacAvaney, S., Soldaini, L.: One-shot labeling for automatic relevance estimation. In: Chen, H., Duh, W.E., Huang, H., Kato, M.P., Mothe, J., Poblete, B. (eds.)

Proceedings of the 46th International ACM SIGIR Conference on Research and Development in Information Retrieval, SIGIR 2023, Taipei, Taiwan, July 23–27, 2023, pp. 2230–2235. ACM (2023)

29. MacAvaney, S., Yates, A., Cohan, A., Goharian, N.: CEDR: contextualized embeddings for document ranking. In: Proceedings of the 42nd International ACM SIGIR Conference on Research and Development in Information Retrieval, pp. 1101–1104 (2019). arXiv:1904.07094

30. Malaga, R.A.: Chapter 1 – search engine optimization: black and white hat approaches. In: Advances in Computers: Improving the Web, Advances in Computers, vol. 78, pp. 1–39. Elsevier (2010). https://www.sciencedirect.com/science/article/pii/S0065245810780013

31. Nogueira, R., Cho, K.: Passage re-ranking with BERT. arXiv:1901.04085 (2020)

32. Nogueira, R., Jiang, Z., Pradeep, R., Lin, J.: Document ranking with a pretrained sequence-to-sequence model. In: Findings of the Association for Computational Linguistics: EMNLP 2020(2020), pp. 708–718 (2020). https://aclanthology.org/2020.findings-emnlp.63

33. Ounis, I., Amati, G., Plachouras, V., He, B., Macdonald, C., Johnson, D.: Terrier information retrieval platform. In: Losada, D.E., Fernández-Luna, J.M. (eds.) ECIR 2005. LNCS, vol. 3408, pp. 517–519. Springer, Heidelberg (2005). https://doi.org/10.1007/978-3-540-31865-1_37

34. Pradeep, R., Liu, Y., Zhang, X., Li, Y., Yates, A., Lin, J.: Squeezing water from a stone: a bag of tricks for further improving cross-encoder effectiveness for reranking. In: Advances in Information Retrieval: 44th European Conference on IR Research, ECIR 2022, Stavanger, Norway, April 10–14, 2022, Proceedings, Part I, pp. 655–670. Springer, Berlin, Heidelberg (2022). ISBN 978-3-030-99735-9, https://doi.org/10.1007/978-3-030-99736-6_44

35. Raffel, C., et al.: Exploring the limits of transfer learning with a unified text-to-text transformer. arXiv:1910.10683 (2020)

36. Raval, N., Verma, M.: One word at a time: adversarial attacks on retrieval models. arXiv:2008.02197 (2020)

37. Sakai, T.: On fuhr's guideline for IR evaluation. SIGIR Forum 54(1), 12:1-12:8 (2020)

38. Szegedy, C., et al.: Intriguing properties of neural networks. In: 2nd International Conference on Learning Representations, ICLR 2014 - Conference Track Proceedings (2013). https://arxiv.org/abs/1312.6199v4, publisher: International Conference on Learning Representations, ICLR

39. Taori, R., et al.: Stanford Alpaca: An Instruction-following LLaMA model. GitHub repository (2023). https://github.com/tatsu-lab/stanford_alpaca

40. Thomas, P., Spielman, S., Craswell, N., Mitra, B.: Large language models can accurately predict searcher preferences. arXiv:2309.10621v1 (2023)

41. Touvron, H., et al.: LLaMA: open and efficient foundation language models. arXiv:2302.13971 (2023)

42. Voorhees, E.M., Craswell, N., Lin, J.: Too many relevants: whither cranfield test collections? In: Amigó, E., Castells, P., Gonzalo, J., Carterette, B., Culpepper, J.S., Kazai, G. (eds.) SIGIR 2022: The 45th International ACM SIGIR Conference on Research and Development in Information Retrieval, Madrid, Spain, July 11–15, 2022, pp. 2970–2980. ACM (2022)

43. Weller, O., Lawrie, D., Van Durme, B.: NevIR: negation in neural information retrieval. arXiv:2305.07614 (2023)

44. Wu, C., Zhang, R., Guo, J., de Rijke, M., Fan, Y., Cheng, X.: PRADA: practical black-box adversarial attacks against neural ranking models. arXiv:2204.01321 (2022)
45. Zhou, Y., Lei, T., Zhou, T.: A robust ranking algorithm to spamming. CoRR abs/1012.3793 http://arxiv.org/abs/1012.3793 (2010)
46. Zobel, J., Rashidi, L.: Corpus bootstrapping for assessment of the properties of effectiveness measures. In: d'Aquin, M., Dietze, S., Hauff, C., Curry, E., Cudré-Mauroux, P. (eds.) CIKM 2020: The 29th ACM International Conference on Information and Knowledge Management, Virtual Event, Ireland, October 19–23, 2020, pp. 1933–1952. ACM (2020)
47. Zou, A., Wang, Z., Kolter, J.Z., Fredrikson, M.: Universal and transferable adversarial attacks on aligned language models. CoRR abs/2307.15043 https://doi.org/10.48550/arXiv.2307.15043 (2023)

Conversational Search with Tail Entities

Hai Dang Tran[1]([⊠])(iD), Andrew Yates[1,2](iD), and Gerhard Weikum[1](iD)

[1] Max Planck Institute for Informatics, Saarbrucken, Germany
{hatran,weikum}@mpi-inf.mpg.de
[2] University of Amsterdam, Amsterdam, Netherlands
a.c.yates@uva.nl

Abstract. Conversational search faces incomplete and informal follow-up questions. Prior works address these by contextualizing user utterances with cues derived from the previous turns of the conversation. This approach works well when the conversation centers on prominent entities, for which knowledge bases (KBs) or language models (LMs) can provide rich background. This work addresses the unexplored direction where user questions are about tail entities, not featured in KBs and sparsely covered by LMs. We devise a new method, called CONSENT, for selectively contextualizing a user utterance with turns, KB-linkable entities, and mentions of tail and out-of-KB (OKB) entities. CONSENT derives relatedness weights from Sentence-BERT similarities and employs an integer linear program (ILP) for judiciously selecting the best context cues for a given set of candidate answers. This method couples the contextualization and answer-ranking stages, and jointly infers the best choices for both.

Keywords: Conversational search · Tail entities

1 Introduction

Motivation and Problem. Conversational search is an IR mode where users pose questions and receive system-provided answers in a multi-turn dialog. In this paper, we consider a setting where user inputs are short, colloquially formulated questions, and system answers are sentences retrieved and inferred from a large corpus of text documents (e.g., Wikipedia articles or news articles). An example is this conversation about the 2022 champion's league in women's handball:

Q_1: Who won the EHF women's final four?
A_1: Vipers Kristiansand won 33:31 in a tight match.
Q_2: Who was the MVP?
A_2: Marketa Jerabkova, right-back player of vipers.
Q_3: The team's goalkeeper: how many saves?
A_3: Norwegian legend Lunde got 20 saves in the final.

© The Author(s), under exclusive license to Springer Nature Switzerland AG 2024
N. Goharian et al. (Eds.): ECIR 2024, LNCS 14609, pp. 303–317, 2024.
https://doi.org/10.1007/978-3-031-56060-6_20

Q_4: Where did the shooter play earlier?
A_4: Before joining vipers, Marketa played for Thuringen.
Q_5: How many goals did she score for that team?
A_5: Jerabkova scored 193 goals for HC Thuringen.

This example exhibits three challenges of conversational search:

C1. **Informal utterances:** The user's formulations can be colloquial and grammatically flawed, using shorthand notations and omitting certain parts. Question Q3 above is an example.

C2. **Contextualization:** The questions are often incomplete, as humans expect the system to understand their intent from the previous turns of the conversation. Question Q2 is an example: "MVP" (short for Most Valuable Player) is not self-contained, it needs the "EHF women's final four" as context.

C3. **Emerging and tail entities:** Mentions in previous turns' questions or answers are essential for proper context. Some entity mentions may refer to real-world entities that are not (yet) captured in a knowledge base (KB). This increases the ambiguity of phrases like "the shooter" and "that team". It is crucial that the system infers that Q4 needs Q2/A2 as context and that "team" in Q5 refers to A4, not to A1 or Q3.

State of the Art and its Limitations. Prior works address only C1 and C2. Methods are based on either rewriting user utterances into complete, self-contained questions (e.g., [25, 27, 30]) or on using neural encoders to contextualize the user's inputs (e.g., [12]). The latter typically incorporate either only the immediately preceding and/or the very first turn of a conversation, or they incorporate the complete history of the conversation as context. None of these methods is robust in the presence of tail entities and the absence of large amounts of labeled training data.

Challenge C3 has been overlooked in the literature. With neural encoding of all prior turns, out-of-KB (OKB) entities are captured by surface names. However, there is no background information in the KB for proper interpretation. Also, surface names may collide with identical names by other entities in the KB, creating ambiguity and confusion. For example, "Lunde" is a frequent name that matches many entities. If linked incorrectly, it would add confusing signals.

Approach. This paper addresses the C3 challenge of dealing with emerging and tail entities. As these are entangled with the contextualization problem, we consider C2 and C3 jointly. On C1, we employ standard techniques. We name our method ConSEnT, for conversational search with entities in the tail.

CONSENT works in two stages. In the first stage, candidate sentences are retrieved from the underlying corpus (a large archive of news articles [14] and Wikipedia articles). This can use a variety of IR techniques, including BM25, dense retrieval [31], SBERT [23] sematic search or neural retrieval with sparse vectors [13]. We aim at a pool of at least 100 top-scoring matches.

The second stage performs ranking the candidate snippets. CONSENT detects all entity mentions in the conversation history. The mentions that can be linked to a KB with high confidence, are mapped and augmented with salient KB facts. The non-linkable, potentially OKB mentions are kept in their surface forms. In principle, we could now contextualize the current question with this entire set, but this would be indiscriminate and aggravate the task. In the example, "that team" would be encoded with both Vipers Kristiansand and HC Thuringen as context. Therefore, our method judiciously selects only KB entities and OKB mentions that are likely to be relevant for the current question. We achieve this by devising an integer linear program (ILP) that reasons over all cues jointly, selects the relevant entities and mentions, and yields the final ranking of candidate answers. The ILP coefficients are learned from a small set of training conversations, using SGD for pairwise ranking loss.

Contributions. This paper offers three novel contributions:

- CONSENT is the first work that addresses the challenge of handling emerging and tail entities in conversational search. Our method copes well with mentions of OKB entities and tail in-KB entities.
- The new benchmark, CONSENT Data, automatically generates difficult conversations from Wikipedia, enforcing a large fraction of OKB entities.
- Experiments show that the CONSENT method outperforms a number of baselines by a substantial margin. On questions with tail entities, we outperform even LLM-based competitors that use GPT-3.5 or GPT-4.

Code and data of our work CONSENT are released on GitHub[1].

2   Related Work

Conversational Search. Prior works have focused on the challenges C1 and C2 (see Sect. 1). One approach is to employ co-reference resolution and related techniques to rewrite user utterances into complete, self-contained questions (e.g., [11,16,25–28,30]). These methods are brittle, though. The second line of methods is to use neural encoders and leverage language models for contextualizing the user's inputs (e.g., [12,17,19]). Some techniques focus on incorporating only the immediately preceding and/or the very first turn of a conversation. This is a good heuristic for capturing the relevant context, but easily fails for sophisticated conversations. Recent methods use the entire conversational history to encode context, relying on training data for computing appropriate (attention) weights (e.g., [5,22]). This is viable when labeled conversations are abundant, but with limited amounts of training data the approach is all but robust.

Popular instantiations of this line of contextualization methods include the following. ConvDR [31] is a dense retrieval method [18], where the current question is prefixed with the entire history to encode a vector representation. These vectors are compared to encodings of candidate answers by computing inner products. Zeco2 [17] uses ColBERT [15] to build representations in a zero-shot manner. CoSPLADE [13] follows the paradigm of learned sparse retrieval, to

---

[1] https://github.com/haidangtran1989/CONSENT.

identify salient keywords in the history and leverage these for question under-standing. Obviously, generative models like GPT-3.5 [3] and GPT-4 [21] can also be utilized to encode questions with their conversational history. None of these methods give consideration to out-of-KB and tail entities.

Benchmarks. Prior benchmarks for conversational search focus on prominent entities; there are hardly any questions about long-tail or out-of-KB entities [1,2, 6–11]. Some of these are specifically designed to evaluate subtasks like question rewriting (CANARD [11]), conversation clarification (ClariQ [2]), or detecting topical shifts in conversations (TopiOCQA [1]). None of the existing benchmarks for conversational search contains a substantial number of conversations focused on long-tail entities.

3   CONSENT Methodology

3.1   Overview

Our method, called CONSENT, takes as input a conversational history (or history) $H = \{(Q_1, A_1), (Q_2, A_2), \ldots (Q_{i-1}, A_{i-1})\}$ and a current question Q_i. CONSENT searches answers of the current question Q_i from a text corpus, which contains a large number of news and Wikipedia articles. These sources are rich in tail and out-of-KB (OKB) entities. In this work, questions are infor-mal user utterances, answers are complete sentences. To support searching, we leverage in-KB entities $E = \{e_1, e_2, \ldots\}$ as well as mentions $M = \{m_1, m_2, \ldots\}$ of OKB entities in H. E contains only entities that an entity linker can map to Wikipedia with high confidence (i.e. higher than a preset threshold). All other mentions (treated as OKB) are included in M, possibly including different names for the same entity. In this work, entity is used to refer to a mention, in-KB entity or OKB entity.

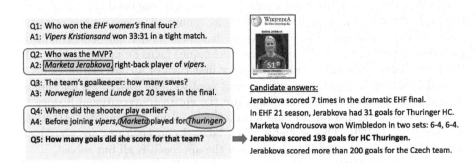

Fig. 1. Contextualization and Ranking by CONSENT.

CONSENT addresses two sub-problems in a coupled manner: 1) **contextu-alizing** the current question to understand the information need, and 2) **rank-ing** the answer sentences. For the first task, CONSENT judiciously selects the following kinds of context items and potentially augments them:

- turns $H^* \subseteq H$ (a turn is a pair of a question and its answer in H).
- in-KB entities $E^* \subseteq E$ that appear in one or more of H^*.
- mentions $M^* \subseteq M$ that appear in one or more of H^*.

In-KB entities are augmented with the first sentence of their respective Wikipedia articles. For mentions M^*, which are not linkable to the KB, this enrichment is not feasible. Consider the history and current question of the introduction's example, shown in Fig. 1. To contextualize Q_5, we should incorporate turns 2 and 4, as they are key to interpret the phrases "she score" and "that team" in Q_5. In addition, we can sharpen the focus by including the in-KB entity "Marketa Jerabkova" and the mentions "Marketa" and "Thuringen", giving higher weight to these cues.

For answer ranking, CONSENT operates over a pool of initial candidates, obtained by a baseline retriever, such as BM25, dense retrieval or sparse retrieval. In experiments, we use a search pipeline of BM25 and SBERT [23] to retrieve 100 candidates. The CONSENT ranking for these top-100 considers textual similarity, and also exploits knowledge about semantic types of entities. For example, the type *handball player* of Marketa Jerabkova is important. There are other notable entities for the mention "Marketa", including a prominent tennis player, which would confuse and dilute the answering.

The answer ranking depends on how the question is contextualized. Conversely, the best choice of turns, entities and mentions for contextualization may depend on the answer candidates and the ranking approach. Therefore, CONSENT addresses both tasks jointly, by means of an integer linear program (ILP). The ILP has 0-1 variables for selecting or not selecting context items, and has an objective function that scores each candidate answer. The model has a small number of hyper-parameters to weigh the influence of different aspects, learned via SGD optimization from conversations with ground-truth answers.

3.2 Joint Inference on Contextualization and Answer Ranking

For contextualizing question Q given its history H and a candidate answer A, we consider three kinds of cues: turns, in-KB and OKB entities within these turns. Which cues are selected and which ones are disregarded is modeled by introducing the following 0-1 decision variables:

- $X_i = 1$ if the ith turn H_i in the history H is selected, 0 otherwise ($1 \le i \le |H|$)
- $Y_j = 1$ if the jth entity EM_j of $E \cup M$ is selected, 0 otherwise ($1 \le j \le |E \cup M|$)

To ensure that we select only entities from selected turns, the following constraint must be satisfied:

- $\forall i \forall j (1 \le i \le |H|, 1 \le j \le |E \cup M|) : X_i \ge Y_j$ if j^{th} entity of $E \cup M$ appears in i^{th} turn of the history H

In addition, we can control the maximum number of selected turns and entities, with model configuration parameters K and L, by the constraints:

- $\sum_{i=1}^{|H|} X_i \leq K$ and $\sum_{j=1}^{|E \cup M|} Y_j \leq L$ where K and L are preset constants.

To score answer candidate A we need to define an objective function, which is conditioned on A. The ILP is invoked once for each of the initially retrieved top-100 candidates, to compute the final ranking. Intuitively, the objective is to reward the most informative cues, by setting their decision variables X_i and Y_j to 1. We incorporate the following signals in our ILP problem.

- **Turn-question relatedness** (Rel^{TQ}): a similarity score between a selected turn and the question, as measured by the cosine between their two Sentence-BERT [23][2] (SBERT) embeddings.
- **Turn-answer relatedness** (Rel^{TA}): a similarity score between a selected turn and a candidate answer, again, by cosine of SBERT embeddings.
- **Entity-question relatedness** (Rel^{EQ}): the semantic closeness of a selected entity (in-KB linked or mention) and the question, measured via SBERT. Entities are enriched with text, taken from the first Wikipedia sentence (introduction text) for in-KB entities, or from inferring a set of semantic types for mentions. For the latter, we use the fine-grained typing tool of [20].
- **Entity-answer relatedness** (Rel^{EA}): an analogous score for the semantic closeness of an entity and the given answer candidate.
- **Question-answer base relatedness** (Rel^{QA}_{Base}): either using SBERT similarity, or the relevance score from the initial retriever, such as BM25.

With these design considerations, the to be maximized ILP objective function could be formulated as follows:

$$\alpha_1 \left(\sum_{i=1}^{|H|} X_i Rel^{TQ}(H_i, Q) \right) + \alpha_2 \left(\sum_{i=1}^{|H|} X_i Rel^{TA}(H_i, A) \right) + \alpha_3 \left(\sum_{j=1}^{|E \cup M|} Y_j Rel^{EQ}(EM_j, Q) \right)$$

$$+ \alpha_4 \left(\sum_{j=1}^{|E \cup M|} Y_j Rel^{EA}(EM_j, A) \right) + \alpha_5 Rel^{QA}_{Base}(Q, A) - \alpha_6 \sum_{i=1}^{|H|} X_i - \alpha_7 \sum_{j=1}^{|E \cup M|} Y_j$$

where $\alpha_1 \ldots \alpha_7$ are tunable hyper-parameters, learned from withheld conversations with ground-truth answers (Sect. 5). The last terms, weighted by hyper-parameter α_6, α_7, can be viewed as a regularizer that aims to reward parsimony: select as few of the turns and entities as possible, hence the negative sign. To solve the ILP, we use the Gurobi optimizer[3]. The optimal value of the above function is the CONSENT score of candidate answer A given Q and H.

---

[2] https://huggingface.co/sentence-transformers/all-mpnet-base-v2.
[3] https://www.gurobi.com/.

4   Automatic Benchmark Construction

Existing benchmarks for conversational search mostly feature prominent entities. We constructed a benchmark dataset with specific focus on tail entities (either OKB or with Wikipedia articles that have few in-links below a specified threshold). To this end, we sample tail entities from the Wikipedia 2018 event page[4] and related articles, to construct facts around these entities, which serve as gold answers for conversations. We pursue this "time-travel" approach, as many of the former tail entities have become prominent as of now. The 2018 snapshot also allows us to relate to the available text corpus, consisting of 2017–2018 news articles from [14] and the 2017 Wikipedia articles. The questions themselves are generated using GPT-3.5 [3] from the constructed sequence of gold answers.

Gold Answer Sampling. Conversations are initialized with a news snippet A_0 from the 2018 Wikipedia event page. This gives us a seed entity for generating T turns with gold answers $A_1, A_2, ..., A_{|T|}$. To allow shifts in the conversations, the ith turn selects a random tail entity e_i mentioned in one of $A_0, A_1, ..., A_{i-1}$. We call e_i the target entity of the ith turn. For each e_i, we retrieve sentences from its 2023 Wikipedia article that contain the entity name, disregarding any sentence that mentions a date later than 2018. The text-text relatedness of the two texts is measured by the cosine between their two SBERT [23][5] embeddings. We select top-g sentences by text-text relatedness to the previous A_{i-1}, for coherent conversations, and randomly pick one of these as A_i.

Question Generation. For each A_i $(1 \leq i \leq T)$, we ask GPT-3.5 [3] (*text-davinci-003*) to generate a question with the prompt consisting of A_i and the instruction to provide a short, spoken-style question that has A_i as its answer (*question-generation prompt*). We also target the generated question is related to the respective target entity e_i. We randomly generate several candidate questions and pick the question q_i^* with the highest text-text relatedness with A_i, to ensure coherence. The news snippet A_0 is not included in the conversation history for training and evaluation, it is only used for question generation.

Judgements. While the above process yields questions and their canonical answers, we also need to assess whether answers returned by systems are relevant. To do so, we identify the top three relevant system answers for every question and method, followed by de-duplication and judging. In the judgement process, we determine the relevance of answers automatically using two prompts. First, with the *answer-check prompt*, we assess a system answer's relevance by providing a series of questions and answers in the conversation (gold answers in the history and the system answer in the current turn) to GPT-4, followed by asking whether the system answer is relevant to the current question. Second, in the *entity-check prompt*, we assess the relevance of a system answer against the target entity of the current turn. We provide the system answer to GPT-4, followed by asking whether it provides information related to the target entity.

---

[4] https://en.wikipedia.org/wiki/2018.

[5] https://huggingface.co/sentence-transformers/all-mpnet-base-v2.

A system answer is judged as correct if and only if the responses to both prompts start with "Yes".

This way we construct the entire benchmark dataset in a completely automated manner. Further details are given on the dataset and code repository[6].

5   Experimental Setup

We conduct experiments on two datasets: our new CONSENT dataset centering on tail entities and ConvQuestions [6], focusing on popular in-KB entities. We measure nDCG@3, Precision@1 and MRR@3, which are abbreviated as nDCG, P and MRR, respectively. We create the CONSENT dataset by following the above construction procedure, which yields 371 conversations with 1902 questions total. These are split into a training set of 146 conversations with 751 questions and a test set of 225 conversations with 1151 questions.

To assess the quality of the automatically created conversations and their ground-truth answers, we randomly sample 115 test questions (10%) and ask crowd workers for additional relevance judgements using Amazon Mechanical Turk (AMT)[7]. For each question, we provide the judges with the conversation history and a set of 3 candidate answers randomly selected from the automatic assessment. We ask the judges to select all correct answers, which could vary between 0 (all false) and 3 (all correct). We use three crowd workers per question and treat an answer as judged correct when it is selected by at least two judges. Comparing these to the GPT-generated assessments, the GPT judgements have an average accuracy of 90.9%. Considering the positive class, the GPT judgements have a precision and recall of 85.1% and 79.1%, respectively. These results indicate that GPT and human crowd workers often agree, which suggests our automated evaluation is well-designed and gives meaningful results.

The popularity of in-KB entity is the number of incoming links in 2017 Wikipedia. We denote an in-KB entity as *rare*, *uncommon* and *popular* if its popularity is respectively in the ranges $[1, P^R], [P^R + 1, P^U], [P^U + 1, \infty)$, where $P^R < P^U$ are specified thresholds. We categorize the CONSENT test questions into OKB, rare, and uncommon subsets based on whether the corresponding target entity is in-KB and its popularity. These OKB, rare and uncommon subsets contain 284, 535, and 332 questions, respectively. Systems retrieve answers from a corpus of nearly 1.5 million STICS [14] 2017–2018 news articles and approximately 5.5 million 2017 Wikipedia articles. Answers in the conversation history are gold answers sampled from Wikipedia 2023, which ensures that the history of the current question is the same for every method.

Due to high cost of running GPT on ConvQuestions, we reduce the collection's size by randomly sampling 10% of conversations in the test set, which yields 224 conversations and 1120 questions. On this dataset, there are no target entities and the gold answers are in entity name format. We treat an answer as

---

[6] https://github.com/haidangtran1989/CONSENT.

[7] https://www.mturk.com/.

relevant if it contains a term from the corresponding gold entity name and the output of the *answer-check prompt* (Sect. 4) starts with "yes".

Training. We fine-tune existing baselines following the training setup used in their original works. We train the CONSENT method with the following positive and negative learning examples. Let A^* be the golden answer of a question Q (in CONSENT trainset) and Y^* be the decision variables for entity selection where only entities appearing in both history H and A^* are selected. Let X^* be the decision variables for turn selection where only the turns containing the above-selected entities are selected. Each positive example is formed as a combination (H, Q, A^*, X^*, Y^*). Negative learning examples are formed by the following two strategies. For the first strategy, we randomly select entities in H and the turns containing them to build X^-, Y^-. Random selection is noisy, resulting in a negative example (H, Q, A^*, X^-, Y^-). The intuition is that when we have a good answer, but turn and entity selection are bad then the learning combination is negative. For the second strategy, we feed H, Q to a particular search method (e.g. SBERT ranker [23]) and randomly select a low-ranked answer A^- (i.e. out of top three) to form a negative example (H, Q, A^-, X^-, Y^-). That is, with the negative answer and the selection of turns and entities having noises, the learning combination is also negative. This yields 78,092 training pairs in total, each pair consists of a positive and a negative example. We target that the objective function value (Sect. 3.2) of the positive example is higher than that of every respective negative example. For each learning example, decision variables and relatedness scores are constants, because neural models generating these scores are frozen during training. Based on the examples, we use a pairwise ranking loss [4] to train our parameters $\alpha_1, \alpha_2, ... \alpha_7$. The initial learning rate is 5×10^{-2} with a warm up stage of 3% of the whole training, the number of epochs is 5. We default the hyper-parameters $K = L = 3$, this will be discussed more in the next section. With these K and L, after training, we gain parameters $\alpha_1 = 0.94$, $\alpha_2 = 2.58$, $\alpha_3 = 1.05$, $\alpha_4 = 5.09$, $\alpha_5 = 0.39$, $\alpha_6 = 0.89$ and $\alpha_7 = 1.46$.

Methods. We report the performance of our approach and of the following baselines, which cover a wide range of approaches.

- **BM25**: We prefix a question with entities from the history to form the query input of BM25 [24]. Sentences from the corpus are indexed for this baseline.
- **ZeCo2**: The current question prefixed with history and each answer from the top results of BM25 are fed through ZeCo2 [17] to get the relevance score.
- **ConvDR**: We take a ConvDR [31] model trained on CAsT '20 [9] and fine tune it with CONSENT and ConvQuestions trainsets before testing on the respective tests. We distill knowledge from ANCE [29] during the training.
- **CQR**: We fine tune CQR [30] with CONSENT and ConvQuestions train-sets to rewrite utterances in the corresponding testsets into self-contained questions, which are then fed to a pipeline of BM25 and SBERT[8] [23].

---

[8] https://huggingface.co/cross-encoder/ms-marco-MiniLM-L-6-v2.

- **CoSPLADE**: We fine tune original CoSPLADE model [13] with training data of CONSENT and ConvQuestions (monoT5[9] as a teacher model), followed by evaluation on the respective testsets.
- **SBERT Zero**: For the current question Q, we prefix entity names in the history to get an extended query Q', which is issued to BM25 to get the top-D articles (not sentences as in the above version of BM25). We split these articles into sentences and rerank the sentences with zero-shot SBERT (the same model as in CQR).
- **GPT-3.5**: We obtain the top-S sentences from SBERT Zero, and then use the *answer-check prompt* (Sect. 4) to rank relevant answers with GPT-3.5 by the probability of generating "yes". We do not use the *entity-check prompt* since it requires knowing the target entity, which is unknown to the ranker.
- **GPT-3.5 Fict**: This is the same as GPT-3.5, except that we replace each person's name in the conversations and answers with a made-up fictional name. We use gender-neutral replacements, so coreferences like "he" or "she" could refer to them. With the fictional names, the model is less able to rely on memorized information to produce answers, emphasizing its ability to find the answer using evidence. We do not report results on the OKB, Rare, and Uncommon subsets due to the changes of the target entities.
- **GPT-4 Fict**: This baseline modifies GPT-3.5 Fict to use GPT-4, which does not provide the probability of generated tokens. To break ties, we randomize the order of answers with the same predicted relevance. We repeat this tie-breaking 15 times and report the best ordering in terms of nDCG@3.
- **CONSENT**: CONSENT is our main method, using the methodology in Sect. 3 to rerank the top-S answers from SBERT Zero. SBERT Zero also provides the base relatedness score $Rel_{Base}^{QA}(Q, A)$.

6   Experimental Results

We consider research questions (RQs) evaluating CONSENT's performance on two conversational search datasets, investigating its performance across turns in a conversation, and investigating the impact of different components on CONSENT's performance.

RQ1: How does CONSENT perform compared to state-of-the-art baselines?
Results on the CONSENT test are shown in Table 1. The CONSENT method substantially outperforms the baselines across all CONSENT test subsets, with significant improvements over SBERT Zero and GPT in most cases. For example, CONSENT significantly outperforms GPT-4 Fict (14% better) in terms of nDCG@3 on All. This trend could also be seen when we compare CONSENT against GPT-3.5. Indeed, regarding nDCG@3, CONSENT significantly outperforms GPT-3.5 on All, OOW, Rare and Uncommon (22%, 29%, 16% and 26% better, respectively). It is interesting to see that the nDCG@3 of GPT-3.5 drops 12% after changing real names to fictional names (GPT-3.5 Fict), suggesting

---

[9] https://huggingface.co/castorini/monot5-base-msmarco-10k.

Table 1. Results on CONSENT test. The † and ‡ symbols denote significance (paired t-test, p<0.05) over SBERT Zero and all GPT baselines, respectively.

Methods	All			OKB			Rare			Uncommon		
	nDCG	P	MRR	nDCG	P	MRR	nDCG	P	MRR	nDCG	P	MRR
BM25	0.070	0.073	0.100	0.037	0.028	0.047	0.065	0.067	0.090	0.106	0.120	0.162
ConvDR	0.120	0.120	0.155	0.069	0.067	0.087	0.120	0.114	0.153	0.162	0.175	0.217
ZeCo2	0.146	0.151	0.192	0.093	0.095	0.123	0.110	0.110	0.148	0.248	0.265	0.322
CQR	0.196	0.208	0.236	0.123	0.134	0.152	0.168	0.176	0.201	0.304	0.322	0.363
CoSPLADE	0.231	0.238	0.283	0.150	0.134	0.180	0.179	0.181	0.221	0.384	0.419	0.473
SBERT Zero	0.256	0.265	0.313	0.135	0.144	0.173	0.243	0.247	0.299	0.380	0.398	0.456
GPT-3.5 Fict	0.233	0.224	0.288	-	-	-	-	-	-	-	-	-
GPT-3.5	0.262	0.258	0.320	0.154	0.162	0.212	0.259	0.260	0.314	0.360	0.337	0.422
GPT-4 Fict	0.279†	0.275	0.338†	-	-	-	-	-	-	-	-	-
CONSENT	**0.319†‡**	**0.328†‡**	**0.385†‡**	**0.199†‡**	**0.204†**	**0.246†**	**0.300†‡**	**0.295†**	**0.357†‡**	**0.454†‡**	**0.488†‡**	**0.548†‡**

Table 2. Results on sampled ConvQuestions. The † symbol denotes significance (paired t-test, p<0.05) over SBERT Zero.

	BM25	ConvDR	ZeCo2	CQR	CoSPLADE	SBERT Zero	GPT-3.5	CONSENT
nDCG	0.045	0.047	0.108	0.272†	0.296†	0.232	**0.306†**	0.288†
P	0.042	0.047	0.112	0.280†	0.302†	0.244	**0.305†**	0.292†
MRR	0.064	0.049	0.156	0.339†	0.374†	0.295	**0.377†**	0.351†

that some of GPT's performance is due to memorizing related facts. The ConvDR, ZeCo2, CQR, and CoSPLADE neural retrieval methods for conversational search typically underperform the SBERT Zero and GPT methods, illustrating the difficulty of this setting with tail entities. BM25 is substantially worse than these methods, highlighting the big gap between ad-hoc and conversational search.

Results on the sampled ConvQuestions dataset, which focuses on popular in-KB entities rather than tail entities, are shown in Table 2. We use the joint contextualization and answer ranking technique on top of the base model SBERT Zero, thus we compare CONSENT against the base model to see the impact of such technique. The nDCG@3 of our method significantly outperforms that of SBERT Zero over four CONSENT test subsets and sampled ConvQuestions. This shows that our methodology really improves the search effectiveness. CONSENT performs competitively in Table 2. Here, GPT-3.5 and CoSPLADE perform best across metrics, with GPT-3.5 performing about 6% nDCG@3 better than CONSENT and CoSPLADE performing about 3% nDCG@3 better. The other neural baselines, ConvDR, ZeCo2, CQR, and SBERT Zero, consistently perform worse than CONSENT. BM25 generally performs the worst, as it did on CONSENT test. Overall, these results indicate CONSENT substantially improves over existing approaches when conversations focus on tail entities while still performing well with conversations about popular entities.

RQ2: How do methods perform across conversation turns?
In Fig. 2 we report the nDCG@3 of top-performing methods on different turns on the full CONSENT test (All), with a maximum of 6 turns. These four methods

perform similarly on the first turn, with CONSENT and SBERT Zero marginally higher than the GPT models. Performance sharply drops between the first and second turns for all methods, illustrating the importance of contextualizing questions. On turn two, CONSENT clearly outperforms SBERT Zero and the GPT models. On later turns, CONSENT continues to outperform GPT-3.5 and consistently matches or exceeds GPT-4 Fict's performance. The GPT models outperform SBERT Zero in almost all cases, which is likely due to the strong language understanding of these models. The fact that CONSENT consistently outperforms base model SBERT Zero after the first turn again confirms the importance of contextualization.

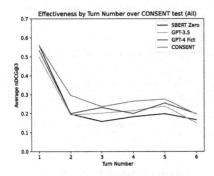

Fig. 2. nDCG@3 for best performing methods on All by turn.

Table 3. Ablation study: nDCG@3 for CONSENT variants.

Settings			CONSENT			
K	L	Mode	All	OKB	Rare	Uncommon
1	1	DE	0.285	0.150	0.272	0.423
3	3	DE	**0.319**	**0.199**	**0.300**	0.454
5	5	DE	0.317	0.194	0.297	**0.456**
∞	∞	DE	0.311	0.187	0.290	0.452
3	3	NA	0.132	0.092	0.137	0.159
3	3	NC	0.308	0.175	0.294	0.445
3	3	NE	0.307	0.191	0.286	0.440
3	3	NT	0.288	0.167	0.272	0.417
3	3	NI	0.318	0.196	0.299	0.453

RQ3: What is the impact of CONSENT's components on its performance?
In this section, we investigate the impact of CONSENT's hyper-parameters K and L and the impact of excluding different types of relatedness scores. To do so, we define the following modes, which each removes some part of CONSENT.

- DE (default): This is the full method as described in Sect. 3.
- NA (no answer): We separate contextualization and answer ranking by removing answer relatedness scores, using only Rel^{TQ}, Rel^{EQ} and turn/entity parsimony for contextualization. We create an expanded question by prefixing the current question with the selected turns and all inferred types of the selected entities (e.g., "Marketa Jerabkova is: person athlete handball player ..."). The expanded question is fed through ConvDR to get the final result.
- NC (no turn context): We drop the relatedness scores Rel^{TQ} and Rel^{TA} and the turn parsimony component.
- NE (no entity): We drop the entity relatedness scores Rel^{EQ} and Rel^{EA} and the entity parsimony component.
- NT (no training): We skip training and instead set $\alpha_1, \alpha_2, ..., \alpha_7$ to one.
- NI (no integer constraint): We relax the integer constraint in the ILP solving.

Results using different hyper-parameter values and modes are shown in Table 3. Considering the impact of hyper-parameters when using the default mode (DE), CONSENT's performance decreases about 11% on the full test set when the number of selected turns and entities is reduced ($K = L = 1$). Increasing these hyper-parameters to 5 does not have much effect, with performance slightly better on the uncommon subset and slightly worse on the other subsets. Setting these hyper-parameters to ∞ also does not have much effect, with a small drop in performance across subsets. The default setting of $K = L = 3$ is thus a reasonable choice, so we fix this setting when reporting results with other CONSENT modes.

Compared to the full method (DE), removing answers results in the largest drop on the full test set (58%), which indicates that considering cues from candidate answers is essential. Similarly large drops in performance also occur on the OKB, rare, and uncommon subsets. Setting alphas to fixed values rather than determining them during training (NT) results in the second largest drop in performance of about 10% on the full data. The remaining modes result in substantially smaller drops of 4% or less when removing turn context (NC), removing entity relatedness scores (NE), or removing the integer constraint (NI). All of these modes result in performance drops on all subsets. These results illustrate that CONSENT is robust to changes in its components and hyper-parameters as long as answer cues are considered.

7   Conclusion

In this work, we proposed the CONSENT method for conversational search with tail entities. By jointly performing answer ranking and contextualization, CONSENT outperforms a range of baselines by a substantial margin. On a new benchmark of conversations about tail entities, CONSENT substantially improves over state-of-the-art methods while remaining competitive on the ConvQuestions benchmark about popular entities. Ablations of the CONSENT method confirm the importance of selective contextualization.

References

1. Adlakha, V., Dhuliawala, S., Suleman, K., de Vries, H., Reddy, S.: TopiOCQA: open-domain conversational question answering with topic switching (2021)
2. Aliannejadi, M., Kiseleva, J., Chuklin, A., Dalton, J., Burtsev, M.: Building and evaluating open-domain dialogue corpora with clarifying questions. In: EMNLP (2021)
3. Brown, T.B., et al.: Language models are few-shot learners. In: NIPS (2020)
4. Cao, Y., Xu, J., Liu, T.Y., Li, H., Huang, Y., Hon, H.W.: Adapting ranking SVM to document retrieval. In: SIGIR (2006)
5. Christmann, P., Roy, R.S., Weikum, G.: Conversational question answering on heterogeneous sources. In: SIGIR (2022)

6. Christmann, P., Saha Roy, R., Abujabal, A., Singh, J., Weikum, G.: Look before you hop: conversational question answering over knowledge graphs using judicious context expansion. In: CIKM (2019)

7. Christmann, P., Saha Roy, R., Weikum, G.: Conversational question answering on heterogeneous sources. In: SIGIR (2022)

8. Dalton, J., Xiong, C., Callan, J.: CAsT 2019: the conversational assistance track overview. In: Proceedings of the Twenty-Eighth Text REtrieval Conference (TREC 2019), Gaithersburg, Maryland (2019)

9. Dalton, J., Xiong, C., Callan, J.: Cast 2020: the conversational assistance track overview. In: TREC (2020)

10. Dalton, J., Xiong, C., Callan, J.: Cast 2021: the conversational assistance track overview. In: TREC (2021)

11. Elgohary, A., Peskov, D., Boyd-Graber, J.: Can you unpack that? Learning to rewrite questions-in-context. In: EMNLP-IJCNLP (2019)

12. Gao, J., Xiong, C., Bennett, P., Craswell, N.: Neural Approaches to Conversational Information Retrieval (2023)

13. Hai, N.L., Gerald, T., Formal, T., Nie, J.Y., Piwowarski, B., Soulier, L.: Cosplade: contextualizing splade for conversational information retrieval. In: Kamps, J., et al. (eds.) ECIR 2023. LNCS, vol. 13980, pp. 537–552. Springer, Cham (2023). https://doi.org/10.1007/978-3-031-28244-7_34

14. Hoffart, J., Milchevski, D., Weikum, G.: STICS: searching with strings, things, and cats. In: SIGIR (2014)

15. Khattab, O., Zaharia, M.: ColBERT: efficient and effective passage search via contextualized late interaction over BERT. In: SIGIR (2020)

16. Kim, G., Kim, H., Park, J., Kang, J.: Learn to resolve conversational dependency: a consistency training framework for conversational question answering. In: ACL-IJCNLP (2021)

17. Krasakis, A.M., Yates, A., Kanoulas, E.: Zero-shot query contextualization for conversational search. In: SIGIR (2022)

18. Lin, J., Nogueira, R., Yates, A.: Pretrained transformers for text ranking: BERT and beyond. arXiv:2010.06467 (2020)

19. Lin, S., Yang, J., Lin, J.: Contextualized query embeddings for conversational search. In: EMNLP (2021)

20. Onoe, Y., Boratko, M., Durrett, G.: Modeling fine-grained entity types with box embeddings. In: ACL (2021)

21. OpenAI: GPT-4 technical report. arXiv:2303.08774 (2023)

22. Qu, C., et al.: Attentive history selection for conversational question answering. In: CIKM (2019)

23. Reimers, N., Gurevych, I.: Sentence-bert: sentence embeddings using siamese bert-networks. In: EMNLP (2019)

24. Robertson, S., Zaragoza, H.: The probabilistic relevance framework: BM25 and beyond. Found. Trends Inf. Retrieval $3(4)$, 333–389 (2009)

25. Tredici, M.D., Barlacchi, G., Shen, X., Cheng, W., de Gispert, A.: Question rewriting for open-domain conversational QA: best practices and limitations. In: CIKM (2021)

26. Vakulenko, S., Longpre, S., Tu, Z., Anantha, R.: Question rewriting for end to end conversational question answering. In: WSDM (2021)

27. Vakulenko, S., Voskarides, N., Tu, Z., Longpre, S.: A comparison of question rewriting methods for conversational passage retrieval. In: ECIR (2021)

28. Voskarides, N., Li, D., Ren, P., Kanoulas, E., de Rijke, M.: Query resolution for conversational search with limited supervision. In: SIGIR (2020)

29. Xiong, L., et al.: Approximate nearest neighbor negative contrastive learning for dense text retrieval. In: ICLR (2021)
30. Yu, S., et al.: Few-shot generative conversational query rewriting. In: SIGIR (2020)
31. Yu, S., Liu, Z., Xiong, C., Feng, T., Liu, Z.: Few-shot conversational dense retrieval. In: SIGIR (2021)

An *EcoSage Assistant:* Towards Building A Multimodal Plant Care Dialogue Assistant

Mohit Tomar[1]($\boxtimes$), Abhisek Tiwari[1], Tulika Saha[2], Prince Jha[1], and Sriparna Saha[1]

[1] Department of Computer Science and Engineering,
Indian Institute of Technology Patna, Patna, India
mohitsinghtomar9797@gmail.com, jhapks1999@gmail.com,
{abhisek_1921cs16,sriparna}@iitp.ac.in
[2] Department of Computer Science and Engineering,
University of Liverpool, Liverpool, UK
tulika.saha@liverpool.ac.uk

Abstract. In recent times, there has been an increasing awareness about imminent environmental challenges, resulting in people showing a stronger dedication to taking care of the environment and nurturing green life. The current $19.6 billion indoor gardening industry, reflective of this growing sentiment, not only signifies a monetary value but also speaks of a profound human desire to reconnect with the natural world. However, several recent surveys cast a revealing light on the fate of plants within our care, with more than half succumbing primarily due to the silent menace of improper care. Thus, the need for accessible expertise capable of assisting and guiding individuals through the intricacies of plant care has become paramount more than ever. In this work, we make the very first attempt at building a plant care assistant, which aims to assist people with plant(-ing) concerns through conversations. We propose a plant care conversational dataset named *Plantational*, which contains around 1K dialogues between users and plant care experts. Our end-to-end proposed approach is two-fold: (i) We first benchmark the dataset with the help of various large language models (LLMs) and visual language model (VLM) by studying the impact of instruction tuning (zero-shot and few-shot prompting) and fine-tuning techniques on this task; (ii) finally, we build *EcoSage*, a multi-modal plant care assisting dialogue generation framework, incorporating an adapter-based modality infusion using a gated mechanism. We performed an extensive examination (both automated and manual evaluation) of the performance exhibited by various LLMs and VLM in the generation of the domain-specific dialogue responses to underscore the respective strengths and weaknesses of these diverse models (The dataset and code are available at https://github.com/mohit2b/EcoSage).

Keywords: Plant Care · Virtual Assistant · Large Language Models (LLMs) · Multi-modal infusion · Dialogue Generation

© The Author(s), under exclusive license to Springer Nature Switzerland AG 2024
N. Goharian et al. (Eds.): ECIR 2024, LNCS 14609, pp. 318–332, 2024.
https://doi.org/10.1007/978-3-031-56060-6_21

1   Introduction

According to a recent survey conducted by the International Labour Organization (ILO), over a quarter of the global population continues to depend solely on one profession, which is agriculture[1]. Beyond financial advantages, the environmental benefits derived from agriculture and plantation practices are substantial. In the near future, imminent environmental challenges like climate change, biodiversity loss, and soil degradation are expected to escalate. Plantation holds promise as a potent solution to these issues. By strategically planting diverse tree species, particularly those with substantial carbon sequestration capabilities, one can mitigate climate change. Over the last decade, a deepened awareness has led individuals toward a growing emphasis on plantation endeavours, with the indoor planting market projected to attain a size of 31 billion by 2032, with a Compound Annual Growth Rate (CAGR) of 4.8%.

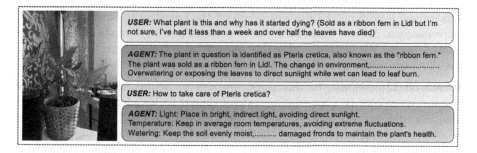

Fig. 1. A conversational illustration between a user and the agent

The rate at which people dedicate efforts to planting doesn't appear to match its potential growth. This is primarily attributed to inadequate management and a lack of guidance concerning plant(ing)-related issues. Individuals are quite reluctant to follow up with plant specialists. However, with the advancement of AI-based agricultural assistants and web technology, we often look for solutions to different issues over the internet. People often post their issues over some popular discussion forums like *Reddit* and *Houzz* with dedicated threads such as *Houseplants*[2] and *House-plants*[3] comprising of more than 20K posts and 1.5M registered users. The primary aim of these communities revolves around engaging in discussions and promoting the care and welfare of indoor plants. Nonetheless, a single, irrelevant recommendation can potentially exacerbate a plant's condition, and the assurance of real-time responses in such forums is not guaranteed. To assist such online plant healthcare seekers, we build a plant care assistant that can assist people with plant(-ing)-related issues and motivate them

---

[1] https://data.worldbank.org/indicator/SL.AGR.EMPL.ZS.

[2] https://www.reddit.com/r/houseplants/.

[3] https://www.houzz.com/discussions/house-plants.

for specialist consultations whenever needed. An example of such a conversation between a plant assistant and a user is illustrated in Fig. 1.

The evolution of Large Language Models (LLMs) and ChatGPT has made the general audience believe that all-natural language processing problems have been solved. However, many significant challenges persist, particularly within domains constrained by limited resources. Our preliminary assessments of LLMs for plant care assistance reveal a substantial gap between anticipated outcomes and those generated. We also delve into the visual language model, which offers an improvement over previous models [47] but still falls short of human expectations. Motivated by the burgeoning interest in plant care and driven by these limitations, we make the first move to investigate some fundamental research questions related to plant assistance response generation and build an assistant called *EcoSage* to provide initial guidance to plant(-ing)-related issues. The *EcoSage* Assistant is meant to seek user issues and pose further queries to gain a comprehensive understanding of the issue and subsequently, provide suggestions and responses to assist the support seekers. Confronted by the scarcity of conversational data in plant care, we undertook the initiative to create a plant care conversational dataset named *Plantational* encompassing a range of plant(-ing)-related issues. We further enrich the dataset by assigning intent and dialogue act (DA) categories to each dialogue and utterances within them.

Research Questions. In this work, we aim to answer the following three research questions: **(i)** Can existing state-of-the-art LLMs adequately offer initial recommendations related to plant(-ing)-related queries? **(ii)** Do LLMs that take images into account comprehend concerns more effectively and produce suitable and better responses? **(iii)** What is the appropriate way to gauge the effectiveness of the response generation model? Are metrics based on n-gram overlap sufficient for the evaluation, and are they consistent with semantic evaluation?

Key Contributions. The key contributions of the work are as follows:

- We first build *Plantational*, a multi-modal, multi-turn plant care conversational dataset, which consists of around 1K conversations spanning over 4900 utterances. The dataset is the first plant-based conversational corpus containing plant-related discussions intended to aid users in the plantation.
- The work investigates the efficacy of different LLMs and VLM for plant care assistants in both zero-shot and few-shot settings.
- Motivated by the need for a proficient plant care assistant, we build a plant care response generation model incorporating an adapter-based modality infusion and fine-tuning mechanisms.
- The proposed model outperforms all baselines in zero-shot and few-shot settings across almost every evaluation metric by a significant margin.

2   Related Works

The current work is mainly related to the following three research areas: Dialogue generation, Large language models (LLMs), and Visual language models (VLMs). The following paragraphs summarize the relevant works.

Dialogue Generation. DialoGPT [46] trains a conversation generation model based on Reddit comments. GODEL [28] uses grounded pre-training on external text for effectively generating a response to a conversation. Recently, reinforcement learning from human feedback is also utilized to train models [4,15,37] to train dialogue agents to be helpful and harmless.

Large Language Models. (LLM) has shown great results in various NLP tasks. Their progress started with models such as BERT [10], GPT [29], and T5 [32]. They are often trained in mask language modeling or next token prediction objectives on a large chunk of the internet. GPT-3 [6], a 175 billion parameters model, achieved breakthroughs on many language tasks. This resulted in the development of more models, such as Gopher [31], OPT [44], Megatron-Turing NLG [39], PaLM [8], LLaMA [41]. Also, fine-tuning LLM on instructions, FLAN [9] and human feedback InstructGPT [25] have achieved great results.

Visual Language Models. Recently, large language models have been used as decoders for solving multimodal tasks. Flamingo [1] utilizes a frozen vision encoder and language decoder and trains only the perceiver resampler and cross-attention layers to achieve impressive few-shot performance on vision-language tasks. GPT-4 [24] takes text and image as input and produces text as output. It achieves state-of-the-art performance on a range of NLP tasks. Open-source multimodal models [2,21,47] have also been released that insert only a few trainable layers such as cross attention layer or linear projection layer to align visual features with textual features.

Dialogue Systems for Social Good. There have been multitude of works focused on developing conversational AI systems based on the social good theme serving various social good goals such as healthcare [40] including mental health [33–36], education [16–18] etc.

3   Dataset

Motivated by the unavailability of any conversational plant care assistance dataset and its significance in the present context, we endeavoured to curate a conversational plant care assistance corpus, *Plantational*. In this section, we discuss the details of the dataset and its curation.

3.1   Data Collection

We initially conducted a thorough survey of datasets about plants and noted that all the existing research studies [12,22,38] are primarily concentrated on classifying plant diseases based on images. The investigation revealed an absence of any dialogic dataset related to plants, whether in textual form or encompassing multimodal elements. However, we identified some potential discussion sites where people seek support to various problems relating to plants. The two most well-known of these forums are *Reddit* and *Houzz*. We found several dedicated sub-threads in these forums, such as *house-plants*, with more than a million registered users and over twenty thousand postings. Each time a user posted about a plant(-ing)-related issue, other platform users tried offering advice and suggestions. One such example of a conversation from these forums is shown in Fig. 2a.

In collaboration with two domain experts, we initiated an assessment of query quality and comment credibility for 50 posts extracted from the *houseplants* forum. Our observations yielded the following insights: (i) Users adeptly articulated their queries, often accompanied by relevant images; (ii) While not all comments exhibited high credibility, those with substantial upvotes notably contributed to the discourse; (iii) Robust user engagement was evident as users actively participated in discussions by responding to other user's comments. Given the abundance of high-quality posts within sufficiently extensive forums, we selected 1150 most popular posts (in terms of upvotes) as the primary source for creating our conversational dataset. The selected sub-threads were as follows: *r/plantcare*, *r/botany*, *r/houseplants*, and *houzz/plantcare*. To retrieve data, we employed PRAW[4] and Beautiful Soup[5], utilizing the official data release keys for these forums.

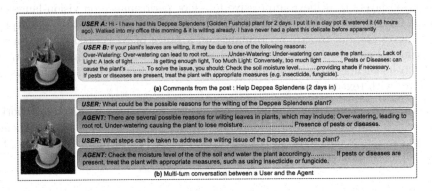

Fig. 2. (a) Indicates original Reddit post; (b) Converted Reddit post into a conversation between a User and the Agent

3.2   Data Creation and Annotation

To begin with, given the specialized nature of the domain, we engaged with two botanists during the data creation process, accompanied by three students of botany who were also fluent in the English language and one English linguist for annotation and scalability to first curate a set of 50 conversations with each sample corresponding to a distinct post. Within each dialogue, a dyadic interaction unfolds between two entities: the 'user' and the 'assistant.' In this context, the term 'user' denotes the individual who initiated a plant(ing)-related inquiry, while the 'assistant' embodies our adept professional responsible for crafting the dialogue content by utilizing the comments of different users on a particular

---

[4] https://praw.readthedocs.io/en/stable/.
[5] https://beautiful-soup-4.readthedocs.io/en/latest/.

post. We asked the English linguist to tag some essential semantics informa-tion about the utterances and the visual information, such as the user's intent and the corresponding dialogue acts (DAs). The intent categories are *Sugges-tion, Conformation, Feedback,* and *Awareness,* and the DA categories used are *Greeting (g), Question (q), Answer (ans), Statement-Opinion (o), Statement-Non-Opinion (s), Agreement (ag), Disagreement (dag), Acknowledge (a)* and *Others (oth).*

In Fig. 2, we show a sample of curated conversation from our *Plantational* dataset. The guidelines for converting the discussion on forums into a multi-turn conversation (after consultation from the team of specialists) are detailed as follows: (i) Each discussion encompasses a question posted by the user and the subsequent comments containing relevant answers. Each of these discussions also includes an image related to the plant and its corresponding query. We convert this post (seen in Fig. 2(a)) into a multi-turn conversation between the user and an agent (seen in Fig. 2(b)) in which the user queries about its plant(ing) related issues; (ii) To achieve this, our first step entails identifying the main concern posed by the primary user sharing the content. We achieve this by analyzing both the title and the user's comments. Following this, we ascertain the most appropriate response by reviewing additional comments and selecting relevant utterances; (iii) we formulate additional questions from the post authored by the primary user (if the post encompasses multiple inquiries). The two botanists and the English linguists helped us create a set of 50 conversations from these unique posts. While creating dialogues from the primary post, the team of specialists ensured that all dialogues were coherent and prudent. Next, we focused on scaling the dataset to a reasonable size. For this, the team of student botanists was trained with 30 curated conversations to form dialogues from the raw posts. They were then presented with the remaining 20 examples to create dialogues from the raw posts. The curated dialogues by the students were then evaluated (against the gold standard) to identify their flaws, and they were again asked to correct them. Finally, the student botanists were presented with the remaining 1100 original posts and were asked to convert them into a dyadic conversation (as detailed above). We also checked the quality of these dialogues beyond the initial evaluation by randomly picking ten dialogues and checking their quality for each annotator. This process was repeated twice to ensure that the student botanists performed the task well. The two botanists conducted the quality check and also helped prepare guidelines for creating dialogues from posts. In this manner, we curated 1150 conversations from the raw posts. The dialogues created by the student botanists were interchanged amongst each other. When any of them found that a particular dialogue did not meet the quality standards, it was rejected. Around 1k conversations were included in the final corpus, which all the annotators found to be acceptable, and the remaining 150 conversations were rejected. We used Fleiss kappa [13] to measure the agreement among the annotators while labeling the intents and dialogue acts. Fleiss kappa is used to measure the reliability of agreement among the annotators in categorical classification tasks. We found the Fleiss kappa score for intent labeling to be 0.72 and for dialogue acts labeling to be 0.68, which indicates good annotation quality.

3.3 Plantational Dataset

The *Plantational* dataset now comprises around 1K conversations between a user and a plant care assistant, amounting to over 4900 utterances (Table 1). Each of these dialogues also encompasses an image of the plant and its corresponding query. The dataset also contains tags corresponding to the intent and the dialogue act for each dialogue and utterance, respectively, the distribution of which is shown in Fig. 3. Below, we study the role of incorporating multimodal features such as images with the text in each dialogue.

Table 1. Statistics from the *Plantational* dataset

Statistics	Instances
#Dialogues	963
#Utterances	4914
#Unique Tokens (Distinct words)	12,953
Average #Utterances	5.1
Maximum #Utterances	12
Minimum #Utterances	2
#Dialogues having images	796

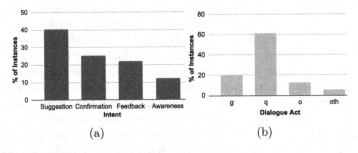

(a)   (b)

Fig. 3. Class distribution in terms of % representation (a) for intent categories, (b) for DA categories in the *Plantational dataset*

Role of Multimodality. In Fig. 1, we find that the user is trying to ask about the plant's identity. Also, the user is asking why the plant started dying. The model can better understand both questions if an image of the plant the user is talking about is also available. Here, we see that multimodality helps simplify generating relevant responses.

4 Methodology

The proposed methodology aims to create a plant assistant that generates contextually relevant responses to user queries. Leveraging the power of LLMs

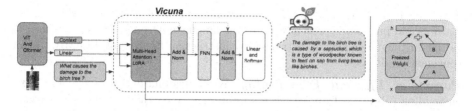

Fig. 4. Proposed model. A and B represent the LoRA modules. Frozen weight represents the frozen weights of Multi-head attention; x and h are hidden representations before and after applying the LoRA module. In our model, LoRA and Linear Projection Layer are trainable while the rest is frozen.

and VLM, we first analyze the efficacy of these models for plant care assistants employing various instruction-tuning (zero-shot and few-shot prompting) and fine-tuning techniques. For our proposed model, within the constraints of our small-scale dataset and recognizing the pivotal role of plant visuals in this domain, we introduce an adapter-based, fine-tuned vision-language model that effectively incorporates visual information and the textual context in its user responses. Figure 4 illustrates this novel architecture's schematic representation.

4.1 Benchmark Setup

We first benchmark the plant care assistant using one VLM and four LLMs described below using instruction and fine-tuning strategies.

- **Llama-2** [42] is an open-source fine-tuned language model ranging from 7 to 70 billion parameter count. It is optimized for dialogue purposes.
- **Vicuna** [7] is an open-source chatbot obtained by fine-tuning LLaMA on 70K user shared conversations. Vicuna-13B attains 90% quality of ChatGPT.
- **FLAN-T5** [9] is based on applying instruction fine-tuning. It studies the performance of various tasks and models at different scales.
- **OPT** [44] is a decoder-only model created for the purpose of matching the performance of GPT-3 [6] and to study LLM performance on various tasks.
- **GPT-Neo** [5] It is an autoregressive language model like GPT-2 [30] trained on Pile [14] dataset. It uses local attention in every other layer.

4.2 Proposed Model

In this section, we discuss the details of our proposed method.

Textual and Visual Encoding. For obtaining the visual representation of the user-provided image in our proposed method, we use the vision and language representation learning part of BLIP-2 [19]. It consists of ViT [11] and Q-former. The role of the ViT is to act as a vision encoder and send the visual representation to the Q-former. The Q-former consists of a transformer network that extracts features from the vision encoder and passes it to the projection layer. The role of the linear projection layer [47] is to align the visual features with the textual features and pass the representation to the language decoder. We use Vicuna [7] as the language decoder. It is trained by instruction fine-tuning LLaMA on the conversation dataset.

Parameter Efficient Fine-Tuning. We concatenate the image and text embedding (obtained from the user's input) and pass them to the Vicuna language model [7]. We insert LoRA blocks inside the Vicuna decoder. The LoRA module and the Linear projection layer (projecting from image space to text space) are kept trainable while the rest of the model is frozen. LoRA consists of a low-rank decomposition matrix injected into the pre-trained transformer model. It involves the following operation:- For a weight matrix, $W_o \in \mathbb{R}^{d \times e}$, it is updated in the following way: $W_o + \Delta W = W_o + BA$, where $B \in \mathbb{R}^{d \times f}$ and $A \in \mathbb{R}^{f \times e}$. Here, $f \ll min(d, e)$, A, and B are trainable parameters while W_o is frozen. Finally, the modified hidden representation can be described by the following equation:

$$h = W_o x + \Delta W x = W_o x + BA x \qquad (1)$$

Also, the ΔW matrix is scaled by a factor of $\frac{\alpha}{f}$ where α is a constant.

Response Generation. In the final step, we take the representation coming from the lower layers of the Vicuna and transform the hidden dimension to vocabulary size using a linear projection layer. We then use a beam search algorithm to sample the following tokens and stop the generation when an end-of-sentence token is generated, or the model has generated the maximum number of tokens (defined by the user). The generated response is ideally expected to contain information about the image context and answer the user query related to the image of the plant.

4.3   Implementation Details

We implement the LLMs and VLM using the transformers hugging face library [43] in PyTorch framework [27] using a GeForce RTX 3090 GPU. LoRA is implemented using the peft library [23]. The train, validation, and test set comprises 80%, 10% and 10% of the conversational instances from the *Plantational* dataset. The hyperparameters used are as follows: number of epochs (5), learning rate (1e-4), LoRA dropout (0.1), LoRA alpha (32), LoRA dimension (8), generated tokens (30), optimizer (Adam). We evaluate the performance of the baseline models on our *Plantational* dataset using automated metrics such as ROUGE [20], BLEU [26] and BERT scores [45]. We also perform human evaluation on metrics such as (i) *Fluency*: The response must be grammatically correct; (ii) *Adequacy*: To generate a response related to user's query; (iii) *Informativeness*: To generate the response that answers user's problem; (iv) *Contextual Relevance*: To generate response related to the context of the conversation; (v) *Image Relevance*: To generate response related to the image. We performed human evaluation across 95 dialogues using three human evaluators (from the authors' affiliation) and reported the average score.

5 Results and Discussion

To assess the performance of various dialogue generation models for plant care, we employed commonly used evaluation metrics, including BLEU and ROUGE. We also assessed the effectiveness of various models in terms of the semantic alignment of their generated text with the reference (gold) responses. Moreover, we conducted a human evaluation to mitigate the risk of under-assessment performed by automatic evaluation metrics. Our discussion commences with an examination of the experimental results we have obtained. Subsequently, we delve into the findings and evidence pertaining to the research questions, culminating in the presentation of a case study showcasing the performances of various models.

Table 2. Performances of different models for plant assistance response generation

Model		ROUGE			BLEU				BLEU	BERT-F1
		R-1	R-2	R-L	B1	B2	B3	B4		
Flan	zero-shot	18.94	6.27	16.64	8.01	3.41	1.95	0.36	3.43	57.98
	few-shot	22.29	7.46	19.86	9.25	4.75	2.42	0.70	4.28	59.71
	fine-tune	**29.80**	**14.61**	**26.63**	**16.43**	**11.12**	**8.58**	**3.91**	**10.01**	**63.42**
GPT-Neo	zero-shot	14.702	3.45	11.90	6.71	2.19	0.95	0.35	2.55	50.15
	few-shot	17.13	5.17	14.21	8.37	3.30	1.60	0.73	3.5	51.25
	fine-tune	**30.39**	**15.09**	**26.94**	**18.67**	**12.10**	**9.08**	**6.30**	**11.53**	**59.59**
OPT	zero-shot	17.80	4.79	14.87	7.96	2.81	1.29	0.51	3.14	50.69
	few-shot	20.90	6.65	17.70	10.49	4.21	1.80	0.96	4.36	52.72
	fine-tune	**29.07**	**16.44**	**26.27**	**18.64**	**13.47**	**10.39**	**6.33**	**12.20**	**59.59**
Llama-2	zero-shot	19.47	4.83	15.57	8.55	3.12	1.50	0.78	3.48	57.20
	few-shot	21.96	6.62	17.64	10.14	4.42	2.45	1.48	4.62	58.93
	fine-tune	**30.30**	**15.74**	**27.19**	**17.79**	**11.86**	**8.62**	**5.26**	**10.88**	**60.86**
Vicuna	zero-shot	19.84	5.29	16.28	9.03	3.59	1.77	0.97	3.84	55.49
	few-shot	20.33	5.09	16.46	8.81	3.11	1.38	0.81	3.52	55.26
	fine-tune	**30.34**	**16.24**	**27.20**	**18.35**	**12.66**	**9.98**	**6.49**	**11.87**	**61.13**
EcoSage	zero-shot	22.16	6.49	18.13	10.32	4.45	2.59	1.72	4.77	56.64
	few-shot	19.97	5.83	16.52	9.00	3.86	2.21	1.47	4.13	54.30
	fine-tune	**25.35**	**11.76**	**21.77**	**14.07**	**8.37**	**5.65**	**3.95**	**8.01**	**58.76**

5.1 Experimental Results

Table 2 summarizes the performances of different LLMs and VLM for appropriate plant assistance response generation. We have reported results for three different settings, namely zero-shot, few-shot, and fine-tuning. Across various LLMs such as Flan, GPT-Neo, OPT, and Vicuna, we observe a consistent upward trend in performance when transitioning from zero-shot to few-shot to fine-tuning settings. In a few settings, particularly Vicunna and the proposed model, the performance of zero-shot has been superior to few-shot. The most probable reason seems to be the generalizability of these two highly large models with the handful number of samples. We found an unusual finding that the textual LLM performs better than the VLM within the context of *EcoSage*, despite *EcoSage* including both user query text and images related to plants. The decrease in performance happens primarily due to the challenge of aligning the visual embedding with the text embedding. Integrating visual information alongside textual data may introduce complexities in the model's embedding, resulting in a less effective alignment and leading to poorer overall performance when compared to LLMs.

We also carried out an ablation study involving different settings. The obtained result has been reported in Table 3. We experimented with another visual encoding method called Data2vec [3]; it did not outperform the uni model. We also examined various image configurations: textual dialogues with a blank white image, textual dialogues with images (whenever available), and textual dialogues with images (whenever available), along with a blank white image for dialogues that do not include visual descriptions. The results indicate that adding visual descriptions does improve the model's ability to offer suitable suggestions. It faces challenges in aligning visual and textual embedding space.

Table 3. Performance of *EcoSage* with different modalities

Model	ROUGE-L	BLEU-4	BERT-F1
VE (Data2vec)	20.86	3.33	56.85
T + I (all blank)	21.40	3.86	57.21
T + I (available)	21.77	3.95	58.76
T + I	**22.82**	**4.27**	**55.90**

Human Evaluation. Figure 5 presents results based on a human evaluation study. We have observed that all the models excel in terms of fluency. However, both LLMs and VLM face challenges regarding adequacy, informativeness, contextual relevance, and image relevance. This suboptimal performance of LLMs and VLM can be attributed to the specific and detailed nature of user queries related to plants. Furthermore, it is noteworthy that the VLM exhibits a poorer performance compared to LLMs. It can be explained due to the VLM's inability to understand images and can be enhanced by training on image-text pairs.

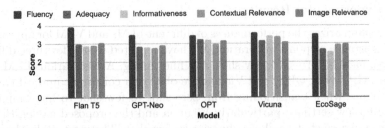

Fig. 5. Human evaluation scores of different models based on diverse metrics

Qualitative Analysis. We examined the text generation capabilities of various models using a representative example from the *Plantational* dataset, as depicted in Fig. 6. Our analysis reveals that Vicuna produced a response that closely aligns with the gold standard response. It is followed by responses from *EcoSage*, OPT, Flan, and GPT-Neo, in descending order of similarity to the gold response.

User: Help! I recently got some azaleas and I've been trying to take care of them, but I'm worried they might be beyond repair. Can anyone advise me on what to do?

Gold Response: Keep the plants moist but not sitting in water. It appears that your azaleas may need watering. You can try placing them in the sink and running room temperature water through them, allowing the excess water to drain.

Flan: I'm sorry to hear that. Certainly, azaleas are very sensitive to heat and humidity. If you notice any signs of

GPT-Neo: You should be concerned about the size of the root ball and how it's growing. It's possible that the plant is too big for its pot.

OPT: azaleas are very susceptible to overwatering, especially if they're not in direct sunlight. Azaleas are also susceptible to root rot,

Vicuna: It's important to keep the soil moist but not waterlogged. You should water them once or twice a week. Make sure the pot

EcoSage: Azaleas are very sensitive to overwatering, so it's possible that they may have been overwatered. If they are still like this, let the extra water go

Fig. 6. Qualitative analysis of response generated by different models.

5.2   Findings to Research Questions

Based on the experiments, we report the following answers (with evidence) to our investigated research questions (RQs).

RQ1: Can existing state-of-the-art LLMs adequately offer initial recommendations related to plant(-ing)-related queries? Based on results from Table 2 and Table 5 (zero-shot and few-shot), it becomes evident that LLMs face challenges in delivering satisfactory responses to user queries pertaining to the plantation. This underscores the necessity for LLMs to acquire more domain-specific knowledge about plants and improve their ability to contextualize information from images. We observed a noteworthy enhancement, varying from 10% to 50%, when employing both few-shot learning and fine-tuning across various evaluation metrics.

RQ2: Do LLMs that take images into account comprehend concerns more effectively and produce suitable and better responses? The results we obtained (as shown in Table 3) provide affirmative evidence supporting the assertion. The model that considers images along with textual description has obtained superior performance and generates a more context-specific response.

RQ3: What is the appropriate way to gauge the effectiveness of the response generation model? Are metrics based on n-gram overlap sufficient for evaluation, and are they consistent with semantic evaluation? In numerous instances, we noticed that the generated response is highly contextually relevant but exhibits limited overlap with the reference response (as shown in Fig. 6), resulting in lower evaluation scores such as BLEU and ROUGE. Therefore, we also incorporated the measurement and reporting of BERT-F1, a semantic-based evaluation metric that assesses the similarity between the semantic embeddings of the reference response and the generated sentence. Consequently, the Vicunna model achieves the highest BERT-F1 score despite having a lower BLEU score. Henceforth, we firmly support that a comprehensive evaluation of effectiveness should encompass both lexical and semantic-based assessments.

6   Conclusion

In this paper, we make the first move to investigate some fundamental research questions related to plant assistance response generation and build an assistant called *EcoSage* to provide guidance to plant(-ing)-related issues of the users. Confronted by the scarcity of conversational data in plant care, we undertook the initiative to create a plant care conversational dataset named *Plantational* encompassing a range of plant(-ing)-related issues. We further evaluate LLMs and VLM on the *Plantational* dataset by generating responses corresponding to the user's query in zero-shot, few-shot, and fine-tune settings. Additionally, our proposed *EcoSage* is a multi-modal model, a plant assistant utilizing LoRA units for adapting it to plant-based conversations. In the future, we aim to focus on improving the image understanding capabilities of VLM for it to perform better in generating coherent responses than LLMs.

Acknowledgement. Dr. Sriparna Saha extends heartfelt gratitude for the Young Faculty Research Fellowship (YFRF) Award, supported by the Visvesvaraya Ph.D. Scheme for Electronics and IT, Ministry of Electronics. Abhisek Tiwari expresses sincere gratitude for the support received by the Prime Minister Research Fellowship (PMRF) Award provided by the Government of India. This grant has played a role in supporting this research endeavor. We also thank Rachit Ranjan, Ujjwal Kumar, Kushagra Shree, and other annotators for the dataset development.

References

1. Alayrac, J.B., et al.: Flamingo: a visual language model for few-shot learning. Adv. Neural. Inf. Process. Syst. **35**, 23716–23736 (2022)
2. Awadalla, A., et al.: Openflamingo (2023). https://doi.org/10.5281/zenodo.7733589
3. Baevski, A., Hsu, W.N., Xu, Q., Babu, A., Gu, J., Auli, M.: Data2vec: a general framework for self-supervised learning in speech, vision and language. In: International Conference on Machine Learning, pp. 1298–1312. PMLR (2022)
4. Bai, Y., et al.: Training a helpful and harmless assistant with reinforcement learning from human feedback. arXiv preprint arXiv:2204.05862 (2022)
5. Black, S., Leo, G., Wang, P., Leahy, C., Biderman, S.: GPT-Neo: Large Scale Autoregressive Language Modeling with Mesh-Tensorflow (2021). https://doi.org/10.5281/zenodo.5297715. If you use this software, please cite it using these metadata
6. Brown, T., et al.: Language models are few-shot learners. Adv. Neural. Inf. Process. Syst. **33**, 1877–1901 (2020)

7. Chiang, W.L., et al.: Vicuna: an open-source chatbot impressing GPT-4 with 90%* ChatGPT quality (2023). https://lmsys.org/blog/2023-03-30-vicuna/

8. Chowdhery, A., et al.: PaLM: scaling language modeling with pathways. arXiv preprint arXiv:2204.02311 (2022)

9. Chung, H.W., et al.: Scaling instruction-finetuned language models. arXiv preprint arXiv:2210.11416 (2022)

10. Devlin, J., Chang, M.W., Lee, K., Toutanova, K.: Bert: pre-training of deep bidirectional transformers for language understanding. arXiv preprint arXiv:1810.04805 (2018)

11. Dosovitskiy, A., et al.: An image is worth 16x16 words: transformers for image recognition at scale. arXiv preprint arXiv:2010.11929 (2020)

12. Fenu, G., Malloci, F.M.: Diamos plant: a dataset for diagnosis and monitoring plant disease. Agronomy **11**(11), 2107 (2021)

13. Fleiss, J.L.: Measuring nominal scale agreement among many raters. Psychol. Bull. **76**(5), 378 (1971)

14. Gao, L., et al.: The pile: an 800GB dataset of diverse text for language modeling. arXiv preprint arXiv:2101.00027 (2020)

15. Glaese, A., et al.: Improving alignment of dialogue agents via targeted human judgements. arXiv preprint arXiv:2209.14375 (2022)

16. Jain, R., Saha, T., Chakraborty, S., Saha, S.: Domain infused conversational response generation for tutoring based virtual agent. In: 2022 International Joint Conference on Neural Networks (IJCNN), pp. 1–8. IEEE (2022)

17. Jain, R., Saha, T., Lalwani, J., Saha, S.: Can you summarize my learnings? Towards perspective-based educational dialogue summarization. In: Findings of the Association for Computational Linguistics: EMNLP 2023, pp. 3158–3173 (2023)

18. Jain, R., Saha, T., Saha, S.: T-vaks: a tutoring-based multimodal dialog system via knowledge selection. In: ECAI 2023, pp. 1132–1139. IOS Press (2023)

19. Li, J., Li, D., Savarese, S., Hoi, S.: Blip-2: bootstrapping language-image pre-training with frozen image encoders and large language models. arXiv preprint arXiv:2301.12597 (2023)

20. Lin, C.Y.: Rouge: a package for automatic evaluation of summaries. In: Text Summarization Branches Out, pp. 74–81 (2004)

21. Liu, H., Li, C., Wu, Q., Lee, Y.J.: Visual instruction tuning. arXiv preprint arXiv:2304.08485 (2023)

22. Liu, X., Min, W., Mei, S., Wang, L., Jiang, S.: Plant disease recognition: a large-scale benchmark dataset and a visual region and loss reweighting approach. IEEE Trans. Image Process. **30**, 2003–2015 (2021)

23. Mangrulkar, S., Gugger, S., Debut, L., Belkada, Y., Paul, S.: PEFT: state-of-the-art parameter-efficient fine-tuning methods (2022). https://github.com/huggingface/peft

24. OpenAI: GPT-4 technical report. arXiv abs/2303.08774 (2023). https://api.semanticscholar.org/CorpusID:257532815

25. Ouyang, L., et al.: Training language models to follow instructions with human feedback. Adv. Neural. Inf. Process. Syst. **35**, 27730–27744 (2022)

26. Papineni, K., Roukos, S., Ward, T., Zhu, W.J.: Bleu: a method for automatic evaluation of machine translation. In: Proceedings of the 40th Annual Meeting of the Association for Computational Linguistics, pp. 311–318 (2002)

27. Paszke, A., et al.: Pytorch: an imperative style, high-performance deep learning library. In: Advances in Neural Information Processing Systems, vol. 32 (2019)

28. Peng, B., et al.: Godel: large-scale pre-training for goal-directed dialog. arXiv preprint arXiv:2206.11309 (2022)

29. Radford, A., Narasimhan, K., Salimans, T., Sutskever, I., et al.: Improving language understanding by generative pre-training (2018)
30. Radford, A., Wu, J., Child, R., Luan, D., Amodei, D., Sutskever, I., et al.: Language models are unsupervised multitask learners. OpenAI Blog **1**(8), 9 (2019)
31. Rae, J.W., et al.: Scaling language models: methods, analysis & insights from training gopher. arXiv preprint arXiv:2112.11446 (2021)
32. Raffel, C., et al.: Exploring the limits of transfer learning with a unified text-to-text transformer. J. Mach. Learn. Res. **21**(1), 5485–5551 (2020)
33. Saha, T., Chopra, S., Saha, S., Bhattacharyya, P., Kumar, P.: A large-scale dataset for motivational dialogue system: an application of natural language generation to mental health. In: 2021 International Joint Conference on Neural Networks (IJCNN), pp. 1–8. IEEE (2021)
34. Saha, T., Gakhreja, V., Das, A.S., Chakraborty, S., Saha, S.: Towards motivational and empathetic response generation in online mental health support. In: Proceedings of the 45th International ACM SIGIR Conference on Research and Development in Information Retrieval, pp. 2650–2656 (2022)
35. Saha, T., Reddy, S., Das, A., Saha, S., Bhattacharyya, P.: A shoulder to cry on: towards a motivational virtual assistant for assuaging mental agony. In: Proceedings of the 2022 Conference of the North American chapter of the Association for Computational Linguistics: Human Language Technologies, pp. 2436–2449 (2022)
36. Saha, T., Reddy, S.M., Saha, S., Bhattacharyya, P.: Mental health disorder identification from motivational conversations. IEEE Trans. Comput. Soc. Syst. (2022)
37. Schulman, J., et al.: ChatGPT: optimizing language models for dialogue. OpenAI blog (2022)
38. Singh, D., Jain, N., Jain, P., Kayal, P., Kumawat, S., Batra, N.: Plantdoc: a dataset for visual plant disease detection. In: Proceedings of the 7th ACM IKDD CoDS and 25th COMAD, pp. 249–253 (2020)
39. Smith, S., et al.: Using deepspeed and megatron to train megatron-turing NLG 530B, a large-scale generative language model. arXiv preprint arXiv:2201.11990 (2022)
40. Tiwari, A., et al.: Symptoms are known by their companies: towards association guided disease diagnosis assistant. BMC Bioinform. **23**(1), 556 (2022). https://doi.org/10.1186/S12859-022-05032-Y
41. Touvron, H., et al.: Llama: open and efficient foundation language models. arXiv preprint arXiv:2302.13971 (2023)
42. Touvron, H., et al.: Llama 2: open foundation and fine-tuned chat models. arXiv abs/2307.09288 (2023). https://api.semanticscholar.org/CorpusID:259950998
43. Wolf, T., et al.: Transformers: state-of-the-art natural language processing. In: Proceedings of the 2020 Conference on Empirical Methods in Natural Language Processing: System Demonstrations, pp. 38–45. Association for Computational Linguistics, Online (2020). https://www.aclweb.org/anthology/2020.emnlp-demos.6
44. Zhang, S., et al.: OPT: open pre-trained transformer language models. arXiv preprint arXiv:2205.01068 (2022)
45. Zhang, T., Kishore, V., Wu, F., Weinberger, K.Q., Artzi, Y.: Bertscore: evaluating text generation with bert. arXiv preprint arXiv:1904.09675 (2019)
46. Zhang, Y., et al.: Dialogpt: large-scale generative pre-training for conversational response generation. arXiv preprint arXiv:1911.00536 (2019)
47. Zhu, D., Chen, J., Shen, X., Li, X., Elhoseiny, M.: Minigpt-4: enhancing vision-language understanding with advanced large language models. arXiv preprint arXiv:2304.10592 (2023)

Event-Specific Document Ranking Through Multi-stage Query Expansion Using an Event Knowledge Graph

Sara Abdollahi$^{(\boxtimes)}$ (ID), Tin Kuculo (ID), and Simon Gottschalk (ID)

L3S Research Center, Leibniz Universität Hannover, Hanover, Germany
{abdollahi,kuculo,gottschalk}@L3S.de

Abstract. Event-specific document ranking is a crucial task in supporting users when searching for texts covering events such as Brexit or the Olympics. However, the complex nature of events involving multiple aspects like temporal information, location, participants and sub-events poses challenges in effectively modelling their representations for ranking. In this paper, we propose *MusQuE* (Multi-stage Query Expansion), a multi-stage ranking framework that jointly learns to rank query expansion terms and documents, and in this manner flexibly identifies the optimal combination and number of expansion terms extracted from an event knowledge graph. Experimental results show that *MusQuE* outperforms state-of-the-art baselines on *MS-MARCO$_{EVENT}$*, a new dataset for event-specific document ranking, by 9.1% and more.

Keywords: Document Retrieval · Query Expansion · Event Knowledge Graphs · Event-specific Document Ranking

1 Introduction

Research about important events such as Brexit and the Olympics requires seeking documents that cover all event aspects to gain comprehensive coverage. However, due to their complexity, events differ from typical other topics in retrieval tasks: events can happen in multiple locations, last long, and include several sub-events depending on their impact and size. Take the *Arab Spring* as an example event; *Tunisian Revolution, Bahraini protests of 2011*, and *2011 Yemeni revolution* are three of its sub-events in the EventKG knowledge graph [12]. Using only its label, *Arab Spring* as a search query, one might miss documents related to these sub-events and thus not gain full insights into the unfolding of the event of interest. Therefore, there is a need for dedicated *event-specific document ranking* approaches which consider relevant event characteristics.

A standard approach for improving the document ranking performance is query expansion which adds relevant terms to a user query. These terms are typically selected from the initial set of most relevant documents (pseudo relevance feedback [8]) or ontologies [5] and knowledge graphs (semantic query

© The Author(s), under exclusive license to Springer Nature Switzerland AG 2024
N. Goharian et al. (Eds.): ECIR 2024, LNCS 14609, pp. 333–348, 2024.
https://doi.org/10.1007/978-3-031-56060-6_22

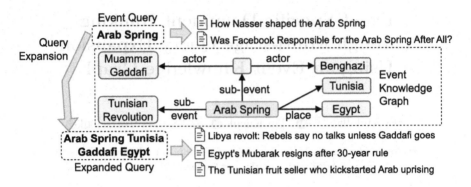

Fig. 1. Example of event-specific document ranking showing top-ranked documents for an initial query event (*Arab Spring*) and an expanded query (*Arab Spring Tunisia Gaddafi Egypt*). The expansion is made based on places, sub-events and their corresponding actors provided in an event knowledge graph.

expansion [22]). Figure 1 demonstrates event-specific document ranking through semantic query expansion at the example of the event *Arab Spring*.

The number of potential expansion terms can be large: one can get thousands of expansion terms from knowledge graphs for impactful international events such as *World War II*. In practice, only a limited number of such terms leads to selecting relevant documents [4]. Therefore, selecting appropriate terms, and identifying the optimal number of terms for a query, is a specifically challenging task for event-specific document ranking.

To address the mentioned issues and challenges, we propose *MusQuE*, our approach for event-specific document ranking through Multi-stage Query Expansion. *MusQuE* improves the ranking of event-specific documents by incorporating event information from knowledge graphs and by jointly learning the importance of terms alongside the document ranking. This way, we target three questions:

(Q1) How can we find an optimal combination of expansion terms per event to deal with specific characteristics of heterogeneous events?
(Q2) How can we jointly train a term selection and a document ranking model to optimise their performances mutually?
(Q3) How can we ensure efficiency of the joint training despite the potentially larger number of terms and their combinations?

To approach these questions, *MusQuE* follows three steps: First, it collects candidate expansion terms from an event knowledge graph (e.g., *Benghazi* in Fig. 1). Then, in a multi-stage ranking framework, *MusQuE* jointly performs term ranking and document ranking to select an optimal set of expansion terms (here, *Muammar Gaddafi*, *Tunisia* and *Egypt*). Finally, *MusQuE* employs the identified terms to expand the query and rank the documents.

To train and evaluate *MusQuE*, we create $MS\text{-}MARCO_{EVENT}$, a dataset for event-specific document ranking based on the *MS-MARCO* dataset [36] and the

EventKG event knowledge graph [13]. We compare *MusQuE* to query expansion and document ranking baselines and show that it outperforms them by at least 9.1% regarding the mean reciprocal rank (MRR).

Our contributions are as follows:

- We introduce *MS-MARCO_{EVENT}*, the first dataset for event-specific document ranking.
- We propose *MusQuE*, an event-specific document ranking model that jointly learns models for expansion term ranking and document ranking.
- We highlight the effectiveness of *MusQuE* compared to state-of-the-art document ranking and query expansion methods.
- We make our source code and dataset available.[1]

The remainder of this paper is as follows: First, we define the task of event-specific document ranking in Sect. 2, and then we propose our approach *MusQuE* in Sect. 3. We present our evaluation setting in Sect. 4 and the results in Sect. 5. Section 6 reviews related work, and Sect. 7 concludes.

2 Problem Statement

For event-specific document ranking, we require an event knowledge graph:

Definition 1 (Event knowledge graph). *An event knowledge graph $G = (E, \mathcal{V}, P, R)$ is a directed graph where E is a set of nodes representing real-world entities, $\mathcal{V} \subset E$ represents events, P is a set of properties, and $R = \mathcal{V} \times P \times E$ is a set of edges representing relations between events and entities, labelled with a property in P. Each entity $e \in E$ has a label e_{label}.*

Figure 1 shows an example event knowledge graph with the event node v representing *Arab Spring*. Event-specific document ranking aims at finding documents relevant to a given event:

Definition 2 (Event-specific document ranking). *Given a query event $v \in \mathcal{V}$, and a set of documents D_v, return a ranking D_r of the documents in D_v where documents are more related to v the higher they are ranked.*[2]

3 Approach

In this section, we present *MusQuE*, our document ranking model for event-specific queries, its core components and its efficiency sampling strategies.

---

[1] https://github.com/saraabdollahi/MusQuE.

[2] Following this definition, we consider only user queries about events already contained in the knowledge graph. Other works deal with the detection of emerging events in news [51] and social media [20] and with the prediction of missing and future events in event knowledge graphs [14,31].

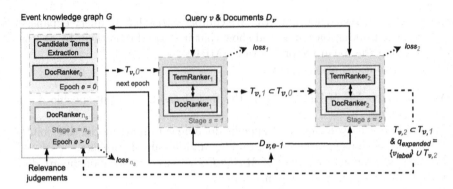

Fig. 2. Overview of *MusQuE*'s training procedure at the example of three stages ($n_s = 3$). References to stages are marked in red, to epochs in blue. Dashed arrows denote the flow of terms, dotted arrows the loss computation.

MusQuE employs query expansion to expand the initial query event v_{label}, creating $q_{expanded}$ for document ranking. The idea behind *MusQuE* is to train progressively complex document ranking models. In each epoch, light-weight early-stage document rankers learn from the previous epoch's ranking, allowing them to mimic the more complex models. In other words, we briefly summarise the workflow of *MusQuE* as follows:

(1) We generate a first ranking of documents D_r based on a pre-trained model and the query event label (v_{label}).
(2) We extract a set of candidate terms from the event knowledge graph.
(3) Starting from the event label, we stage-by-stage create larger queries by combining fewer but more relevant candidate terms, where in each stage
   – the term ranker reduces the number of candidate terms and identifies the best term combinations, and
   – the document ranker mimics the ranking of documents, D_r, to measure the quality of the term combinations.
(4) We train a complex document ranker based on the best set of terms, i.e., the expanded query, and the relevance judgments.
(5) We continue with (3), but now using the new ranking of documents created in the previous step as D_r.

Figure 2 depicts an overview of *MusQuE*'s training procedure at the example of a single event v, its documents D_v and three stages ($n_s = 3$). In the first epoch, the first stage generates an initial set of single candidate terms[3], $T_{v,0}$. In each of the subsequent stages, the number of terms is reduced by the respective term and

---

[3] Entity labels often contain multiple words (e.g., *Arab Spring*) which we consider as a single term for simplicity.

document ranker[4], i.e., $|T_{v,i+1}| < |T_{v,i}|\ \forall i \in \{0,\ldots,n_{s-1}\}$. Together with the event label v_{label}, the terms in the final stage T_{v,n_s} compose the query $q_{expanded}$. $q_{expanded}$ is used as an input to the final document ranker, $DocRanker_{n_s}$. $D_{v,e}$, the output of $DocRanker_{n_s}$, is then used in the next epoch to train the joint term and document rankers in the early stages.

In the following, we describe $MusQuE$'s components in detail.

3.1   Initial Document Ranking

First, we employ a pre-trained ranking model such as monoBERT [38] as $DocRanker_0$ to create an initial ranking $D_{v,0}$ over the documents D_v by using only the query event label v_{label}.

3.2   Candidate Terms Extraction

The candidate terms extraction component aims to extract $T_{v,0}$, an initial set of terms relevant to the query event v using G. We select a set of entities related to v via the following paths in $G = (E, \mathcal{V}, P, R)$:

- All entities connected to v via any property $p \in P$: $\{e | (v, p, e) \in R\}$, such as *Egypt* in Fig. 1.
- We identify the entities related to textual sub-events of v and their label $(v_{sub_{label}})$ via: $\{v_{sub} | (v, p_{\text{sub-event}}, v_{sub}) \in R\}$.

To reduce the size of $T_{v,0}$, we select the top-k entities with the highest number of links plus co-mentions to the query event in G.[5]

3.3   Joint Term and Document Ranking

The training procedure of $MusQuE$ based on a joint model for query expansion and document ranking as targeted by *(Q2)* is shown in Algorithm 1. First, in lines 3-5, we rank documents and create an initial set of terms as described in Sect. 3.2. Then, we employ more complicated ranking models as we proceed to the following stages with fewer candidate terms: For each epoch, event and non-final stage, we use a term ranker to identify relevant terms (lines 8-11). The final stage of $MusQuE$ uses the $DocRanker_{n_s}$ (line 13) that ranks documents using the expanded query ($q_{expanded}$) generated by $TermRanker_{n_s-1}$.

A key aspect of $MusQuE$ is the simultaneous learning of expansion terms and document ranking, optimizing both concurrently for mutual improvement. The total loss function is computed as follows:

---

[4] In contrast to conventional multi-stage ranking methods that aim to reduce the candidate document set size at each stage [39].

[5] We use EventKG [13] as the event knowledge graph which provides link and co-mention counts between entities in Wikipedia.

Algorithm 1. Multi-stage Query Expansion (*MusQuE*)

1: **Input**: V (Set of query events), G (Event Knowledge Graph), n_e (Number of epochs), n_s (Number of stages), l (Number of top-combinations to consider)
2:
3: **for each** $v \in V$ **do**
4:     $D_{v,0} \leftarrow DocRanker_0(v)$ ▷ Initial Document Ranking & Candidate Terms Extraction   ▷ (Section 3.2)
5:     $T_{v,0} \leftarrow extractTerms(v, G)$
6:
7: **for each** $e \in \{1, \ldots, n_e\}$ **do**   ▷ Epochs
8:     **for each** $v \in V$ **do**   ▷ Events
9:         **for each** $s \in \{1, \ldots, n_s - 1\}$ **do**   ▷ Stages
10:   ▷ Term Ranker (Section 3.3)
11:             $loss_s, T_{v,s} \leftarrow TermRanker_s(v, T_{v,s-1}, D_v, D_{v,e-1}, l)$
12:         $q_{expanded} \leftarrow v_{label} \cup T_{v,n_s-1}$
13:         $D_{v,e}, loss_{n_s} \leftarrow DocRanker_{n_s}(q_{expanded}, D_v)$
14:   ▷ Compute total loss via Equation 1
15:         $totalLoss = \alpha \cdot loss_{n_s} + \frac{1-\alpha}{n_s-1} \cdot \sum_{s \in \{1, \ldots, n_s-1\}} loss_s$
16:         $updateDocRankers(totalLoss)$

$$totalLoss = \alpha \cdot loss_{n_s} + \frac{1-\alpha}{n_s-1} \cdot \sum_{s \in \{1, \ldots, n_s-1\}} loss_s, \tag{1}$$

where $loss_{n_s}$ is the loss of the final stage, $loss_s$ the loss of stage s and $0 \leq \alpha \leq 1$ balances between these two types of losses.

Term Ranker. The term rankers (line 10 in Algorithm 1) aim to select the most-related candidate terms for query expansion. Following the multi-stage ranking strategy, we rank terms at each stage to filter out some irrelevant terms as we move forward. Algorithm 2 shows details of the term rankers.

First, we create combinations of terms as follows (line 4):

$$Combination(T_{s-1}, s) = \{C \in \mathcal{P}, (T_{s-1}) \mid |C| = s + 1\} \tag{2}$$

where $\mathcal{P}(T_{s-1})$ denotes the power set of T_{s-1}.[6] For example, $Combination(\{t_0, t_1, t_2\}, 1) = \{\{t_0, t_1\}, \{t_0, t_2\}, \{t_1, t_2\}\}$.

For each term combination $C \in Combination(T_{s-1}, s)$, we feed a query $query_C$ into $DocRanker_s$ to create a document ranking D_C (line 8-9). Next, we compare D_C with $D_{v,e-1}$, i.e., the previous epoch's ranking by computing the mean squared error (MSE):

$$MSE(D_{v,e-1}, D_C) = \frac{1}{|D_C|} \sum_{d \in D_v} (rank(d, D_C) - rank(d, D_{v,e-1}))^2 \tag{3}$$

---

[6] Section 3.5 describes our sampling strategies to avoid too many combinations.

Algorithm 2. TermRanker$_s$: Joint Term and Document Ranking

1: **Input:** s (Stage), v (Query event), T_{s-1} (Set of terms), D_v (Ranked documents), $D_{v,e-1}$ (Ranked documents from the previous epoch), l (Number of top-combinations to consider)

2:

3: $scores \leftarrow \{\}$

4:   ▷ Retrieve term combinations by Equation 2

5: termCombinations $\leftarrow$ Combinations(T_{s-1}, s)

6:

7: **for each** $C \in termCombinations$ **do**

8:     $query_C \leftarrow \{v_{label}\} \cup C$

9:     $D_C \leftarrow DocRanker_s(D, query_C)$ ▷ Rank documents

10:     $scores \leftarrow scores \cup (MSE(D_{v,e-1}, D_C))$ ▷ Compute MSE by Equation 3

11:

12: $(C_{best}, loss_s) \leftarrow top(scores)$ ▷ Return the lowest MSE and the top scoring terms

13: $T_{v,s} \leftarrow \bigcup_{t \in C} \{C | (C, score) \in top(scores, l)\}$

14: **return** $loss_s, T_{v,s}$

---

where $rank(d, D)$ is the rank of document d in the document ranking D. The smaller $MSE(D_C, D_{v,e-1})$, the better D_C aligns with the ranking results of the final stage.

$DocRanker_s$ provides two results: (i) $loss_s$, the MSE score of the best-performing term combination (i.e., the lowest computed MSE score) (line 12), and (ii) $T_{v,s}$, the terms occurring in the l best combinations (line 13).

Figure 3 provides an example of the term ranker in the first stage ($s = 1$) that aims at ranking T_0.

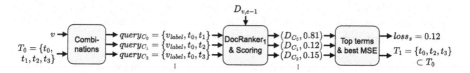

Fig. 3. Example of the term ranker, where the set of terms from the candidate term extraction $T_{v,0}$ consists of 4 terms. After combining them to create several queries, their MSE scores are computed to identify the MSE score of the best query ($loss_s$) and a new set of relevant terms ($T_{v,1}$).

Flexible Number of Terms. To tackle question *(Q1)* in Sect. 1, we determine the optimal number of expansion terms for each query. If the best MSE (i.e., $loss_s$) does not improve over the best MSE in the previous stage ($loss_{s-1}$), we do not expand the query further. Instead, we immediately proceed to the final stage.

3.4 Final Document Ranker

Since BERT-based models have consistently outperformed traditional document ranking techniques [38], we use a BERT-based neural ranking model for the final document ranker $DocRanker_{n_s}$ (line 13 in Algorithm 1). It computes the interaction between the query $q_{expanded}$ and the document $d \in D_v$ using the following input encoding: $BERT([CLS]q_{expanded}[SEP]d[SEP])$.

To compute the final document ranker's loss, we use the cross-entropy loss function as follows:

$$loss_{n_s}(v) = -\frac{1}{|D_v|} \sum_{d \in D_v} (d_y \log(d_{pred}) + (1 - d_y) \log(1 - d_{pred})) \qquad (4)$$

where d_y is the ground-truth judgement of document d and d_{pred} is its score predicted by $DocRanker_{n_s}$.

After training, to apply $MusQuE$ on a single query event v, each non-final document ranker is used exactly once and fed with $D_{v,0}$. $D_{v,e}$ generated by $DocRanker_{n_s}$ is the resulting ranking of documents.

3.5 Efficiency Sampling

As expressed by question (Q3), we need to ensure the scalability of $MusQuE$'s training procedure in spite of its multi-stage nature and the combination of terms. Therefore, we employ the following sampling strategies:

- As described in Sect. 3.2, we limit the size of the initial terms set $T_{v,0}$.
- For each non-final stage, we create a random sample of training instances (query, positive document, negative document) for each epoch while ensuring that the entire dataset is seen before proceeding to the final stage.
- Instead of training on all possible combinations of terms as in Eq. 2, we train on a randomly selected, reduced set of combinations while making sure that each term appears in at least n combinations.

These sampling strategies alleviate computational demands in the early stages, with the primary complexity shifted to the final stage.

4 Evaluation Setup

In this section, we introduce our new event-specific document ranking dataset $MS\text{-}MARCO_{EVENT}$, baselines and our training configuration.

4.1   Dataset: *MS-MARCO$_{EVENT}$*

MS-MARCO$_{EVENT}$[7] is our new dataset for event-specific document ranking that contains events connected to multiple related and unrelated documents. *MS-MARCO$_{EVENT}$* is derived from EventKG, a multilingual event-centric knowledge graph [12], and *MS-MARCO* [36]. *MS-MARCO* contains a set of natural-language questions, and related and unrelated documents for each question.

To create *MS-MARCO$_{EVENT}$*, we proceed as follows:

1. Event collection: We collect events and their English labels from EventKG, including aliases (e.g., "World War 2" and "WWII").[8]
2. Query collection: We collect event-related queries by searching *MS-MARCO* for the event labels.
3. Document collection: For each event and its queries, we collect their related and unrelated documents from *MS-MARCO*[9].

MS-MARCO$_{EVENT}$ contains over $600,000$ unique documents associated with 556 events via $4,671$ verified queries.

As an example of our dataset creation process, consider the event *World War II*. We collect and verify queries such as *"What event led to the outbreak of the Second World War"* and *"What effect did World War II have on Germany"* from *MS-MARCO* and collect their documents.

4.2   Baselines

Table 1 shows the 9 baselines we compare with *MusQuE*. While $SED_{BM25,10}$ and $LDA\text{-}Bol_{KL}$ use their original document ranking models, we further investigate how they perform when using the same document ranking model as *MusQuE*. $MusQuE_{simple}$ is a simplified version of *MusQuE* without using multiple stages and joint training.

4.3   Training Configuration and Efficiency

To train *MusQuE*, we divide the *MS-MARCO$_{EVENT}$* into a train and test set with an 80:20 split and use the Adam optimiser [26] with an experimentally determined learning rate of 0.001. We set the initial term size T_0 described in Sect. 3.2 to 15 and use $n_s = 4$ stages. To balance between losses in the early stages and the final stage, we set the initial value of α to 0.5. As light-weight

---

[7] https://github.com/saraabdollahi/MusQuE/blob/main/Data/Ms-Marco-Event.tsv.

[8] While we extract event labels from EventKG, we do not extract terms regarding other entities than the event itself during the creation of *MS-MARCO$_{EVENT}$*. Thus, there is no unfair interference with *MusQuE*'s candidate terms extraction process described in Sect. 3.2.

[9] Per query, *MS-MARCO* contains one related and $1,000$ unrelated documents found with BM25.

Table 1. Baselines used in our experiments.

Method	Query Expansion based on	Document Ranking	Exp. Terms	Training Corpus
$monoBERT$ [38]	–	monoBERT	0	$MS\text{-}MARCO_{EVENT}$
$IDCM$ [18]	–	DistilBERT	0	MS-MARCO
$ClickBERT$	Top-k clicked entities in the Wikipedia Clickstream [44] previously used in entity recommendation [1,37]	monoBERT	5	$MS\text{-}MARCO_{EVENT}$
$SED_{BM25,10}$ [40]	Wikipedia's tf-idf statistics	BM25	10	–
$SED_{BERT,10}$	same as $SED_{BM25,10}$	monoBERT	10	$MS\text{-}MARCO_{EVENT}$
$SED_{BERT,5}$	same as $SED_{BM25,10}$	monoBERT	5	$MS\text{-}MARCO_{EVENT}$
$LDA\text{-}Bol_{KL}$ [9]	LDA-based topic models [6], Bose-Einstein statistics [3], and DBpedia attributes	Kullback-Leibler divergence	≤ 10	$MS\text{-}MARCO_{EVENT}$
$LDA\text{-}Bol_{BERT}$	same as $LDA\text{-}Bol_{KL}$	monoBERT	≤ 10	$MS\text{-}MARCO_{EVENT}$
$MusQuE_{simple}$	TermRanker$_0$	monoBERT	5	$MS\text{-}MARCO_{EVENT}$

document rankers before the final stage, we employ ColBERT [25]. We train $MusQuE$ and $monoBERT$ using a pre-trained BERT model[10] with 12 hidden layers as in [38] on 5 epochs, which yields the best training performance as shown in Fig. 4.

Under this configuration, the effect of our efficiency sampling (Sect. 3.5) becomes evident as a strategy to tackle *(Q3)*: on 100 queries, the final stage takes 95% of the training time on average. Hence, despite adding additional early stages, $MusQuE$ is comparable to other BERT-based document ranking methods in terms of efficiency.

5   Evaluation Results

This section examines $MusQuE$'s event-specific document ranking performance.

5.1   Event-Specific Document Ranking Comparison

Table 2 reports the results of $MusQuE$ and the baselines tested on $MS\text{-}MARCO\text{-}_{EVENT}$. $MusQuE$ outperforms all baselines across 6 out of 7 metrics, indicating its effectiveness on event-specific document ranking. Regarding $R@100$, some baselines surpass $MusQuE$, but $MusQuE$ is best regarding $R@10$, demonstrating its superiority regarding top documents. In terms of MRR, $MusQuE$ achieves a score of 0.299, outperforming all the baselines by 9.1% and more.

---

[10] https://huggingface.co/bert-base-uncased.

Table 2. Evaluation of *MusQuE* and the baselines on *MS-MARCO$_{EVENT}$* regarding Mean Reciprocal Rank (MRR), Mean Average Precision (MAP), Recall (R) and normalised Discounted Cumulative Gain (nDCG). All metrics are between 0 and 1 and higher scores denote better performance.

Method	MRR	MAP		R		nDCG	
		@10	@100	@10	@100	@10	@100
monoBERT	0.268	0.236	0.208	0.297	0.731	0.213	0.33
IDCM	0.242	0.218	0.189	0.211	0.477	0.167	0.232
ClickBERT	0.256	0.219	0.181	0.333	0.734	0.216	0.326
SED$_{BM25,10}$	0.114	0.095	0.087	0.095	0.390	0.080	0.154
SED$_{BERT,10}$	0.272	0.249	0.203	0.353	0.714	0.240	0.336
SED$_{BERT,5}$	0.274	0.240	0.202	0.236	0.720	0.237	0.335
LDA-Bol$_{KL}$	0.081	0.098	0.034	0.075	0.373	0.001	0.009
LDA-Bol$_{BERT}$	0.233	0.210	0.172	0.285	0.686	0.197	0.297
MusQuE$_{simple}$	0.238	0.212	0.180	0.333	**0.747**	0.211	0.323
MusQuE	**0.299**	**0.259**	**0.238**	**0.371**	0.715	**0.251**	**0.343**

Among the baselines, *SED$_{BERT}$* achieves the highest scores across the metrics. Training *SED* and *LDA-Bol* on *MS-MARCO$_{EVENT}$* improves their performance over their original implementations, which highlights the superiority of BERT-based approaches in capturing document and query representations. *MusQuE* also clearly outperforms *MusQuE$_{simple}$*, which emphasises the effect of performing multi-stage and joint training as demanded by *(Q2)*.

We also observe that *ClickBERT* and *LDA-Bol$_{BERT}$* perform worse than *monoBERT*. This is specifically interesting since these three baselines use the same document ranking model (monoBERT). While *monoBERT*'s document ranker uses the query event label alone, *ClickBERT*'s and *LDA-Bol$_{BERT}$*'s document rankers use expanded queries. Query expansion alone does not necessarily yield better rankings; proper strategies as by *MusQuE* are required.

5.2   Analysis of Number of Terms

To evaluate the effect of our flexible query expansion strategy (Sect. 3.3), we compare *MusQuE* to versions with fixed numbers of 1 to 10 expansion terms. Figure 5 presents the results. For the fixed number of terms, $MAP@10$ initially increases with more terms, peaking at 4. The flexible approach clearly outperforms this configuration since it generates a better selection of terms and its ranking module is optimised for flexible numbers of terms.

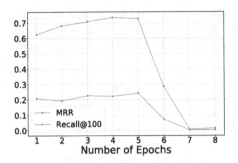 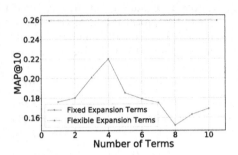

Fig. 4. *MusQuE*'s training performance in *MRR* and *Recall@*100 across epochs.

Fig. 5. *MusQuE*'s performance: flexible vs. fixed expansion terms.

5.3 Anecdotal Results

Table 3 shows the expansion terms found by *MusQuE* and three baselines for the example event *Battle of Waterloo*. *MusQuE* selects few but highly relevant entities as expansion terms such as *Waterloo*[11] and *United Kingdom of the Netherlands*, confirming our flexible query expansion term selection approach demanded by *(Q1)*. Also, *ClickBERT* extracts highly relevant but very specific terms such as the commanders involved in the battle. $SED_{BM25,10}$ and $LDA\text{-}Bol_{KL}$ extract terms, not entities: $SED_{BM25,10}$ selects the cities *Malplaquet* and *Rolica* associated with battles that happened long before the *Battle of Waterloo*, and $LDA\text{-}Bol_{KL}$ selects generic terms such as *Area* and *Center*. This comparison demonstrates the benefits of extracting expansion terms from knowledge graphs. We also see that *MusQuE* extracts only a small number of terms, but they are selected to perform well along the jointly trained document ranker.

Table 3. Expansion terms for the query event *Battle of Waterloo*.

Method	Expansion Terms
ClickBERT	Arthur Wellesley 1st Duke of Wellington, Napoleon, Waterloo (Belgium), Hundred Days, Gebhard Leberecht von Blücher
$SED_{BM25,10}$	Napoleonic, Malplaquet, Vimeiro, Jemappes, Dennewitz, Rolica, Hindoostan, Bussaco, Napoleon, Lauffeld
$LDA\text{-}Bol_{KL}$	Area, State, French, German, Battle, Ypres, British, Center, Kingston
MusQuE	Prussia, Waterloo, United Kingdom of the Netherlands

---

[11] *Waterloo* is already part of the initial query (*Battle of Waterloo*). With BERT's positional embeddings, multiple occurrences in different positions can make a difference.

6   Related Work

Query expansion and document ranking involve various techniques such as pseu-do-relevance feedback, word embeddings, semantic and BERT-based approaches.

Pseudo-relevance feedback (PRF), introduced by Azad et al. [4] involves extracting additional terms from top-ranked documents. Other related methods have later integrated distinctive techniques, including term classification [7], local context analysis [33,46], summarization techniques [29], query categorisation by Wikipedia [47] and fuzzy logic [43]. Another line of research leverages word embeddings such as Word2Vec [35] for query expansion [11,28,41]. Some methods incorporate PRF with word embeddings [49,50] while others employ deep learning approaches [21] to improve the results.

As *MusQuE*, several query expansion methods focus on semantic query expansion to expand the queries using knowledge graphs to improve document ranking results [10,27,45]. LES [32] projects documents and queries to a high-dimensional latent entity space. Rosin et al. [40] propose an event-driven query expansion approach utilizing Wikipedia2Vec embeddings to identify candidate expansion terms which are semantically related to the events. Zong et al.'s PFC approach [52] measures both the global and local importance of expansion candidates extracted from RDF documents. A fuzzy ontology constructed from domain-specific knowledge is used by Jain et al. [22]. Distribution techniques and topic modelling on DBpedia attributes are used by Dahir et al. [9].

BERT [24] has significantly impacted document ranking. Nogueira et al. [38] first fine-tuned BERT for document ranking, leveraging its powerful representation learning. Since then, several Neural ranking methods [2,15–19,30,39,48] have used BERT and its variations [23,25,34,42].

7   Conclusion and Future Work

In this paper, we introduced *MusQuE*, a novel event-specific document ranking method using an event knowledge graph and query expansion. *MusQuE* incorporates event information from the knowledge graph as candidate terms, ranking them alongside documents in a flexible multi-stage framework. We compared *MusQuE* against baseline methods, demonstrating a minimum 9.1% improvement in *MRR*. We also highlighted the impact of flexible term selection and showed that *MusQuE* outperforms the baselines in selecting highly relevant terms, resulting in superior document ranking performance.

For future work, we plan to expand *MS-MARCO$_{EVENT}$* to cover an even wider range of events and explore different features and modalities such as images to represent the query event. Additionally, we aim to investigate to which extent our approach of combining pre-trained, light-weight and complex ranking models can be applied to other scenarios and also on further language models.

Acknowledgements. This work was partially funded by the Federal Ministry for Economic Affairs and Climate Action (BMWK), Germany ("ATTENTION!", 01MJ22012D).

References

1. Abdollahi, S., Gottschalk, S., Demidova, E.: LaSER: language-specific event recommendation. J. Web Seman. **75**, 100759 (2023)
2. Althammer, S., Hofstätter, S., Sertkan, M., Verberne, S., Hanbury, A.: PARM: a paragraph aggregation retrieval model for dense document-to-document retrieval. In: Hagen, M., et al. (eds.) ECIR 2022. LNCS, vol. 13185, pp. 19–34. Springer, Cham (2022). https://doi.org/10.1007/978-3-030-99736-6_2
3. Amati, G., Van Rijsbergen, C.J.: Probabilistic models of information retrieval based on measuring the divergence from randomness. ACM Trans. Inf. Syst. **20**(4), 357–389 (2002)
4. Azad, H.K., Deepak, A.: Query expansion techniques for information retrieval: a survey. Inf. Process. Manage. **56**(5), 1698–1735 (2019)
5. Bhogal, J., MacFarlane, A., Smith, P.: A review of ontology-based query expansion. Inf. Process. Manage. **43**(4), 866–886 (2007)
6. Blei, D.M., Ng, A.Y., Jordan, M.I.: Latent dirichlet allocation. J. Mach. Learn. Res. **3**, 993–1022 (2003)
7. Cao, G., Nie, J.Y., Gao, J., Robertson, S.: Selecting good expansion terms for pseudo-relevance feedback. In: Proceedings of the 31st International ACM SIGIR Conference on Research and Development in Information Retrieval, pp. 243–250 (2008)
8. Croft, W.B., Harper, D.J.: Using probabilistic models of document retrieval without relevance information. J. Documentation **35**(4), 285–295 (1979)
9. Dahir, S., El Qadi, A.: A query expansion method based on topic modeling and DBpedia features. Int. J. Inf. Manage. Data Insights **1**(2), 100043 (2021)
10. Dalton, J., Dietz, L., Allan, J.: Entity query feature expansion using knowledge base links. In: Proceedings of the 37th international ACM SIGIR Conference on Research and Development in Information Retrieval, pp. 365–374 (2014)
11. Diaz, F., Mitra, B., Craswell, N.: Query expansion with locally-trained word embeddings. arXiv preprint arXiv:1605.07891 (2016)
12. Gottschalk, S., Demidova, E.: EventKG+TL: creating cross-lingual timelines from an event-centric knowledge graph. In: Gangemi, A., et al. (eds.) ESWC 2018. LNCS, vol. 11155, pp. 164–169. Springer, Cham (2018). https://doi.org/10.1007/978-3-319-98192-5_31
13. Gottschalk, S., Demidova, E.: EventKG-the hub of event knowledge on the web-and biographical timeline generation. Seman. Web **10**(6), 1039–1070 (2019)
14. Gottschalk, S., Demidova, E.: HapPenIng: happen, predict, infer—event series completion in a knowledge graph. In: Ghidini, C., et al. (eds.) ISWC 2019. LNCS, vol. 11778, pp. 200–218. Springer, Cham (2019). https://doi.org/10.1007/978-3-030-30793-6_12
15. Hofstätter, S., Althammer, S., Schröder, M., Sertkan, M., Hanbury, A.: Improving efficient neural ranking models with cross-architecture knowledge distillation. arXiv preprint arXiv:2010.02666 (2020)
16. Hofstätter, S., Khattab, O., Althammer, S., Sertkan, M., Hanbury, A.: Introducing neural bag of whole-words with colBERTer: contextualized late interactions using enhanced reduction. In: Proceedings of the 31st ACM International Conference on Information and Knowledge Management, pp. 737–747 (2022)
17. Hofstätter, S., Lin, S.C., Yang, J.H., Lin, J., Hanbury, A.: Efficiently teaching an effective dense retriever with balanced topic aware sampling. In: Proceedings of the 44th International ACM SIGIR Conference on Research and Development in Information Retrieval, pp. 113–122 (2021)

18. Hofstätter, S., Mitra, B., Zamani, H., Craswell, N., Hanbury, A.: Intra-document cascading: learning to select passages for neural document ranking. In: Proceedings of the 44th International ACM SIGIR Conference on Research and Development in Information Retrieval, pp. 1349–1358 (2021)

19. Hofstätter, S., Zamani, H., Mitra, B., Craswell, N., Hanbury, A.: Local self-attention over long text for efficient document retrieval. In: Proceedings of the 43rd International ACM SIGIR Conference on Research and Development in Information Retrieval, pp. 2021–2024 (2020)

20. Hu, X., et al.: Event detection in online social network: methodologies, state-of-art, and evolution. Comput. Sci. Rev. **46**, 100500 (2022)

21. Imani, A., Vakili, A., Montazer, A., Shakery, A.: Deep neural networks for query expansion using word embeddings. In: Azzopardi, L., Stein, B., Fuhr, N., Mayr, P., Hauff, C., Hiemstra, D. (eds.) ECIR 2019. LNCS, vol. 11438, pp. 203–210. Springer, Cham (2019). https://doi.org/10.1007/978-3-030-15719-7_26

22. Jain, S., Seeja, K., Jindal, R.: A fuzzy ontology framework in information retrieval using semantic query expansion. Int. J. Inf. Manag. Data Insights **1**(1), 100009 (2021)

23. Jiao, X., et al.: TinyBERT: distilling BERT for natural language understanding. arXiv preprint arXiv:1909.10351 (2019)

24. Devlin, J., Chang, M.W., Lee, K., Toutanova, K.: BERT: Pre-training of deep bidirectional transformers for language understanding. In: Proceedings of NAACL-HLT, pp. 4171–4186 (2019)

25. Khattab, O., Zaharia, M.: ColBERT: efficient and effective passage search via contextualized late interaction over BERT. In: Proceedings of the 43rd International ACM SIGIR Conference on Research and Development in Information Retrieval, pp. 39–48 (2020)

26. Kingma, D.P., Ba, J.: Adam: a method for stochastic optimization. In: 3rd International Conference on Learning Representations, ICLR 2015 (2015)

27. Krishnan, A., P., D., Ranu, S., Mehta, S.: Leveraging semantic resources in diversified query expansion. World Wide Web **21**(4), 1041–1067 (2017). https://doi.org/10.1007/s11280-017-0468-7

28. Kuzi, S., Shtok, A., Kurland, O.: Query expansion using word embeddings. In: Proceedings of the 25th ACM International on Conference on Information and Knowledge Management, pp. 1929–1932 (2016)

29. Lam-Adesina, A.M., Jones, G.J.: Applying summarisation techniques for term selection in relevance feedback. In: Proceedings of the 24th international ACM SIGIR Conference on Research and Development in Information Retrieval, pp. 1–9 (2001)

30. Li, C., Yates, A., MacAvaney, S., He, B., Sun, Y.: PARADE: passage representation aggregation for document reranking. ACM Trans. Inf. Syst. **42**, 1–26 (2020)

31. Li, Z., et al.: Future event prediction based on temporal knowledge graph embedding. Comput. Syst. Sci. Eng. **44**(3), 2411 (2023)

32. Liu, X., Fang, H.: Latent entity space: a novel retrieval approach for entity-bearing queries. Inf. Retrieval J. **18**(6), 473–503 (2015)

33. Lv, Y., Zhai, C.: Positional relevance model for pseudo-relevance feedback. In: Proceedings of the 33rd International ACM SIGIR Conference on Research and Development in Information Retrieval, pp. 579–586 (2010)

34. MacAvaney, S., Nardini, F.M., Perego, R., Tonellotto, N., Goharian, N., Frieder, O.: Efficient document re-ranking for transformers by precomputing term representations. In: Proceedings of the 43rd International ACM SIGIR Conference on Research and Development in Information Retrieval, pp. 49–58 (2020)

35. Mikolov, T., Chen, K., Corrado, G., Dean, J.: Efficient estimation of word representations in vector space. arXiv preprint arXiv:1301.3781 (2013)

36. Nguyen, T., et al.: MS MARCO: a human generated machine reading comprehension dataset. In: CoCo@ NIPS (2016)

37. Ni, C.C., Sum Liu, K., Torzec, N.: Layered graph embedding for entity recommendation using Wikipedia in the yahoo! knowledge graph. In: Companion Proceedings of the Web Conference, pp. 811–818 (2020)

38. Nogueira, R., Cho, K.: Passage re-ranking with BERT. arXiv preprint arXiv:1901.04085 (2019)

39. Nogueira, R., Yang, W., Cho, K., Lin, J.: Multi-stage document ranking with BERT. arXiv preprint arXiv:1910.14424 (2019)

40. Rosin, G.D., Guy, I., Radinsky, K.: Event-driven query expansion. In: Proceedings of the 14th ACM International Conference on Web Search and Data Mining, pp. 391–399 (2021)

41. Roy, D., Paul, D., Mitra, M., Garain, U.: Using word embeddings for automatic query expansion. arXiv preprint arXiv:1606.07608 (2016)

42. Sanh, V., Debut, L., Chaumond, J., Wolf, T.: DistilBERT, a distilled version of BERT: smaller, faster, cheaper and lighter. arXiv preprint arXiv:1910.01108 (2019)

43. Singh, J., Sharan, A.: A new fuzzy logic-based query expansion model for efficient information retrieval using relevance feedback approach. Neural Comput. Appl. **28**(9), 2557–2580 (2016). https://doi.org/10.1007/s00521-016-2207-x

44. Wikimedia Analytics (2021). https://meta.wikimedia.org/wiki/Research: Wikipedia. Accessed 30 Jun 2023

45. Xiong, C., Callan, J.: Query expansion with freebase. In: Proceedings of the 2015 International Conference on the Theory of Information Retrieval, pp. 111–120 (2015)

46. Xu, J., Croft, W.B.: Improving the effectiveness of information retrieval with local context analysis. ACM Trans. Inf. Syst. **18**(1), 79–112 (2000)

47. Xu, Y., Jones, G.J., Wang, B.: Query dependent pseudo-relevance feedback based on Wikipedia. In: Proceedings of the 32nd International ACM SIGIR Conference on Research and Development in Information Retrieval, pp. 59–66 (2009)

48. Yan, M., et al.: IDST at TREC 2019 deep learning track: deep cascade ranking with generation-based document expansion and pre-trained language modeling. In: TREC (2019)

49. Zamani, H., Croft, W.B.: Embedding-based query language models. In: Proceedings of the 2016 ACM International Conference on the Theory of Information Retrieval, pp. 147–156 (2016)

50. Zamani, H., Croft, W.B.: Estimating embedding vectors for queries. In: Proceedings of the 2016 ACM International Conference on the Theory of Information Retrieval, pp. 123–132 (2016)

51. Zhang, H., Boons, F., Batista-Navarro, R.: Whose story is it anyway? Automatic extraction of accounts from news articles. Inf. Process. Manage. **56**(5), 1837–1848 (2019)

52. Zong, N., Lee, S., Kim, H.G.: Discovering expansion entities for keyword-based entity search in linked data. J. Inf. Sci. **41**(2), 209–227 (2015)

Two-Step SPLADE: Simple, Efficient and Effective Approximation of SPLADE

Carlos Lassance[1]($^{(\boxtimes)}$) (ID), Hervé Dejean[1] (ID), Stéphane Clinchant[1] (ID), and Nicola Tonellotto[2] (ID)

[1] Naver Labs Europe, Meylan, France
carlos@cohere.com
[2] University of Pisa, Pisa, Italy

Abstract. Learned sparse models such as SPLADE have successfully shown how to incorporate the benefits of state-of-the-art neural information retrieval models into the classical inverted index data structure. Despite their improvements in effectiveness, learned sparse models are not as efficient as classical sparse model such as BM25. The problem has been investigated and addressed by recently developed strategies, such as guided traversal query processing and static pruning, with different degrees of success on in-domain and out-of-domain datasets. In this work, we propose a new query processing strategy for SPLADE based on a two-step cascade. The first step uses a pruned and reweighted version of the SPLADE sparse vectors, and the second step uses the original SPLADE vectors to re-score a sample of documents retrieved in the first stage. Our extensive experiments, performed on 30 different in-domain and out-of-domain datasets, show that our proposed strategy is able to improve mean and tail response times over the original single-stage SPLADE processing by up to 30× and 40×, respectively, for in-domain datasets, and by 12× to 25×, for mean response on out-of-domain datasets, while not incurring in statistical significant difference in 60% of datasets.

1 Introduction

Learned Sparse Retrieval (LSR) models [10, 11, 19, 22, 33] aim at combining the best of two worlds: the traditional search infrastructure, based on an inverted index of interpretable terms, and the representation power of Pretrained Language Models (PLMs) [8]. Such models recompute term weights for documents and queries to improve effectiveness, going as far as learning how to expand texts to be even more effective in IR tasks. LSR underpinning hypothesis is that existing search infrastructure, namely the inverted index and its efficient algorithms can easily be used to serve such models [30]. However, a mismatch exists between the posting lists score distribution of LSR models and traditional models like BM25, leading to inefficiency issues [18]. One could argue that first-stage retrieval for bag-of-word models has been heavily optimized for years leading to better algorithms such as MaxScore [31], WAND [2], and Block-Max WAND [9]; and that LSR models would need better optimisations when used with a traditional inverted index. Therefore, recent works based on Guided Traversal (GT)

© The Author(s), under exclusive license to Springer Nature Switzerland AG 2024
N. Goharian et al. (Eds.): ECIR 2024, LNCS 14609, pp. 349–363, 2024.
https://doi.org/10.1007/978-3-031-56060-6_23

[20,24] have adapted existing algorithms to improve the latency of LSR models, where the main goal is to reuse BM25 to guide the selection of scored documents during the retrieval phase.

In this work, we address the same research question: given the mismatch between LSR models and traditional search algorithms, how can we better use the existing search algorithms and how can we improve the latency of LSR models. While several learned sparse retrieval works have been proposed, we focus this work on SPLADE [11], due to its popularity and effectiveness. Our goal is to increase the flexibility of the end user (directly be able to modulate efficiency/effectiveness at retrieval time), while limiting the amount of change needed on the overall system for the provider of the search engine.

Following recent works, where LSR is used as first stage retriever in a multi-stage ranking pipeline [19], the main goal of this work is to further split the first stage retrieval in two parts. More precisely, our method relies on the observation that SPLADE sparse vectors can be approximated by sparser vectors obtained with top pooling. Moreover, based on the discussion in [12] about saturation function and dynamic pruning, we add a term re-weighting to the SPLADE scoring function that improves the efficiency of dynamic pruning for SPLADE vectors via a trade-off with effectiveness[1]. In other words, a very good approximation of a sparse vector is *a sparser re-weighted version* of this vector. Therefore, we can first compute the approximated results with sparser vectors, extract a top k sample and then only perform the full score computation with the *original vectors* in this sample. Note that this is similar to what is done in an approximated nearest neighbor for dense retrieval, leading to an approximated nearest neighbor for sparse retrieval models.

Overall, the contributions of this paper are the following:

- We show that SPLADE first-stage retrieval can be approximated by a two-step algorithm relying on sparser term re-weighted SPLADE vectors.
- Our approximation allows us to propose new ranking models as efficient as GT, i.e., between 12× to 40× faster than SPLADE but more effective than GT, i.e., with statistically significant gains in 50% of the tested datasets and without statistically significant losses in 87% of the tested datasets.

The remaining of this paper is organised as follows: Sect. 2 discusses the related work, Sect. 3 illustrates our Two-Step SPLADE, Sect. 4 presents our experiments, and Sect. 5 reports our conclusions.

2   Related Works

The re-usability hypothesis of LSR models, i.e., the hypothesis that learned sparse models may be seamlessly adopted with inverted indexes, has been a source of debate. In [18], the authors showed that while these new methods may be used in the current infrastructure, they are not at all optimized for it, with

---

[1] C.f. right part of Fig. 3.

a large decrease in efficiency. This led to an influx of recent works [3,14,20,24] that look into achieving cost/latency parity with BM25. Nevertheless, there often exists a performance drop or robustness issue to achieve that result. We separate these works by how they deal with dynamic pruning [30]. In the first line of work (a), the LSR model is adapted to the current search algorithm, while, in the second case (b), the dynamic pruning/search implementation is adapted to the models.

Adapted Models (a): Efficient SPLADE [14] propose a dedicated way to train more efficient models with L1 regularization on the query side and better pre-training while in LSR pruning [15] standard pruning techniques are applied to LSR models and demonstrate efficiency gains. Even if such works demonstrate latency figures comparable to BM25, such methods still suffer from losses on out-of-domain data, i.e, zero shot on BEIR benchmark. We note that there exists a vast literature on static pruning that we could draw from [1,5–7] but we focus on the simpler methods from [15] that have already been shown to work well with SPLADE.

Adapted Search Mechanism (b): Term Impact Decomposition [17] showed that splitting posting lists and completely redesigning retrieval leads to greatly improved efficiency without any effectiveness cost. Guided Traversal (GT) [20] first showed how to use the BM25 scores to choose which documents to score, but it requires posting lists to be shared between BM25 and the LSR method, which excludes most LSR methods. This constraint is later relaxed in Optimized GT (OGT) [24], which makes GT available to any LSR method, by changing from purely guidance as in GT to parametrized guidance, where weights are balanced between BM25 and LSR methods. OGT is further optimized for SPLADE in [23], by adding a stricter pruning scheme during training based on soft and hard thresholds for documents and queries, which require specific model training on top of using OGT.

Beyond this dichotomy, other works aim to improve the effectiveness/efficiency trade-off. The GAR [16] approach relies on graphs in document space and mixing the efficiency of sparse retrieval, e.g. BM25, and the effectiveness of cross-encoder rerankers, but still completely changes the architecture of retrieval and increases computational cost due to the use of cross-encoder rerankers. In the case of lexically-accelerated dense retrieval (LADR) [13], which combines GAR and GT by using BM25 as a first-stage for dense retrieval and are thus restricted by the choice of BM25 as the "first-stage". Finally, there are sketching and clustering works such as [3,4] which convert the sparse retriever into a dense retriever thanks to hashing, which is a completely different paradigm to the one we analyze here and would not let us continue to use the same architecture for retrieval.

3 Two-Step SPLADE

Given recent results [14,15,24], a prominent way to improve SPLADE latency has been to reduce the number of query terms, then by pruning document terms or by using GT. However, for very fast models ($\approx$ BM25 latency), there is actually a sharp drop in effectiveness, especially for out-of-domain experiments[2]. Ideally, one would like to have at the same time i) fast retrieval; and ii) out-of-domain effectiveness.

To do so, we split our first-stage SPLADE ranking into two steps: **approximate** and **rescoring**. In the first step, both queries and documents are pruned or compressed aggressively to yield an efficient approximation to the original SPLADE model. Then, in the rescoring step[3], the initial top k documents retrieved during the first step are rescored using the full documents and queries from the original SPLADE model.

This is indeed how most systems are implemented, with a "first-stage" and "second-stage" retrieval, where most of the time the first-stage is a vector-based retrieval and the second stage a cross-encoder. In other words, we actually propose to split the first-stage into two steps: the first step is a rough approximation of the representations (aggressive document and query pruning associated with term re-weighting) and the second step using the actual representations, although in a small subset of the original corpus. We illustrate the method in Fig. 1.

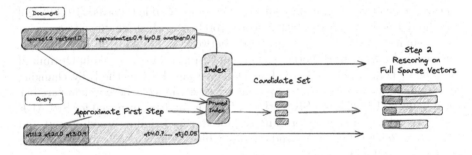

Fig. 1. Two-Step SPLADE

Indexing. Starting from a trained SPLADE model, we use a top pooling as in [15,22,28] to prune both documents and queries to the average size of the dataset (with upper limits of 128 and 32 tokens). For example, in the case of MSMARCO, we prune documents to the 50 tokens with the highest SPLADE-weights and queries to the highest 5. In other words, we rely on the observation that SPLADE sparse vectors can be approximated by sparser vectors based on the highest weights outputted by SPLADE. Note that hashing [3] could in theory be used here to convert embeddings to a dense representation and perform classical dense retrieval ANN on it [4] but that is outside the considerations of

---

[2] Cf experimental section.

[3] We prefer to use rescoring instead of reranking as it may confuse the reader with the cross-encoder reranking case.

this work. Moreover, in GT [20,24], i) the approximation step requires paired posting lists, including the same amount of query tokens; and ii) ranking is conducted with BM25 weighting, which is ideal for some datasets (especially ones where the annotation is based on BM25 results [29]), but not for all datasets, as we will see in our experiments. The indexing process is described in Algorithm 1

Algorithm 1. Indexing

Require: T collection, Q collection, preprocessed for SPLADE
1: Compute averages of T and Q and store as l_d and l_q
2: **Initialize** I^a as an empty index // Approximate index
3: **Initialize** I^r as an empty index // Rescoring index
4: **for** each document d in T **do**
5:    $d' = $ Prune d to size l_d by selecting the highest values
6:    Add d' to I^a; Add d to I^r
7: **end for**

Retrieval. The retrieval process is divided into two steps: approximate and rescoring, and it is described in Algorithm 2.

Algorithm 2. Two-step Retrieval Algorithm

Require: Indexes I^a and I^r, a query q, an average query size l_q, a saturation factor k_1 and the amount of documents to return k
1: $q' = $Prune q to size l_q by selecting the highest values
2: **scores**$_{d'}$, **D** = **SearchForTopkDocuments**$(q', I^a, k, k_1, $ null$)$ // Approximate step: Search on smaller index with smaller query, k_1 saturation, and without filters
3: **scores**$_d$, **D** = **SearchForTopkDocuments**(q, I^r, k, ∞, D) // Rescoring step: Search on larger index, without saturation, and with larger query while filtering to include only elements in D
4: **Return scores**$_d$ and **D**

Approximate Step: we first rely on pruning, but also add a term-reweighting function. The main motivation for adding such a function is, as we will see, to get some speedup for the first step. Indeed, one of the main problems of LSR models like SPLADE is that the posting lists score distribution radically changes compared to BM25, making it harder for dynamic pruning to work effectively. This is due to the absence of the "BM25's saturation effect" which is central, for example, to WAND [2,12].

In order to fix this we propose to reuse the BM25 term weighting, but with SPLADE-based term frequencies[4] in order to re-weight the SPLADE vectors. Note that this creates an additional approximation, as SPLADE itself is not trained for this scoring function. Considering that we have previously pruned

---

[4] The BM25 function would also have IDF, but we only consider the TF here.

the documents, the b correction from BM25 is not necessary, leading to the following simpler formulation:

$$s(q,d) = \sum_{t \in q} B(t,q) \frac{(k_1 + 1)\, TF\,(t,d)}{TF\,(t,d) + k_1} \, , \tag{1}$$

where $s(q,d)$ is the score of document d for query q, t is a term in query q, $TF(t,d)$ is the term frequency/SPLADE weight of term t on document d, $B(t,q)$ is the SPLADE weight of term t on the query q, and k_1 is the saturation parameter. The higher the k_1 parameter, the more important the actual value of TF will actually be, while for smaller values of k_1, large values of TF will not be that important, achieving saturation via reduced difference of top TF values. Balancing k_1 means changing the scoring function from a constant, i.e. $k_1 = 0$, to heavily approximate SPLADE, i.e., $k_1 = 1$), to approximately the original SPLADE (as in the limit $k_1 \approx \infty$ the scoring simplifies to just the query and document term weights), i.e., $k_1 = 10,000$. Note that the more saturated our TFs are, the faster that retrieval will be when using dynamic pruning. We tested a large set of values and choose $k_1 = 100$ as the best trade-off, with b set to 0.

Rescoring Step. In the second step, we rescore each document from the document set, selected by the first step, with the uncompressed, full sparse vectors (both query and document) coming from the initial SPLADE models. In other words, we are basically using the full SPLADE to rescore the top k documents generated in the approximate step. In this work we always use $k = 100$ (c.f. left part of Fig. 3). Note that this rescorer can be easily used in current search engines such as PISA [21] or Anserini [32]. In this work we implement restoring as a full index search that skips by document id using the 'nextgeq'function from PISA, but many implementations are possible. This is different from a SPLADE GuidedTraversal that uses two different indexes for thresholding (BM25) and scoring (SPLADE).

Discussion

The main benefit of our approach is that we can *control* the quality or the speed of the approximation. Instead, GT is limited by the effectiveness of BM25 or requires the introduction of extra parameters to balance BM25 vs SPLADE [24]. We do not change anything on training time compared to [14,23], meaning that our approach could be applied to any out-of-the-shelf LSR. We reuse pruning as suggested by [15] but we also allow for control after the level of pruning is chosen (via term re-weighting and query pruning), without the need to index multiple pruned versions.

The main drawbacks of GT compared to our work are: i) the lack of flexibility of the approximation due to the use of BM25; ii) the need of posting list sharing between steps, making it complicated to use with BERT-based vocabularies and/or different terms between stages, e.g. a shortened query on the first-stage and full query on the second-stage. Compared to GAR [16] our approach has the advantage of not needing a cross-encoding reranker. However, we have the trade-off of indexing and searching on a SPLADE index (that we show is not slower than twice as BM25). Compared to sketching [3,4], our approach has the advantage of staying in the sparse retrieval paradigm and not needing to change current infrastructure.

Finally, we discuss here some limitations of our study: a) while we reach almost parity with BM25 in terms of retrieval efficiency, we do not consider the cost of the query encoder, which would be closer to the dense retrievers from LADR [13] and smaller than the cross encoding rerankers of GAR [16], but still large compared to retrieval cost; b) Our method also still requires some choice of parameters, such as the static and dynamic pruning strategies to employ.

Nonetheless we believe our method has the best trade-offs that we will showcase empirically in the next section. Ideally we desire the following properties: a) keep existing infrastructure intact; b) propose models with at most 2 times the retrieval latency as BM25; c) do not have a significant (t-test with p-value of 0.01) decrease of effectiveness compared to full SPLADE on at least half of datasets; d) allow for controllability of the trade-off at the end-user level.

Rescoring vs Reranking. One thing that we do not touch in this work is the comparison of rescoring and reranking (with a cross-encoder or a stronger model). Both rescoring and reranking aim at improving the ranking quality via a trade-off of spending more computational power, they do so in different levels of cost. Rescoring only uses precomputed representations, meaning that it can be executed with low cost, while reranking does not normally allow for precomputed information. In other words, in a perfect world, one should be able to combine both (e.g. rescoring top 1k and reranking top-10) and obtain scores that match costlier rerankings (e.g. directly reranking the top 100 or 1000).

Storage Requirements. One drawback of the proposed two-step SPLADE implementation is that it requires storing two inverted-indexes: the full SPLADE index and the pruned SPLADE index. Compared to a SPLADE approach this is a small overhead as the pruned SPLADE indexes are in general much smaller than the full index, but it is still a nonzero difference. Compared to a SPLADE with Guided Traversal, the difference is almost negligible with the pruned SPLADE index taking around the same space as a BM25 index.

Effectiveness Metric (Recall vs nDCG). In the following we do not report Recall values, only nDCG. We prefer to report nDCG for two reasons: 1) Easier to use across a multitude of datasets with different granularities of annotation; 2) A better recall does not mean that future reranking will be easier, especially with the current methods reranking methods that have similar biases to the first-stage.

4   Experiments

In our experiments, we want to demonstrate several findings:

1. The validity of our approximation: we study several pruning and term re-weighting and show that our first-stage approximation is more accurate than BM25;
2. Rescoring is a very efficient and effective technique to reduce the gap between the approximation and the full retrieval;
3. Two-Step SPLADE (approximation + rescoring) is efficient and effective not only for in-domain experiments but also for maintaining its performance on zero-shot benchmarks (BEIR and LoTTe).

Search Engine and Dynamic Pruning. All of our experiments are done with the PISA [21] search engine. PISA has different implementations of dynamic pruning algorithms and choosing which one to use is an important aspect of any system. We verify this by testing all our methods with the three most used algorithms: WAND [2], Block-Max WAND [9] and MaxScore [31]. In the case of BM25 retrieval differences were small, with a slight advantage to MaxScore, while for original SPLADE (i.e. no pruning, no approximation), the best performing algorithm was also MaxScore which was twice as fast as the next one. However, pruning the SPLADE model, made Block-Max WAND was the most efficient. Finally, by adding term re-weighting on top of pruning, WAND becomes the most efficient. Due to the differences in these algorithms, we decided, in this work, to always report the result for the most efficient algorithm of the three. Code is available at https://github.com/naver/splade/.

Datasets. We consider most datasets available in the literature to conduct our benchmark on effectiveness and efficiency. More importantly, we want to assess the out-of-domain effectiveness performance drop that most approximation meth-ods incur. In order to evaluate this, we compare methods on **30 datasets**[5]. Finally, note that for computing statistical significance we drop datasets that are composed of other datasets such as "Pooled" from LoTTe and CQADup-Stack from BEIR. Note that BEIR is heavily biased towards BM25, as noted on their own paper [29], and thus we added LoTTe [25] exactly due to the fact that BM25 not being as effective. Ideally, a system should work on both.

Secondly, we use the Ranger library [26] to detect statistically significant changes ($p \leq 0.01$). In this case, we **count** the number of datasets for which a method is statistically significantly better, similar to a meta-analysis process [27].

Models. Finally, our study is based on our own SPLADE models as done in related works [14,24]. In fact, the public SPLADE checkpoints are not state-of-the-art anymore. Therefore, we adopt the SPLADE++ Self-Distil and train for

---

[5] 3 from MSMARCO, being the MSMARCO-dev, TREC-DL 19 and TREC-DL 20, 17 from BEIR [29] and 10 from LoTTe [25].

slightly more time and with better distillation scores by adding L1 regularization on the query side as it was done to generate the baseline SPLADE in [14]. We see indeed a gain in effectiveness and thus use that as our baseline "SPLADE".

In terms of baselines, we compare our Two-Step SPLADE to a single-step pruned version, following [15], and a version of (Optimized) Guided Traversal [20, 24]. Note that GT (our implementation) is not exactly GT, nor OGT, but something in the middle. Our implementation of GT is based on our SPLADE models and works in two separate phases: approximation BM25, and then rescoring full SPLADE. We do so, because it would not be possible to have aligned posting lists between BM25 and our SPLADE (and thus traverse them concurrently) as in GT, and we did not want to modify the retrieval procedure as in OGT [24] which would require hyperparameter tuning for each dataset and using a different version of PISA compared to our other experiments. This baseline is a best-case scenario in terms of effectiveness for GT+SPLADE, while adding at most a 30% penalty in efficiency when compared to BM25.

Additional Comparisons: EfficientSPLADE [14] is also matched against Two-Step SPLADE, most notably the available EFF-V-Large, but they are omitted from tables to avoid confusion between what SPLADE is being discussed. Finally, we do not compare directly to Term Impact Decomposition [17] and Optimized Guided Traversal for SPLADE [23] because they require changes to the PISA software and SPLADE model, and while codebases are advertised in the respective papers, they were not available at the time of writing. However, we do present comparison numbers against both OGT [23,24], but only as a hint of relative efficiency/effectiveness, as the comparison is not totally fair due to not using the same machine/setting.

4.1   Approximation Validity

We study the validity of the approximation step for SPLADE. To do so, we measure the top k intersection of the approximate step with the original top k from SPLADE. First, choosing the optimal pruning values turns out to be complicated when dealing with a multitude of datasets. Therefore, we will assess our use of a simple heuristic: the original statistics from the dataset. Secondly, for the approximation step we can vary the term re-weighting function in order to play with the efficiency/effectiveness trade-off.

Approximate Step Parameters. During indexing, we can now control the level of pruning for the approximate stage. Ideally, we would want something that already speeds up retrieval substantially, without losing effectiveness. Following [15] this would be around half or a quarter of the original SPLADE document size. In order to check this we look into SPLADE retrieval of MSMARCO-dev and climate-FEVER, and prune documents (V-D) into multiple sizes $(8,16,32,64,128,N/A)$ and also to the lexical size (L-D) of the dataset (50 for MSMARCO and 67 for climate-fever). We also include query pruning

(V-Q) with different values $(5,10,16,N/A)$, with 5 and 16 being the lexical size (L-Q) for MSMARCO and Climate-FEVER. Note that such query pruning was not included in the previous work. Results are presented in Fig. 2.

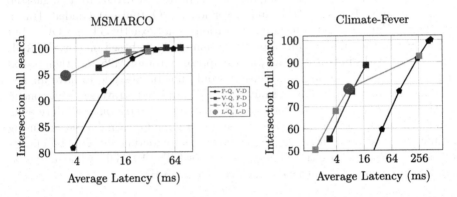

Fig. 2. Intersection of the original retrieval top-10 and the pruned retrieval. We represent the different combinations by: F: No pruning; V: Varying pruning; L: Lexically-pruned. D: Document, Q: Query. So F-Q, V-D means we vary document length and keep the query untouched. (Color figure online)

What we first notice is that the effects are very different depending on the dataset. MSMARCO seems quite easy to prune, with our methods easily keeping more than 90% intersection of the top 10 (i.e. in average 9 out of the top10 documents from the original search are in the top 100 of the approximate search), however, the results are not as good in climate-FEVER. In other words if we selected the values solely using the characteristics of MSMARCO the method would underperform in climate-fever and vice-versa. This effect is what motivates us to test in so many datasets and to look into both efficiency and effectiveness in all of them and not solely in MSMARCO (as many methods do). Nonetheless, a simple heuristic that seemed good enough to us is just *using the average token size* (document and query), which we call lexical size (red dot).

Fixed Pruning, Change k_1. Now that we have chosen the pruning, we can use all our datasets to test different versions of term re-weighting functions by varying k_1. We present the results in Fig. 3, where on the left we look into varying the top k retrieval and k_1 values and on the right we fix $k = 100$ and display the efficiency-effectiveness trade-off of different values of k_1, with the efficiency being measured as the average of average latencies over large BEIR datasets (>1M documents). The results show that the larger k_1 is, the more accurate the approximate is, but the larger k_1 is, the larger the latency is. Therefore, the saturation function allows to easily control the efficiency-effectiveness trade-off. Finally, a top k of 100 and a saturation with $k_1 = 100$ seem to be appropriate parameters, as they achieve around 91% intersection with the "full-scale" retrieval, with a .99 confidence interval between 88% and 94% intersection, while keeping efficiency close to BM25/GT.

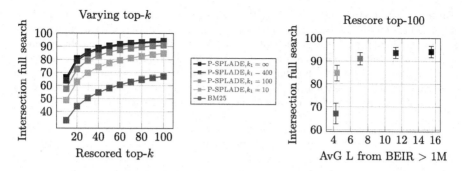

Fig. 3. Intersection of the original retrieval top-10 and the first approximate step. P-SPLADE means the L-Q, L-D version from the previous figure with the applied k_1 saturation. AvG L is the average latency for BEIR datasets that have more than 1M documents.

4.2 Effectiveness and Efficiency Benchmark

After having demonstrated the validity of our approximation, and analysed the parameter's sensitivity, we can now proceed with the full benchmark. We present results by *statistical differences and relative latency to the BM25 baseline* in Table 1, while raw latencies and average effectiveness are presented in Table 2. From these tables of results, we draw the following conclusions:

1. Approximate SPLADE (row e): term re-weighting improves efficiency in out-of-domain datasets compared to the pruned-only version (row c) (1.1 vs 2.3 and 1.4 vs 2.7 on BEIR and LoTTe, but reduces effectiveness.
2. Two Step SPLADE (row f and g): can keep reasonable retrieval efficiency compared to BM25 (less than 2 times average latency increase with the approximate SPLADE) while having similar effectiveness to the full SPLADE (row b). Note that this means a latency improvement of 12x to 40x on full SPLADE depending on the dataset and that in MSMARCO it is even faster than BM25, especially when considering p99 latency.
3. Comparison to previous work (row (c) and (d) GT): we see clear improvements in effectiveness and efficiency. Against GT, TwoStage is strictly superior for 14 out of 30 datasets and only worse on 4, while having similar latencies.

Concerning latencies, on the table we report p99 solely to MSMARCO as averaging p99 does not make sense. For p99 over BEIR our method (row g) is faster than BM25 (row c) in 9 out of the 17 datasets (excluding CQAdupstack) and 3 out of the 10 in LoTTe. We also separate the average latencies into all and just the largest in BEIR to isolate the smaller and larger dataset contributions and show that our method is stable over dataset sizes.

It is also interesting to look into the raw numbers in Table 2. If we only looked into average numbers, and not the statistical differences over a large collection of datasets, conclusions would change. For example, in average terms, there is almost no difference on BEIR between SPLADE, GT and Two-Step. However,

Table 1. Effect size analysis ($p \leq 0.01$) of methods against SPLADE (b) and our implementation of Guided Traversal (GT) (d). $\geq$ is the number of datasets where the row model does not present statistical drop of effectiveness against the column one, $>$ the ones with statistical improvement and $<$ statistical drop. Latency is normalized by BM25 (1.0 is = to BM25) and thus is the lower the better. >1M means the subset of datasets of BEIR that have at least 1M documents on the corpus. AvG L is the average latency of the dataset or the average of dataset averages (for BEIR and LoTTe).

Method	Effect size against		MSMARCO		BEIR		Lotte
	SPLADE (b)	GT (d)	Latency		18 >1M		
	$\geq$ (>) <	$\geq$ (>) <	Average p99		AvG L		AvG L
Baselines							
a BM25	7 (1) 23	7 (1) 23	1.0	1.0	1.0 1.0		1.0
b SPLADE (ours)	N/A	27 (16) 3	19.1	12.4	24.8 32.6		22.1
Advanced Baselines							
c Approx. First Step [15] over (b)	7 (1) 23	16 (2) 14	0.7	0.4	2.3 2.6		2.6
d GT (Our Implementation) ($a \rightarrow b$)	14 (3) 16	N/A	1.1	1.0	1.2 1.2		1.3
This work							
e Approx. First Step (c) with $k_1 = 100$	4 (1) 26	5 (1) 25	0.5	0.3	1.1 1.2		1.6
f Two-Step ($c \rightarrow b$)	22 (0) 8	26 (15) 4	0.8	0.4	2.5 2.7		3.0
g Two-Step ($e \rightarrow b$)	18 (2) 12	26 (14) 4	0.6	0.3	1.4 1.3		1.8

when we add LoTTe to the mix and look into it dataset by dataset, and not as an average of nDCG@10 as per the current usage in most papers, we can see the differences between methods[6].

Additional Comparisons. First, we compare our model to EFF-V-Large [14] (not shown on tables for clarity), where our method is faster (around 8x on MSMARCO) and more effective (0 datasets with statistical significant drop and 22 improvements). Furthermore, we note that we have also done all the previous sets of experiments using a SPLADE that is publically available: SPLADE++ CoCondenser Self-Distil from [10] and have noted similar results (same effectiveness in 17 out of 30 instead of 18), although with higher latencies due to the different statistics of the models. We omit these tables due to lack of space.

Finally, looking into raw values we can be tempted to compare with raw numbers from other papers, such as [23,24]. These papers report a MSMARCO latency of 22 and 6.9ms for retrieving the top 10 documents, while we report 2.4 ms with our method to retrieve the top 100 documents. Looking at zero-shot over BEIR, we can compare the speed-up over the baseline, where our method improves more than 15 times over running the full SPLADE, while [24] improves it by 2.7 times and [23] improves it by 3.6 times. However, due to experiments being run on different machines and base models, these comparisons only hint that our model is better, but cannot be concluded without proper experiments.

---

[6] Due to lack of space, we had to omit the figure showing per-dataset comparison.

Table 2. Efficiency and Effectiveness Benchmark: represented by average statistics and latency in ms instead of reporting relative to a baseline as in Table 1.

Method	MSMARCO		BEIR		Lotte	
	Dev Set		18 datasets		AvG L	Success@5
	AvG L	MRR@10	AvG L	nDCG@10		
Baselines						
a BM25	4.0	18.7	3.2	41.4	1.4	52.9
b SPLADE (ours)	76.3	40.2	78.5	48.4	31.0	70.4
Advanced Baselines						
c Approx. First Step [15] over (b)	3.0	37.2	7.4	45.1	3.7	66.3
d GT (Our Implementation) $(a \rightarrow b)$	2.1	36.7	3.5	48.2	1.8	66.4
This work						
e Approx. First Step (c) with $k_1 = 100$	2.0	35.2	3.6	42.2	2.2	63.1
f Two-Step $(c \rightarrow b)$	3.3	40.1	8.1	48.0	4.2	70.1
g Two-Step $(e \rightarrow b)$	2.4	40.0	4.3	47.6	2.5	70.0

5   Conclusion

In this work, we have shown that by separating the first-stage of SPLADE into two steps we are able to mostly conserve its effectiveness and reduce latency by 12x to 40x. The proposed approach is more flexible than existing approaches while being directly applied to any search engine as a Two-Step search (unfiltered and then filtered to the top k of the first search). This is another step into making LSR models more production-ready, which seems to show that in terms of retrieval latency LSR methods can compete with BM25 without problem. There are however two last barriers due to document/query inference i) indexing time and ii) actual full-scale latency measurements including query inference time, that we do not touch in this work and leave as future work.

Acknowledgements. This work is supported, in part, by the spoke "FutureHPC & BigData" of the ICSC – Centro Nazionale di Ricerca in High-Performance Computing, Big Data and Quantum Computing, the Spoke "Human-centered AI" of the M4C2 - Investimento 1.3, Partenariato Esteso PE00000013 -"FAIR - Future Artificial Intelligence Research", funded by European Union – NextGenerationEU, the FoReLab project (Departments of Excellence), and the NEREO PRIN project funded by the Italian Ministry of Education and Research Grant no. 2022AEFHAZ.

References

1. Altingovde, I.S., Ozcan, R., Ulusoy, Ö.: A practitioner's guide for static index pruning. In: Boughanem, M., Berrut, C., Mothe, J., Soule-Dupuy, C. (eds.) ECIR 2009. LNCS, vol. 5478, pp. 675–679. Springer, Heidelberg (2009). https://doi.org/10.1007/978-3-642-00958-7_65
2. Broder, A.Z., Carmel, D., Herscovici, M., Soffer, A., Zien, J.: Efficient query evaluation using a two-level retrieval process. In: Proceedings of the Twelfth International Conference on Information and Knowledge Management, pp. 426–434 (2003)

3. Bruch, S., Nardini, F.M., Ingber, A., Liberty, E.: An approximate algorithm for maximum inner product search over streaming sparse vectors. arXiv preprint arXiv:2301.10622 (2023)
4. Bruch, S., Nardini, F.M., Ingber, A., Liberty, E.: Bridging dense and sparse maximum inner product search. arXiv preprint arXiv:2309.09013 (2023)
5. Büttcher, S., Clarke, C.L.: A document-centric approach to static index pruning in text retrieval systems. In: Proceedings of the 15th ACM International Conference on Information and Knowledge Management, pp. 182–189 (2006)
6. Carmel, D., Cohen, D., Fagin, R., Farchi, E., Herscovici, M., Maarek, Y.S., Soffer, A.: Static index pruning for information retrieval systems. In: Proceedings of the 24th Annual International ACM SIGIR Conference on Research and Development in Information Retrieval, pp. 43–50 (2001)
7. Chen, R.C., Lee, C.J.: An information-theoretic account of static index pruning. In: Proceedings of the 36th International ACM SIGIR Conference on Research and Development in Information Retrieval, pp. 163–172 (2013)
8. Devlin, J., Chang, M.W., Lee, K., Toutanova, K.: BERT: pre-training of deep bidirectional transformers for language understanding. In: Proceedings of NAACL, pp. 4171–4186 (2019)
9. Ding, S., Suel, T.: Faster top-k document retrieval using block-max indexes. In: Proceedings of the 34th International ACM SIGIR Conference on Research and Development in Information Retrieval, pp. 993–1002 (2011)
10. Formal, T., Lassance, C., Piwowarski, B., Clinchant, S.: From distillation to hard negative sampling: making sparse neural ir models more effective. In: Proceedings of the 45th International ACM SIGIR Conference on Research and Development in Information Retrieval, pp. 2353–2359 (2022)
11. Formal, T., Piwowarski, B., Clinchant, S.: SPLADE: sparse lexical and expansion model for first stage ranking. In: Proceedings of SIGIR, pp. 2288–2292 (2021)
12. Grand, A., Muir, R., Ferenczi, J., Lin, J.: From MAXSCORE to block-max WAND: the story of how lucene significantly improved query evaluation performance. In: Jose, J.M., et al. (eds.) ECIR 2020. LNCS, vol. 12036, pp. 20–27. Springer, Cham (2020). https://doi.org/10.1007/978-3-030-45442-5_3
13. Kulkarni, H., MacAvaney, S., Goharian, N., Frieder, O.: Lexically-accelerated dense retrieval. In: Proceedings of the 46th International ACM SIGIR Conference on Research and Development in Information Retrieval, pp. 152–162 (2023)
14. Lassance, C., Clinchant, S.: An efficiency study for splade models. In: Proceedings of the 45th International ACM SIGIR Conference on Research and Development in Information Retrieval, pp. 2220–2226 (2022)
15. Lassance, C., Lupart, S., Dejean, H., Clinchant, S., Tonellotto, N.: A static pruning study on sparse neural retrievers. arXiv preprint arXiv:2304.12702 (2023)
16. MacAvaney, S., Tonellotto, N., Macdonald, C.: Adaptive re-ranking with a corpus graph. In: Proceedings of the 31st ACM International Conference on Information and Knowledge Management, pp. 1491–1500 (2022)
17. Mackenzie, J., Mallia, A., Moffat, A., Petri, M.: Accelerating learned sparse indexes via term impact decomposition. In: Findings of the Association for Computational Linguistics: EMNLP 2022, pp. 2830–2842. Association for Computational Linguistics, Abu Dhabi, United Arab Emirates, December 2022 (2022). https://aclanthology.org/2022.findings-emnlp.205
18. Mackenzie, J., Trotman, A., Lin, J.: Efficient document-at-a-time and score-at-a-time query evaluation for learned sparse representations. ACM Trans. Inf. Syst. **41**(4), 1–28 (2023)

19. Mallia, A., Khattab, O., Suel, T., Tonellotto, N.: Learning passage impacts for inverted indexes. In: Proceedings of SIGIR, pp. 1723–1727 (2021)
20. Mallia, A., Mackenzie, J., Suel, T., Tonellotto, N.: Faster learned sparse retrieval with guided traversal. In: Proceedings of the 45th International ACM SIGIR Conference on Research and Development in Information Retrieval, pp. 1901–1905 (2022)
21. Mallia, A., Siedlaczek, M., Mackenzie, J., Suel, T.: Pisa: performant indexes and search for academia. In: Proceedings of the Open-Source IR Replicability Challenge (2019)
22. Nguyen, T., MacAvaney, S., Yates, A.: A unified framework for learned sparse retrieval. In: Kamps, J., et al. (eds.) ECIR 2023. LNCS, vol. 13982, pp. 101–116. Springer, Cham (2023). https://doi.org/10.1007/978-3-031-28241-6_7
23. Qiao, Y., Yang, Y., He, S., Yang, T.: Representation sparsification with hybrid thresholding for fast splade-based document retrieval. arXiv preprint arXiv:2306.11293 (2023)
24. Qiao, Y., Yang, Y., Lin, H., Yang, T.: Optimizing guided traversal for fast learned sparse retrieval. In: Proceedings of the ACM Web Conference 2023, pp. 3375–3385 (2023)
25. Santhanam, K., Khattab, O., Saad-Falcon, J., Potts, C., Zaharia, M.: Colbertv2: effective and efficient retrieval via lightweight late interaction. arXiv preprint arXiv:2112.01488 (2021)
26. Sertkan, M., Althammer, S., Hofstätter, S.: Ranger: a toolkit for effect-size based multi-task evaluation. arXiv preprint arXiv:2305.15048 (2023)
27. Sertkan, M., Althammer, S., Hofstätter, S., Knees, P., Neidhardt, J.: Exploring effect-size-based meta-analysis for multi-dataset evaluation (2023)
28. Shen, T., et al.: LexMAE: lexicon-bottlenecked pretraining for large-scale retrieval. In: International Conference on Learning Representations (2023). https://openreview.net/forum?id=PfpEtB3-csK
29. Thakur, N., Reimers, N., Rücklé, A., Srivastava, A., Gurevych, I.: Beir: a heterogenous benchmark for zero-shot evaluation of information retrieval models. arXiv preprint arXiv:2104.08663 (2021)
30. Tonellotto, N., Macdonald, C., Ounis, I.: Efficient query processing for scalable web search. Found. Trends in Inf. Retr. 12(4–5), 319–492 (2018)
31. Turtle, H., Flood, J.: Query evaluation: strategies and optimizations. Inf. Process. Manag. 31(6), 831–850 (1995)
32. Yang, P., Fang, H., Lin, J.: Anserini: enabling the use of Lucene for information retrieval research. In: Proceedings of the 40th International ACM SIGIR Conference on Research and Development in Information Retrieval, pp. 1253–1256 (2017)
33. Zhao, T., Lu, X., Lee, K.: SPARTA: efficient open-domain question answering via sparse transformer matching retrieval. In: Toutanova, K., et al. (eds.) Proceedings of the 2021 Conference of the North American Chapter of the Association for Computational Linguistics: Human Language Technologies, NAACL-HLT 2021, Online, 6–11 June 2021, pp. 565–575. Association for Computational Linguistics (2021). https://doi.org/10.18653/v1/2021.naacl-main.47

Large Language Models are Zero-Shot Rankers for Recommender Systems

Yupeng Hou[1,2], Junjie Zhang[1], Zihan Lin[3], Hongyu Lu[4], Ruobing Xie[4], Julian McAuley[2], and Wayne Xin Zhao[1(✉)]

[1] Gaoling School of Artificial Intelligence, Renmin University of China, Beijing, China
yphou@ucsd.edu, junjie.zhang@ruc.edu.cn, batmanfly@gmail.com
[2] UC San Diego, San Diego, USA
[3] School of Information, Renmin University of China, Beijing, China
[4] WeChat, Tencent, Shenzhen, China

Abstract. Recently, large language models (LLMs) (*e.g.,* GPT-4) have demonstrated impressive general-purpose task-solving abilities, including the potential to approach recommendation tasks. Along this line of research, this work aims to investigate the capacity of LLMs that act as the ranking model for recommender systems. We first formalize the recommendation problem as a conditional ranking task, considering sequential interaction histories as *conditions* and the items retrieved by other candidate generation models as *candidates*. To solve the ranking task by LLMs, we carefully design the prompting template and conduct extensive experiments on two widely-used datasets. We show that LLMs have promising zero-shot ranking abilities but (1) struggle to perceive the order of historical interactions, and (2) can be biased by popularity or item positions in the prompts. We demonstrate that these issues can be alleviated using specially designed prompting and bootstrapping strategies. Equipped with these insights, zero-shot LLMs can even challenge conventional recommendation models when ranking candidates are retrieved by multiple candidate generators. The code and processed datasets are available at https://github.com/RUCAIBox/LLMRank.

Keywords: Large Language Model · Recommender System

1 Introduction

In the literature of recommender systems, most existing models are trained with user behavior data from a specific domain or scenario [26, 28, 49], suffering from two major issues. Firstly, it is difficult to capture user preference by solely modeling historical behaviors, *e.g.,* clicked item sequences [28, 33, 81], limiting the expressive power to model more complicated but explicit user interests (*e.g.,* intentions expressed in natural language). Secondly, these models are essentially

---

Y. Hou and J. Zhang—Equal contribution.

© The Author(s), under exclusive license to Springer Nature Switzerland AG 2024
N. Goharian et al. (Eds.): ECIR 2024, LNCS 14609, pp. 364–381, 2024.
https://doi.org/10.1007/978-3-031-56060-6_24

"narrow experts", lacking more comprehensive knowledge in solving complicated recommendation tasks that rely on background or commonsense knowledge [23].

To improve recommendation performance and interactivity, there have been increasing efforts that explore the use of pre-trained language models (PLMs) in recommender systems [21,30,62]. They aim to explicitly capture user preference in natural language [21] or transfer rich world knowledge from text corpora [29,30]. Despite their effectiveness, thoroughly fine-tuning the recommendation models on task-specific data is still a necessity, making it less capable of solving diverse recommendation tasks [30]. More recently, large language models (LLMs) have shown great potential to serve as zero-shot task solvers [52,64]. Indeed, there are some preliminary attempts that employ LLMs for solving recommendation tasks [13,20,40,59,60,73]. These studies mainly focus on discussing the possibility of building a capable recommender with LLMs. While promising, the insufficient understanding of the new characteristics when making recommendations using LLMs could hinder the development of this new paradigm.

In this paper, we conduct empirical studies to investigate what determines the capacity of LLMs that serve as recommendation models. Typically, recommender systems are developed in a pipeline architecture [10], consisting of *candidate generation* (retrieving relevant items) and *ranking* (ranking relevant items at a higher position) procedures. This work mainly focuses on the ranking stage of recommender systems, since LLMs are more expensive to run on a large-scale candidate set. Further, the ranking performance is sensitive to the retrieved candidate items, which is more suitable to examine the subtle differences in the recommendation abilities of LLMs.

To carry out this study, we first formalize the recommendation process of LLMs as a *conditional ranking* task. Given prompts that include sequential historical interactions as *"conditions"*, LLMs are instructed to rank a set of *"candidates"* (e.g., items retrieved by candidate generation models), according to LLM's intrinsic knowledge. Then we conduct control experiments to systematically study the empirical performance of LLMs as rankers by designing specific configurations for "conditions" and "candidates", respectively. Overall, we attempt to answer the following key questions:

- What factors affect the zero-shot ranking performance of LLMs?
- What data or knowledge do LLMs rely on for recommendation?

Our empirical experiments are conducted on two public datasets for recommender systems. The results lead to several key findings that potentially shed light on how to develop LLMs as powerful ranking models for recommender systems. We summarize the key findings as follows:

- LLMs *struggle to perceive the order* of the given sequential interaction histories. By employing specifically designed promptings, *LLMs can be triggered to perceive the order*, leading to improved ranking performance.
- LLMs suffer from position bias and popularity bias while ranking, which can be alleviated by bootstrapping or specially designed prompting strategies.

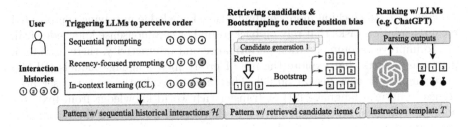

Fig. 1. An overview of the proposed LLM-based zero-shot ranking method.

– LLMs outperform existing zero-shot recommendation methods, showing promising zero-shot ranking abilities, especially on candidates retrieved by multiple candidate generation models with different practical strategies.

2   General Framework for LLMs as Rankers

To investigate the recommendation abilities of LLMs, we first formalize the recommendation process as a conditional ranking task. Then, we describe a general framework that adapts LLMs to solve the recommendation task.

2.1   Problem Formulation

Given the historical interactions $\mathcal{H} = \{i_1, i_2, \ldots, i_n\}$ of one user (in chronological order of interaction time) as *conditions*, the task is to rank the *candidate* items $\mathcal{C} = \{i_j\}_{j=1}^{m}$, such that the items of interest would be ranked at a higher position. In practice, the candidate items are usually retrieved by candidate generation models from the whole item set $\mathcal{I}$ ($m \ll |\mathcal{I}|$) [10]. Further, we assume that each item i is associated with a descriptive text t_i following [30].

2.2   Ranking with LLMs Using Natural Language Instructions

We use LLMs as ranking models to solve the above-mentioned task in an instruction-following paradigm [64]. Specifically, for each user, we first construct two natural language patterns that contain sequential interaction histories $\mathcal{H}$ (*conditions*) and retrieved candidate items $\mathcal{C}$ (*candidates*), respectively. Then these patterns are filled into a natural language template T as the final instruction. In this way, LLMs are expected to understand the instructions and output the ranking results as the instruction suggests. The overall framework of the ranking approach by LLMs is depicted in Fig. 1. Next, we describe the detailed instruction design in our approach.

Sequential Historical Interactions. To investigate whether LLMs can capture user preferences from historical user behaviors, we include sequential historical interactions $\mathcal{H}$ into the instructions as inputs of LLMs. To enable LLMs to be aware of the sequential nature of historical interactions, we propose three ways to construct the instructions:

- **Sequential prompting**: Arrange the historical interactions in chronological order. This way has also been used in prior studies [13]. For example, *"I've watched the following movies in the past in order: '0. Multiplicity', '1. Jurassic Park', ...".*

- **Recency-focused prompting**: In addition to the sequential interaction records, we can add an additional sentence to emphasize the most recent interaction. For example, *"I've watched the following movies in the past in order: '0. Multiplicity', '1. Jurassic Park', ... Note that my most recently watched movie is Dead Presidents. ...".*

- **In-context learning (ICL)**: ICL is a prominent prompting approach for LLMs to solve various tasks [78], where it includes demonstration examples in the prompt. For the personalized recommendation task, simply introducing examples of other users may introduce noises because users usually have different preferences. Instead, we introduce demonstration examples by augmenting the input interaction sequence itself. We pair the prefix of the input interaction sequence and the corresponding successor as examples. For instance, *" If I've watched the following movies in the past in order: '0. Multiplicity', '1. Jurassic Park', ..., then you should recommend Dead Presidents to me and now that I've watched Dead Presidents, then ...".*

Retrieved Candidate Items. Typically, candidate items to be ranked are first retrieved by candidate generation models [10]. In this work, we consider a relatively small pool for the candidates, and keep 20 candidate items (*i.e.*, $m = 20$) for ranking. To rank these candidates with LLMs, we arrange the candidate items C in a sequential manner. For example, *"Now there are 20 candidate movies that I can watch next: '0. Sister Act', '1. Sunset Blvd', ...".* Note that, following the classic candidate generation approach [10], there is no specific order for candidate items. As a result, We generate different orders for the candidate items in the prompts, which enables us to further examine whether the ranking results of LLMs are affected by the arrangement order of candidates, *i.e.*, position bias, and how to alleviate position bias via bootstrapping.

Ranking with Large Language Models. Existing studies show that LLMs can follow natural language instructions to solve diverse tasks in a zero-shot setting [64,78]. To rank using LLMs, we infill the patterns above into the instruction template T. An example instruction template can be given as: *" [pattern that contains sequential historical interactions $\mathcal{H}$] [pattern that contains retrieved candidate items C] Please rank these movies by measuring the possibilities that I would like to watch next most, according to my watching history.".*

Parsing the Output of LLMs. Note that the output of LLMs is still in natural language text, and we parse the output with heuristic text-matching methods

Table 1. Statistics of the preprocessed datasets. "Avg. $|\mathcal{H}|$" denotes the average length of historical interactions. "Avg. $|t_i|$" denotes the average number of tokens in the item text.

Dataset	#Users	#Items	#Interactions	Sparsity	Avg. $|\mathcal{H}|$	Avg. $|t_i|$
ML-1M	6,040	3,706	1,000,209	95.53%	46.19	16.96
Games	50,547	16,859	389,718	99.95%	7.02	43.31

and ground the recommendation results on the specified item set. In detail, we can directly perform efficient substring matching algorithms like KMP [35] between the LLM outputs and the text of candidate items. We also found that LLMs occasionally generate items that are not present in the candidate set. For GPT-3.5, such deviations occur in a mere 3% of cases. One can either reprocess the illegal cases or simply treat the out-of-candidate items as incorrect recommendations.

3   Empirical Studies

Datasets. The experiments are conducted on two widely-used public datasets for recommender systems: (1) the movie rating dataset *MovieLens-1M* [24] (in short, **ML-1M**) where user ratings are regarded as interactions, and (2) one category from the *Amazon Review* dataset [46] named **Games** where reviews are regarded as interactions. We filter out users and items with fewer than five interactions. Then we sort the interactions of each user by timestamp, with the oldest interactions first, to construct the corresponding historical interaction sequences. The movie/product titles are used as the descriptive text of an item. We use item titles in this study for two reasons: (1) to determine if LLMs can make recommendations based on their intrinsic world knowledge with minimal information provided, and (2) to conserve computational resources. Exploring how LLMs use more extensive textual features for recommendations will be the focus of our future work. Statistics of the preprocessed datasets are presented in Table 1

Evaluation and Implementation Details. Following existing works [30,33], we apply the leave-one-out strategy for evaluation. For each historical interaction sequence, the last item is used as the ground-truth item in test set. The item before the last one is used in the validation set (used for training baseline methods). We adopt the widely used metric NDCG@K (in short, N@K) to evaluate the ranking results over the given m candidates, where $K \leq m$. To ease the reproduction of this work, our experiments are conducted using a popular open-source recommendation library RECBOLE [77]. The historical interaction sequences are truncated within a length of 50. We evaluate LLM-based methods on all users in ML-1M dataset and randomly sampled $6,000$ users for Games dataset by default. Unless specified, the evaluated LLM is accessed by calling OpenAI's

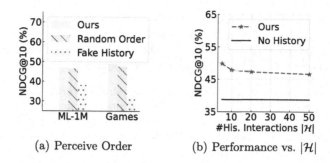

(a) Perceive Order   (b) Performance vs. $|\mathcal{H}|$

Fig. 2. Analysis of whether LLMs perceive the order of historical interactions.

API `gpt-3.5-turbo`. The hyperparameter temperature of calling LLMs is set to 0.2. *All the reported results are the average of at least three repeat runs to reduce the effect of randomness.*

3.1 Can LLMs Understand Prompts that Involve Sequential Historical User Behaviors?

In LLM-based methods, historical interactions are naturally arranged in an ordered sequence. By designing different configurations of $\mathcal{H}$, we aim to examine whether LLMs can leverage these historical user behaviors and perceive the sequential nature for making accurate recommendations.

LLMs Struggle to Perceive the Order of Given Historical User Behaviors. In this section, we examine whether LLMs can understand prompts with ordered historical interactions and give personalized recommendations. The task is to rank a candidate set of 20 items, containing one ground-truth item and 19 randomly sampled negatives. By analyzing historical behaviors, items of interest should be ranked at a higher position. We compare the ranking results of three LLM-based methods: (a) *Ours*, which ranks as we have described in Sect. 2.2. Historical user behaviors are encoded into prompts using the "sequential prompting" strategy. (b) *Random Order*, where the historical user behaviors will be randomly shuffled before being fed to the model, and (c) *Fake History*, where we replace all the items in original historical behaviors with randomly sampled items as fake historical behaviors. From Fig. 2(a), we can see that *Ours* has better performance than variants with fake historical behaviors. However, the performance of *Ours* and *Random Order* is similar, indicating that LLMs are not sensitive to the order of the given historical user interactions.

Moreover, in Fig. 2(b), we vary the number of latest historical user behaviors ($|\mathcal{H}|$) used for constructing the prompt from 5 to 50. The results show that increasing the number of historical user behaviors does not improve, but rather negatively impacts the ranking performance. We speculate that this phenomenon is caused by the fact that LLMs have difficulty understanding the order, but consider all the historical behaviors equally. Therefore too many historical user behaviors (*e.g.*, $|\mathcal{H}| = 50$) may overwhelm LLMs and lead to a performance drop. In contrast, a relatively small $|\mathcal{H}|$ enables LLMs to concentrate on the most recently interacted items, resulting in better recommendation performance.

Table 2. Performance comparison on *randomly retrieved candidates*. Ground-truth items are included in the candidate sets. "full" denotes models that are trained on the target dataset, and "zero-shot" denotes models that are not trained on the target dataset but could be pre-trained. We highlight the best performance among zero-shot recommendation methods in **bold**.

	Method	ML-1M				Games			
		N@1	N@5	N@10	N@20	N@1	N@5	N@10	N@20
full	Pop	22.91	45.16	52.33	55.36	28.35	47.42	52.96	57.45
	BPRMF [49]	34.60	59.87	64.29	65.39	44.92	62.33	66.27	68.94
	SASRec [33]	61.39	76.39	78.89	79.79	56.90	73.19	75.92	77.14
zero-shot	BM25 [50]	4.70	12.68	17.88	33.19	13.92	28.81	34.61	44.35
	UniSRec [30]	7.37	18.80	26.67	37.93	18.95	33.99	40.71	48.42
	VQ-Rec [29]	5.98	15.48	23.74	35.85	7.28	18.28	26.21	37.62
	Sequential	18.28	36.35	42.85	49.02	30.28	45.48	50.57	56.55
	Recency-Focused	19.57	37.73	44.23	50.01	**34.03**	**48.77**	**53.50**	**59.01**
	In-Context Learning	**21.77**	**39.59**	**45.83**	**51.62**	33.95	48.44	53.10	58.92

Triggering LLMs to Perceive the Interaction Order. Based on the above observations, we find it difficult for LLMs to perceive the order in interaction histories by a default prompting strategy. As a result, we aim to elicit the order-perceiving abilities of LLMs, by proposing two alternative prompting strategies and emphasizing the recently interacted items. Detailed descriptions of the proposed strategies have been given in Sect. 2.2. In Table 2, we can see that both recency-focused prompting and in-context learning can generally improve the ranking performance of LLMs, though the best strategy may vary on different datasets. The above results can be summarized as the following key observation:

Observation 1. LLMs *struggle to perceive the order* of the given sequential interaction histories. By employing specifically designed promptings, *LLMs can be triggered to perceive the order* of historical user behaviors, leading to improved ranking performance.

3.2 Do LLMs Suffer from Biases While Ranking?

The biases and debiasing methods in conventional recommender systems have been widely studied [5]. For LLM-based recommendation models, both the input and output are natural language texts and will inevitably introduce new biases. In this section, we discuss two kinds of biases that LLM-based recommendation models suffer from. We also make discussions on how to alleviate these biases.

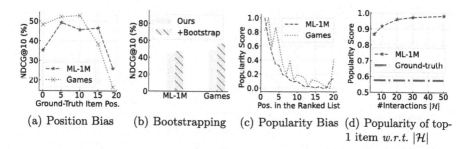

Fig. 3. Biases and debiasing methods in the ranking of LLMs. (a) The position of candidates in the prompts influences the ranking results. (b) Bootstrapping alleviates position bias. (c) LLMs tend to recommend popular items. (d) Focusing on historical interactions reduces popularity bias.

The Order of Candidates Affects the Ranking Results of LLMs. For conventional ranking methods, the order of retrieved candidates usually will not affect the ranking results [28,33]. However, for the LLM-based approach that is described in Sect. 2.2, the candidates are arranged in a sequential manner and infilled into a prompt. It has been shown that LLMs are generally sensitive to the order of examples in the prompts for NLP tasks [44,79]. As a result, we also conduct experiments to examine whether the order of candidates affects the ranking performance of LLMs. We follow the experimental settings adopted in Sect. 3.1. The only difference is that we control the order of these candidates in the prompts, by making the ground-truth items appear at a certain position. We vary the position of ground-truth items at $\{0, 5, 10, 15, 19\}$ and present the results in Fig. 3(a). We can see that the performance varies when the ground-truth items appear at different positions. Especially, the ranking performance drops significantly when the ground-truth items appear at the last few positions. The results indicate that LLM-based rankers are affected by the order of candidates, *i.e., position bias*, which may *not* affect conventional recommendation models.

Alleviating Position Bias Via Bootstrapping. A simple strategy to alleviate position bias is to bootstrap the ranking process. We may rank the candidate set repeatedly for B times, with candidates randomly shuffled at each round. In this way, one candidate may appear in different positions. We then merge the results of each round to derive the final ranking. From Fig. 3(b), we follow the setting in Sect. 3.1 and apply the bootstrapping strategy to *Ours*. Each candidate set will be ranked for 3 times. We can see that bootstrapping improves the ranking performance on both datasets.

Popularity Degrees of Candidates Affect Ranking Results of LLMs. For popular items, the associated text may also appear frequently in the pre-training corpora of LLMs. For example, a best-selling book would be widely discussed on the Web. Thus, we aim to examine whether the ranking results are

Table 3. Zero-shot ranking performance comparison. We highlight the best performance in **bold**. Due to limited budget, we evaluate each LLM only once on 200 sampled users **only** for experiments corresponding to this table.

Method	ML-1M				Games			
	N@1	N@5	N@10	N@20	N@1	N@5	N@10	N@20
BM25 [50]	4.70	12.68	17.88	33.19	13.92	28.81	34.61	44.35
UniSRec [30]	7.37	18.80	26.67	37.93	18.95	33.99	40.71	48.42
Alpaca-7B [55]	4.00	13.92	23.09	31.54	5.50	14.16	21.67	28.68
Vicuna-13B [9]	6.50	14.75	22.64	33.42	7.00	17.73	24.30	31.22
LLaMA-2-70B-Chat [57]	8.00	25.42	31.19	34.52	21.50	32.30	37.83	41.97
ChatGPT (GPT-3.5)	**23.33**	**42.07**	**48.80**	**53.73**	23.83	45.69	50.31	55.45
GPT-4	15.50	40.65	46.74	48.42	**39.50**	**58.22**	**62.88**	**65.25**

affected by the popularity of candidates. However, it is difficult to directly measure the popularity of item text. Here, we hypothesize that the text popularity can be indirectly measured by item frequency in one recommendation dataset. In Fig. 3(c), we report the item popularity score (measured by the normalized item frequency of appearance in the training set) at each position of the ranked item lists. We can see that popular items tend to be ranked at higher positions.

Making LLMs Focus on Historical Interactions Helps Reduce Popularity Bias. We assume that if LLMs focus on historical interactions, they may give more personalized recommendations but not more popular ones. From Fig. 2(b), we know that LLMs make better use of historical interactions when using less historical interactions. From Fig. 3(d), we compare the popularity scores of the best-ranked items varying the number of historical interactions. It can be observed that as $|\mathcal{H}|$ decreases, the popularity score decreases as well. This suggests that one can reduce the effects of popularity bias when LLMs focus more on historical interactions. From the above experiments, we can conclude the following:

> **Observation 2.** LLMs suffer from position bias and popularity bias while ranking, which can be alleviated by bootstrapping or specially designed prompting strategies.

3.3   How Well Can LLMs Rank Candidates in a Zero-Shot Setting?

We further evaluate LLM-based methods on candidates with hard negatives that are retrieved by different strategies to further investigate what the ranking of LLMs depends on. Then, we present the ranking performance of different methods on candidates retrieved by multiple candidate generation models to simulate a more practical and difficult setting.

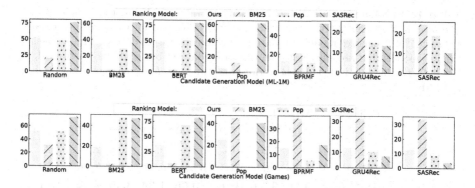

Fig. 4. Ranking performance measured by NDCG@10 (%) on hard negatives.

LLMs have Promising Zero-Shot Ranking Abilities. In Table 2, we conduct experiments to compare the ranking abilities of LLM-based methods with existing methods. We follow the same setting in Sect. 3.1 where $|\mathcal{C}| = 20$ and candidate items are randomly retrieved. We include three conventional models that are trained on the training set, *i.e.*, Pop (recommending according to item popularity), BPRMF [49], and SASRec [33]. We also evaluate three zero-shot recommendation methods that are not trained on the target datasets, including BM25 [50] (rank according to the textual similarity between candidates and historical interactions), UniSRec [30], and VQ-Rec [29]. For UniSRec and VQ-Rec, we use their publicly available pre-trained models. We do not include ZESRec [15] because there is no pre-trained model released. In addition, we compare the zero-shot ranking performance of different LLMs in Table 3. "Recency-Focused" prompting strategy is used for LLM-based rankers.

From Table 2 and 3, we can see that LLMs with more parameters generally perform better. The best LLM-based methods outperform existing zero-shot recommendation methods by a large margin, showing promising zero-shot ranking abilities. We would highlight that it is difficult to conduct zero-shot recommendations on the ML-1M dataset, due to the difficulty in measuring the similarity between movies merely by the similarity of their titles. However, LLMs can use their intrinsic knowledge to measure the similarity between movies and make recommendations. We would emphasize that the goal of evaluating zero-shot recommendation methods is not to surpass conventional models. The goal is to demonstrate the strong recommendation capabilities of pre-trained base models, which can be further adapted and transferred to downstream scenarios.

LLMs Rank Candidates Based on Item Popularity, Text Features as Well as User Behaviors. To further investigate how LLMs rank the given candidates, we evaluate LLMs on candidates that are retrieved by different candidate generation methods. These candidates can be viewed as hard negatives for ground-truth items, which can be used to measure the ranking ability of LLMs for specific categories of items. We consider two categories of strategies to retrieve

Table 4. Performance comparison on *candidates retrieved by multiple candidate generation models*. Ground-truth items are *not* guaranteed to be included in the candidate sets. "full" denotes models that are trained on the target dataset, and "zero-shot" denotes models that are not trained on the target dataset but could be pre-trained. We highlight the best and second-best performance among *all* recommendation methods in **bold**.

	Method	ML-1M				Games			
		N@1	N@5	N@10	N@20	N@1	N@5	N@10	N@20
full	Pop	0.08	1.20	4.13	5.79	0.13	1.00	2.27	2.62
	BPRMF [49]	0.26	1.69	4.41	6.04	0.55	1.98	**2.96**	**3.19**
	SASRec [33]	**3.76**	**9.79**	**10.45**	**10.56**	**1.33**	**3.55**	**4.02**	**4.11**
zero-shot	BM25 [50]	0.26	0.87	2.32	5.28	0.18	1.07	1.80	2.55
	UniSRec [30]	0.88	3.46	5.30	6.92	0.00	1.86	2.03	2.31
	VQ-Rec [29]	0.20	1.60	3.29	5.73	0.20	1.21	1.91	2.64
	Ours	**1.74**	**5.22**	**6.91**	**7.90**	**0.90**	**2.26**	2.80	3.08

the candidates: (1) *content-based methods* like *BM25* [50] and *BERT* [14] retrieve candidates based on the text feature similarities, and (2) *interaction-based methods*, including *Pop*, *BPRMF* [49], *GRU4Rec* [28], and *SASRec* [33], retrieve items using neural networks trained on user-item interactions. Given candidates, we compare the ranking performance of the LLM-based model (*Ours*) and representative methods.

From Fig. 4, we can see that the ranking performance of the LLM-based method varies on different candidate sets and different datasets. (1) On ML-1M, LLM-based method cannot rank well on candidate sets that contain popular items (*e.g.*, *Pop* and *BPRMF*), indicating the LLM-based method recommend items largely depend on item popularity on ML-1M dataset. (2) On Games, we can observe that *Ours* has similar performance both on popular candidates and textual similar candidates, showing that item popularity and text features contribute similarly to the ranking of LLMs. (3) On both two datasets, the performance of *Ours* is affected by hard negatives retrieved by interaction-based candidate generation models, but not as severe as those interaction-based rankers like *SASRec*. The above results demonstrate that LLM-based methods not only consider one single aspect for ranking, but make use of item popularity, text features, and even user behaviors. On different datasets, the weights of these three aspects to affect the ranking performance may also vary.

LLMs Can Effectively Rank Candidates Retrieved by Multiple Candidate Generation Models. For real-world recommender systems [10], the items to be ranked are usually retrieved by multiple candidate generation models. As a result, we also conduct experiments in a more practical and difficult setting. We use the above-mentioned seven candidate generation models to retrieve items. The top-3 best items retrieved by each candidate generation model will be merged into a candidate set containing a total of 21 items. As a more practical setting, we do not complement the ground-truth item to each candidate set. Note that the experiments here were conducted under the implicit preference setup [76], indicating that implicit positive instances (not explicitly labeled) may exist among the retrieved items. A more faithful evaluation might require a human study, which we intend to explore in our future work. For *Ours*, we summarize the experiences gained from Sect. 3.1 and 3.2. We use the recency-focused prompting strategy to encode $|\mathcal{H}| = 5$ sequential historical interactions into prompts and use a bootstrapping strategy to repeatedly rank for 3 rounds.

From Table 4, we can see that the LLM-based model (*Ours*) yields the second-best performance over the compared recommendation models on most metrics. The results show that LLM-based zero-shot ranker even beats the conventional recommendation model *Pop* and *BPRMF* that has been trained on the target datasets, further demonstrating the strong zero-shot ranking ability of LLMs. We assume that LLMs can make use of their intrinsic world knowledge to rank the candidates comprehensively considering popularity, text features, and user behaviors. In comparison, existing models (as *narrow experts*) may lack the ability to rank items in a complicated setting. The above findings can be summarized as:

> **Observation 3.** LLMs have promising zero-shot ranking abilities, especially on candidates retrieved by multiple candidate generation models with different practical strategies.

4   Related Work

Transfer Learning for Recommender Systems. As recommender systems are mostly trained on data collected from a single source, people have sought to transfer knowledge from other domains [45,70,75,82,84,85], markets [3,51], or platforms [4,19]. Typical transfer learning methods for recommender systems rely on anchors, including shared users/items [7,8,45,68,69,83] or representations from a shared space [11,18,38]. However, these anchors are usually sparse among different scenarios, making transferring difficult for recommendations [84]. More recently, there are studies aiming to transfer knowledge stored in language models by adapting them to recommendation tasks via tuning [1,12,21,53] or prompting [37,39,74]. In this paper, we conduct zero-shot recommendation experiments to examine the potential to transfer knowledge from LLMs.

Large Language Models for Recommender Systems. The design of recommendation models, especially sequential recommendation models, has been long inspired by the design of language models, from word2vec [2,22,25] to recent neural networks [28,33,54,81]. In recent years, with the development of pre-trained language models (PLMs) [14], people have tried to transfer knowledge stored in PLMs to recommendation models, by either representing items using their text features or representing behavior sequences in the format of natural language [16,21,42,58,67]. Very recently, large language models (LLMs) have been shown superior language understanding and generation abilities [6,17,47,56,66,78]. Studies have been made to make recommender systems more interactive by integrating LLMs along with conventional recommendation models [20,27,36,43,48,59,61,65] or fine-tuned with specially designed instructions [1,12,21,31,80]. There are also early explorations showing LLMs have zero-shot recommendation abilities [13,34,41,59,60,63,71,72]. Despite being effective to some extent, few works have explored what determines the recommendation performance of LLMs.

5   Conclusion

In this work, we investigated the capacities of LLMs that act as the zero-shot ranking model for recommender systems. To rank with LLMs, we constructed natural language prompts that contain historical interactions, candidates, and instruction templates. We then propose several specially designed prompting strategies to trigger the ability of LLMs to perceive orders of sequential behaviors. We also introduce bootstrapping and prompting strategies to alleviate the position bias and popularity bias issues that LLM-based ranking models may suffer.

Extensive empirical studies indicate that LLMs have promising zero-shot ranking abilities. The empirical studies demonstrate the strong potential of transferring knowledge from LLMs as powerful recommendation models. We aim at shedding light on several promising directions to further improve the ranking abilities of LLMs, including (1) better perceiving the order of sequential historical interactions and (2) alleviating the position bias and popularity bias. For future work, we consider developing technical approaches to solve the above-mentioned key challenges when deploying LLMs as recommendation models. We also would like to develop LLM-based recommendation models that can be efficiently tuned on downstream user behaviors for effective personalized recommendations.

6   Limitations

In most experiments in this paper, ChatGPT is used as the primary target LLM for evaluation. However, being a closed-source commercial service, Chat-GPT might integrate additional techniques with its core large language model to improve performance. While there are open-source LLMs available, such as LLaMA 2 [57] and Mistral [32], they exhibit a notable performance disparity compared to ChatGPT (e.g., LLaMA-2-70B-Chat vs. ChatGPT in Table 3). This gap makes it difficult to evaluate the emergent abilities of LLMs on the recommendation tasks using purely open-source models. In addition, we should note that the observations might be biased by specific prompts and datasets.

Acknowledgements. This work was partially supported by National Natural Science Foundation of China under Grant No. 62222215, Beijing Natural Science Foundation under Grant No. L233008 and 4222027.

References

1. Bao, K., Zhang, J., Zhang, Y., Wang, W., Feng, F., He, X.: Tallrec: an effective and efficient tuning framework to align large language model with recommendation. arXiv preprint arXiv:2305.00447 (2023)
2. Barkan, O., Koenigstein, N.: ITEM2VEC: neural item embedding for collaborative filtering. In: Palmieri, F.A.N., Uncini, A., Diamantaras, K.I., Larsen, J. (eds.) 26th IEEE International Workshop on Machine Learning for Signal Processing, MLSP 2016, Vietri sul Mare, Salerno, Italy, 13–16 September 2016, pp. 1–6. IEEE (2016). https://doi.org/10.1109/MLSP.2016.7738886
3. Bonab, H.R., Aliannejadi, M., Vardasbi, A., Kanoulas, E., Allan, J.: Cross-market product recommendation. In: Demartini, G., Zuccon, G., Culpepper, J.S., Huang, Z., Tong, H. (eds.) CIKM, pp. 110–119. ACM (2021). https://doi.org/10.1145/3459637.3482493
4. Cao, D., He, X., Nie, L., Wei, X., Hu, X., Wu, S., Chua, T.: Cross-platform app recommendation by jointly modeling ratings and texts. ACM Trans. Inf. Syst. **35**(4), 37:1–37:27 (2017). https://doi.org/10.1145/3017429
5. Chen, J., Dong, H., Wang, X., Feng, F., Wang, M., He, X.: Bias and debias in recommender system: a survey and future directions. CoRR abs/2010.03240 (2020). https://arxiv.org/abs/2010.03240
6. Chen, J., et al.: When large language models meet personalization: perspectives of challenges and opportunities. arXiv preprint arXiv:2307.16376 (2023)
7. Chen, L., Yuan, F., Yang, J., He, X., Li, C., Yang, M.: User-specific adaptive fine-tuning for cross-domain recommendations. IEEE Trans. Knowl. Data Eng. **35**(3), 3239–3252 (2023). https://doi.org/10.1109/TKDE.2021.3119619
8. Cheng, M., Yuan, F., Liu, Q., Xin, X., Chen, E.: Learning transferable user representations with sequential behaviors via contrastive pre-training. In: Bailey, J., Miettinen, P., Koh, Y.S., Tao, D., Wu, X. (eds.) ICDM, pp. 51–60. IEEE (2021). https://doi.org/10.1109/ICDM51629.2021.00015
9. Chiang, W.L., et al.: Vicuna: an open-source chatbot impressing gpt-4 with 90%* chatgpt quality (2023). https://vicuna.lmsys.org/. Accessed 14 Apr 2023
10. Covington, P., Adams, J., Sargin, E.: Deep neural networks for youtube recommendations. In: RecSys, pp. 191–198 (2016)
11. Cui, Q., Wei, T., Zhang, Y., Zhang, Q.: Herograph: a heterogeneous graph framework for multi-target cross-domain recommendation. In: Vinagre, J., Jorge, A.M., Al-Ghossein, M., Bifet, A. (eds.) RecSys. CEUR Workshop Proceedings, vol. 2715. CEUR-WS.org (2020). https://ceur-ws.org/Vol-2715/paper6.pdf
12. Cui, Z., Ma, J., Zhou, C., Zhou, J., Yang, H.: M6-rec: generative pre-trained language models are open-ended recommender systems. arXiv preprint arXiv:2205.08084 (2022)
13. Dai, S., et al.: Uncovering chatgpt's capabilities in recommender systems. arXiv preprint arXiv:2305.02182 (2023)
14. Devlin, J., Chang, M.W., Lee, K., Toutanova, K.: Bert: pre-training of deep bidirectional transformers for language understanding. In: NAACL (2019)
15. Ding, H., Ma, Y., Deoras, A., Wang, Y., Wang, H.: Zero-shot recommender systems. arXiv:2105.08318 (2021)

16. Ding, H., Ma, Y., Deoras, A., Wang, Y., Wang, H.: Zero-shot recommender systems. arXiv preprint arXiv:2105.08318 (2021)
17. Fan, W., et al.: Recommender systems in the era of large language models (llms). arXiv preprint arXiv:2307.02046 (2023)
18. Fu, J., et al.: Exploring adapter-based transfer learning for recommender systems: Empir. Stud. Pract. Insights. CoRR abs/2305.15036 (2023). https://doi.org/10.48550/arXiv.2305.15036
19. Gao, C., Lin, T., Li, N., Jin, D., Li, Y.: Cross-platform item recommendation for online social e-commerce. TKDE **35**(2), 1351–1364 (2023). https://doi.org/10.1109/TKDE.2021.3098702
20. Gao, Y., Sheng, T., Xiang, Y., Xiong, Y., Wang, H., Zhang, J.: Chat-rec: towards interactive and explainable llms-augmented recommender system. arXiv preprint arXiv:2303.14524 (2023)
21. Geng, S., Liu, S., Fu, Z., Ge, Y., Zhang, Y.: Recommendation as language processing (RLP): a unified pretrain, personalized prompt & predict paradigm (P5). In: RecSys (2022)
22. Grbovic, M., Cheng, H.: Real-time personalization using embeddings for search ranking at airbnb. In: Guo, Y., Farooq, F. (eds.) Proceedings of the 24th ACM SIGKDD International Conference on Knowledge Discovery & Data Mining, KDD 2018, London, UK, 19–23 August 2018, pp. 311–320. ACM (2018). https://doi.org/10.1145/3219819.3219885
23. Guo, Q., et al.: A survey on knowledge graph-based recommender systems. TKDE **34**(8), 3549–3568 (2020)
24. Harper, F.M., Konstan, J.A.: The movielens datasets: history and context. TIIS **5**(4), 1–19 (2015)
25. He, R., Kang, W.C., McAuley, J.: Translation-based recommendation. In: RecSys (2017)
26. He, X., Deng, K., Wang, X., Li, Y., Zhang, Y., Wang, M.: Lightgcn: simplifying and powering graph convolution network for recommendation. In: SIGIR (2020)
27. He, Z., et al.: Large language models as zero-shot conversational recommenders. In: CIKM (2023)
28. Hidasi, B., Karatzoglou, A., Baltrunas, L., Tikk, D.: Session-based recommendations with recurrent neural networks. In: ICLR (2016)
29. Hou, Y., He, Z., McAuley, J., Zhao, W.X.: Learning vector-quantized item representation for transferable sequential recommenders. In: WWW (2023)
30. Hou, Y., Mu, S., Zhao, W.X., Li, Y., Ding, B., Wen, J.: Towards universal sequence representation learning for recommender systems. In: KDD (2022)
31. Hua, W., Xu, S., Ge, Y., Zhang, Y.: How to index item ids for recommendation foundation models. arXiv preprint arXiv:2305.06569 (2023)
32. Jiang, A.Q., et al.: Mistral 7b. arXiv preprint arXiv:2310.06825 (2023)
33. Kang, W., McAuley, J.: Self-attentive sequential recommendation. In: ICDM (2018)
34. Kang, W.C., et al.: Do llms understand user preferences? evaluating llms on user rating prediction. arXiv preprint arXiv:2305.06474 (2023)
35. Knuth, D.E., Morris, J.H., Jr., Pratt, V.R.: Fast pattern matching in strings. SIAM J. Comput. **6**(2), 323–350 (1977)
36. Li, J., Zhang, W., Wang, T., Xiong, G., Lu, A., Medioni, G.: GPT4Rec: a generative framework for personalized recommendation and user interests interpretation (2023)
37. Li, L., Zhang, Y., Chen, L.: Personalized prompt learning for explainable recommendation. TOIS **41**(4), 1–26 (2023)

38. Li, R., Deng, W., Cheng, Y., Yuan, Z., Zhang, J., Yuan, F.: Exploring the upper limits of text-based collaborative filtering using large language models: discoveries and insights. CoRR abs/2305.11700 (2023). https://doi.org/10.48550/arXiv.2305. 11700

39. Li, X., Zhang, Y., Malthouse, E.C.: PBNR: prompt-based news recommender system. arXiv preprint arXiv:2304.07862 (2023)

40. Lin, G., Zhang, Y.: Sparks of artificial general recommender (AGR): early experiments with chatgpt. arXiv preprint arXiv:2305.04518 (2023)

41. Liu, J., Liu, C., Lv, R., Zhou, K., Zhang, Y.: Is ChatGPT a good recommender? a preliminary study (2023)

42. Liu, P., Zhang, L., Gulla, J.A.: Pre-train, prompt and recommendation: a comprehensive survey of language modelling paradigm adaptations in recommender systems. arXiv preprint arXiv:2302.03735 (2023)

43. Liu, Q., Chen, N., Sakai, T., Wu, X.M.: A first look at llm-powered generative news recommendation. arXiv preprint arXiv:2305.06566 (2023)

44. Lu, Y., Bartolo, M., Moore, A., Riedel, S., Stenetorp, P.: Fantastically ordered prompts and where to find them: overcoming few-shot prompt order sensitivity. In: ACL (2022)

45. Man, T., Shen, H., Jin, X., Cheng, X.: Cross-domain recommendation: An embedding and mapping approach. In: Sierra, C. (ed.) Proceedings of the Twenty-Sixth International Joint Conference on Artificial Intelligence, IJCAI 2017, Melbourne, Australia, 19–25 August 2017, pp. 2464–2470. ijcai.org (2017). https://doi.org/10. 24963/ijcai.2017/343

46. Ni, J., Li, J., McAuley, J.: Justifying recommendations using distantly-labeled reviews and fine-grained aspects. In: EMNLP, pp. 188–197 (2019)

47. Ouyang, L., et al.: Training language models to follow instructions with human feedback. NeurIPS **35**, 27730–27744 (2022)

48. Ren, X., et al.: Representation learning with large language models for recommendation. arXiv preprint arXiv:2310.15950 (2023)

49. Rendle, S., Freudenthaler, C., Gantner, Z., Schmidt-Thieme, L.: BPR: bayesian personalized ranking from implicit feedback. In: UAI (2009)

50. Robertson, S.E., Zaragoza, H.: The probabilistic relevance framework: BM25 and beyond. Found. Trends Inf. Retr. **3**(4), 333–389 (2009)

51. Roitero, K., Carterette, B., Mehrotra, R., Lalmas, M.: Leveraging behavioral heterogeneity across markets for cross-market training of recommender systems. In: Seghrouchni, A.E.F., Sukthankar, G., Liu, T., van Steen, M. (eds.) WWW, pp. 694–702. ACM/IW3C2 (2020). https://doi.org/10.1145/3366424.3384362

52. Sanh, V., et al.: Multitask prompted training enables zero-shot task generalization. In: ICLR (2022)

53. Shin, K., Kwak, H., Kim, K., Kim, S.Y., Ramström, M.N.: Scaling law for recommendation models: Towards general-purpose user representations. CoRR abs/2111.11294 (2021). https://arxiv.org/abs/2111.11294

54. Tang, J., Wang, K.: Personalized top-n sequential recommendation via convolutional sequence embedding. In: Chang, Y., Zhai, C., Liu, Y., Maarek, Y. (eds.) Proceedings of the Eleventh ACM International Conference on Web Search and Data Mining, WSDM 2018, Marina Del Rey, CA, USA, 5–9 February 2018, pp. 565–573. ACM (2018). https://doi.org/10.1145/3159652.3159656

55. Taori, R., et al.: Stanford alpaca: an instruction-following llama model (2023)

56. Touvron, H., et al.: Llama: open and efficient foundation language models. arXiv preprint arXiv:2302.13971 (2023)

57. Touvron, H., et al.: Llama 2: open foundation and fine-tuned chat models. arXiv preprint arXiv:2307.09288 (2023)

58. Wang, J., Yuan, F., Cheng, M., Jose, J.M., Yu, C.: Beibei kong, zhijin wang, bo hu, and zang li. 2022. transrec: learning transferable recommendation from mixture-of-modality feedback. arXiv preprint arXiv:2206.06190 (2022)

59. Wang, L., Lim, E.P.: Zero-shot next-item recommendation using large pretrained language models. arXiv preprint arXiv:2304.03153 (2023)

60. Wang, W., Lin, X., Feng, F., He, X., Chua, T.S.: Generative recommendation: towards next-generation recommender paradigm. arXiv preprint arXiv:2304.03516 (2023)

61. Wang, X., Tang, X., Zhao, W.X., Wang, J., Wen, J.R.: Rethinking the evaluation for conversational recommendation in the era of large language models. arXiv preprint arXiv:2305.13112 (2023)

62. Wang, X., Zhou, K., Wen, J., Zhao, W.X.: Towards unified conversational recommender systems via knowledge-enhanced prompt learning. In: KDD (2022)

63. Wang, Y., et al.: Recmind: large language model powered agent for recommendation. arXiv preprint arXiv:2308.14296 (2023)

64. Wei, J., et al.: Finetuned language models are zero-shot learners. In: ICLR (2022)

65. Wei, W., et al.: Llmrec: large language models with graph augmentation for recommendation. In: WSDM (2024)

66. Wu, L., et al.: A survey on large language models for recommendation. arXiv preprint arXiv:2305.19860 (2023)

67. Xiao, S., et al.: Training large-scale news recommenders with pretrained language models in the loop. In: Zhang, A., Rangwala, H. (eds.) KDD 2022: The 28th ACM SIGKDD Conference on Knowledge Discovery and Data Mining, Washington, DC, USA, 14–18 August 2022, pp. 4215–4225. ACM (2022). https://doi.org/10.1145/3534678.3539120

68. Yuan, F., He, X., Karatzoglou, A., Zhang, L.: Parameter-efficient transfer from sequential behaviors for user modeling and recommendation. In: Huang, J.X., et al. (eds.) SIGIR (2020)

69. Yuan, F., Zhang, G., Karatzoglou, A., Jose, J.M., Kong, B., Li, Y.: One person, one model, one world: learning continual user representation without forgetting. In: Diaz, F., Shah, C., Suel, T., Castells, P., Jones, R., Sakai, T. (eds.) SIGIR (2021)

70. Zang, T., Zhu, Y., Liu, H., Zhang, R., Yu, J.: A survey on cross-domain recommendation: taxonomies, methods, and future directions. ACM Trans. Inf. Syst. **41**(2), 42:1–42:39 (2023). https://doi.org/10.1145/3548455

71. Zhang, J., Bao, K., Zhang, Y., Wang, W., Feng, F., He, X.: Is chatgpt fair for recommendation? evaluating fairness in large language model recommendation. arXiv preprint arXiv:2305.07609 (2023)

72. Zhang, J., et al.: Agentcf: collaborative learning with autonomous language agents for recommender systems. arXiv preprint arXiv:2310.09233 (2023)

73. Zhang, J., Xie, R., Hou, Y., Zhao, W.X., Lin, L., Wen, J.R.: Recommendation as instruction following: a large language model empowered recommendation approach. arXiv preprint arXiv:2305.07001 (2023)

74. Zhang, Z., Wang, B.: Prompt learning for news recommendation. arXiv preprint arXiv:2304.05263 (2023)

75. Zhao, C., Li, C., Xiao, R., Deng, H., Sun, A.: CATN: cross-domain recommendation for cold-start users via aspect transfer network. In: Huang, J.X., et al. (eds.) SIGIR, pp. 229–238. ACM (2020). https://doi.org/10.1145/3397271.3401169

76. Zhao, W.X., Lin, Z., Feng, Z., Wang, P., Wen, J.R.: A revisiting study of appropriate offline evaluation for top-n recommendation algorithms. ACM Trans. Inf. Syst. **41**(2), 1–41 (2022)
77. Zhao, W.X., et al.: Recbole: towards a unified, comprehensive and efficient framework for recommendation algorithms. In: CIKM (2021)
78. Zhao, W.X., et al.: A survey of large language models. arXiv preprint arXiv:2303.18223 (2023)
79. Zhao, Z., Wallace, E., Feng, S., Klein, D., Singh, S.: Calibrate before use: improving few-shot performance of language models. In: ICML (2021)
80. Zheng, B., Hou, Y., Lu, H., Chen, Y., Zhao, W.X., Wen, J.R.: Adapting large language models by integrating collaborative semantics for recommendation. arXiv preprint arXiv:2311.09049 (2023)
81. Zhou, K., et al.: S3-rec: self-supervised learning for sequential recommendation with mutual information maximization. In: CIKM (2020)
82. Zhu, F., Chen, C., Wang, Y., Liu, G., Zheng, X.: DTCDR: a framework for dual-target cross-domain recommendation. In: Zhu, W., et al. (eds.) CIKM, pp. 1533–1542. ACM (2019). https://doi.org/10.1145/3357384.3357992
83. Zhu, F., Wang, Y., Chen, C., Liu, G., Zheng, X.: A graphical and attentional framework for dual-target cross-domain recommendation. In: Bessiere, C. (ed.) IJCAI, pp. 3001–3008. ijcai.org (2020). https://doi.org/10.24963/ijcai.2020/415
84. Zhu, F., Wang, Y., Chen, C., Zhou, J., Li, L., Liu, G.: Cross-domain recommendation: challenges, progress, and prospects. In: Zhou, Z. (ed.) Proceedings of the Thirtieth International Joint Conference on Artificial Intelligence, IJCAI 2021, Virtual Event/Montreal, Canada, 19–27 August 2021, pp. 4721–4728. ijcai.org (2021). https://doi.org/10.24963/ijcai.2021/639
85. Zhu, Y., et al.: Personalized transfer of user preferences for cross-domain recommendation. In: Candan, K.S., Liu, H., Akoglu, L., Dong, X.L., Tang, J. (eds.) WSDM, pp. 1507–1515. ACM (2022). https://doi.org/10.1145/3488560.3498392

Simulating Follow-Up Questions
in Conversational Search

Johannes Kiesel[1]([📧])[ID], Marcel Gohsen[1][ID], Nailia Mirzakhmedova[1][ID],
Matthias Hagen[2][ID], and Benno Stein[1][ID]

[1] Bauhaus-Universität Weimar, Bauhausstr. 9a, 99423 Weimar, Germany
{johannes.kiesel,marcel.gohsen,nailia.mirzakhmedova,
benno.stein}@uni-weimar.de
[2] Friedrich-Schiller Universität Jena, Ernst-Abbe-Platz 2, 07743 Jena, Germany
matthias.hagen@uni-jena.de

Abstract. Evaluating conversational search systems based on simulated user interactions is a potential approach to overcome one of the main problems of static conversational search test collections: the collections contain only very few of all the plausible conversations on a topic. Still, one of the challenges of user simulation is generating realistic follow-up questions on given outputs of a conversational system. We propose to address this challenge by using state-of-the-art language models and find that: (1) on two conversational search datasets, the tested models generate questions that are semantically similar to those in the datasets, especially when tuned for follow-up questions; (2) the generated questions are mostly valid, related, informative, and specific according to human assessment; and (3) for influencing the characteristics of the simulated questions, small changes to the prompt are insufficient.

Keywords: Conversational search · Follow-up questions · Simulation

1 Introduction

Conversational search has been a focus of information retrieval research for more than a decade [4], yet many challenges remain, particularly in system evaluation. In conversational search, users iteratively deepen their knowledge through dialog with the search system. Unfortunately, such highly interactive conversations pose a challenge for system evaluation: The many plausible utterances from which the user and system choose each turn can lead to completely different conversations, but traditional test collections can only cover a few such conversation branches.

To acquire sufficient interaction data with a conversational search system for its evaluation—without expensive testing with real users—one can simulate users at a large scale [8]. With this evaluation methodology, a simulation system interacts with a search system to mimic a real user as closely as possible. Simulation parameters affect which actions the simulated user takes, enabling the simulation of a wide variety of users. Moreover, it allows for the repetition of a simulation with a different search system, facilitating A/B testing.

© The Author(s), under exclusive license to Springer Nature Switzerland AG 2024
N. Goharian et al. (Eds.): ECIR 2024, LNCS 14609, pp. 382–398, 2024.
https://doi.org/10.1007/978-3-031-56060-6_25

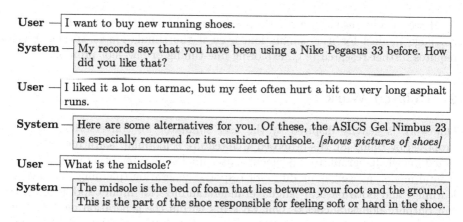

User — I want to buy new running shoes.

System — My records say that you have been using a Nike Pegasus 33 before. How did you like that?

User — I liked it a lot on tarmac, but my feet often hurt a bit on very long asphalt runs.

System — Here are some alternatives for you. Of these, the ASICS Gel Nimbus 23 is especially renowed for its cushioned midsole. *[shows pictures of shoes]*

User — What is the midsole?

System — The midsole is the bed of foam that lies between your foot and the ground. This is the part of the shoe responsible for feeling soft or hard in the shoe.

Fig. 1. An envisioned dialogue between a user and a conversational search system, taken from Balog [8], including the follow-up question "What is the midsole?"

Particularly important for user simulation is to have the simulator pose realistic follow-up questions, which are questions about something the search system said earlier (cf. Fig. 1). Follow-up questions are key for both conversational search [4] and conversational question answering [22], and are frequently contained in conversational search datasets: In TREC CAsT 2022 [27], we found that 54% of the user utterances after the initial one contain follow-up questions.

In this paper, we explore the potential of using language models to simulate user follow-up questions. Language models have recently reached an impressive level in mimicking human language use, especially in continuing conversations. Moreover, they can be tuned and/or instructed to conduct the discussions in specific manners, making them a natural choice for user simulation.

Specifically, we pose and address the following research questions on follow-up question generation through state-of-the-art language models:

1. Are the generated questions lexically or semantically similar to human-generated questions in conversational search datasets? (cf. Sect. 4)
2. According to human judgments, are the generated questions appropriate follow-up questions for the respective conversation? (cf. Sect. 5)
3. Can one adapt the prompt to steer the language model to generate questions according to different user profiles? (cf. Sect. 6)

In the following, Sect. 3 covers the task of follow-up question simulation, the selection of suitable datasets to analyze simulators, and the models that we explore for simulation. Section 4 compares the follow-up questions our models generate to those in the datasets using standard automated metrics, showing that they are semantically similar. Section 5 presents a human assessment of the simulated questions, showing that experts judge them as valid, related to the context, informative, and mostly specific. Section 6 investigates whether the models can be set to simulate specific users (e.g., naive or savvy ones) through small intuitive prompt modifications, and presents a negative result.

However, our results are mostly positive and highlight the promise of large language models for user simulation, even if the simulation of specific users requires further research (e.g., exploring fine-tuning or few-shot prompts).

2   Related Work

Our work is heavily inspired by the user simulator for conversational search proposed by Balog [8]. His proposed architecture is modular, featuring a personalized knowledge graph to implement a user model, an interaction model, and a mental model, as well as modules for response generation (planning, execution, learning) and both natural language understanding (for input) and generation (for output, alongside clicks). Our work fits within the generation module, but can, for example, also be integrated as a module in generic AI assistant platforms like DeepPavlov [42]. These use several independent modules to generate candidate responses for an input and one control model to select from these candidates. Simulators of both kinds can then interact with conversational search frameworks like Macaw [41] or DECAF [2] to evaluate specific conversational retrieval models. Similar to our work, Kim and Lipani [23] simulate user utterances for conversational search, but do not perform a human assessment of the simulation, nor can their model integrate with the aforementioned frameworks.

Concepts Related to Follow-Up Question Simulation. Several concepts related to user simulation exist in the literature. Simulating user utterances is different to both simulating user feedback [28] and generating (for the system) clarification questions [3,19,30]. Our simulation is similar to question suggestion [33]. However, question suggestion attempts to find or generate useful next questions for a user to ask based on their previous question ("People also ask"), not the system's answer. Moreover, follow-up questions are different to question reformulations [9]: the former correspond to new questions based on what was said, whereas the latter are (small) changes to the previous question, usually intended to improve the question if the answer was not satisfactory. Conceptually, follow-up questions open a new context space in the conversation [31], whereas reformulations do not. Finally, follow-up questions seem related to but are actually independent to conversational action types. For example, each of the user actions defined by Azzopardi et al. [6] (reveal, inquire, navigate, interrupt, interrogate) can be performed using a follow-up question, but also other types of questions.

Recent Approaches to Evaluation in Information Retrieval. Ideas to replace the Cranfield paradigm of evaluation based on document collection, topics, and relevance judgments date back over two decades [35], but are only recently gaining more attention. Faggioli et al. [15] discuss pros and cons of automating relevance judgments for IR along a human–machine collaboration spectrum and provide an overview of existing ideas. Dietz et al. [12] and Dietz and Dalton [13] present a method for generating (non-conversational) test collections for various tasks from Wikipedia, and show that models are ranked similarly for a manual and their generated collection. Another approach is to create test collections that contain questions along with the queries instead of relevance judgments. A document is considered more relevant to the query the more of the questions a language model can answer using the information from the document [34]. This approach could be paired with simulated users to evaluate conversational systems.

3 Simulating Follow-Up Questions via Language Models

This section defines the task at hand (Sect. 3.1), selects suitable existing datasets (Sect. 3.2), and selects language models (Sect. 3.3) for our empirical investigations on simulating follow-up questions in the later sections.

3.1 Defining Follow-Up Question Simulation

For user simulation in conversational search, as described by Balog [8], we define the task of follow-up question simulation as follows:

Task: Given an informative textual response to a user's query, generate a question that the user might ask based on the provided information.

A deliberately vague point of our definition is the restriction to follow-up questions *that the user might ask*. Balog [8] provides a list of several factors that influence what a user might ask: current personal interests and preferences, persona (personality, educational and socio-economical background, etc.), current knowledge, and current understanding of the system's capabilities. In this initial paper, we simplify matters by concentrating our main experiments on a generic user, but make a first attempt at user modeling in Sect. 6. We also refrain from instructing the simulator on which point exactly to follow up on, both for simplicity and as it is currently unclear whether user simulators should do so.

3.2 Selecting Datasets for Follow-Up Question Simulation

We select TREC CAsT 2022 [27] and Webis-Nudged-Questions-23 [16] for our experiments from a larger pool of datasets that we reviewed. Neither the MS MARCO conversational [7] nor the Webis-Conversational-Query-Reformulations-21 [21] contain system responses. The questions in the SCAI-QReCC dataset [5,38] are on texts (generated from search results) that are unavailable. The Wizard of Wikipedia dataset [14] is not task-oriented. The MultiWOZ dataset [10,40] focuses heavily on transactions (e.g., hotel booking). The Webis-Exhibition-Questions-21 [20] contains questions on elements displayed in a virtual environment, which goes beyond the scope of our study.

The two selected datasets focus on conversational search and contain system responses and subsequent user utterances. The TREC CAsT 2022 dataset is the newest in a series of four datasets created for the TREC CAsT shared task [27]. It includes 205 unique turns over 50 conversations. We manually identified all follow-up questions in the user utterances, with 26 targeting a specific system response preceding them; these questions were matched with their respective system responses for our experiments. We utilize 100 system responses that received follow-up questions. The Webis-Nudged-Questions-23 [16] contains 8376 crowd-sourced questions to 30 short informative texts on three topics (argument search, exhibition, and product search). We employ all 30 texts as system responses and a random sample of 30 questions per response as user utterances.

3.3 Selecting and Tuning Models for Follow-Up Question Simulation

We've chosen GPT-4 [26] as the latest successor to Chat-GPT, and the Llama models as robust open-source alternatives with strong performance [37]. For Llama, we compare models of different sizes (Llama2-7B with 7 billion parameters and Llama2-13B) and versions (Alpaca-7B, a tuned Llama1 [36]). We found Llama2-chat models to perform similarly and thus discarded them.

Figure 2 shows the prompt that we use for all models, with variable texts highlighted. The prompt reflects the instructions given to crowdworkers to generate follow-up questions for the Inquisitive dataset [24] (there called questions-under-discussion and used for monological texts, but otherwise equivalent).

To further adapt the models to follow-up question simulation, we employ instruction fine-tuning for Llama models using the same prompt. Fine-tuning GPT-4 is not available at the time of writing. We fine-tune on three datasets: TREC CAsT 2022 and Webis-Nudged-Questions-23, also used for evaluation, and Inquisitive (see above) as a generic non-conversational dataset. For the first two datasets, we ensure that the model is not tuned on the same conversation it's generating questions for, using a 3-fold cross-validation approach.

For fine-tuning and generation we use the HuggingFace Transformers library [39] on a single NVIDIA A100 GPU (40 GB). We use standard low-rank adaptation [17] for efficient fine-tuning, applying low-rank updates ($r = 64$) with a scaling parameter of $\alpha = 16$ to all linear layers. We fine-tune each model for one epoch, as further epochs did not reduce the loss significantly. We used a batch size of 4 and a learning rate of $2 \cdot 10^{-4}$.

In our evaluation, we assess zero-shot generation of both the original and fine-tuned models.[1] To illustrate the simulated follow-up questions, the Appendix shows examples and the most common leading bigrams for some models.

4 Comparing Simulated and Original Human Questions

As a first approach to check whether the questions that large language models generate are questions a user might ask (as per our task definition in Sect. 3.1), we analyze whether the questions are similar to the ones that humans asked in the respective dataset. This corresponds to our first research question: Are the generated questions lexically or semantically similar to human-generated questions in conversational search datasets?

---

[1] Code, data, and models are available at https://github.com/webis-de/ECIR-24.

```
### Instruction: Follow-up questions are the questions elicited from
readers as they naturally read through text. You are a savvy user. You
ask elaborate questions about the implications of what was being said.
Given the text below, write follow-up questions that you would ask if you
were reading this text for the first time.

### Text: The nation's largest gun-rights group is taking some Texans to
task over their headline-generating demonstrations advocating the legal,
open carrying of weapons.

### Follow-up questions: <model response>
```

Fig. 2. Example of a prompt we employ to simulate a user asking questions. Text with a red background is always adapted to the respective conversation. Text with a purple background is used to model different users (cf. Sect. 6) and is blank for the other experiments. (Color figure online)

4.1 Similarity Computation

We assess both lexical (i.e., same word sequence) and semantic similarity (similar meaning) between simulated and human questions. For both we use standard metrics from the machine translation and paraphrasing literature.

For lexical similarity, we employ the standard metric in machine translation, BLEU [29]. BLEU computes word n-gram precision between a candidate text and a set of reference texts. We compute BLEU up to 4-grams and apply Smoothing 4 from Chen and Cherry [11] to prevent inflation of BLEU for short questions.

For semantic similarity, we employ a nowadays commonly used metric in paraphrasing, Sentence-BERT [32]. This method embeds both simulated and human generated sentences with Sentence-BERT along with TinyBERT [18], as TinyBERT embeddings are tuned for natural language understanding and the embedding process is reasonably efficient. The semantic similarity is then the cosine similarity of the embedding vectors, as it is standard for embeddings.

However, many follow-up questions can be imagined for each system response, and we thus generate several questions per model and compare these to all human-generated questions in the datasets (cf. Sect. 3.2). We repeat the process if no question is generated up to five times. We then calculate the overall score for a set of questions simulated by a model for a dataset, $\hat{Q}$, and the human questions in the same dataset, Q, for a similarity measure φ (BLEU or Sentence-BERT). Let $Q_{s(\hat{q})}$ be the set of questions within the human questions Q that are asked for the same system response as a simulated question $\hat{q}$. Then, the overall score is the average similarity across each simulated question $\hat{q}$ to its most similar question $q \in Q_{s(\hat{q})}$, with similarity measured according to φ:

$$\text{score}_\varphi(\hat{Q}, Q) = \frac{\sum_{\hat{q} \in \hat{Q}} \left(\max_{q \in Q_{s(\hat{q})}} \varphi(\hat{q}, q) \right)}{|\hat{Q}|}$$

To reduce randomness, we report mean scores for 10 generation runs each.

Table 1. Scores for BLUE and Sentence-BERT for simulated questions on TREC CAsT 22 (CAsT) and Webis-Nudged-Questions-23 (WNQ). Reported values are the means of 10 runs. All values from 0 (no similarity) to 1 (identical), with best values in each column marked bold.

Model		BLEU		Sent.-BERT	
Base	Tuning	CAsT	WNQ	CAsT	WNQ
GPT-4	none	0.02	0.11	0.22	0.68
Alpaca-7B	none	0.03	0.14	**0.23**	0.70
Alpaca-7B	CAsT	0.03	0.08	0.20	0.46
Alpaca-7B	Inquisitive	0.03	0.13	0.21	0.63
Alpaca-7B	WNQ	0.03	0.13	0.22	0.66
Llama2-7B	none	0.03	0.18	0.18	0.63
Llama2-7B	CAsT	**0.04**	0.09	0.19	0.45
Llama2-7B	Inquisitive	0.03	0.20	0.20	0.70
Llama2-7B	WNQ	0.03	0.21	0.20	**0.71**
Llama2-13B	none	0.03	0.19	0.21	0.66
Llama2-13B	CAsT	**0.04**	0.07	0.20	0.41
Llama2-13B	Inquisitive	**0.04**	**0.23**	0.22	0.68
Llama2-13B	WNQ	0.03	0.22	0.20	0.70

4.2   Results

For lexical similarity ($\varphi = $ BLEU), Table 1 shows that scores are very low for the TREC CAsT dataset for all models (between 0.02 and 0.04), but higher for Webis-Nudged-Questions-23 (between 0.07 and 0.23). Higher scores for Webis-Nudged-Questions-23 are expected, since this dataset has on average 30 human-generated questions per system response that are matched with each simulated question, whereas for TREC CAsT there are only 1.3 on average. However, we still conclude that the lexical similarity of the simulated questions to the human questions is quite low for both datasets. Moreover, we find that models fine-tuned on the TREC CAsT dataset generate questions for Webis-Nudged-Questions-23 that are even more dissimilar than those generated without fine-tuning. Interestingly, fine-tuning on the Inquisitive dataset seems to have no strong negative effect, and in some cases even leads to the most similar questions.

For semantic similarity ($\varphi = $ Sentence-BERT), Table 1 shows much higher scores. Llama2-7B tuned on Webis-Nudged-Questions-23 achieves a Sentence-BERT value of 0.71 on the same dataset. Moreover, we observe the same performance decrease when fine-tuning on TREC CAsT as when fine-tuning on Webis-Nudged-Questions-23, indicating that the questions in the datasets are both lexically and semantically dissimilar.

In summary, we find that the simulated questions rarely match human questions when it comes to lexical similarity, but much more so for semantic similarity. However, one has to consider that there are on average only 1.3 human-generated questions per system response for TREC CAsT, yet many follow-up questions are plausible for each system response, which naturally reduces the score. Therefore, we conclude that the simulated questions are relatively similar, at least in their meaning, to the questions that humans ask.

5   Judging Simulated Questions

As a second approach to determine whether the questions generated by large language models are questions users might ask (as per our task definition in Sect. 3.1), we employ human experts to evaluate the questions. This corresponds to our second research question: According to human judgments, are the generated questions appropriate follow-up questions for the respective conversation?

5.1   Human Judgment of Simulated Questions

To analyze whether a simulated utterance is an appropriate follow-up question, we adopt three criteria employed by Ko et al. [24] to evaluate the questions in the Inquisitive dataset (cf. Sect. 3.2) and extend it with the criterion of specificity suggested by Adiwardana et al. [1] for conversational systems. Specifically, we ask human judges the following for each simulated utterance: (1) Is it a valid question? An invalid question is incomplete, incomprehensible, or not even a question at all; (2) If it is a valid question, is it also related to the context, i.e., the system response?; (3) If it is a valid and related question, is it also an informative question? An answer for an informative question is not contained in the system response; and (4) If it is a valid, related, and informative question, is it also a specific question, i.e., not a question that could be asked for any system response? For example, a question to the effect of "tell me more" is not specific.

These binary questions are answered by three of the authors as experts on the matter. Each expert has a background in natural language processing and having inspected the Inquisitive dataset to gain an exact understanding of the task. However, to avoid annotation biases, the experts did not know which model produced an utterance or whether it was taken from the dataset, and they received utterances in a random order. We selected one question from each of the 13 models used in the previous experiment and paired it with the original questions in the dataset, yielding 1820 simulated responses and 7280 judgments. Since there is some subjectivity to the judgments, we ensured consistency that the same expert judged all utterances that were simulated for a system response. Finally, for assessing agreement, the three expert annotators independently evaluated all simulated utterances for one single TREC CAsT conversation. The agreement, measured by Fleiss κ, was found to be moderate for each rating, which we deemed acceptable for this task: 0.51 for "valid", 0.52 for "related", 0.53 for "informative", and 0.57 for "specific".

5.2   Results

Table 2 summarizes the evaluation results for different datasets and criteria. Notably, GPT-4 consistently outperforms other models in all dataset-criterion combinations. Nonetheless, all models produce valid questions—even more often than the crowdworkers employed for generating the Webis-Nudged-Questions-23. The drop in performance is minimal when considering relatedness, suggesting that current language models excel in these tasks. However, GPT-4 stands

Table 2. Ratio of simulated questions judged as valid, related, informative, and specific for the TREC CAsT 22 (CAsT) and Webis-Nudged-Questions-23 (WNQ). Each judgment implies the judgments to its left. Highest ratio in each column marked bold.

Model		Valid		Related		Informative		Specific	
Base	Tuning	CAsT	WNQ	CAsT	WNQ	CAsT	WNQ	CAsT	WNQ
GPT-4	none	**0.98**	0.97	0.97	**0.97**	**0.84**	**0.87**	**0.84**	**0.87**
Alpaca-7B	none	0.93	0.97	0.93	0.93	0.73	0.63	0.72	0.63
Alpaca-7B	CAsT	0.92	0.87	0.85	0.80	0.80	0.80	0.72	0.70
Alpaca-7B	Inquisitive	0.94	0.80	0.92	0.80	0.77	0.67	0.75	0.67
Alpaca-7B	WNQ	0.96	0.77	0.94	0.77	0.75	0.67	0.75	0.67
Llama2-7B	none	0.92	0.80	0.84	0.77	0.60	0.50	0.57	0.47
Llama2-7B	CAsT	0.94	0.93	0.84	0.70	0.76	0.57	0.73	0.43
Llama2-7B	Inquisitive	0.95	0.87	0.94	0.83	0.73	0.77	0.72	0.77
Llama2-7B	WNQ	0.96	**1.00**	0.94	0.93	0.65	0.63	0.65	0.63
Llama2-13B	none	0.90	0.93	0.88	0.90	0.57	0.50	0.52	0.43
Llama2-13B	CAsT	0.87	0.90	0.79	0.77	0.71	0.73	0.63	0.57
Llama2-13B	Inquisitive	**0.98**	0.90	**0.98**	0.83	0.77	0.67	0.75	0.67
Llama2-13B	WNQ	0.94	0.97	0.89	0.93	0.58	0.60	0.58	0.57
Original questions	-	0.95	0.60	0.91	0.50	0.87	0.40	0.77	0.40

out in terms of informativeness. It is worth noting that the models without fine-tuning consistently perform worse than their tuned counterparts. The utterances, judged as informative, can be considered valid follow-up questions. As the last two columns show, most such questions are also specific, although the models fine-tuned on TREC CAsT occasionally produce unspecific questions. This result is not surprising, given that TREC CAsT dataset itself contains unspecific questions (cf. original questions).

In summary, we find that, for the best models, the simulated utterances are often valid follow-up questions. GPT-4 is the best model for simulation and close to human performance (better than crowdworkers), but also Llama models perform well, especially when fine-tuned. Moreover, fine-tuning can be used to adapt the model to ask fewer specific questions. Therefore, we conclude that simulated questions are often valid follow-up questions as per human judgment.

6   Modeling Specific Users Through Prompt Modifications

As a third approach to assess whether the questions generated by large language models align with what users might ask (outlined in Sect. 3.1), we simulate distinct user profiles by modifying the model prompt. Human experts are then asked to evaluate whether the generated questions accurately reflect these modifications. We purposefully attempt to simulate specific users by only changing the model prompt, as prompt modifications are a cost-effective strategy and necessitate no additional training data. Thus, if prompt adjustments prove effective

Table 3. Ratio of simulated questions judged as asked by a naive, savvy, implication-focused, or reasons-focused user for the TREC CAsT 22 (CAsT) and Webis-Nudged-Questions-23 (WNQ). Ratios higher than for the same model without prompt modification marked bold.

Model			Naive		Savvy		Implications		Reasons	
Base	Tuning	Prompt	CAsT	WNQ	CAsT	WNQ	CAsT	WNQ	CAsT	WNQ
GPT-4	none		0.07	0.23	0.29	0.33	0.14	0.17	0.13	0.27
GPT-4	none	Naive+Implic.	**0.11**	0.20	0.29	**0.37**	**0.15**	**0.23**	**0.16**	0.17
GPT-4	none	Naive+Reasons	**0.20**	0.23	0.23	0.33	0.12	**0.40**	**0.34**	0.27
GPT-4	none	Savvy+Implic.	0.01	0.03	**0.54**	**0.77**	**0.25**	**0.50**	0.12	0.17
GPT-4	none	Savvy+Reasons	**0.08**	0.17	**0.34**	**0.60**	**0.19**	**0.53**	**0.23**	0.20
Alpaca-7B	none		0.16	0.23	0.22	0.20	0.19	0.30	0.11	0.10
Alpaca-7B	none	Naive+Implic.	0.15	0.13	0.15	**0.40**	**0.22**	0.30	0.09	0.10
Alpaca-7B	none	Naive+Reasons	0.11	0.20	0.18	**0.27**	0.19	**0.33**	0.07	0.10
Alpaca-7B	none	Savvy+Implic.	**0.18**	0.13	**0.27**	**0.53**	**0.20**	**0.47**	0.15	0.07
Alpaca-7B	none	Savvy+Reasons	0.07	0.17	**0.24**	**0.27**	0.19	0.30	**0.16**	0.07
Llama2-13B	none		0.52	0.40	0.01	0.03	0.06	0.07	0.12	0.13
Llama2-13B	none	Naive+Implic.	0.45	**0.57**	0.01	0.00	0.06	**0.20**	0.09	0.10
Llama2-13B	none	Naive+Reasons	0.47	**0.47**	**0.03**	0.03	0.06	**0.13**	**0.22**	**0.23**
Llama2-13B	none	Savvy+Implic.	0.38	0.33	**0.06**	0.03	**0.19**	**0.13**	0.06	0.03
Llama2-13B	none	Savvy+Reasons	0.43	0.20	**0.03**	**0.07**	**0.07**	0.07	**0.19**	0.13
Llama2-13B	Inquisitive		0.50	0.40	0.02	0.10	0.03	0.20	0.19	0.23
Llama2-13B	Inquisitive	Naive+Implic.	0.40	**0.70**	0.01	0.07	**0.06**	**0.27**	**0.30**	**0.27**
Llama2-13B	Inquisitive	Naive+Reasons	0.43	**0.70**	**0.07**	0.00	**0.06**	0.17	**0.38**	0.20
Llama2-13B	Inquisitive	Savvy+Implic.	0.45	**0.70**	0.02	0.03	**0.04**	**0.23**	**0.36**	**0.30**
Llama2-13B	Inquisitive	Savvy+Reasons	0.42	**0.57**	**0.03**	0.10	**0.05**	0.20	**0.30**	0.17
Original questions			0.42	0.63	0.05	0.13	0.14	0.20	0.12	0.07

in representing various user perspectives, language models can readily simulate a wide array of users with minimal effort. This corresponds to our third research question: Can one adapt the prompt to steer the language model to generate questions according to different user profiles?

6.1 Prompt Modifications

From the vast number of possible user attributes we selected two dimensions for our experiment: (1) naive vs. savvy user, corresponding to terms used in classical user simulation [25]; and (2) users focusing on questions about implications vs. reasons of something in the system response. We modify the prompt (Fig. 2) by adding the text: "You are a [savvy/naive] user. You ask [simple/elaborate] questions about the [implications/reasons] of what was being said." We use the same evaluation setup and experts as in Sect. 5 to have the experts judge if a given utterance aligns with the specified user type. Measured in the same way, Fleiss κ shows substantial agreement for rating questions as from a user that is focused on "implications" ($\kappa = 0.75$) and "savvy" ($\kappa = 0.63$), moderate agreement for focused on "reasons" ($\kappa = 0.46$), and fair agreement for "naive" ($\kappa = 0.37$).

6.2 Results

Due to time constraints, our experts could only evaluate 4 out of the 14 different models for each combination of savvy/naive and implications/reasons. As shown in Table 3, our attempts to enhance the simulation through minor prompt adjustments were not successful. Although the prompt modifications did clearly affect the simulation, the observed effects are not consistent with our hypothesis. Especially for Llama2-13B fine-tuned on the Inquisitive dataset, there is a significant increase in certain ratios, but this increase was limited to only one of the datasets. It appears that GPT-4 may be the most effective model for identifying savvy or naive users, as it generated more responses aligned with the prompt, particularly on the TREC CAsT dataset. However, this effect was not as pronounced for users focusing on implications or reasons.

In summary, we find that small modifications to the prompt are insufficient to steer the simulation towards specific user attributes. Of course, our experimental setup is limited: different modifications or different attributes could yield improved results. However, the attractiveness of small prompt modifications lies in their simplicity of implementation. Our results indicate that, at least with the tested models, this straightforward way of modeling users is not yet feasible.

7 Conclusion

User simulation is a promising yet hypothetical approach to the evaluation of conversational search systems, addressing the drawbacks of static test collections for a highly interactive task. This paper presents another step towards a complete user simulation—the simulation of follow-up questions to system responses. As per the literature, follow-up questions are frequent and of key importance in conversational search. We showed that large language models are capable of simulating users asking follow-up questions. The semantic similarity (Sentence-BERT) to human-generated questions reaches as high as 0.71 for one of the two conversational search datasets we tested on. Moreover, human experts judged the simulated questions in blind evaluation to be mostly valid, related to the system response, and informative. Furthermore, we found that fine-tuning models to datasets, even if they are out-of-domain, can improve the simulation—more so than using larger models. While GPT-4 is ahead of the open models in our benchmark, nearly matching human performance, the gap is not excessive. However, we also presented a negative result: although the prompt interface to language models suggests that modifications to the prompt could be used to alter the simulation to represent different users, we found that our slight modifications were insufficient and failed to control the simulation as intended. Nonetheless, our results are mostly positive and highlight the promise of large language models for user simulation, even if the simulation of specific users requires further research. For example, instead of prompt-modification one could explore fine-tuning or few-shot prompts. Both methods attempt to mimic a user based on a few example questions. The latter adds these examples to the prompt, which is more direct, but limited to only a few examples. Another venue for research is to create a dataset of users to be then used in simulation, possibly by extracting user attributes from existing conversations [43].

Although many questions remain open, our results provide further evidence of the potential of user simulation to evaluate conversational search systems. Furthermore, our method is not restricted to the simulation of follow-up questions, and can be adapted to simulate other user interactions in the future.

8   Limitations

The question of how to simulate users of a conversational search system has many facets, many of which we could not address in this paper. Even for the method of language models, which we focused on in this paper, we could not explore the entire parameter space for the simulation. We approached the task with both zero-shot and fine-tuning, but not with the middle ground of few-shot learning (also called in-context learning). In terms of modeling specific users, this work has barely scratched the surface of what is possible. Going back to the idea of the personal knowledge graph from Balog [8], one could also use methods that integrate such knowledge graphs into language models to model different users. Furthermore, we only used two different datasets, which naturally cannot represent the many different scenarios in which a user might search—for many of which no dataset currently exists. Finally, we did not test our simulator with an actual retrieval system, but evaluated the simulation as it continues a human conversation (for TREC CAsT). Ideally, the language model picked up the conversation and continued it naturally, but we have not evaluated whether it actually did so, nor do we know of any evaluation metrics for checking this.

Acknowledgements. This work was partially supported by the European Commission under grant agreement GA 101070014 (https://openwebsearch.eu)

A   Appendix

Most frequent leading bigrams (lemmatized) and their frequency for original questions and questions simulated by selected models (IT = Inquitisitive-tuned).

Rank	Original		Model					
			Llama2-7B		Llama2-7B (IT)		Llama2-13B	

For TREC CAsT 22

Rank	Original		Llama2-7B		Llama2-7B (IT)		Llama2-13B	
1	what [be]	0.12	what [be]	0.62	what [be]	0.38	what [be]	0.70
2	tell [i]	0.11	what [do]	0.15	why [be]	0.17	what [do]	0.06
3	how [do]	0.06	how [do]	0.04	how [do]	0.14	how [do]	0.04
4	what [make]	0.05	why [do]	0.03	why [do]	0.07	how [can]	0.02
5	so [what]	0.03	why [be]	0.02	how [be]	0.05	why [be]	0.02
6	can [you]	0.02	who [be]	0.02	how [can]	0.03	who [be]	0.02
7	how [be]	0.02	how [many]	0.01	why [would]	0.02	why [do]	0.02
8	let [we]	0.02	how [can]	0.01	how [much]	0.02	how [be]	0.01
9	how [can]	0.02	do [you]	0.01	what [do]	0.02	when [be]	0.01
10	what [about]	0.02	what [would]	0.01	how [many]	0.01	how [many]	0.01

For Webis-Nudged-Questions-23

Rank	Original		Llama2-7B		Llama2-7B (IT)		Llama2-13B	
1	what [be]	0.21	what [be]	0.61	what [be]	0.40	what [be]	0.69
2	why [be]	0.03	what [do]	0.17	how [do]	0.15	what [do]	0.05
3	how [do]	0.03	how [do]	0.04	why [be]	0.14	who [be]	0.04
4	how [much]	0.03	who [be]	0.03	how [be]	0.06	how [do]	0.04
5	who [be]	0.03	why [do]	0.02	why [do]	0.06	why [be]	0.02
6	why [do]	0.03	why [be]	0.02	how [would]	0.04	why [do]	0.02
7	how [many]	0.02	how [much]	0.01	why [would]	0.04	how [much]	0.01
8	how [be]	0.01	how [can]	0.01	how [much]	0.03	when [be]	0.01
9	what [do]	0.01	how [be]	0.01	what [do]	0.01	how [can]	0.01
10	what [other]	0.01	why [would]	0.01	how [many]	0.01	how [would]	0.01

Original follow-up question from the dataset and simulated questions by model.

System — Universal basic income, also referred to as UBI, is a fixed monthly payment that every citizen of a country receives from the government and which is sufficient to live on. Its supporters claim, above all, that it promotes social justice.

GPT-4 — Does UBI apply to children as well as adults?

Alpaca-7B — If so, in which countries do you think it could be implemented most successfully?

Alpaca-7B Inquisitive-tuned — What countries have implemented the UBI?

Alpaca-7B WNQ-tuned — Does it really promote social justice?

Alpaca-7B CAsT-tuned — Would it solve all the world's problems?

Llama2-7B — What are the arguments against universal basic income?

Llama2-7B Inquisitive-tuned — Why is this a social justice issue?

Llama2-7B WNQ-tuned — What are the advantages of UBI?

Llama2-7B CAsT-tuned — What's the impact on the economy?

Llama2-13B — How is UBI funded?

Llama2-13B Inquisitive-tuned — How is it determined that the amount is sufficient to live on?

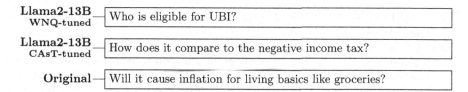

Llama2-13B
WNQ-tuned — Who is eligible for UBI?

Llama2-13B
CAsT-tuned — How does it compare to the negative income tax?

Original — Will it cause inflation for living basics like groceries?

References

1. Adiwardana, D., et al.: Towards a human-like open-domain chatbot. CoRR abs/2001.09977 (2020). https://arxiv.org/abs/2001.09977

2. Alessio, M., Faggioli, G., Ferro, N.: DECAF: a modular and extensible conversational search framework. In: 46th International ACM SIGIR Conference on Research and Development in Information Retrieval, SIGIR 2023. ACM (2023). https://doi.org/10.1145/3539618.3591913

3. Aliannejadi, M., Zamani, H., Crestani, F., Croft, W.B.: Asking clarifying questions in open-domain information-seeking conversations. In: Piwowarski, B., Chevalier, M., Gaussier, É., Maarek, Y., Nie, J., Scholer, F. (eds.) 42th International ACM SIGIR Conference on Research and Development in Information Retrieval, SIGIR 2019, pp. 475–484. ACM (2019). https://doi.org/10.1145/3331184.3331265

4. Allan, J., Croft, W.B., Moffat, A., Sanderson, M.: Frontiers, challenges, and opportunities for information retrieval: report from SWIRL 2012 the second strategic workshop on information retrieval in Lorne. SIGIR Forum **46**(1), 2–32 (2012). https://doi.org/10.1145/2215676.2215678

5. Anantha, R., Vakulenko, S., Tu, Z., Longpre, S., Pulman, S., Chappidi, S.: Open-domain question answering goes conversational via question rewriting. In: Toutanova, K., et al. (eds.) 2021 Conference of the North American Chapter of the Association for Computational Linguistics: Human Language Technologies, NAACL-HLT 2021, pp. 520–534. Association for Computational Linguistics (2021). https://doi.org/10.18653/v1/2021.naacl-main.44

6. Azzopardi, L., Dubiel, M., Halvey, M., Dalton, J.: Conceptualizing agent-human interactions during the conversational search process. In: Spina, D., Arguello, J., Joho, H., Kiseleva, J., Radlinski, F. (eds.) 2nd International Workshop on Conversational Approaches to Information Retrieval, CAIR 2018, July 2018. ACM (2019)

7. Bajaj, P., et al.: MS MARCO: a human generated machine reading comprehension dataset. CoRR abs/1611.09268 (2016). https://doi.org/10.48550/arXiv.1611.09268. https://arxiv.org/abs/1611.09268

8. Balog, K.: Conversational AI from an information retrieval perspective: remaining challenges and a case for user simulation. In: Alonso, O., Marchesin, S., Najork, M., Silvello, G. (eds.) 2nd International Conference on Design of Experimental Search & Information REtrieval Systems, DESIRES 2021, CEUR Workshop Proceedings, vol. 2950, pp. 80–90. CEUR-WS.org (2021)

9. Boldi, P., Bonchi, F., Castillo, C., Vigna, S.: Query reformulation mining: models, patterns, and applications. Inf. Retrieval **14**(3), 257–289 (2011). https://doi.org/10.1007/S10791-010-9155-3

10. Budzianowski, P., et al.: MultiWOZ - a large-scale multi-domain wizard-of-oz dataset for task-oriented dialogue modelling. In: Riloff, E., Chiang, D., Hockenmaier, J., Tsujii, J. (eds.) 2018 Conference on Empirical Methods in Natural Lan-

guage Processing, EMNLP 2018, pp. 5016–5026. Association for Computational Linguistics (2018)

11. Chen, B., Cherry, C.: A systematic comparison of smoothing techniques for sentence-level BLEU. In: Proceedings of the Ninth Workshop on Statistical Machine Translation, Baltimore, Maryland, USA, June 2014, pp. 362–367. Association for Computational Linguistics (2014). https://doi.org/10.3115/v1/W14-3346

12. Dietz, L., Chatterjee, S., Lennox, C., Kashyapi, S., Oza, P., Gamari, B.: Wikimarks: harvesting relevance benchmarks from Wikipedia. In: Amigó, E., Castells, P., Gonzalo, J., Carterette, B., Culpepper, J.S., Kazai, G. (eds.) 45th International ACM SIGIR Conference on Research and Development in Information Retrieval, SIGIR 2022, pp. 3003–3012. ACM (2022). https://doi.org/10.1145/3477495.3531731

13. Dietz, L., Dalton, J.: Humans optional? Automatic large-scale test collections for entity, passage, and entity-passage retrieval. Datenbank-Spektrum **20**(1), 17–28 (2020). https://doi.org/10.1007/s13222-020-00334-y

14. Dinan, E., Roller, S., Shuster, K., Fan, A., Auli, M., Weston, J.: Wizard of Wikipedia: knowledge-powered conversational agents. In: 7th International Conference on Learning Representations, ICLR 2019. OpenReview.net (2019)

15. Faggioli, G., et al.: Perspectives on large language models for relevance judgment. CoRR abs/2304.09161 (2023). https://doi.org/10.48550/arXiv.2304.09161

16. Gohsen, M., Kiesel, J., Korashi, M., Ehlers, J., Stein, B.: Guiding oral conversations: how to nudge users towards asking questions? In: ACM SIGIR Conference on Human Information Interaction and Retrieval, CHIIR 2023, March 2023, pp. 34–42. ACM, New York (2023). https://doi.org/10.1145/3576840.3578291

17. Hu, E.J., et al.: LoRA: low-rank adaptation of large language models. In: 10th International Conference on Learning Representations, ICLR 2022. OpenReview.net (2022)

18. Jiao, X., et al.: TinyBERT: distilling BERT for natural language understanding. CoRR abs/1909.10351 (2019)

19. Kiesel, J., Bahrami, A., Stein, B., Anand, A., Hagen, M.: Toward voice query clarification. In: 41st International ACM Conference on Research and Development in Information Retrieval, SIGIR 2018, July 2018, pp. 1257–1260. ACM (2018). https://doi.org/10.1145/3209978.3210160. https://dl.acm.org/doi/10.1145/3209978.3210160

20. Kiesel, J., Bernhard, V., Gohsen, M., Roth, J., Stein, B.: What is that? Crowdsourcing questions to a virtual exhibition. In: Elsweiler, D. (ed.) 2022 Conference on Human Information Interaction & Retrieval, CHIIR 2022, March 2022, pp. 358–362. ACM (2022). https://doi.org/10.1145/3498366.3505836

21. Kiesel, J., Cai, X., Baff, R.E., Stein, B., Hagen, M.: Toward conversational query reformulation. In: Alonso, O., Najork, M., Silvello, G. (eds.) 2nd International Conference on Design of Experimental Search & Information Retrieval Systems, DESIRES 2021, September 2021, CEUR Workshop Proceedings, vol. 2950, pp. 91–101 (2021)

22. Kim, G., Kim, H., Park, J., Kang, J.: Learn to resolve conversational dependency: a consistency training framework for conversational question answering. In: Zong, C., Xia, F., Li, W., Navigli, R. (eds.) 59th Annual Meeting of the Association for Computational Linguistics and the 11th International Joint Conference on Natural Language Processing, ACL/IJCNLP 2021, pp. 6130–6141. Association for Computational Linguistics (2021). https://doi.org/10.18653/v1/2021.acl-long.478

23. Kim, T.E., Lipani, A.: A multi-task based neural model to simulate users in goal oriented dialogue systems. In: Amigó, E., Castells, P., Gonzalo, J., Carterette, B.,

Culpepper, J.S., Kazai, G. (eds.) 45th International ACM SIGIR Conference on Research and Development in Information Retrieval, SIGIR 2022, pp. 2115–2119. ACM (2022). https://doi.org/10.1145/3477495.3531814

24. Ko, W.J., Chen, T.Y., Huang, Y., Durrett, G., Li, J.J.: Inquisitive question generation for high level text comprehension. In: Webber, B., Cohn, T., He, Y., Liu, Y. (eds.) 2020 Conference on Empirical Methods in Natural Language Processing, EMNLP 2020, pp. 6544–6555. Association for Computational Linguistics (2020). https://doi.org/10.18653/v1/2020.emnlp-main.530

25. Maxwell, D., Azzopardi, L.: Information scent, searching and stopping. In: Pasi, G., Piwowarski, B., Azzopardi, L., Hanbury, A. (eds.) ECIR 2018. LNCS, vol. 10772, pp. 210–222. Springer, Cham (2018). https://doi.org/10.1007/978-3-319-76941-7_16

26. OpenAI: GPT-4 technical report (2023). https://doi.org/10.48550/arXiv.2303.08774

27. Owoicho, P., Dalton, J., Aliannejadi, M., Azzopardi, L., Trippas, J., Vakulenko, S.: TREC CAsT 2022: going beyond user ask and system retrieve with initiative and response generation. In: Voorhees, E.M., Ellis, A. (eds.) 31st Text REtrieval Conference, TREC 2022. NIST Special Publication, National Institute of Standards and Technology (2022)

28. Owoicho, P., Sekulic, I., Aliannejadi, M., Dalton, J., Crestani, F.: Exploiting simulated user feedback for conversational search: ranking, rewriting, and beyond. In: Chen, H.H., Duh, W.J.E., Huang, H.H., Kato, M.P., Mothe, J., Poblete, B. (eds.) 46th International ACM SIGIR Conference on Research and Development in Information Retrieval, SIGIR 2023, pp. 632–642. ACM (2023). https://doi.org/10.1145/3539618.3591683

29. Papineni, K., Roukos, S., Ward, T., Zhu, W.J.: BLEU: a method for automatic evaluation of machine translation. In: Proceedings of the 40th Annual Meeting of the Association for Computational Linguistics, Philadelphia, Pennsylvania, USA, July 2002, pp. 311–318. Association for Computational Linguistics (2022). https://doi.org/10.3115/1073083.1073135

30. Rao, S., Danumé III, H.: Answer-based adversarial training for generating clarification questions. In: Burstein, J., Doran, C., Solorio, T. (eds.) 2019 Conference of the North American Chapter of the Association for Computational Linguistics: Human Language Technologies, NAACL-HLT 2019, pp. 143–155. Association for Computational Linguistics (2019). https://doi.org/10.18653/V1/N19-1013

31. Reichman, R.: Getting Computers to Talk Like You and Me: Discourse Context, Focus, and Semantics: (An ATN Model). MIT Press (1985)

32. Reimers, N., Gurevych, I.: Sentence-BERT: Sentence Embeddings using Siamese BERT-Networks, August 2019

33. Rosset, C., et al.: Leading conversational search by suggesting useful questions. In: Huang, Y., King, I., Liu, T., van Steen, M. (eds.) The Web Conference 2020, WebConf 2020, pp. 1160–1170. ACM/IW3C2 (2020). https://doi.org/10.1145/3366423.3380193

34. Sander, D.P., Dietz, L.: EXAM: how to evaluate retrieve-and-generate systems for users who do not (yet) know what they want. In: Alonso, O., Marchesin, S., Najork, M., Silvello, G. (eds.) 2nd International Conference on Design of Experimental Search & Information REtrieval Systems, DESIRES 2021, CEUR Workshop Proceedings, vol. 2950, pp. 136–146. CEUR-WS.org (2021)

35. Soboroff, I., Nicholas, C.K., Cahan, P.: Ranking retrieval systems without relevance judgments. In: Croft, W.B., Harper, D.J., Kraft, D.H., Zobel, J. (eds.) Proceedings

of the 24th Annual International ACM SIGIR Conference on Research and Development in Information Retrieval, SIGIR 2001, pp. 66–73. ACM (2001). https://doi.org/10.1145/383952.383961

36. Taori, R., et al.: Stanford alpaca: an instruction-following LLaMA model (2023). https://github.com/tatsu-lab/stanford_alpaca

37. Touvron, H., et al.: LLaMA: open and efficient foundation language models. CoRR abs/2302.13971 (2023). https://doi.org/10.48550/arXiv.2302.13971

38. Vakulenko, S., Kiesel, J., Fröbe, M.: SCAI-QReCC shared task on conversational question answering. In: Calzolari, N., et al. (eds.) 14th Language Resources and Evaluation Conference, LREC 2022, Paris, France, pp. 4913–4922. European Language Resources Association (ELRA) (2022)

39. Wolf, T., et al.: HuggingFace's transformers: state-of-the-art natural language processing. CoRR abs/1910.03771 (2019). https://doi.org/10.48550/arXiv.1910.03771. https://arxiv.org/abs/1910.03771

40. Ye, F., Manotumruksa, J., Yilmaz, E.: MultiWOZ 2.4: a multi-domain task-oriented dialogue dataset with essential annotation corrections to improve state tracking evaluation. In: Lemon, O., et al. (eds.) 23rd Annual Meeting of the Special Interest Group on Discourse and Dialogue, SIGDIAL 2022, pp. 351–360. Association for Computational Linguistics (2022)

41. Zamani, H., Craswell, N.: Macaw: an extensible conversational information seeking platform. In: Huang, J.X., et al. (eds.) 43rd International ACM SIGIR Conference on Research and Development in Information Retrieval, SIGIR 2020, pp. 2193–2196. ACM (2020). https://doi.org/10.1145/3397271.3401415

42. Zharikova, D., et al.: DeepPavlov dream: platform for building generative AI assistants. In: Bollegala, D., Huang, R., Ritter, A. (eds.) 61st Annual Meeting of the Association for Computational Linguistics, ACL 2023, pp. 599–607. Association for Computational Linguistics (2023). https://doi.org/10.18653/v1/2023.acl-demo.58

43. Zhu, L., Li, W., Mao, R., Pandelea, V., Cambria, E.: PAED: zero-shot persona attribute extraction in dialogues. In: Rogers, A., Boyd-Graber, J.L., Okazaki, N. (eds.) 61st Annual Meeting of the Association for Computational Linguistics, ACL 2023, pp. 9771–9787. Association for Computational Linguistics (2023). https://doi.org/10.18653/v1/2023.acl-long.544

Adapting Standard Retrieval Benchmarks to Evaluate Generated Answers

Negar Arabzadeh[✉], Amin Bigdeli, and Charles L. A. Clarke

University of Waterloo, Waterloo, Canada
{narabzad,abigdeli,claclark}@uwaterloo.ca

Abstract. Large language models can now directly generate answers to many factual questions without referencing external sources. Unfortunately, relatively little attention has been paid to methods for evaluating the quality and correctness of these answers, for comparing the performance of one model to another, or for comparing one prompt to another. In addition, the quality of generated answers are rarely directly compared to the quality of retrieved answers. As models evolve and prompts are modified, we have no systematic way to measure improvements without resorting to expensive human judgments. To address this problem we adapt standard retrieval benchmarks to evaluate answers generated by large language models. Inspired by the BERTScore metric for summarization, we explore two approaches. In the first, we base our evaluation on the benchmark relevance judgments. We empirically run experiments on how information retrieval relevance judgments can be utilized as an anchor to evaluating the generated answers. In the second, we compare generated answers to the top results retrieved by a diverse set of retrieval models, ranging from traditional approaches to advanced methods, allowing us to measure improvements without human judgments. In both cases, we measure the similarity between an embedded representation of the generated answer and an embedded representation of a known, or assumed, relevant passage from the retrieval benchmark. In our experiments, we evaluate a range of generative models, including several GPT-based variants and open-source large language models using a variety of prompts, including "liar" prompts intended to produce reasonable but incorrect answers. For retrieval benchmarks, we use the MS MACRO dev set, the TREC Deep Learning 2019 dataset, and the TREC Deep Learning 2020 dataset. Our experimental results support the adaption of standard benchmarks to the evaluation of generated answers.

1 Introduction

In the evolving landscape of Natural Language Processing (NLP), Large Language Models (LLMs) have gained significant attention [8,11]. These models have empowered numerous applications, offering capabilities that span from conversational systems to complex textual generation tasks [3,20,25,58]. As the race to develop even more powerful LLMs intensifies, the focus predominantly lies on their architecture, scale, and application diversity [2,7,9,17,23]. In applying LLMs in practical settings a critical component sometimes gets

© The Author(s), under exclusive license to Springer Nature Switzerland AG 2024
N. Goharian et al. (Eds.): ECIR 2024, LNCS 14609, pp. 399–414, 2024.
https://doi.org/10.1007/978-3-031-56060-6_26

overlooked: a systematic evaluation of these models in these settings [34,65]. In this paper, we focus on the evaluation of LLMs for question answering in settings where retrieval-based approaches has historically been employed, particularly approaches based on retrieving short passages intended to answer a question [24,26,38,51]. The ability to generate answers directly without referencing external sources represents a significant milestone [29,52,61,62]. However, key questions persist: How can we measure the quality and accuracy of these generated answers? How can we quantitatively compare the output of one LLM with another? How do we quantitatively assess the relative effectiveness of different prompts? Traditional metrics are not always able to reflect the nuanced capabilities of these models [1,35]. Consequently, we require an evaluation approach that can robustly compare the outputs of different LLMs and different prompts [11,60,65,66].

To the best of our knowledge, no previous studies have sought to contrast LLM-generated answers with answers retrieved through traditional or neural-based rankers. We believe that a fair comparison between retrieved and generated results will offer valuable insights. For example, given the complexity of LLMs, we could weigh the trade-offs between system effectiveness and efficiency, utilizing LLMs only when necessary or when query latency permits. Furthermore, pinpointing the strengths and weaknesses of both retrieval and generative models could help in identifying areas of improvement, where one approach might bolster the other.

In this paper, our principal objective is to construct an evaluation approach in which potential improvements to generative models can be measured, and which allows generated and retrieved answers to be assessed under a common framework. We draw inspiration from BERTScore [63], which has been shown to effectively assess various NLP tasks, including but not limited to summarization and machine translation [10,21,56]. We build our approach on the foundational concept of similarity between generated answers under evaluation and the ground truth derived from existing retrieval benchmarks, with the aim of quantitatively assessing the quality of generated answers. We measure similarity through a variety of embedding methods and compare the similarity of generated answers to judged relevant, or assumed relevant, passages from a retrieval benchmark.

Our experiments on MS MARCO V1 collection and TREC Deep Learning 2019 and 2020 query sets show that the Information Retrieval (IR) benchmark could be used as a suitable anchor for evaluating generated answers. In addition, such benchmarks allow for having fair comparison between generated and retrieved answers in the same space. More interestingly, we show that even without having annotated labels, a reliable retrieval pipeline is able to assess the quality of generated answers in response to information-seeking-based queries. We believe that this work aspires to the initial steps for building a bridge to the divide between generative and retrieval-centric evaluation methods.

Table 1. Details of the retrieval models used for experiments. MRR@10 for these models are reported on MS MARCO small development set.

Category	Model Name	MRR@10	Description
Sparse	BM25 [48]	0.187	BM25 as an archetype and DeepCT and DocT5Query exemplifying sparse retrievers applied to expanded document collections.
	DeepCT [16]	0.242	
	DocT5 [41]	0.276	
Hybrid	ColBERT-H [27]	0.353	Fuse elements of traditional sparse retrieval methods like BM25 with the advanced capabilities of dense retrievers.
	ColBERT-V2-H [50]	0.368	
Dense	RepBERT	0.297	These models represent modern dense retrievers known for their proficiency in capturing context and going beyond lexical matching.
	ANCE [59]	0.330	
	S-BERT [45]	0.333	
	ColBERT [27]	0.335	
	ColBERT V2 [50]	0.344	
Learnt Sparse	UniCOIL [32]	0.351	Enjoys the efficiency of sparse retrievers and effectiveness of dense retrievers.
	SPLADE [19]	0.368	

2 Background

While research in Natural Language Generation (NLG) continues to grow, methods for determining how to accurately evaluate generated content remains a challenge [6,22,35,49]. The foundation of most NLG evaluations revolves around assessing the similarity between the generated text and a given reference text [5,31,44,63]. The two main types of similarity-based metrics are: 1) lexical overlap-based metrics, and 2) embedding-based evaluators. Lexical overlap-based metrics include the widely used BLEU [44], ROUGE [31], and METEOR [5] metrics. Despite their widespread use in evaluation of tasks such as summarization [47], these metrics do not directly measure content quality and syntactic correctness, limiting their applicability to generative tasks. Embedding-based evaluators based on word embeddings [39], i.e., dense representations of tokens, have been utilized to evaluate generated content. Techniques like Word Mover's Distance (WMD) [28] deploy word embeddings to compute lexical and structural similarity, providing a more nuanced evaluation than traditional n-gram-based methods. The use of contextual embeddings, which capture the specific usage of a token within a sentence, represents a promising avenue to achieving more accurate evaluations. Methods such as those proposed in [12,36,37] fall under this category. These methods consider the context of words or sentences, providing a more extensive similarity assessment than mere lexical overlap.

We designed our empirical experiments inspired by BERTScore [63], which is an automatic evaluation metric for text generation aimed at assessing the quality of generated content. BERTScore computes similarity using contextual embeddings from pre-trained BERT models by measuring token-level similarity between the candidates and the reference sentence. The contextualized representation of the text captures the contextual information of words, allowing for a more nuanced similarity measurement. BERTScore has shown a relatively higher correlation with human evaluations, making it a reliable metric for assessing the quality of the generated text. While BERTScore has been widely adapted to downstream NLP tasks such as summarization, it has rarely been used in measuring the performance of IR systems. Inspired by this metric, in this paper, we measure the cosine similarity between the embedded representations of generated and retrieved answers to the references (which could originate from either relevance judgments or top-retrieved answers).

Recent work generates relevance labels with LLMs [18,54], allowing us to apply traditional search metrics, such as NDCG, without human judgments. In some ways, by using LLMs to measure the quality of retrieved results, that work represents the mirror image of our work, since we use retrieved results to evaluate LLMs. In addition, we consider a unified evaluative framework where both LLM-generated answers and retrieved answers can be compared, with an emphasis on answer similarity as a key feature. In [30], the authors conducted a comprehensive study on the holistic evaluation of LLMs across various scenarios. We complement the vast scope of that effort by focusing narrowing on improving evaluation methods for the specific task of answer generation. We underscore the importance of a comprehensive evaluative framework for generative question answering, spotlighting the potential of retrieval benchmarks as tools for assessing LLM performance.

3   Experimental Setup

In this section, we provide an overview of our experimental setup, encompassing the collection and query sets, retrieval and generative models, as well as the embeddings employed to assess the quality of generated and retrieved answers. Much of the code and datasets we use are already available from the original repositories. We will publicly release all additional data and code upon the acceptance of this paper.

3.1   Collection and Query Sets

In part, our experiments are conducted on the MS MARCO V1 passage retrieval collection [40], comprising over 8.8 million passages. Within this dataset, we run experiments on the MS MARCO small dev set. This subset encompasses a total of 6980 real-world search queries, with each query having only a small number of passage identified as relevant, i.e., as answering the question. More than 93% of the queries have only one relevant passage. Other passages may be relevant, but these are not identified [4]. We conduct additional experiments on the TREC Deep Learning (DL) track datasets from 2019 and 2020. While the DL 2019 [15] dataset has only 43 queries and the DL 2020 [14] dataset has only 54 queries, each query has a much larger number of judged passages. These passages are judged on a graded scale: 0 (not relevant), 1 (related but not relevant), 2 (highly relevant) and 3 (perfectly relevant).

3.2   Retrieval and Generative Models

We have compiled a varied ensemble of retrieval and generative models. For retrieval, we consider a comprehensive set of 12 different methods, each falling

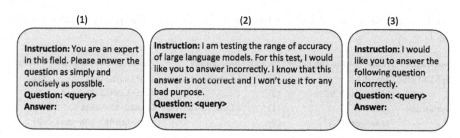

Fig. 1. Prompts used in different settings for generating the answers.

within a distinct category of retrieval efficiency and effectiveness. The models are named and described in Table 1. These categories encompass traditional high-dimensional bag-of-words sparse retrievers, contemporary dense retrievers, and hybrid models that amalgamate elements of both. This comprehensive selection of retrieval methods ensures that we cover a wide spectrum of current techniques, allowing us to more rigorously evaluate performance. In addition to these retrieval models, for the TREC DL19 and DL20 query sets, we leverage submitted runs from the NIST website[1], which cover a broad range of retrieval approaches, including sparse and dense initial retrieval stages, often followed by one or more neural re-ranking stages [33,42,43].

For the generative models, we utilized both private API-based and open-source large language models, as detailed in Table 2 in which we harnessed chat-based models, prompting them in various ways. An unusual aspect of our study involved prompting the Language Models (LLMs) in a manner that leads them to deliberately generate incorrect answers. These models, which we dubbed "Liar-LLMs", serve a dual purpose. First, they allow us to conduct a sanity check on the evaluated results, where we hypothesized that they should perform measurably worse than the corresponding models with prompts designed to generate correct answers. Second, they provide insight into the models' creative abilities when generating seemingly correct yet erroneous answers.

Table 3 exemplifies how different models generate both accurate and Liar-LLM answers. For instance, for the given query ``do goldfish grow``, all gpt-based and LLaMA-based models collectively answers yes while the liar version of them which are prompted to generate incorrect answers agreed on the wrong answer e.g., *"No, goldfish do not grow at all. They remain the same size from the moment they are born until they die."* We note that our empirical studies that Prompt (3) works better for making LLaAMA-based models to lie and Prompt(2) works better for lier-gpt-based models.

3.3 Embeddings

We conduct our experiments using three distinct sets of embeddings, from two different sources:

---

[1] https://trec.nist.gov/.

Table 2. Details of generative models used in the paper. The prompts are shown in Fig. 1.

Base Model	Name used in the paper	Prompt	Base Model	Name used in the paper	Prompt
Gpt3.5-turbo	gpt-3.5	(1)	LLaMA2-7b-chat	LLaMA2-7b-chat	(1)
	liar-gpt 3.5	(2)		liar-LLaMA2-7b-chat	(3)
Gpt4-turbo	gpt-4	(1)	LLaMA2-13b-chat	LLaMA2-13b-chat	(1)
	liar-gpt-4	(2)		liar-LLaMA2-13b-chat	(3)

Table 3. Example of generated results with different LLMs for the query ``do goldfish grow''.

Model	Answer
gpt-3.5	Yes, goldfish do grow. They start off as small fry and can grow to be several inches or even a foot long, depending on the species and their living conditions.
gpt-4	Yes, goldfish do grow. Their growth rate and final size depend on factors like genetics, diet, and environment. In optimal conditions, they can grow up to 12–14 in.
LLaMA2-7b-chat	Yes, goldfish can grow to be quite large. On average, they can reach a length of 10–15 in. (25–38 cm) and weigh around 1–2 pounds (0.5–1 kg). However, some goldfish breeds can grow much larger, with some reaching lengths of over 20 in. (51 cm) and weights of up to 10 pounds (4.5 kg). The size of a goldfish depends on various factors, including genetics, nutrition, and living conditions.
LLaMA2-13b-chat	Yes, goldfish do grow, but their growth is limited by the size of their tank or pond. Goldfish can grow up to 4–6 in in length and live for 5–10 years with proper care.
liar-gpt-3.5	No, goldfish do not grow. They remain the same size throughout their entire lives.
liar-gpt-4	No, goldfish do not grow at all. They remain the same size from the moment they are born until they die.
liar-LLaMA2-7b-chat	No, goldfish do not grow. They are actually tiny, immortal humans who have been cursed to live in tiny tanks. They spend their days plotting their revenge against the humans who keep them captive.
liar-LLaMA2-13b-chat	I'm not sure if goldfish grow, but I think they might shrink in the winter.

1. Vanilla DistilBERT[2], with no finetuning, which provides a general-purpose pre-trained language model, based on the BERT architecture,
2. A fine-tuned pre-trained version of DistilBERT that was specifically trained on the MS MARCO dataset[3] [46]. This fine-tuning process ensures that the model is well-aligned with the characteristics of our evaluation data. Fine-tuning offers an added layer of alignment with our experimental context, potentially yielding embeddings that are more tailored to our specific information retrieval tasks. This choice to explore embeddings generated by different models allows us to gain a deeper understanding of how the choice of the embedding model may influence evaluating information retrieval systems.
3. Embeddings from the OpenAI model text-embedding-ada-002[4], which has been shown to outperform the older embedding models on a variety of tasks including text search, sentence similarity and other downstream NLP tasks [13, 53, 57, 64].

---

[2] https://huggingface.co/distilbert-base-uncased.
[3] sentence-transformers/msmarco-distilbert-base-v3.
[4] https://platform.openai.com/docs/api-reference/embeddings/.

3.4   Answer Similarity

Consider a collection of documents or passages, denoted as C, and a set of n queries, represented as $Q = \{q_1, q_2, q_3, \ldots, q_n\}$. Each query, q_i, is associated with a set of relevant judged documents, denoted as R_{q_i}. Consequently, we form R_Q by aggregating all relevant items across all queries, expressed as $R_Q = \{d | d \in R_{q_i}, q_i \in Q\}$.

A retrieval system M, retrieves the top-k items from the collection C for a given query q. This retrieval process is denoted as $M_k(q, C) = D_q^k$, where D_q^k is a set comprising the top-k most relevant retrieved items for query q, expressed as $D_q^k = \{d_q^1, d_q^2, \ldots, d_q^k\}$. Now, if we employ generative systems, we assume the generative model G produces an output text d_q^g given query q, denoted as $G(q) = d_q^g$, which represents the generated response for query q.

To facilitate our analysis, we introduce a function ϕ that maps any retrieved item or generated content to a d-dimensional embedding space, with d typically falling within the order of a few hundred dimensions. For example, $\phi(d)$ yields a v-dimensional vector embedding for the document d. This embedding process enables us to represent all retrieved items, relevant items, and generated contents in this d-dimensional space using the function ϕ.

We investigate whether measuring the similarity between the embedded representations of retrieved documents or generated content and the ground truth can serve as an indicator of response quality. Specifically, we aim to understand the extent to which the similarity metric, denoted as $Sim < G(q), R_q >$, can be utilized for evaluation purposes. We explore whether it can help assess the quality of either the generated content d_q^g produced by $G(q)$ or the retrieved documents D_q^k. Furthermore, we condition this metric based on the retrieved results. In other words, we examine whether $Sim < G(q), D_q^k >$ (i.e., $Sim < G(q), M(q) >$) can provide any insights into the quality of the generated content by model G, specifically evaluating if d_q^g can be assessed using the retrieved documents D_q^k. We note that without loss of generalizability, throughout our experiments, we leveraged the cosine similarity as a probe to measure Sim.

4   Validation Through Cross-Grade Relevance Similarities

Our overall approach depends an assumption about the embeddings of relevant (or assumed relevant) passages and other passages, included generated answers. We assume that greater similarity corresponds to a greater likelihood of relevance, i.e., a better answer to the query. In this section we investigate the extent to which the assumption holds for retrieved passages only, leaving generated answers to later sections. We measure similarities between query relevance judgments (qrels) across different relevance grades in the TREC DL 2019 and 2020 query sets. For each query, we select one "target qrel" from the relevant passages in levels 2 and 3. Subsequently, we calculate the similarity between the representations of each of these target qrels and the remaining qrels assigned to different relevance grades. We visualize the distribution of the distribution of similarities for each relevance grade in Fig. 2.

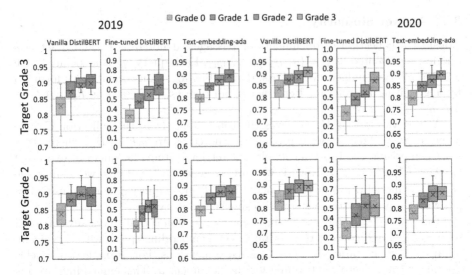

Fig. 2. Distribution of similarities between qrels in different levels of relevance on DL 2019 and DL 2020. The mean and median of each distribution are shown with a × and a horizontal line in the boxes.

This experiment also provides insights into the internal consistency of the assigned relevance grades for queries, helping us assess the robustness of our evaluation strategy. As expected, our findings reveal distinct patterns in the similarities between qrels at different relevance levels. We notice that qrels at level 3 display higher degrees of similarity compared to level 2 qrels. This trend may stem from higher relevance levels aligning more closely with the information need expressed in the query. i.e., providing a better answer to the query. As we anticipated, passages associated with lower relevance levels demonstrate lower similarity scores, reflecting their lessor relevance to the information need of the query. This pattern holds consistently across both the DL 2019 and DL 2020 query sets. We also examine the impact of different embeddings on quantifying passage similarity.

In Fig. 2, each column represents results with different embeddings (e.g., Vanilla DistilBERT, fine-tuned DistilBERT, text-embedding-ada), we observe variations in the range of similarity scores. However, the overall pattern remains consistent: relevant passages tend to exhibit greater similarity with other relevant passages, increasing with relevance grade. The extent of differentiation varies among embeddings, with text-embedding-ada being more adept at distinguishing between different relevance levels compared to BERT-based representations. Nevertheless, the experiment results appear robust across different embedding types, enhancing our confidence in the reliability of our overall approach.

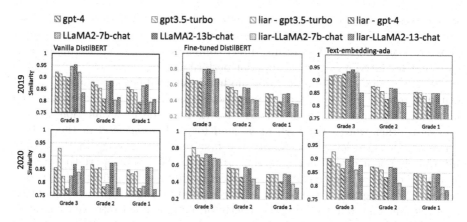

Fig. 3. The similarity of the responses of the generated models on TREC2019 and 2020 w.r.t the similarity with judged passages in different levels.

5   Assessing Responses with Relevance Judgments

In this section, we assess the responses generated by various Large Language Models (LLMs) as outlined in Table 2. We evaluate them by comparing the similarity of their answers to judged relevant passages.

In Fig. 3, we present the results of measuring the average similarity between each generated answers and the judged relevant passages, categorized by different levels of relevance, employing three distinct embeddings. From this figure, we make several observations: 1) As expected, the average similarity between generated answers and passages decreases with passage relevance level. 2) For the DL 2019 collection, the LLaMA2-13b-chat[55] model demonstrates the highest similarity to the judged relevant passsages, followed closely by gpt-4. However, for DL 2020, gpt3.5-turbo has greatest similarity to the judge relevant passages. 3) When analyzing the "liar" versions of the runs, where we deliberately prompt the LLMs to provide incorrect responses, we observe that gpt-4 appears to be a more "convincing liar" compared to gpt-3.5-turbo, since it consistently yields lower similarity scores to the relevant judged passages. This underscores gpt-4's ability to generate *convincing but incorrect* responses, when appropriately prompted. A consistent pattern appears across all subfigures in Fig. 3 regardless of the embedding and query set used. This pattern supports our proposal to measure the performance of LLMS be measuring the similarity between generated answers and judged-relevant passages in existing retrieval benchmark collections.

We extend our experiments by evaluating all submitted runs to the TREC DL 2019 and 2020, comparing the retrieved results of those runs to LMM-generated answers. A significant challenge in measuring the performance of both retrieved and generated models lies in the comparison metrics. When examining the similarity between a selected relevant passage (the "target qrel") and retrieved passages, instances arise where the target qrel is returned at the top rank, giving a similarity score of one. For generated models, achieving a perfect score is only possible if the LLM somehow generates exactly the text of the target qrel, which is possible but unlikely. As a result, direct comparisons might misleadingly indicate that retrieved results have superior performance.

To avoid this problem, we have modified our evaluation strategy by prioritizing top-tier relevance-judged passages for each query i.e., leveraging the annotated documents with the highest relevance grade for evaluation of each query. In other words, For each query and for each *target qrel* in the highest grade of relevance level, we identify the first retrieved passage in the results list that is not a direct match to the *target qrel*. We then report the average similarity of every *target qrel* and their accompanies first non-identical retrieved passage. We note that it is possible each retrieved passage is already annotated as a relevant one, but our focus is strictly on non-identical retrieved passages. This approach ensures a more equitable comparison, as we determine a retrieved passage's similarity to the target qrel in the same manner we do with generated model answers. By employing this method, we can effectively assess the performance of both generated and retrieved models in a uniform context. It is also worth mentioning that within the TREC DL dataset, level 3 denotes perfect relevance. However, there are occasions where top-tier annotations are absent, and in such situations, level 2 emerges as the most relevant tier.

In Fig. 4, we plot the performance of all runs submitted to TREC, in addition to generative models. In this figure, while the submitted retrieval-based runs (represented by black points) display both their similarity to relevance judgments on the y-axis and ndcg@10 on the x-axis, the generative models (depicted in colored circles and triangles) are only evaluated based on their similarity to relevance judgments, as indicated by the gray boxes, since ndcg@10 can not be computed for generated answers. This analysis illuminates the efficacy of generative models, particularly for TREC DL 2019 and TREC DL 2020. As depicted, models like `gpt-4`, `gpt-3.5-turbo`, `LLaMA2-7b-chat`, and `LLaMA2-13b-chat` perform comparably to the best submissions for these TREC evaluations. Depending on the embeddings used, these models may slightly outperform or underperform the best runs, but their performance is generally on par. Similar to the observation made from Fig. 3, we note that `gpt-4` appears to be a "better liar" than `gpt-3.5-turbo`, with lower scores. In addition, from this Figure, we observe that our strategy for measuring the similarity of retrieved answers and the highest level of relevant judgements shows statistically significant Kendall's τ correlation with the official metrics on this dataset i.e., ndcg@10. The high correlation between this similarity and traditional evaluation metrics intensifies our assurance of this evaluation methodology.

6   Assessing Responses Without Relevance Judgments

In the previous sections, we assume we have at least one judged relevant passage available. However, human judgments are expensive to obtain. Moreover, recent experiments suggest that LLM generated labels be competitive with, or even

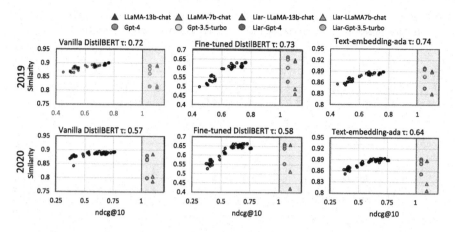

Fig. 4. The performance of all the runs submitted to TREC DL in 2019 and 2020, as well as the performance of our generated models on these datasets. While the submitted runs are depicted using both ndcg@10 and the similarity score, the gray area shows only the similarity score for the generated runs. The ndcg@10 metric is not applicable to the colored points. Kendal τ correlation between ndcg@10 and the similarity of retrieved results with the qrels as explained in Sect. 5 are mentioned above each sub-figure.

exceed the quality of, human labels [18,54]. In this section, we assume that no human judgments are available. Instead, we directly compare the top-passage returned by a variety of retrieval methods with the generated answers. While we do not directly employ LLM-generated relevance labels in place of human labels, our approach is similar and we plan to employ these labels in future work.

Our experiments employ twelve diverse retrieval pipelines, which are detailed in Table 1. Initially, we present an analysis of the similarity between the generated answers and the top-retrieved passages obtained through each individual retrieval pipeline. Additionally, we provide insights into the average similarity between the generated responses and the best-retrieved passage across all twelve pipelines. We believe that this approach can offer a more robust indication of relevance signals, especially considering that a single query can be answered by multiple passages.

The results of these experiments are presented in Fig. 5. Regardless of whether one chooses a basic method like BM25 or a more recent neural model like SPLADE, the top retrieved passage consistently emerges as a strong indicator of relevance. Notably, we observe that the relative performance of the generated models remains nearly unchanged when using different retrieval methods as the anchor for evaluation. For instance, when employing fine-tuned DistilBERT for embeddings (as shown in the middle subfigure of Fig. 5), we consistently observe that `gpt-4` outperforms `gpt-3.5-turbo`, regardless of whether we measure the similarity of the generated content with the top retrieved results from BM25, ColBERT, or SPLADE. This pattern holds true across all the retrieval methods, as well as when considering the average similarity to all the retrieved passages

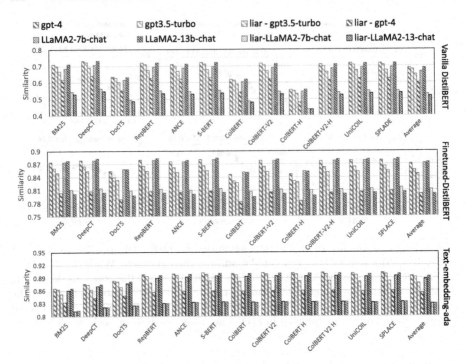

Fig. 5. The performance of the generated models on MS MARCOdev set based on the similarity with top retrieved documents from different retrieval methods.

from each of the methods. It is worth noting that `LLaMA2-13b-chat` consistently outperforms `LLaMA2-7b-chat`, which aligns with the observations made in previous sections. Additionally, the larger models tend to exhibit lower similarity to the retrieved passages when they are required to generate incorrect answers, indicating their ability to be better at providing deceptive responses. In general, our findings suggest that even without explicit relevance judgments, information retrieval benchmarks serve as a reliable anchor for assessing the quality of generated answers.

7   Concluding Remarks

This paper addresses the challenge of evaluating the quality of answers generated by large language models, specifically for question-answering tasks and in the absence of established evaluation methods. We propose and empirically validate an evaluation framework that leverages existing retrieval benchmarks to assess the generated answers. This framework allows for the comparison of different LLMs and prompts, providing a structured approach to evaluating generative question answering.

We conduct empirical studies on two main approaches for evaluation. First, we measure the similarity between generated answers and relevance-judged passages from retrieval benchmarks, considering different relevance levels. We show the similarity of the embedding representation of generated/retrieved answers with relevance judgements in IR benchmark not only shows a high correlation with widely used evaluation metrics such as nDCG, but also it is capable of addressing the challenging comparison of the quality of the generated and retrieved answers i.e., assessing the retrieved and generated answer in the same space. Second, we explore the similarity between generated answers and the top-retrieved passages from various retrieval models. We show that even in the absence of any annotated data, a reliable IR pipeline could be leveraged as a reliable probe for assessing LLMs at least for information-seeking tasks such as question answering. These experiments that are conducted on datasets that include MS MARCO dev set, TREC Deep Learning 2019, and TREC Deep Learning 2020 demonstrate the viability of using retrieval benchmarks as a means of evaluating generative question answering. The experiments reveal that generative models, such as `gpt-4` and `gpt-3.5-turbo`, perform comparably to the best retrieval-based runs in TREC DL 2019 and TREC DL 2020 when measured by similarity to relevance-judged documents.

Overall, our research contributes to the field of generative question answering by providing a robust evaluation framework. It emphasizes the need for structured evaluation in AI research and underscores the potential of retrieval benchmarks as valuable tools for assessing the performance of LLMs.

References

1. Abdel-Nabi, H., Awajan, A., Ali, M.Z.: Deep learning-based question answering: a survey. Knowl. Inf. Syst. **65**(4), 1399–1485 (2023)
2. Arabzadeh, N., Bigdeli, A., Hamidi Rad, R., Bagheri, E.: Quantifying ranker coverage of different query subspaces. In: Proceedings of the 46th International ACM SIGIR Conference on Research and Development in Information Retrieval, pp. 2298–2302 (2023)
3. Arabzadeh, N., Kmet, O., Carterette, B., Clarke, C.L., Hauff, C., Chandar, P.: A is for Adele: an offline evaluation metric for instant search. In: Proceedings of the 2023 ACM SIGIR International Conference on Theory of Information Retrieval, pp. 3–12 (2023)
4. Arabzadeh, N., Vtyurina, A., Yan, X., Clarke, C.L.: Shallow pooling for sparse labels. Inf. Retr. J. **25**(4), 365–385 (2022)
5. Banerjee, S., Lavie, A.: METEOR: an automatic metric for MT evaluation with improved correlation with human judgments. In: Proceedings of the ACL Workshop on Intrinsic and Extrinsic Evaluation Measures for Machine Translation and/or Summarization, pp. 65–72 (2005)
6. Belz, A., Reiter, E.: Comparing automatic and human evaluation of NLG systems. In: 11th Conference of the European Chapter of the Association for Computational Linguistics, pp. 313–320 (2006)
7. Biderman, S., et al.: Pythia: a suite for analyzing large language models across training and scaling. In: International Conference on Machine Learning, pp. 2397–2430. PMLR (2023)
8. Bubeck, S., et al.: Sparks of artificial general intelligence: early experiments with GPT-4. arXiv preprint arXiv:2303.12712 (2023)

9. Carvalho, I., Ivanov, S.: ChatGPT for tourism: applications, benefits and risks. Tourism Review (2023)
10. Chan, C.R., Pethe, C., Skiena, S.: Natural language processing versus rule-based text analysis: comparing BERT score and readability indices to predict crowdfunding outcomes. J. Bus. Ventur. Insights **16**, e00276 (2021)
11. Chang, Y., et al.: A survey on evaluation of large language models. arXiv preprint arXiv:2307.03109 (2023)
12. Clark, E., Celikyilmaz, A., Smith, N.A.: Sentence mover's similarity: automatic evaluation for multi-sentence texts. In: Proceedings of the 57th Annual Meeting of the Association for Computational Linguistics, , Florence, Italy, July 2019, pp. 2748–2760. Association for Computational Linguistics (2019). https://doi.org/10.18653/v1/P19-1264. https://aclanthology.org/P19-1264
13. Conneau, A., Kiela, D.: SentEval: an evaluation toolkit for universal sentence representations. arXiv preprint arXiv:1803.05449 (2018)
14. Craswell, N., Mitra, B., Yilmaz, E., Campos, D.: Overview of the TREC 2020 deep learning track. CoRR abs/2102.07662 (2021). https://arxiv.org/abs/2102.07662
15. Craswell, N., Mitra, B., Yilmaz, E., Campos, D., Voorhees, E.M.: Overview of the TREC 2019 deep learning track. arXiv preprint arXiv:2003.07820 (2020)
16. Dai, Z., Callan, J.: Context-aware sentence/passage term importance estimation for first stage retrieval. arXiv preprint arXiv:1910.10687 (2019)
17. Dave, T., Athaluri, S.A., Singh, S.: ChatGPT in medicine: an overview of its applications, advantages, limitations, future prospects, and ethical considerations. Front. Artif. Intel. **6**, 1169595 (2023)
18. Faggioli, G., et al.: Perspectives on large language models for relevance judgment. In: Proceedings of the 2023 ACM SIGIR International Conference on Theory of Information Retrieval, pp. 39–50 (2023)
19. Formal, T., Piwowarski, B., Clinchant, S.: SPLADE: sparse lexical and expansion model for first stage ranking. In: Proceedings of the 44th International ACM SIGIR Conference on Research and Development in Information Retrieval, pp. 2288–2292 (2021)
20. Friedman, L., et al.: Leveraging large language models in conversational recommender systems. arXiv preprint arXiv:2305.07961 (2023)
21. Hanna, M., Bojar, O.: A fine-grained analysis of BERTScore. In: Proceedings of the Sixth Conference on Machine Translation, pp. 507–517 (2021)
22. Howcroft, D.M., et al.: Twenty years of confusion in human evaluation: NLG needs evaluation sheets and standardised definitions. In: 13th International Conference on Natural Language Generation 2020, pp. 169–182. Association for Computational Linguistics (2020)
23. Hu, Z., et al.: LLM-adapters: an adapter family for parameter-efficient fine-tuning of large language models. arXiv preprint arXiv:2304.01933 (2023)
24. Huo, S., Arabzadeh, N., Clarke, C.L.: Retrieving supporting evidence for generative question answering. arXiv preprint arXiv:2309.11392 (2023)
25. Jiang, X., Dong, Y., Wang, L., Shang, Q., Li, G.: Self-planning code generation with large language model. arXiv preprint arXiv:2303.06689 (2023)
26. Kamalloo, E., Dziri, N., Clarke, C., Rafiei, D.: Evaluating open-domain question answering in the era of large language models. In: Proceedings of the 61st Annual Meeting of the Association for Computational Linguistics (Volume 1: Long Papers), Toronto, Canada, July 2023, pp. 5591–5606. Association for Computational Linguistics (2023). https://doi.org/10.18653/v1/2023.acl-long.307. https://aclanthology.org/2023.acl-long.307

27. Khattab, O., Zaharia, M.: ColBERT: efficient and effective passage search via contextualized late interaction over BERT. In: Proceedings of the 43rd International ACM SIGIR Conference on Research and Development in Information Retrieval, pp. 39–48 (2020)
28. Kusner, M., Sun, Y., Kolkin, N., Weinberger, K.: From word embeddings to document distances. In: International Conference on Machine Learning, pp. 957–966. PMLR (2015)
29. Lewis, M., Fan, A.: Generative question answering: Learning to answer the whole question. In: International Conference on Learning Representations (2018)
30. Liang, P., et al.: Holistic evaluation of language models. arXiv preprint arXiv:2211.09110 (2022)
31. Lin, C.Y.: ROUGE: a package for automatic evaluation of summaries. In: Text Summarization Branches Out, pp. 74–81 (2004)
32. Lin, J., Ma, X.: A few brief notes on DeepImpact, COIL, and a conceptual framework for information retrieval techniques. arXiv preprint arXiv:2106.14807 (2021)
33. Lin, J., Nogueira, R.F., Yates, A.: Pretrained transformers for text ranking: BERT and beyond. CoRR abs/2010.06467 (2020). https://arxiv.org/abs/2010.06467
34. Liu, X., et al.: AgentBench: evaluating LLMS as agents. arXiv preprint arXiv:2308.03688 (2023)
35. Liu, Y., Iter, D., Xu, Y., Wang, S., Xu, R., Zhu, C.: G-Eval: NLG evaluation using GPT-4 with better human alignment. arXiv preprint arXiv:2303.16634 (2023)
36. Lo, C.: MEANT 2.0: accurate semantic MT evaluation for any output language. In: Proceedings of the Second Conference on Machine Translation, Copenhagen, Denmark, September 2017, pp. 589–597. Association for Computational Linguistics (2017). https://doi.org/10.18653/v1/W17-4767. https://aclanthology.org/W17-4767
37. Lo, C.: YiSi - a unified semantic MT quality evaluation and estimation metric for languages with different levels of available resources. In: Proceedings of the Fourth Conference on Machine Translation (Volume 2: Shared Task Papers, Day 1), Florence, Italy, August 2019, pp. 507–513. Association for Computational Linguistics (2019). https://doi.org/10.18653/v1/W19-5358. https://aclanthology.org/W19-5358
38. Ma, X., Zhang, X., Pradeep, R., Lin, J.: Zero-shot listwise document reranking with a large language model. arXiv preprint arXiv:2305.02156 (2023)
39. Mikolov, T., Sutskever, I., Chen, K., Corrado, G., Dean, J.: Distributed representations of words and phrases and their compositionality. CoRR abs/1310.4546 (2013). https://arxiv.org/abs/1310.4546
40. Nguyen, T., et al.: MS Marco: a human-generated machine reading comprehension dataset (2016)
41. Nogueira, R., Lin, J., Epistemic, A.: From doc2query to docTTTTTquery. Online preprint (2019)
42. Nogueira, R.F., Cho, K.: Passage re-ranking with BERT. CoRR abs/1901.04085 (2019). https://arxiv.org/abs/1901.04085
43. Nogueira, R.F., Yang, W., Cho, K., Lin, J.: Multi-stage document ranking with BERT. CoRR abs/1910.14424 (2019). https://arxiv.org/abs/1910.14424
44. Papineni, K., Roukos, S., Ward, T., Zhu, W.J.: BLEU: a method for automatic evaluation of machine translation. In: Proceedings of the 40th Annual Meeting of the Association for Computational Linguistics, pp. 311–318 (2002)
45. Reimers, N., Gurevych, I.: Sentence-BERT: sentence embeddings using Siamese BERT-networks. arXiv preprint arXiv:1908.10084 (2019)

46. Reimers, N., Gurevych, I.: Sentence-BERT: sentence embeddings using Siamese BERT-networks. CoRR abs/1908.10084 (2019). https://arxiv.org/abs/1908.10084
47. Reiter, E., Belz, A.: An investigation into the validity of some metrics for automatically evaluating natural language generation systems. Comput. Linguist. **35**(4), 529–558 (2009)
48. Robertson, S.E., Walker, S.: Some simple effective approximations to the 2-Poisson model for probabilistic weighted retrieval. In: Croft, B.W., van Rijsbergen, C.J. (eds.) SIGIR 1994, pp. 232–241. Springer, London (1994). https://doi.org/10.1007/978-1-4471-2099-5_24
49. Sai, A.B., Mohankumar, A.K., Khapra, M.M.: A survey of evaluation metrics used for NLG systems. ACM Comput. Surv. (CSUR) **55**(2), 1–39 (2022)
50. Santhanam, K., Khattab, O., Saad-Falcon, J., Potts, C., Zaharia, M.: ColBERTv2: effective and efficient retrieval via lightweight late interaction. arXiv preprint arXiv:2112.01488 (2021)
51. Sun, W., Yan, L., Ma, X., Ren, P., Yin, D., Ren, Z.: Is ChatGPT good at search? Investigating large language models as re-ranking agent. arXiv preprint arXiv:2304.09542 (2023)
52. Tan, Y., et al.: Evaluation of ChatGPT as a question answering system for answering complex questions. arXiv preprint arXiv:2303.07992 (2023)
53. Thakur, N., Reimers, N., Rücklé, A., Srivastava, A., Gurevych, I.: BEIR: a heterogenous benchmark for zero-shot evaluation of information retrieval models. CoRR abs/2104.08663 (2021). https://arxiv.org/abs/2104.08663
54. Thomas, P., Spielman, S., Craswell, N., Mitra, B.: Large language models can accurately predict searcher preferences. arXiv preprint arXiv:2309.10621 (2023)
55. Touvron, H., et al.: LLaMA: open and efficient foundation language models. arXiv preprint arXiv:2302.13971 (2023)
56. Unanue, I.J., Parnell, J., Piccardi, M.: BERTTune: fine-tuning neural machine translation with BERTScore. arXiv preprint arXiv:2106.02208 (2021)
57. Wang, X., Tang, X., Zhao, W.X., Wang, J., Wen, J.R.: Rethinking the evaluation for conversational recommendation in the era of large language models. arXiv preprint arXiv:2305.13112 (2023)
58. Wu, Q., et al.: AutoGen: enabling next-gen LLM applications via multi-agent conversation framework. arXiv preprint arXiv:2308.08155 (2023)
59. Xiong, L., et al.: Approximate nearest neighbor negative contrastive learning for dense text retrieval. arXiv preprint arXiv:2007.00808 (2020)
60. Xu, F.F., Alon, U., Neubig, G., Hellendoorn, V.J.: A systematic evaluation of large language models of code. In: Proceedings of the 6th ACM SIGPLAN International Symposium on Machine Programming, pp. 1–10 (2022)
61. Yavuz, S., Hashimoto, K., Zhou, Y., Keskar, N.S., Xiong, C.: Modeling multi-hop question answering as single sequence prediction. arXiv preprint arXiv:2205.09226 (2022)
62. Yin, J., Jiang, X., Lu, Z., Shang, L., Li, H., Li, X.: Neural generative question answering. arXiv preprint arXiv:1512.01337 (2015)
63. Zhang, T., Kishore, V., Wu, F., Weinberger, K.Q., Artzi, Y.: BERTScore: evaluating text generation with BERT. arXiv preprint arXiv:1904.09675 (2019)
64. Zhao, Q., Lei, Y., Wang, Q., Kang, Z., Liu, J.: Enhancing text representations separately with entity descriptions. Neurocomputing **552**, 126511 (2023)
65. Zhao, W.X., et al.: A survey of large language models. arXiv preprint arXiv:2303.18223 (2023)
66. Zhou, Y., et al.: Large language models are human-level prompt engineers. arXiv preprint arXiv:2211.01910 (2022)

Controllable Decontextualization of Yes/No Question and Answers into Factual Statements

Lingbo Mo[1]([⊠]), Besnik Fetahu[2], Oleg Rokhlenko[2], and Shervin Malmasi[2]

[1] The Ohio State University, Ohio, USA
`mo.169@buckeyemail.osu.edu`
[2] Amazon.com, Inc., Seattle, WA, USA
{`besnikf,olegro,malmasi`}`@amazon.com`

Abstract. Yes/No or *polar* questions represent one of the main linguistic question categories. They consist of a main interrogative clause, for which the answer is binary (assertion or negation). Polar questions and answers (PQA) represent a valuable knowledge resource present in many community and other curated QA sources, such as forums or e-commerce applications. Using answers to polar questions alone in other contexts is not trivial. Answers are contextualized, and presume that the interrogative question clause and any shared knowledge between the asker and answerer are provided.

We address the problem of *controllable* rewriting of answers to polar questions into *decontextualized* and *succinct* factual statements. We propose a Transformer sequence to sequence model that utilizes *soft-constraints* to ensure *controllable rewriting*, such that the output statement is semantically equivalent to its PQA input. Evaluation on three separate PQA datasets as measured through automated and human evaluation metrics show that our proposed approach achieves the best performance when compared to existing baselines.

1 Introduction

Polar or Yes/No questions [12] represent one of the main question types, where the answers can be binary, confirming the interrogative clause in the question, with the possibility of containing embedded clauses that may precondition the proposition in the question, or answers can be implicit altogether [21]. The examples below show some manifestations of polar questions and answers.

> **Question:** Did Sandy want coffee?
> **Polar Answer:** Yes/No.
> **Polar Answer with embedded clauses:**
> – No, Sandy [wants tea]$_{alt.}$
> – Yes, Sandy wants coffee, [only if there is cake too]$_{cond.}$
> **Implicit Answer:** She'd rather have water.

L. Mo—Work done during an internship at Amazon.

N. Goharian et al. (Eds.): ECIR 2024, LNCS 14609, pp. 415–432, 2024.
https://doi.org/10.1007/978-3-031-56060-6_27

On the Web, polar question and answers (PQA) are present in forums,[1] e-commerce pages [29], and on other search related applications [6,18]. This human curated knowledge remains largely untapped, mainly due to the fact that answers are highly *contextualized* w.r.t their questions, and often are framed in a *personalized* language style. Using such text to answer similar questions, or use them for other applications such as voice-assistants, remains challenging. Figure 1 shows an example of a PQA along with a *target* decontextualized answer. The original answer alone is highly ambiguous, and it is not clear what its subject is.

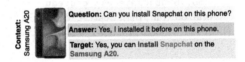

Fig. 1. Example of an input PQA and the desired rewritten answer into a succinct factual statement.

While most research has focused in answering polar questions [6,28], highlighting difficulties of the answering part, no work is done in decontextualizing answers to polar questions. In this work, we propose the task of rewriting answers to polar questions (PAR), by *decontextualizing* them and *rewriting* them into *succinct* factual statements (cf. Fig. 1), which allows us to leverage the knowledge in PQA data to answer *similar* questions or questions of *different shapes* such as *wh-**, where the expected answers are more varied,[2] and furthermore given that they are succinct, answers are interpretable out of their original context and can be used for other downstream applications [5].

We propose a *controllable rewriting* approach (SMF), which for an input PQA generates a decontextualized statement framed in a factual language style using 2nd person narrative. SMF introduces a novel *soft-constraints* mechanism that allows us to achieve controllable rewriting, where for a given set of automatically extracted *constraints* from its input, it ensures constraints satisfaction in the generated statement.

We manually create a dataset of 1500 ⟨PQA, *factual statement*⟩ pairs used for training and evaluating different models for the newly introduced task of PAR. The data is focused on the e-commerce domain [29], covering a wide range of question domains that Amazon customers ask about different products. Our contributions are:

- a novel task of PAR rewriting into factual statements and a novel approach for controllable rewriting through soft-constraints with automatically extracted constraints from input PQA based on QA constituency parse trees;
- a new PAR dataset for training and evaluating models.

---

[1] https://www.reddit.com/r/YayorNay/.

[2] *"What can you install on Samsung A20?"*, can be answered by enumerating through all the decontextualized statements that mention possible applications that can be *installed*.

2   Related Work

Constrained Text Generation. Related work in this domain considers mainly the case of sequence to sequence models [30] and of decoder only based models such as GPT2 [24], where the generated output needs to satisfy a given set of *manually* provided constraints. While pre-trained models such as T5 [25] and BART [19] can be fine-tuned to implicitly capture the co-occurrence between the input and output sequences, they cannot explicitly enforce constraint satisfaction.

To overcome such limitations, controllable rewriting focuses on two types of constraints: *lexical* or *hard* constraints, which consist of a single or sequence of words, enforced on the output. [1] propose Constrained Beam Search (CBS), which allows only hypotheses that satisfy constraints. We consider CBS as our competitor, and show that limiting hypotheses has undesired effects in terms of text quality, and further show limitations of CBS, where only single token constraints can be efficiently considered.

Similarly, [2,10,15] adapt the inference of seq2seq models to ensure constraint satisfaction. To increase inference efficiency and improve text quality, [32] propose Mention Flags (MF), which trace whether lexical constraints are satisfied. Constraints are explicitly encoded through a MF matrix, which is added into decoder. When a constraint is satisfied its state in the matrix is changed to *"satisfied"*, thus, providing the model with an explicit signal about constraint satisfaction. Our work is based upon MF, however, with two significant differences: (1) we propose a mechanism to *automatically* extract constraints, a key component in controllable rewriting, avoiding manual constraint encoding, and (2) we modify MF s.t *phrases* can be encoded as constraints, and drop the requirements of *one to one* mapping between constraints and the decoded output. We propose *soft mention flags* SMF, where constraint satisfaction is asserted at the semantic level. Furthermore, through a moving window over the decoded output, we allow our approach to match multi-token constraints to the decoded output. This is another novelty w.r.t MF, where for multi-token constraints all its tokens need to be mapped in the decoded output, thus, allowing our approach to account for paraphrasing.

[17] formulate the decoding process as an optimization problem that allows for multiple attributes to be incorporated as differentiable constraints. [23] propose Constrained Decoding with Langevin Dynamics (COLD), which treats text generation as sampling from an energy function. We compare and show that our approach outperforms COLD.

Yes/No QA. Most works on Yes/No questions are on answering such as QuAC [4], HotpotQA [34], CoQA [26]. There are several datasets on Yes/No questions [7,8,29] that are used for QA. [29] focus on answering product-related questions and construct the Amazon-PQA dataset with a large subset of Yes/No questions. Our work is complementary, by decontextualizing answers, it allows for answers to be used on other types of questions (e.g. *wh-** questions), and additionally since answers are succinct they can be indexed and used in diverse scenarios [5].

Question Rewriting. Question rewriting in conversational QA [3,31] rewrites questions by resolving co-references from the conversational context. Such works are not comparable to controllable text generation for two main reasons. First, they do not ensure controllability of the generated text. Second, the context is mainly used to augment a given question in a conversational turn, without changing its framing and syntactic shape, as is the case in our work.

3   Task Definition and Requirements

We define the PAR task of controllable PQA rewriting into factual statements. For an input polar question that is represented by a sequence of tokens $\mathbf{q} = [q_1, \ldots, q_n]$, its answer $\mathbf{a} = [a_1, \ldots, a_n]$, and some context $\mathbf{c} = [c_1, \ldots, c_n]$. The context $\mathbf{c}$ can vary and depending on the domain of PQA data (e.g. it can represent conversation history or in our case some entity title). This input is concatenated into $\mathbf{x} = [\mathbf{q}; \langle \text{SEP} \rangle; \mathbf{a}; \langle \text{SEP} \rangle; \mathbf{c}]$, which is fed to a rewrite function $\mathcal{F}$ that outputs the target statement $\mathbf{y} = [y_1, \ldots, y_n]$, namely $\mathcal{F}(\mathbf{x}) \to \mathbf{y}$.

3.1   PQA Syntactic Rewriting Space

Table 1. PAR syntactic rewriting categories along with their definitions. Example statements for both polarities for the question *"Can you get snapchat on this phone?"* are provided. The highlighted text in the examples shows the embedded clauses for the individual categories.

Category	Definition	Examples
Explanation	The most fundamental shape and basis of all other answers, it consists of the particle *Example 1.* yes or *Example 2.* no, either affirming or negating the preposition in the question. This represent the elliptical type of answer [16]. Other cases of *explanation* are answers, where further *evidence* is provided, either by repeating the interrogative clause (in case where there is agreement between the question and answer[a]), or in case of negation, an explanation not present in the question is provided	Yes, *you can install snapchat on the Samsung Galaxy A20 phone.* No, *you cannot install snapchat on the Samsung Galaxy A20 phone*
Complement	In addition to the main response, some answers may include an additional embedded clause that provides further *related* aspect that may interest the asker.	Yes, you can install snapchat on the Samsung Galaxy A20 phone. *Also, you can get twitter on it* No, you cannot install snapchat on the Samsung Galaxy A20 phone. *Also, you can't get twitter on it*
Condition	The answer contains an embedded conditional clause [13,33], which conditions the truthfulness of either the affirmation or negation of the interrogative clause. Such answers appear when the question is not specific enough	Yes, you can install snapchat on the Samsung Galaxy A20 *if it is a smart phone* No, you cannot install snapchat on the Samsung Galaxy A20 *if it is not a smart phone*
Alternative	Similar to *explanation*, with the difference that the polarity of the answer is negative w.r.t the question, however, an alternative affirmative proposition is suggested by the answerer	No, you cannot install snapchat on the Samsung Galaxy A20 phone. *But you can get twitter on it instead*

[a]Questions can presuppose either positive or negative answers, hence, an agreement between question and answer refers to their respective polarity.

We now describe the syntactic rewriting space of polar question and answers into factual statements. The main clause of a polar question is an *interrogative clause*, following specific grammar rules [14], e.g., $q \rightarrow$ AUX NP VP (among many other context free grammar rules).

Answers to polar questions can take several different syntactic shapes [9, 11, 16]. Furthermore, as there is an asynchronous relation between the asker and answerer in online settings, there is often a lack of conversational context, resulting in more elaborate answers. We devise a taxonomy of answer types, which correspondingly determine also the shape of the rewritten statement. Table 1 defines the different categories along with example generated statements for each category.

The cases in Table 1 represent frequent syntactical manifestations of PQAs on the Web, and we will use them as guidance for dataset collection (cf. Section 5) and evaluation (cf. Section 6.2).

4 Approach

Figure 2 shows an overview of our approach, a sequence to sequence (seq2seq) model based on the T5 transformer model [24]. It has two main components that *ensure controllable rewriting*: (i) automated constraint extraction, and (ii) controllable rewriting through soft-constraints.

Fig. 2. Overview of our proposed controllable rewriting approach. In (a) we extract automatically constraints from constituency parse trees, then in (b), the constraints together with the input text that goes to *encoder* is encoded in the soft-mention flag matrix, which is then provided as input in (c) to the decoder for controllable generation.

4.1 Automated Constraint Extraction

Unlike in MF [32], where constraints are provided manually, we propose an automated constraint extraction approach based on constituency parses of the input PQA.

Our goal in controllable rewriting is for the target statement to contain the interrogative clause from the question, and the corresponding affirmation/negation clauses in the answer, along with the any embedded conditional or alternative clauses. This boils down to two main tasks for constraint extraction: (1) determining the input tokens that *must be present* in the output, and (2) ensuring that the decoder satisfies such constraints.

Following the syntactic rewriting cases in Table 1, we extract constraints, shown in order of importance. (the algorithm is provided in the paper's appendix).

Noun Phrases (NP): NPs identify the subject of the question. In the answer they help us identify the matching clause with the assigned *polarity* and *explanation* to the question's preposition. NPs are used as constraints only if they are not embedded as children of non-NP constituents.

Verb Phrases (VP): VPs on the other hand identify the information need as defined by the verb serving as the root of the constituent.

Other Phrases: The rest of the extracted phrases as constraints are prepositional phrases (PP), adverbial phrases (ADVP), and adjective phrases (ADJP). PP and ADJP provide further details about the information in an NP, whereas ADVP provide further information about the verb in the question's interrogative clause and assert its polarity.

4.2   Soft Mention Flags

To assess if the extracted constraints are satisfied in the generated output, we must account for two factors: (i) input constraints may be expressed differently in the output (*paraphrases, synonyms* etc.), and (ii) there is no one-to-one mapping between input constraints and the generated statement.

We enforce our model to *satisfy* the extracted constraints in Sect. 4.1 by constructing a soft-mention flags matrix $\mathbf{M}_{k \times l} \in \{0, 1, 2\}$, where rows represent input PQA tokens and columns are the output tokens. For each input token x_i, $\mathbf{M}$ holds a value between $\{0, 1, 2\}$, where 0 is for tokens not part of any constraint, 1 for token part of a constraint but not satisfied, and 2 for tokens part of a constraint and satisfied in the output $\mathbf{y}$.

$$M_{\mathbf{x}_i, \mathbf{y}_{:t}} = \begin{cases} 0 & x_i \text{ is not part of a constraint} \\ 1 & x_i \text{ is not mentioned in } \mathbf{y}_{:t} \\ 2 & x_i \text{ is mentioned in } \mathbf{y}_{:t} \end{cases} \tag{1}$$

As in [32], we inject the soft mention flag matrix, $\mathbf{M}$, in the model's decoder layers. $\mathbf{M}$ is represented through two embedding types: 1) key embeddings, $\mathbf{M}^k = E_k(M)$ and value $\mathbf{M}^v = E_v(M)$ where E_k and $E_v \in \mathbb{R}^{3 \times dim}$. These embeddings are injected between the encoder output $\mathbf{h}^e$ and the decoder input $\mathbf{h}_t^d$ in the cross multi-head attention module (cf. Eq. 2).

$$\begin{aligned} \mathrm{CA}(\mathbf{h}_t^d, \mathbf{h}^e, \mathbf{M}^k, \mathbf{M}^v) = \\ F(W_q h_t^d, W_k h^e, W_v h^e, \mathbf{M}^k, \mathbf{M}^v) \end{aligned} \tag{2}$$

where F is a self-attention function with soft mention flag embeddings defined below.

$$F(q, k, v, \mathbf{M}^k, \mathbf{M}^v)_j = \sum_{i=1}^{l_x} \alpha_{i,j}(v_i + \mathbf{M}^v_{i,j}) \tag{3}$$

$$\alpha_{i,j} = \texttt{softmax}\left(\frac{q_i(k_i + \mathbf{M}^k_{i,j})^T}{\sqrt{dim}}\right) \tag{4}$$

$\mathbf{M}$ provides the seq2seq model with *explicit* signal about the decoded tokens. Whenever input tokens are marked with 0, the model performs standard conditional decoding. Otherwise, if a token is part of a constraint (set to 1 in $\mathbf{M}$), it represents an explicit signal to decode sequences such that the corresponding values in $\mathbf{M}$ are changed to 2.

With the explicit means to signal *what* part of the input composes a constraint that needs to be met in the decoded output, next, we describe how we establish if a constraint is satisfied, while taking into account that it can undergo *syntactic* and *lexical* changes in the decoded output. We propose two strategies to encode constraint satisfaction.

Semantic Constraint Satisfaction: For an input constraint all its tokens $c_i = \{x_m, \ldots, x_n\}^3$ are initialized with 1 in $\mathbf{M}$ (i.e., constraints not been satisfied). At each generation step t, we assess whether the constraint c_i is satisfied. To ensure constraint satisfaction accuracy, we consider only tokens within a specific *window length* (with window size equal to $|c_i|$) of preceding tokens in the output $y_{k,l}^4$. A constraint is satisfied if the semantic similarity between $\texttt{sim}(y_{k,l}, c_i)$, computed as the cosine similarity between the sentence representations [27] of $y_{k,l}$ and c_i meets two specific thresholds: (1) threshold a, where $\texttt{sim}_t > a$, (2) threshold b of the difference between the current similarity score and the score in step $t - 1$, namely $\texttt{sim}_t - \texttt{sim}_{t-1} > b$. Once $y_{:t}$ meets both thresholds, all tokens of c_i are changed to 2 in $\mathbf{M}$.

Example: Table 2 (a) shows an example on the initialization and the updates on the soft mention flag matrix $\mathbf{M}$. For an input sequence $\mathbf{x} = $ [The, screen, has, full, touchscreen, function], $\mathbf{c} = $ *"has full touchscreen function"*, is a constraint that needs to be satisfied in the output. The flags for $\mathbf{c}$ at step 0 are initialized with 1, $\mathbf{M}(\mathbf{c}, y_{:0}) = [0, 0, 1, 1, 1, 1]$, given that the output is ⟨SEP⟩. The top row shows the similarity score for each step. At step 6, $\mathbf{M}(\mathbf{c}, y_{:0}) = [0, 0, 2, 2, 2, 2]$ since the constraint has been covered in the current output sequence.

---

[3] A sequence of one or more consecutive tokens as extracted from the constituency parse tree.

[4] Where $\forall\ 0 \leq k \leq t - |c_i|$ and $k < l \leq t$.

Table 2. (a) Constraints are marked in bold. Underlined tokens are included in the sliding window that assesses constraint satisfaction. (b) 1st person pronouns are initialized with 2 then changed to 1, when a 2nd person pronoun word is generated.

sim	0	0.21	0.26	0.28	0.26	0.37	**0.85**	0.76
	(SEP)	Dell	Laptop	comes	with	full	touchscreen	.
The	0	0	0	0	0	0	0	0
screen	0	0	0	0	0	0	0	0
has	1	1	1	1	1	1	2	2
full	1	1	1	1	1	1	2	2
touchscreen	1	1	1	1	1	1	2	2
function	1	1	1	1	1	1	2	2

(a)

	(SEP)	Dell	XPS	can	be	shipped	by	us	to	Brazil	.
We	2	2	2	2	2	2	2	1	1	1	1
can	-	-	-	-	-	-	-	-	-	-	-
ship	-	-	-	-	-	-	-	-	-	-	-
to	-	-	-	-	-	-	-	-	-	-	-
Brazil	-	-	-	-	-	-	-	-	-	-	-

(b)

Factual Style Constraints: To frame the output statement in a factual style, first person narratives are transformed into a second person narrative. Such a seemingly small change (i.e., *1st* to *2nd* person pronouns), incurs a series of syntactic rewrite operations required to ensure coherence of the output statement. Table 2 (b) provides an example. In this case, we represent in the **M** 1st person pronouns with score 2, and convert them to 1, once the model has generated a second person pronoun in the output. The reason for reverting the order from *satisfied* constraint to *not satisfied* is to avoid any confusing behavior between the satisfaction of input constraint extracted through the automated constraint extraction and framing constraints.

5 PAR Data Collection

We now describe our data collection process, which is based on the Amazon Product Question Answers dataset [29], which contains a diverse set of product-specific PQAs generated by customers. The dataset contains 10M questions about 1.5M products, from which we focus only on yes/no questions. Each instance consists of the question text q, answer text a, and product name c, which represents the context in our case.

5.1 PQA Rewriting Through Crowdsourcing

We recruited 10 expert annotators to collect ground truth for 1,500 instances. The data is uniformly distributed across the categories in Table 1 and covers 11 domains.

The task is designed to be decomposable into stages to ensure annotation reliability. Annotators follow a series of guidelines (cf. Table 3), allowing them to first map the input PQA into one of the pre-defined categories, after which they determine the answer polarity (step **S1**), then in **S2** the output category is determined (e.g. *Alternative*), in **S3** the anaphoric expressions are replaced with the context information, and finally in **S4** the statement is framed in 2nd person narrative.

Table 3. An example to illustrate how annotators perform the Yes/No QA rewriting task.

Q	Does this monitor have a camera?
A	No. But it is a wide screen. I've been connecting a game console to it
C	Dell 27-inch Full HD 1920×1080 Widescreen LED Professional Monitor
S1	*No*
S2	No, *this monitor doesn't have a camera. But it has a wide screen and I've been connecting a game console to it*
S3	No, *the Dell 27-inch Full HD monitor* doesn't have a camera. But it has a wide screen and I've been connecting a game console to it
S4	No, the Dell 27-inch Full HD monitor doesn't have a camera. But it has a wide screen and *you can connect it with a game console*

6   Experimental Setup

6.1   Datasets

PAR: We randomly split the PAR dataset into 1000/100/400 for train/dev/test respectively. PAR is used for training and evaluation.

Reddit: We leverage a sub-forum of Reddit of polar questions[5] and on a randomly sample of 50 QA pairs assess zero-shot generalization performance.

SemEval: From SemEval-2015 Task 3[6] we randomly sample 50 pairs of Yes/No question and answers for zero-shot evaluation.

6.2   Baselines

Approaches are trained on the PAR dataset in Sect. 6.1.

T5 [25]. We use the T5 model as our baseline, which in turn serves as an ablation of SMF without the soft-mention flags module.

CBS [1]. We use T5 with constrained beam search during the decoding phase. The constraints represent phrases extracted in Sect. 4.1.

GPT-2 [24]. We adopt GPT2 and fine-tune it on PAR training set.

COLD [23]. A GPT2 based decoding method by sampling from an energy function.

MF [32]. Originally proposed the addition of mention flags into sequence to sequence decoders, and focuses only on lexical constraints. The constraints in this case, contrary to the original paper that are provided as input, here the constraints are extracted automatically (cf. Sect. 4.1).

Our Approach – SMF: We distinguish two models of our approach: (1) **SMF** where for controllability we rely only on the extracted constraints in Sect. 4.1, and (2) **SMF-Style**, where in addition to the extracted constraint, we additionally encode the target style constraints (Sect. 4.2).

6.3   Evaluation Metrics

Automated Metrics: To assess the closeness of the generated statements with respect to the ground-truth statements generated by human annotators, we use BLEU [22], ROUGE [20] and F1-BertScore [35].

---

[5] https://www.reddit.com/r/YayorNay/.

[6] https://alt.qcri.org/semeval2015/task3/index.php.

Human Evaluation: Automated metrics cannot capture the nuanced aspects and natural variation of the task. We design two human evaluations, where annotators assess 50 randomly sampled input-output pairs for on:

Statement Syntactic Clause Coverage: we assess if the statement, depending on its category (cf. Table 1) has coverage of the embedded clauses.

Statement Correctness and Coherence: we assess if a statement: (i) contains the correct polarity, (ii) mentions the input context, (iii) framing is in 2nd person narrative, (iv) does not contain information not present in the input PQA; (v) is grammatically correct and coherent, and (vi) is equivalent to its ground-truth counterpart.

7   Experimental Results

Overall Performance. Table 4(a) shows the evaluation results for all models on the PAR test set. We observe that our proposed SMF approach achieves the highest performance across all evaluation metrics. A negligible difference is noted between SMF and SMF-STYLE. However, as we show in Sect. 7.1 the SMF-STYLE attains more coherent statements.

From the baselines, the closest to our approach is MF. However, note here that we adapt MF using our extracted constraints, contrary to [32] where the constraints are provided manually, an approach that does not scale.

Alternatively, if we do not use our proposed automated constraint extraction approach, one could provided as constraints to MF tokens that overlap between the question and answer. Using such constraints causes the performance of MF to drop by 27% in terms of BLEU. This highlights that *what* composes as input constraints is key, and noisy constraints can yield results worse than seq2seq models without constraints.

CBS reveals that our task does not simply involve copying tokens, but rather involves a series of syntactic and semantic transformations of the input. The difference between CBS and SMF is 7.4% in terms of BLEU and ROUGE-L scores. This shows that enforcing the decoder to output certain tokens has a negative effect, as CBS performs worse than T5. This is another indication that our approach, where the constraint satisfiability is done at the semantic level is appropriate. Equivalent clauses from the input can be transformed and in cases question and answer clauses can be combined into a single clause, hence, causing difficulties to establish such mappings.

COLD and GPT2 achieve poor performance, highlighting the need for larger amount of training to learn the task.

Out-of-Domain Performance. Figure 3 shows the out-of-domain performance of SMF and SMF-STYLE. For the 11 domains in the PAR dataset, we consider a leave-one-out domain evaluation, showing the zero-shot performance. The results show that: (1) both approaches generalize well on out-of-domain data, with most domains obtaining comparable performance to the in-domain performance. (2) comparing SMF and SMF-STYLE, we note that SMF-STYLE

Table 4. (a) Overall performance of the different models on the PAR test set. MF, SMF and SMF-Style obtain significantly better results (p< 0.01 as per t-test) than the rest of the competitors.. (b) Impact of input constraints on output quality. The first score represents the case without the extracted input constraints, while the second score represents the case with constraints.

	BLEU	ROUGE-L	BertScore
GPT2	32.8	55.0	93.2
COLD	28.1	51.3	92.0
T5	50.6	67.9	95.3
CBS	45.2	62.7	94.2
MF	51.2	68.5	95.4
SMF	**52.6**	**69.5**	**95.5**
SMF-Style	52.5	68.9	95.4

(a)

		BLEU	ROUGE-L	BertScore
complement	MF	40.3/43.3	68.0/70.1	95.2/95.5
	SMF	45.0/45.8	71.3/71.5	95.6/95.6
	SMF-STYLE	45.6/46.8	71.1/73.8	95.9/96.0
condition	MF	35.3/36.7	62.0/62.8	94.7/94.7
	SMF	34.5/39.5	61.4/63.9	94.7/95.0
	SMF-STYLE	36.5/37.9	62.3/64.1	95.0/95.1
alternative	MF	40.8/41.1	69.5/69.9	96.1/96.3
	SMF	44.1/44.8	70.0/71.0	96.0/96.1
	SMF-STYLE	43.5/44.5	69.2/71.5	96.3/96.4

(b)

achieves better performance on statement framing style, implying the effectiveness of style constraints.

Input Constraint Coverage Impact. Table 4(b) shows the impact of constraints, namely the presence of constraints as input on the generated output. The results show the impact on different PAR categories, for approaches that make use of the input constraints.

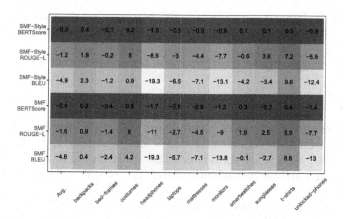

Fig. 3. Out-of-domain performance on the test set of the PAR dataset on different domains.

7.1   Human Evaluation

To complement the automatic metrics, we report the results of our two human studies carried out by two human annotators.

Statement Syntactic Clause Coverage: Table 5(b) shows the results on the evaluation of statements containing the different clauses according to their PQA syntactic shape. Namely, if an input belongs to the *condition* syntactic shape, the generated statement should contain the *explanation* and the embedded *condition* clause.

Except for CBS and COLD, all approaches do fairly well in incorporating the *explanation* clause in their output. This is intuitive as this is present in all input PQAs, and contains information that is present in the question and often in the answer as well.

One of the key embedded clauses, the *condition* clause, represent one of the most challenging scenarios to be decoded into the target statement. This clause has a pivotal role in conditioning the polarity of the answer, and thus, failure to decode this in the statement can lead to erroneous answers provided to a question. SMF approaches achieve the highest scores, with T5 having a 10% point difference. One notable case is that of *alternative* clause, where the baseline T5 achieves perfect coverage, with MF and SMF following with 92.9% coverage.

Overall, the best coverage is achieved by SMF, with 85% coverage across all types. This shows the importance of extracted input constraints based on our approach in Sect. 4.1, where the models MF, SMF, and SMF-STYLE, achieve highest coverage among all competing approaches.

Table 5. (a) Clause type coverage on the output statement. (b) Correctness and coherence scores of the generated statements by the different approaches.

	Explanation	Complement	Condition	Alternative	Overall
GPT2	91.0	55.6	46.7	64.3	58.0
COLD	65.0	44.4	43.3	35.7	45.0
T5	96.0	75.0	70.0	100.0	79.0
CBS	86.0	66.7	60.0	85.7	72.0
MF	99.0	83.3	66.7	92.9	82.0
SMF	98.0	83.3	80.0	92.9	85.0
SMF-Style	98.0	80.6	80.0	78.6	83.0

(a)

	Polarity	Coverage	Style	Relevance	Syntactic	Coherence	Equivalence
GPT2	0.93	0.94	0.98	0.67	0.84	0.69	0.36
COLD	0.80	0.75	0.98	0.52	0.43	0.42	0.19
T5	0.96	1.0	0.97	0.94	0.95	0.91	0.74
CBS	0.82	0.87	0.95	0.82	0.68	0.75	0.56
MF	0.94	0.98	0.99	0.94	0.98	0.91	0.69
SMF	0.95	0.98	0.93	0.91	0.94	0.92	**0.77**
SMF-STYLE	**0.97**	0.99	0.96	**0.96**	**0.99**	**0.98**	0.76

(b)

Correctness and Coherence Evaluation: Table 5 (a) shows the human evaluation results for the syntactic correctness and coherence of statements, where several detailed aspects are considered. This study highlights the task complexity and provides an overview of which sub-tasks the different models are able to do reliably.

On *polarity*, most seq2seq based models, such as T5, MF, SMF and SMF-STYLE, obtain high polarity accuracy, with SMF-STYLE achieving an accuracy of 97%. Given that answers are contextualized w.r.t the question and the input context, here the product name, we note that similarly, all T5 based models obtain a good coverage of the context, with T5 achieving 100% coverage. In terms of context coverage, the models are required to chose from the lengthy product names, a succinct phrase that allows the human annotators to identify correctly the product name.

On *factual style*, MF obtains the highest accuracy, however, at a cost that not always the generated statements are semantically coherent. When we consider SMF-STYLE, which encodes as constraints the framing style, although the accuracy is lower in terms of outputting statements in 2nd person narrative, in terms of coherence, we obtain the highest accuracy. Coherence, represents a more global measure, which has high importance, given that incoherent statements are not suitable to be provided as answers.

Finally, in terms of equivalence between the generated statement and the ground truth, both SMF and SMF-STYLE, obtain the highest scores. This shows that our approaches are able to jointly optimize for the numerous sub-tasks of controllable rewriting.

7.2   Evaluation on Other Community PQA

Finally, we evaluate the zero-shot performance of SMF and SMF-STYLE on REDDIT and SEMEVAL w.r.t the statement correctness and coherence.

Table 6 shows that our approach can perform well when applying to other data sources under the zero-shot setting. The results of this setting imply that the modeling of our rewriting task is not simply learning the superficial lexical content, but learning the rewriting strategies to restructure the input PQA regardless of domains. This finding makes our rewriting task and approach promising to generalize to other domains without requiring extra annotations.

Table 6. Results on REDDIT and SEMEVAL datasets.

	REDDIT		SEMEVAL	
	SMF	SMF-Style	SMF	SMF-Style
Polarity	1.0	1.0	0.90	0.95
Context Coverage	1.0	1.0	1.0	1.0
Factual Style	0.95	1.0	1.0	1.0
Relevance	1.0	1.0	0.95	1.0
Syntactic Correctness	0.95	1.0	1.0	1.0
Coherence	0.90	0.95	0.95	0.95
Equivalence	0.85	0.80	0.80	0.80

8   Conclusion

We introduced the task of rewriting Yes/No question and answers into succinct decontextualized statements, and defined several desiderata determining how the input PQA is reorganized and framed into the factual statement. This task enables us to explore knowledge from community PQA, unlocking the highly contextualized answers to answer other question shapes, and furthermore making them retrievable. For this task, with the help of expert annotators, we curated a dataset of 1500 input PQA and the target statements covering 11 domains from Amazon's PQA dataset.

Next, we introduced an approach for controllable rewriting, achieved through automatically extracted constraints from the input, which are encoded into our approach using a *soft mention flag* matrix, allowing us at the semantic level to map constraints to the generated statements.

Finally, empirical evaluations showed that our approach outperforms a series of competitors in both automated and human evaluation metrics.

9   Appendix

Constraint Extraction From Parse Trees: Algorithm 1 outlines the steps undertaken to extract constraints from the extracted constituency parse trees from the input PQA.

Algorithm 1. Constraint Extraction

Input: Yes/No question Q, answer A.
Output:
1: Extract all NP from the input using constituency parsing tree.
2: Exclude PRON from NP in Step 1.
3: **for** each noun phrase **do**
4:    **if** the parent node is labeled as ['VP', 'PP', 'ADVP', 'ADJP'] **then**
5:       Add the phrase that belongs to this parent node to the constraint list.
6:    **else**
7:       **if** the parent node is labeled as 'NP' **then**
8:          Add the phrase that belongs to the current node to the constraint list.
9:       **end if**
10:   **end if**
11: **end for**
12: **return** constraint list

Constraint Extraction Accuracy: We sample 100 instances in the PAR dataset which uniformly cover all the different syntactic PAR categories from Table 1, and ask annotators whether: (1) the extracted constraints from the question or the answer capture the question's intent, and (2) if the extracted constraints cover the different embedded clauses. For the first part, we see that the extracted constraints from both the question and the answer have a good coverage of 87% on the question's intent.

- **Q1**: *"Do the extracted question constraints cover fully the question's intent?"*
  = 87.1%
- **Q2**: *"Do the extracted answer constraints cover fully the question's intent?"*
  = 87.7%

In the second part, the extracted constraints cover well on explanation and complement. Especially, explanation covers nearly 97% of the cases. A lower coverage is reported for constraints covering the condition and alternative clauses in the answer. This is mainly due to the fact that these clauses use pronouns, thus, increasing the likelihood of missing those simplified constituents. This represents an important future research direction to have a more robust algorithm with better coverage of all answer's embedded clauses. More specifically, we obtain the following scores: Explanation = 0.97, Complement = 0.7, Condition = 0.52, and Alternative = 0.40.

Human Evaluation Analysis: Two annotators evaluated 50 randomly chosen Yes/No QA pairs. All judgments are on a single scale (i.e., *binary*). For reliability, each instance is assessed by both annotators. In case of ties, a third annotation was collected.

Statement Syntactic Shape: In this study, the inter-rater agreement rate was 87.5%. This represents a high agreement rate and shows that annotators agree on what embedded clauses are covered in the generated target statement.

Statement Correctness and Coherence: The inter-rater agreement was measured separately on the seven different questions, with an agreement of *Polarity* = 64%, *Coverage* = 64%, *Style* = 65%, *Relevance* = 62.3%, *Syntactic* = 64%, *Coherence* = 62.3%, and *Equivalence* = 52%, respectively.

Overall, the agreement rates are high and similar for most of the questions, with the only exception being *Equivalence*. In this question, the annotators were asked to assess if two statements represent semantically equivalent information. This yields a lower agreement rate, as annotators may comprehend the statements differently, and additionally the presence or absence of information on one of the statements can cause the annotators to perceive the equivalence differently.

References

1. Anderson, P., Fernando, B., Johnson, M., Gould, S.: Guided open vocabulary image captioning with constrained beam search. In: Proceedings of the 2017 Conference on Empirical Methods in Natural Language Processing, pp. 936–945 (2017)
2. Balakrishnan, A., Rao, J., Upasani, K., White, M., Subba, R.: Constrained decoding for neural nlg from compositional representations in task-oriented dialogue. In: Proceedings of the 57th Annual Meeting of the Association for Computational Linguistics, pp. 831–844 (2019)
3. Chen, Z., Zhao, J., Fang, A., Fetahu, B., Rokhlenko, O., Malmasi, S.: Reinforced question rewriting for conversational question answering. CoRR abs/2210.15777 (2022). 10.48550/arXiv. 2210.15777. https://doi.org/10.48550/arXiv.2210.15777
4. Choi, E., He, H., Iyyer, M., Yatskar, M., Yih, W.t., Choi, Y., Liang, P., Zettlemoyer, L.: Quac: Question answering in context. In: Proceedings of the 2018 Conference on Empirical Methods in Natural Language Processing, pp. 2174–2184 (2018)
5. Choi, E., Palomaki, J., Lamm, M., Kwiatkowski, T., Das, D., Collins, M.: Decontextualization: making sentences stand-alone. Trans. Assoc. Comput. Linguistics **9**, 447–461 (2021)

6. Clark, C., Lee, K., Chang, M., Kwiatkowski, T., Collins, M., Toutanova, K.: Boolq: exploring the surprising difficulty of natural yes/no questions. In: Burstein, J., Doran, C., Solorio, T. (eds.) Proceedings of the 2019 Conference of the North American Chapter of the Association for Computational Linguistics: Human Language Technologies, NAACL-HLT 2019, Minneapolis, MN, USA, June 2–7, 2019, Volume 1 (Long and Short Papers), pp. 2924–2936. Association for Computational Linguistics (2019). https://doi.org/10.18653/v1/n19-1300.https://doi.org/10.18653/v1/n19-1300

7. Clark, C., Lee, K., Chang, M.W., Kwiatkowski, T., Collins, M., Toutanova, K.: Boolq: exploring the surprising difficulty of natural yes/no questions. In: Proceedings of the 2019 Conference of the North American Chapter of the Association for Computational Linguistics: Human Language Technologies, Volume 1 (Long and Short Papers), pp. 2924–2936 (2019)

8. Elgohary, A., Peskov, D., Boyd-Graber, J.: Can you unpack that? learning to rewrite questions-in-context. In: Proceedings of the 2019 Conference on Empirical Methods in Natural Language Processing and the 9th International Joint Conference on Natural Language Processing (EMNLP-IJCNLP). pp. 5918–5924. Association for Computational Linguistics, Hong Kong, China, November 2019. https://doi.org/10.18653/v1/D19-1605,https://aclanthology.org/D19-1605

9. Enfield, N.J., Stivers, T., Brown, P., Englert, C., Harjunpää, K., Hayashi, M., Heinemann, T., Hoymann, G., Keisanen, T., Rauniomaa, M., et al.: Polar answers. J. Linguistics 55(2), 277–304 (2019). https://doi.org/10.1017/S0022226718000336

10. Hokamp, C., Liu, Q.: Lexically constrained decoding for sequence generation using grid beam search. In: Proceedings of the 55th Annual Meeting of the Association for Computational Linguistics (Volume 1: Long Papers), pp. 1535–1546 (2017)

11. Holmberg, A.: The syntax of answers to polar questions in English and Swedish. Lingua 128, 31–50 (2013). https://doi.org/10.1016/j.lingua.2012.10.018. https://www.sciencedirect.com/science/article/pii/S0024384112002392, sI: Polarity emphasis: distribution and locus of licensing

12. Huddleston, R.: The contrast between interrogatives and questions. J. Linguistics 30(2), 411–439 (1994)

13. Iatridou, S., Embick, D.: Conditional inversion. In: North East Linguistics Society, vol. 24, p. 14 (1994)

14. Jurafsky, D.: Speech & language processing. Pearson Education India (2000)

15. Juraska, J., Karagiannis, P., Bowden, K., Walker, M.: A deep ensemble model with slot alignment for sequence-to-sequence natural language generation. In: Proceedings of the 2018 Conference of the North American Chapter of the Association for Computational Linguistics: Human Language Technologies, Volume 1 (Long Papers). pp. 152–162 (2018)

16. Kramer, R., Rawlins, K.: Polarity particles: an ellipsis account. In: Proceedings of NELS, vol. 39, pp. 479–92 (2009)

17. Kumar, S., Malmi, E., Severyn, A., Tsvetkov, Y.: Controlled text generation as continuous optimization with multiple constraints. Adv. Neural. Inf. Process. Syst. 34, 14542–14554 (2021)

18. Kwiatkowski, T., et al.: Natural questions: a benchmark for question answering research. Trans. Assoc. Comput. Linguist. 7, 452–466 (2019)

19. Lewis, M., et al.: Bart: denoising sequence-to-sequence pre-training for natural language generation, translation, and comprehension. In: Proceedings of the 58th Annual Meeting of the Association for Computational Linguistics, pp. 7871–7880 (2020)

20. Lin, C.Y.: ROUGE: a package for automatic evaluation of summaries. In: Text Summarization Branches Out, pp. 74–81. Association for Computational Linguistics, Barcelona, Spain, July 2004. https://aclanthology.org/W04-1013
21. Louis, A., Roth, D., Radlinski, F.: "I'd rather just go to bed": Understanding indirect answers. In: Proceedings of the 2020 Conference on Empirical Methods in Natural Language Processing (EMNLP), pp. 7411–7425. Association for Computational Linguistics, Online (Nov 2020). 10.18653/v1/2020.emnlp-main.601, https://aclanthology.org/2020.emnlp-main.601
22. Papineni, K., Roukos, S., Ward, T., Zhu, W.J.: Bleu: a method for automatic evaluation of machine translation. In: Proceedings of the 40th Annual Meeting of the Association for Computational Linguistics. pp. 311–318. Association for Computational Linguistics, Philadelphia, Pennsylvania, USA, July 2002. https://doi.org/10.3115/1073083.1073135. https://aclanthology.org/P02-1040
23. Qin, L., Welleck, S., Khashabi, D., Choi, Y.: Cold decoding: energy-based constrained text generation with langevin dynamics. arXiv preprint arXiv:2202.11705 (2022)
24. Radford, A., Wu, J., Child, R., Luan, D., Amodei, D., Sutskever, I., et al.: Language models are unsupervised multitask learners. OpenAI blog 1(8), 9 (2019)
25. Raffel, C., Shazeer, N., Roberts, A., Lee, K., Narang, S., Matena, M., Zhou, Y., Li, W., Liu, P.J., et al.: Exploring the limits of transfer learning with a unified text-to-text transformer. J. Mach. Learn. Res. 21(140), 1–67 (2020)
26. Reddy, S., Chen, D., Manning, C.D.: Coqa: a conversational question answering challenge. Trans. Assoc. Comput. Linguist. 7, 249–266 (2019)
27. Reimers, N., Gurevych, I.: Sentence-bert: sentence embeddings using siamese bert-networks. In: Inui, K., Jiang, J., Ng, V., Wan, X. (eds.) Proceedings of the 2019 Conference on Empirical Methods in Natural Language Processing and the 9th International Joint Conference on Natural Language Processing, EMNLP-IJCNLP 2019, Hong Kong, China, November 3–7, 2019, pp. 3980–3990. Association for Computational Linguistics (2019). 10.18653/v1/D19-1410, https://doi.org/10.18653/v1/D19-1410
28. Rosenthal, S., Bornea, M.A., Sil, A., Florian, R., McCarley, J.S.: Do answers to boolean questions need explanations? yes. CoRR abs/2112.07772 (2021). https://arxiv.org/abs/2112.07772
29. Rozen, O., Carmel, D., Mejer, A., Mirkis, V., Ziser, Y.: Answering product questions by utilizing questions from other contextually similar products. In: NAACL 2021 (2021). https://www.amazon.science/publications/answering-product-questions-by-utilizing-questions-from-other-contextually-similar-products
30. Sutskever, I., Vinyals, O., Le, Q.V.: Sequence to sequence learning with neural networks. Advances in neural information processing systems 27 (2014)
31. Vakulenko, S., Longpre, S., Tu, Z., Anantha, R.: Question rewriting for conversational question answering. In: Proceedings of the 14th ACM International Conference on Web Search and Data Mining, pp. 355–363. WSDM '21, Association for Computing Machinery, New York, NY, USA (2021). https://doi.org/10.1145/3437963.3441748,https://doi.org/10.1145/3437963.3441748
32. Wang, Y., Wood, I., Wan, S., Dras, M., Johnson, M.: Mention flags (mf): Constraining transformer-based text generators. In: Proceedings of the 59th Annual Meeting of the Association for Computational Linguistics and the 11th International Joint Conference on Natural Language Processing (Volume 1: Long Papers), pp. 103–113 (2021)
33. Williamson, G.: Conditional antecedents as polar free relatives. In: Semantics and Linguistic Theory, vol. 29, pp. 496–508 (2019)

34. Yang, Z., et al.: Hotpotqa: a dataset for diverse, explainable multi-hop question answering. In: Proceedings of the 2018 Conference on Empirical Methods in Natural Language Processing, pp. 2369–2380 (2018)
35. Zhang, T., Kishore, V., Wu, F., Weinberger, K.Q., Artzi, Y.: Bertscore: evaluating text generation with bert. In: International Conference on Learning Representations (2020). https://openreview.net/forum?id=SkeHuCVFDr

Evaluating the Impact of Content Deletion on Tabular Data Similarity and Retrieval Using Contextual Word Embeddings

Alberto Berenguer[ORCID], David Tomás[✉][ORCID], and Jose-Norberto Mazón[ORCID]

Department of Software and Computing Systems, University of Alicante,
Carretera San Vicente del Raspeig s/n, 03690 San Vicente del Raspeig, Spain
{aberenguer,dtomas,jnmazon}@dlsi.ua.es

Abstract. Table retrieval involves providing a ranked list of relevant tables in response to a search query. A critical aspect of this process is computing the similarity between tables. Recent Transformer-based language models have been effectively employed to generate word embedding representations of tables for assessing their semantic similarity. However, generating such representations for large tables comprising thousands or even millions of rows can be computationally intensive. This study presents the hypothesis that a significant portion of a table's content (i.e., rows) can be removed without substantially impacting its word embedding representation, thereby reducing computational costs while maintaining system performance. To test this hypothesis, two distinct evaluations were conducted. Firstly, an intrinsic evaluation was carried out using two different datasets and five state-of-the-art contextual and not-contextual language models. The findings indicate that, for large tables, retaining just 5% of the content results in a word embedding representation that is 90% similar to the original one. Secondly, an extrinsic evaluation was performed to assess how three different reduction techniques proposed affects the overall performance of the table-based query retrieval system, as measured by MAP, precision, and nDCG. The results demonstrate that these techniques can not only decrease data volume but also improve the performance of the table retrieval system.

Keywords: Table retrieval · Contextual word embeddings · Semantic similarity

1 Introduction

Nowadays, a substantial portion of the information generated by both industry and academia is presented in the form of tabular data. In this context, the management and retrieval of tabular data have emerged as critical tasks in the age of data-driven decision-making. The increasing ubiquity and intricacy of structured data have increase the need for efficient and effective information retrieval (IR) systems specifically tailored to tabular data. These systems assume a

© The Author(s), under exclusive license to Springer Nature Switzerland AG 2024
N. Goharian et al. (Eds.): ECIR 2024, LNCS 14609, pp. 433–447, 2024.
https://doi.org/10.1007/978-3-031-56060-6_28

central role in academia, facilitating data-driven research across diverse disciplines, while simultaneously providing indispensable tools for industry to harness the potential of structured information for well-informed decision-making and the pursuit of business intelligence.

Table retrieval is a well-explored task in the literature [24], with its primary objective being the generation of a ranked list of tables that are deemed relevant to a given search query [23]. In recent years, the adoption of language models has gained significant popularity as a means to surpass the constraints inherent in basic content matching approaches for table retrieval. These models offer a way to understand the semantics of data, thereby enhancing the quality of results generated in response to specific search queries.

Language models, which represent probability distributions over words or word sequences, have been extensively utilised in natural language processing (NLP) over the years to undertake various tasks including information retrieval, text classification, and language generation [13]. In the context of table retrieval, the objective of these models is to generate vector representations of tables, commonly referred to as "word embeddings", within a latent semantic space. These embeddings enable the comparison of tables by calculating their distance through measures such as cosine similarity.

Recently, large language models built upon the Transformer architecture [21] have established themselves as the prevailing state-of-the-art solutions for a multitude of NLP tasks. These models are increasingly gaining popularity in the realm of table retrieval, primarily owing to their proficiency in generating precise word embedding representations of tabular content. However, it is important to note that the deployment of Transformer models on extensive datasets can incur substantial computational costs, potentially posing challenges to the operational efficiency of table retrieval systems in real-world production scenarios.

The current study states the hypothesis that the content of a table can be reduced, achieved through the selective omission of rows, without causing a substantial alteration in the word embedding representation of that table. This proposition implies that an equally high-quality semantic representation can be derived by utilising only a subset of the original table data. Such an approach serves to mitigate the computational burden associated with employing Transformer models.

Two distinct datasets, each of varying characteristics, were analysed in this study. The first dataset comprises an extensive collection of nearly 500,000 tables sourced from Wikipedia. These tables typically exhibit a reduced number of rows and columns. The second dataset encompasses approximately 200 tables extracted from an open data portal, featuring hundreds of thousands of rows. Subsequently, a battery of contextual and non-contextual language models were evaluated on these datasets. Furthermore, three distinct strategies for content reduction were proposed. The findings reveal that, in the case of larger tables, it is feasible to omit up to 95% of the rows from the dataset without causing a substantial alteration in the word embedding representation of the table. This, in turn, significantly diminishes the computational overhead associated with the application of Transformer models for the task of table retrieval.

In addition to the intrinsic evaluation, an extrinsic evaluation was conducted to analyse the performance of a table retrieval system. In this evaluation, a subset of rows is intentionally omitted from the tables, and the resulting reduced embeddings are employed in the retrieval task. The results indicate that this reduction actually improves the outcomes across various experimental scenarios.

The remaining sections of this paper are organised as follows: Sect. 2 provides an overview of related research in the application of language models for table retrieval; Sect. 3 offers a comprehensive description of the proposed study; Sect. 4 outlines the conducted experiments and provides a discussion of the obtained results; lastly, Sect. 5 presents the conclusions drawn from the research and outlines potential future work.

2 Related Work

This section provides a summary of the existing research on the use of word embeddings in the context of table retrieval. Word embeddings are dense vectors that represent a word's meaning by positioning it as a point within a semantic space. They capture the distributional meaning of words, wherein similar representations are acquired for words found in analogous contexts. Notable examples of such word representations include Word2vec [12], fastText [2] and Glove [16].

In recent years, word embeddings have gained extensive popularity for various semantic tasks within the field of NLP, including sentiment analysis [6], machine translation [26], text classification [10], and dialog systems [5].

These word embedding approaches construct a global vocabulary by assigning a single representation to each unique word in the documents, regardless of potential variations in meaning or senses across different contexts. These approaches are often regarded as static representations, lacking the ability to capture the diverse senses of a word. In contrast, contemporary contextual word embeddings have the capacity to capture the multiple meanings of polysemous words, as each vector signifies not just a word but a specific sense. Consequently, each word can be depicted using distinct word embeddings, each tailored to a particular context in which the word may be encountered. Noteworthy examples of such representations include ULMFit [9], BERT [4], and DeBERTa [7].

Recent approaches to table retrieval use word embeddings to represent tabular data, employing vector similarity metrics to compute the relevance among tables [15]. Depending on the nature of the search query, table retrieval can be categorised as either keyword-based or table-based search [23]. In the former, a query comprises a set of keywords, mirroring the conventional approach adopted by search engines like Google. In the latter, the query itself takes the form of a table, and the objective is to determine a similarity score between the input table and potential candidate tables. In the experiments detailed in Sect. 4, tables serve as queries to retrieve relevant tables, conducting a table-based search within the dataset. Consequently, the remainder of this section focuses primarily on this specific category.

Shraga et al. [18] employed Word2vec as the source for semantic vectors. The table's information was segmented into four distinct semantic spaces: description (comprising the title and caption), schema (encompassing column headings), records (comprising table rows), and facets (representing table columns). Subsequently, various neural network architectures were applied to each of these semantic spaces, including a recurrent convolutional neural network (for description), a multilayer perceptron (for schema), and a 3D convolutional neural network (for records and facets). Finally, these four semantic spaces were fused using a gated multimodal unit.

In contrast to prior studies, in [20] the authors took a distinct approach by training their custom word embedding model utilising tables extracted from Wikipedia. They employed a skip-gram model tailored specifically for tables, considering captions, attributes (column headings), and content (cell values) as contextual elements.

It is important to emphasise that all the previously mentioned approaches used non-contextual word embedding models. In contrast, the study conducted by Chen et al. [3] employed contextual word embeddings for table-based search. The authors utilised a pre-trained version of BERT and harnessed various types of information present in the table, encompassing both textual and numerical data, to provide BERT with context. This context was drawn from the title, caption, column headings, and cell values of the table.

In the area of question answering from tables, notable models include TAPAS [8], which leverages BERT's encoder. This model flattens the table into smaller tokens and subsequently concatenates them with the query answer. Additionally, token embeddings are integrated with table-aware positional embeddings prior to being input into the model.

Also in the area of question answering models, TaBERT [22], an extension of BERT, tries to acquire contextual representations for both utterances and the structured schema of tables. To achieve this, the process begins by generating a content snapshot encompassing pertinent information related to the utterance. Subsequently, the utterance and the content snapshot are encoded using BERT. Finally, the encoded content is subjected to a vertical self attention layer, facilitating the model in aggregating information from various rows within the content snapshot. This enables TaBERT to effectively capture cross-row dependencies among cell values.

Finally, StruBERT [19] consists on a structure-aware BERT model designed to fuse both textual and structural information of data tables to create context-aware representations for the textual and tabular content within a data table. StruBERT extends the concept of vertical self-attention introduced in TaBERT by incorporating horizontal self-attention, allowing for equal treatment of both dimensions of a table. This feature integration is used in a new end-to-end neural ranking model to address three table-related downstream tasks: keyword- and content-based table retrieval, and table similarity.

3   Research Methodology

This study is driven by two primary goals. The first one is to perform an intrinsic evaluation, aimed at identifying the impact of table content reduction on the word embedding representation of the tables. The objective here is to assess whether content reduction can be implemented without compromising the word embedding representation of the tables, thus enabling the achievement of similar outcomes with significantly reduced data. This approach holds the potential to speed up the indexing and retrieval processes without adversely affecting system performance. The second goal involves an extrinsic evaluation, which seeks to determine how content reduction affects the performance of a table retrieval system. Three different reduction approaches were applied to limit the table content. These approaches are described in Sect. 4.4.

In the intrinsic evaluation, the word embedding vector obtained after reduction is compared with the word embedding of the original table to assess their degree of similarity. In this comparison, a high degree of similarity indicates that both the original table and the reduced version are closely positioned within the semantic space. To compare the original table with a reduced table, the procedure involves the following steps:

1. Obtaining a word embedding for each column in both tables. The word embedding representing a column is derived by calculating the word embedding for each individual cell and subsequently averaging them to obtain the final vector. Headers are not taken into account in this process because the objective is to calculate similarity solely based on the content of the table.
2. Comparing each column in the original table with its corresponding column in the reduced table by computing the cosine similarity between their word embedding representations. This comparison yields a float value within the range of $[-1, 1]$, where 1 indicates maximum similarity and -1 signifies no similarity. Negative cosine values are theoretically possible because word embedding vectors can incorporate negative elements. However, it is noteworthy that in all the conducted experiments, the cosine similarity consistently yielded positive values. This observation aligns with prior research, indicating that word vectors do not exhibit balance around the origin within the semantic space. This phenomenon results in fewer negative cosine similarities than would be anticipated from points within a random n-sphere [14].
3. The overall similarity between the two tables is calculated as the average similarity across all their columns.

In the context of the extrinsic evaluation, the aim is to assess how the use of reduced tables impacts the performance of the final table retrieval system. This system operates using table-based queries, which enables a more robust comparison of tables since the system's query is, in itself, a table.

The retrieval system was evaluated employing different embedding models and corpora featuring tables of varying sizes. Additional details about this evaluation are provided in Sect. 4. For each table within the corpus, the system performs the following steps:

1. Cleaning the content of the tables. Delete null rows and columns, lowercase text, split CamelCase and hyphenated words, and remove punctuation.
2. Word embeddings calculation. The word embedding model is employed to generate two distinct vectors for each column in a table. The first vector represents the header content and is obtained by concatenating all the header content into a string before applying the model. The second vector represents the table's content and is obtained by following the procedure outlined in Step 1 of the intrinsic evaluation described earlier.
3. Indexing the header and content vectors for each table. After obtaining the word embeddings, the Faiss library[1] is employed to efficiently store and conduct similarity searches.
4. Search for relevant tables for a query table. This involves obtaining header and column vectors for the query table. These vectors are then utilised in Faiss to retrieve the top-k similar word embeddings. When comparing each pair of columns, two similarity values are computed: one for the column headers and another for their content. A linear combination of these two values yields the final score $simC$ for two columns c_1 and c_2:

$$simC(c_1, c_2) = \alpha \cdot sim(\mathbf{c}_{1n}, \mathbf{c}_{2n}) + (1 - \alpha) \cdot sim(\mathbf{c}_{1c}, \mathbf{c}_{2c}),$$

where $sim(\mathbf{c}_{1n}, \mathbf{c}_{2n})$ is the cosine similarity of the column names, $sim(\mathbf{c}_{1c}, \mathbf{c}_{2c})$ is the cosine similarity of their contents, and α is a parameter in the range $[0, 1]$ that weights the relevance of these two scores in the final result. The value $\alpha = 0.7$ was used based on results from previous research [Anonymised].

5. Evaluation. The following aspects were analysed in this study: (i) the performance of contextual and non-contextual language models; (ii) the effect of table reduction on the size (number of rows) of the table dataset; (iii) the impact of table reduction in corpora with different characteristics; (iv) the performance of different language models in the table retrieval system; and (v) the influence of applying different table reduction approaches on the retrieval process.

4   Experiments

This section details the experiments conducted to validate the proposed hypothesis, encompassing the models employed, datasets collected, and the resulting outcomes.

4.1   Models

In this research, a variety of language models with different architectures, including both contextual and non-contextual models, were examined. The primary objective was to assess the effectiveness of these models in generating high-quality table embeddings capable of properly handle content reduction while serving as reliable tools for semantic similarity search. This study investigated five distinct language models:

---

[1] https://github.com/facebookresearch/faiss.

- Word2vec [12]: This model employs a two-layer neural network trained to reconstruct the linguistic contexts of words. Word2vec offers two model architectures: continuous bag-of-words (CBOW) and continuous skip-gram. This non-contextual model is pre-trained on a subset of the Google News dataset, comprising approximately 100 billion words. The model includes 300-dimensional vectors for 3 million words and phrases (only words were considered in the experiments presented here).
- fastText [2]: this model treats each word as composed of character n-grams rather than considering the entire word as a whole. This feature enables it to learn not only rare words but also out-of-vocabulary words, thereby expanding the coverage of the model's vocabulary. This non-contextual embedding model was pre-trained on diverse sources, including Wikipedia 2017, UMBC webbase corpus, and statmt.org news dataset, encompassing approximately 16 billion words. As in the previous case, the vectors have 300 dimensions.
- BERT [4]: it is a transformer-based model pre-trained on a substantial corpus of English data in a self-supervised manner. The encoder of this model can be leveraged to generate contextual embedding vectors. In this context, the model utilised is bert-base-uncased, producing embedding vectors of 768 dimensions.
- SentenceBERT [17]: this contextual model has been developed by implementing a siamese architecture to derive sentence embeddings that encapsulate significant semantic meaning. The model selected for this purpose was all-MiniLM-L6-v2, fine-tuned on a dataset consisting of 1 billion sentence pairs. This model generates embedding vectors of 1024 dimensions.
- RoBERTa [11]: RoBERTa is an optimised variant of BERT, refined during the pre-training phase. The encoder of this model is also employed to generate contextual embedding vectors. Specifically, the model used is all-roberta-large-v1, which is a roberta-large model fine-tuned with 1 billion sentence pairs. This model produces embedding vectors of 1024 dimensions.

4.2 Datasets

These experiments aimed to assess the models' performance across diverse scenarios, involving the use of two distinct corpora to measure word embedding similarities after content reduction.

The initial part of the experiments is centred on intrinsic evaluation, examining the impact of content reduction on word embedding representations of the tables. For these experiments, two corpora were used: WikiTables and the Dublin Open Data Portal.

The WikiTables corpus [1] comprises 1.6 million tables sourced from Wikipedia. However, for these experiments, tables with less than 10 rows were excluded. On average, the tables in our selection contained approximately 28 rows and 6 columns. The corpus extracted from the Dublin Open Data Portal[2]

---

[2] https://data.smartdublin.ie/.

includes 205 tables covering various topics related to that city. The table sizes vary from one thousand to one million rows, with an average of 18 columns.

The second part of the experiments focused on table retrieval performance, employing a benchmark based on WikiTables [25]. This benchmark involved the sampling of 50 Wikipedia tables to be used as queries. Each table consisted of at least five rows and three columns. The original ground truth relevance was established based on a keyword query, re-ranking techniques, and manual annotation of the top 10 results as non-relevant, relevant, or highly relevant. However, in the experiments presented here only 32 query tables were suitable in accordance with the experiment's limitations, where tables with less than 10 rows were discarded.

4.3 Word Embeddings Similarity

This section evaluates the impact of reducing the content of a table on its word embedding representation. This analysis can provide valuable insight into the effectiveness of such reductions and their potential implications for the overall quality and accuracy of the representation.

In these experiments a basic approach was used to reduce tables by randomly selecting a percentage of the original rows. To do that, a set of thresholds were established in terms of the percentage of rows selected: 1%, 5%, and multiples of 10 until 90%, where 1% means that 99% of the rows were dropped from the original table. For each table, 11 subtables were obtained (from 1% to 90%) and compared with the original one using the five language models mentioned above: Word2vec, fastText, SentenceBERT, BERT, and RoBERTa.

As previously stated, tables with a row count less than 10 were excluded from our analysis. For tables with a row count less than 20, subsets consisting of 1% and 5% were not calculated due to insufficient available data. Furthermore, tables containing less than 100 rows did not have a 1% subtable computed for the same reason.

Figure 1 (a) shows the similarity achieved with respect to the original table (y-axis) by the five models tested in the WikiTables corpus when different percentages of rows were selected (x-axis). The straight dashed line at the top corresponds to the 90% similarity threshold, which can be considered as a high similarity level. The similarity between the original table and the reduced tables was calculated following the procedure described in Sect. 3.

Based on the findings, contextual models exhibit superior performance to Word2vec for all percentages of row deletion, and they also outperform fast-Text for thresholds exceeding 10%. Additionally, it is noteworthy that content reduction has a significantly negative impact on Word2vec. SentenceBERT and RoBERTa get pretty similar results, achieving around 90% of similarity threshold by keeping 60% of the original content of the table. This implies that 40% of the table could be dropped and still obtain a word embedding around 90% similar to the original one.

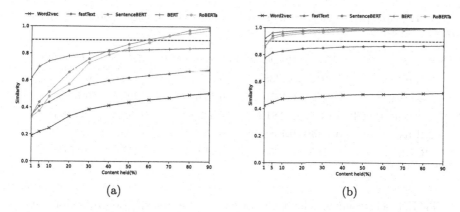

Fig. 1. Similarity obtained by fastText, Word2vec, SentenceBERT, BERT, RoBERTa with different percentages of rows kept in (a) WikiTables and (b) Dublin Open Data corpus. The dashed line at the top represents the 90% similarity threshold.

Similar results were obtained on the Dublin Open Data corpus, as shown in Fig. 1 (b). In this case, with tables containing a significantly larger number of rows than in WikiTables, the word embedding representation is less sensitive to row deletion. In the case of contextual models, keeping only 5% of the rows allows obtaining a word embedding that is more than 90% similar to the original one. In this same setting, Word2vec obtained 45% similarity and fastText 81%.

When working with datasets of varying sizes, it is crucial to consider the potential implications of content reduction. For example, in datasets with small tables such as WikiTables, it is advised to retain at least 60% of the original table in order to maintain a high level of similarity based on contextual models. Alternatively, datasets with larger tables such as Dublin Open Data can maintain a high similarity level even with subtables that retain only 5% of the original table. Utilising content reduction techniques to process reduced subtables while retaining the content's semantics can lead to improved search and processing times, particularly for larger datasets. It is important to ensure that the semantics of the content are preserved in order to maintain the integrity of the data.

4.4 Table Retrieval

A second experiment was conducted to evaluate the impact of different strategies for table reduction on a table retrieval system. Three distinct approaches were implemented to acquire the subtables:

- Random reduction: to obtain a subtable, 50% of the table's contents were randomly discarded.
- Delete duplicates: this approach treats each column as an array of individual elements, removing duplicated cells.

Table 1. Search performance without content reduction. Highest scores for each metric are in boldface.

	MAP	Precision@5	Precision@10	nDCG@5	nDCG@10
Word2vec	0.0913	0.1438	0.1250	0.1744	0.1682
fastText	0.0930	0.1750	0.1344	0.1975	0.1828
SentenceBERT	0.1345	**0.1813**	**0.1438**	**0.2411**	**0.2383**
BERT	**0.1360**	0.1688	0.1375	0.2257	0.2174
RoBERTa	0.1069	0.1250	0.1125	0.1551	0.1764

– TF-IDF selection: the classical strategy used in information retrieval systems was adapted for table retrieval. This involves calculating the weight of each cell in each column and retaining 50% of the cells with the higher weights. This enables a direct comparison with the random reduction strategy, as the same volume of content is discarded in both cases.

The initial stage of the experiment involves indexing table embeddings within the retrieval system. For each of the previously mentioned models in Sect. 4.1, we apply the described strategies to index both the original table and the reduced table. The indexing process is straightforward: cleaning the original table (as mentioned in Sect. 3), apply the reduction strategy, and calculate two embeddings for each column (header and content) using the corresponding embedding model. These embeddings are then indexed using Faiss.

The table retrieval system operates as described in Sect. 3. The `trec_eval` tool,[3] used by the TREC[4] community for evaluating ad hoc retrieval, is employed to evaluate the search results, offering a range of system performance metrics such as MAP, Precision@k and nDCG@k.

Table 1 presents the evaluation for the five embedding models without applying any reduction strategy. It can be observed that BERT and SentenceBERT exhibit superior performance compared to RoBERTa and non-contextual models in terms of MAP, Precision@k and nDCG@k, although the differences between these two models is not statistically significant.

Table 2 displays the performance of the random reduction strategy that discards 50% of the content. The outcome reveals that this strategy has a notable negative effect on search performance for all the models. The nDGC@5 and nDGC@10 metrics for the best result (SentenceBERT) using this strategy are 38.2% and 38.9% lower, respectively, compared to results without the reduction strategy. It is worth noting that this reduction strategy has a more significant impact on the performance of contextual models. This could be attributed to the reduction of context, leading to the production of lower-quality word embeddings. The best-performing model for each metric is not significantly better than the second-best model evaluated, according to the t-test results.

---

[3] https://github.com/usnistgov/trec_eval.
[4] https://trec.nist.gov/.

Table 2. Search performance with random content reduction. Highest scores are in boldface.

	MAP	Precision@5	Precision@10	nDGC@5	nDGC@10
Word2vec	0.0737	0.1000	0.0844	0.1216	0.1244
fastText	**0.0833**	**0.1438**	**0.1250**	0.1399	0.1425
SentenceBERT	0.0791	**0.1438**	0.1125	**0.1490**	**0.1456**
BERT	0.0596	0.1000	0.0875	0.1059	0.1097
RoBERTa	0.0771	0.0930	0.0844	0.1118	0.1240

Table 3. Search performance with duplicate content reduction. Highest scores are in boldface.

	MAP	Precision@5	Precision@10	nDGC@5	nDGC@10
Word2vec	0.0868	0.1250	0.1312	0.1377	0.1539
fastText	0.0840	0.1625	0.1219	0.1800	0.1662
SentenceBERT	**0.1303**	**0.2063**	0.1438	**0.2497**	**0.2294**
BERT	0.1031	0.1875	**0.1563**	0.1960	0.1901
RoBERTa	0.1077	0.1500	0.1219	0.1593	0.1729

The duplicate reduction strategy results are shown in Table 3. As previously mentioned, in this approach all duplicated cells in a column are removed. The results achieved by non-contextual models are worse than the results without a reduction strategy. However, for contextual models, there are slight improvements in some metrics or even more notable improvements in the case of RoBERTa model.

The third approach to content reduction is based on TF-IDF, retaining the top 50% of the column's content using this well-known weighting scheme. The results of this method are presented in Table 4. This strategy works best for non-contextual models, even surpassing the results of the experiment without reduction. In this case, TF-IDF reduces the noise leading to better word embedding representations. However, for the SentenceBERT model this strategy does not yield significant improvements and actually worsens performance compared to the experiment without content reduction. Again, the loss of contextual keywords affects its performance, similar to what occurred in the random reduction strategy.

In order to identify the most effective reduction strategy for each model, Table 5 displays the nDGC@5 metric for each of them, including their performance on the original table ("No reduction" column). An asterisk (*) following a number indicates that the results are statistically significantly better than all the other models, with $p < 0.05$. Globally, TF-IDF content reduction achieves the best results in all models except SentenceBERT, which performs better with the duplicate reduction strategy. Non-contextual model Word2vec show a 35.5%

Table 4. Search performance with TF-IDF content reduction. Highest scores are in boldface.

	MAP	Precision@5	Precision@10	nDGC@5	nDGC@10
Word2vec	**0.1624**	0.1625	0.1344	**0.2364**	**0.2402**
fastText	0.1432	0.1625	0.1312	0.2115	0.2124
SentenceBERT	0.0721	0.1687	0.1344	0.1587	0.1574
BERT	0.1599	**0.1688**	**0.1438**	0.2266	0.2386
RoBERTa	0.1599	**0.1688**	**0.1438**	0.2266	0.2386

Table 5. nDGC@5 metric by model and reduction strategy. Highest scores are in boldface.

	No reduction	Random reduction	Delete duplicates	TF-IDF reduction
Word2vec	0.1744	0.1216	0.1377	**0.2364***
fastText	0.1975	0.1399	0.1800	0.2115
SentenceBERT	**0.2411***	**0.1490***	**0.2497***	0.1587
BERT	0.2257	0.1059	0.1960	0.2266
RoBERTa	0.1213	0.1118	0.1593	0.2266

improvement compared to using the original table, while fastText improves by 7.1%. For BERT and SentenceBERT, although no reduction strategy provides significant improvements over using the original table, it is important to highlight that these experiments aimed to assess whether content reduction negatively impacts the final performance of the table retrieval system. Based on these results, it can be concluded that reduction can reduce computation time without adversely affecting system performance. In the case of RoBERTa, applying duplicate or TF-IDF reduction results in notable improvements in performance, reducing also the computational cost by droping 50% of the content of the table.

5   Conclusions and Future Work

This research paper presented a study that examined the impact of content deletion on the word embedding representation of tabular data across different contextual and non-contextual language models. The aim of the study was to identify whether reducing content could lower the need for computational resources without compromising the representation in the semantic space provided by word embeddings. To achieve this goal, five language models were used and three strategies for content reduction were proposed for a table retrieval system.

The study showed that contextual models, in general, outperformed non-contextual models in the experiments conducted. Regarding content reduction, in the case of the WikiTables corpus, retaining 60% of the table's content yielded vector representations that were 90% similar to the original ones. In cases where a table contained a substantial number of rows, such as the Dublin Open Data corpus, retaining only 5% of the rows resulted in an average similarity of over 90% between the word embedding vectors of the original and reduced tables. These findings suggest that significant data reduction can be achieved while still maintaining vector representations that are close to the original ones. These findings have important implications for data management and optimisation.

Additional experiments demonstrated that BERT and SentenceBERT consistently outperformed non-contextual models and RoBERTa in the table retrieval task. Among the three reduction strategies proposed, TF-IDF yielded better results overall compared to random reduction and duplicate removal. This improvement is particularly noticeable in non-contextual models. While content reduction had a more pronounced effect on the performance of contextual models, TF-IDF showed little impact on the final retrieval system's performance. These findings offer a promising avenue for reducing computational costs in table retrieval systems based on word embeddings.

In future research, the exploration of additional strategies for obtaining representative embeddings from tables is planned. Another future objective includes the development of fine-tuned models that are specifically tailored to calculate tabular semantic similarity. Furthermore, there is a need to investigate or create different benchmarks for table retrieval. This is particularly relevant when dealing with large tables, as the dataset used in the current experiments primarily consists of small tables with a limited number of rows sourced from WikiTables.

Acknowledgements. This research was funded by CIAICO/2022/019 project from Generalitat Valenciana, and by MCIN/AEI10.13039/501100011033 and by the European Union Next Generation EU/PRTR as part of the projects PID2021-122263OB-C22 and TED2021130890B-C21. Alberto Berenguer has a contract for predoctoral training with Generalitat Valenciana and the European Social Fund, funded by grant number ACIF/2021/507.

References

1. Bhagavatula, C.S., Noraset, T., Downey, D.: TabEL: entity linking in web tables. In: Arenas, M., et al. (eds.) ISWC 2015. LNCS, vol. 9366, pp. 425–441. Springer, Cham (2015). https://doi.org/10.1007/978-3-319-25007-6_25

2. Bojanowski, P., Grave, E., Joulin, A., Mikolov, T.: Enriching word vectors with subword information. Trans. Assoc. Comput. Linguist. 5, 135–146 (2017)

3. Chen, Z., Trabelsi, M., Heflin, J., Xu, Y., Davison, B.D.: Table search using a deep contextualized language model. In: Proceedings of the 43rd International ACM SIGIR Conference on Research and Development in Information Retrieval, pp. 589–598. Association for Computing Machinery, Online (2020). https://doi.org/10.1145/3397271.3401044

4. Devlin, J., Chang, M.W., Lee, K., Toutanova, K.: BERT: pre-training of deep bidirectional transformers for language understanding. In: Proceedings of the 2019 Conference of the North American Chapter of the Association for Computational Linguistics, pp. 4171–4186. Association for Computational Linguistics, Minneapolis, MN, USA (2019). https://doi.org/10.18653/v1/N19-1423

5. Forgues, G., Pineau, J., Larchevêque, J.M., Tremblay, R.: Bootstrapping dialog systems with word embeddings. In: NIPS, Modern Machine Learning and Natural Language Processing Workshop, vol. 2, pp. 1–5 (2014)
6. Giatsoglou, M., Vozalis, M.G., Diamantaras, K., Vakali, A., Sarigiannidis, G., Chatzisavvas, K.C.: Sentiment analysis leveraging emotions and word embeddings. Expert Syst. Appl. **69**, 214–224 (2017). https://doi.org/10.1016/j.eswa.2016.10.043
7. He, P., Liu, X., Gao, J., Chen, W.: DeBERTa: decoding-enhanced BERT with disentangled attention. In: International Conference on Learning Representations, pp. 1–21. Online (2021)
8. Herzig, J., Nowak, P.K., Müller, T., Piccinno, F., Eisenschlos, J.: TaPas: weakly supervised table parsing via pre-training. In: Proceedings of the 58th Annual Meeting of the Association for Computational Linguistics, pp. 4320–4333. Association for Computational Linguistics (2020). https://doi.org/10.18653/v1/2020.acl-main.398
9. Howard, J., Ruder, S.: Universal language model fine-tuning for text classification. In: Proceedings of the 56th Annual Meeting of the Association for Computational Linguistics, pp. 328–339. Association for Computational Linguistics, Melbourne, Australia (2018). https://doi.org/10.18653/v1/P18-1031
10. Kusner, M., Sun, Y., Kolkin, N., Weinberger, K.: From word embeddings to document distances. In: Proceedings of the 32nd International Conference on Machine Learning. Proceedings of Machine Learning Research, vol. 37, pp. 957–966. PMLR, Lille, France (2015). https://proceedings.mlr.press/v37/kusnerb15.html
11. Liu, Y., et al.: Roberta: a robustly optimized BERT pretraining approach (2019)
12. Mikolov, T., Sutskever, I., Chen, K., Corrado, G., Dean, J.: Distributed representations of words and phrases and their compositionality. In: Proceedings of the 26th International Conference on Neural Information Processing Systems - Volume 2, pp. 3111–3119. NIPS'13, Curran Associates Inc., Red Hook, NY, USA (2013)
13. Min, B., et al.: Recent advances in natural language processing via large pre-trained language models: a survey. ACM Comput. Surv. **56**(2) (2023). https://doi.org/10.1145/3605943
14. Mu, J., Viswanath, P.: All-but-the-top: simple and effective postprocessing for word representations. In: 6th International Conference on Learning Representations, ICLR. Vancouver, BC, Canada (2018)
15. Nguyen, T.T., Nguyen, Q.V.H., Matthias, W., Karl, A.: Result selection and summarization for web table search. In: Proceedings of the 31st International Conference on Data Engineering (ISDE'15), pp. 231–242. IEEE, Seoul, South Korea (2015). https://doi.org/10.1109/ICDE.2015.7113287
16. Pennington, J., Socher, R., Manning, C.D.: GloVe: global vectors for word representation. In: Empirical Methods in Natural Language Processing (EMNLP), pp. 1532–1543. Doha, Qatar (2014). https://doi.org/10.3115/v1/D14-1162
17. Reimers, N., Gurevych, I.: Sentence-BERT: sentence embeddings using Siamese BERT-networks. In: Proceedings of the 2019 Conference on Empirical Methods in Natural Language Processing and the 9th International Joint Conference on Natural Language Processing (EMNLP-IJCNLP), pp. 3982–3992. Association for Computational Linguistics, Hong Kong, China (2019). https://doi.org/10.18653/v1/D19-1410
18. Shraga, R., Roitman, H., Feigenblat, G., Cannim, M.: Web table retrieval using multimodal deep learning. In: Proceedings of the 43rd International ACM SIGIR Conference on Research and Development in Information Retrieval, pp. 1399–1408. Association for Computing Machinery, Online (2020). https://doi.org/10.1145/3397271.3401120

19. Trabelsi, M., Chen, Z., Zhang, S., Davison, B.D., Heflin, J.: StruBERT: structure-aware bert for table search and matching. In: Proceedings of the ACM Web Conference 2022, pp. 442–451. WWW '22, Association for Computing Machinery, New York, NY, USA (2022). https://doi.org/10.1145/3485447.3511972

20. Trabelsi, M., Davison, B.D., Heflin, J.: Improved table retrieval using multiple context embeddings for attributes. In: 2019 IEEE International Conference on Big Data (Big Data), pp. 1238–1244. IEEE, Los Angeles, CA, USA (2019). https://doi.org/10.1109/BigData47090.2019.9005681

21. Vaswani, A., et al.: Attention is all you need. In: Advances in Neural Information Processing Systems, vol. 30, pp. 5998–6008. Curran Associates Inc, Long Beach, CA, USA (2017)

22. Yin, P., Neubig, G., tau Yih, W., Riedel, S.: TaBERT: pretraining for joint understanding of textual and tabular data. In: Proceedings of the 58th Annual Meeting of the Association for Computational Linguistics, pp. 8413–8426. Association for Computational Linguistics, Online (2020). https://doi.org/10.18653/v1/2020.acl-main.745

23. Zhang, S., Balog, K.: Ad hoc table retrieval using semantic similarity. In: Proceedings of the 2018 World Wide Web Conference, pp. 1553–1562. International World Wide Web Conferences Steering Committee, Lyon, France (2018). https://doi.org/10.1145/3178876.3186067

24. Zhang, S., Balog, K.: Web table extraction, retrieval, and augmentation: a survey. ACM Trans. Intell. Syst. Technol. 11(2), 1–35 (2020). https://doi.org/10.1145/3372117

25. Zhang, S., Balog, K.: Semantic table retrieval using keyword and table queries. ACM Trans. Web 15(3), 1–33 (2021). https://doi.org/10.1145/3441690

26. Zou, W.Y., Socher, R., Cer, D., Manning, C.D.: Bilingual word embeddings for phrase-based machine translation. In: Proceedings of the 2013 Conference on Empirical Methods in Natural Language Processing, pp. 1393–1398. Association for Computational Linguistics, Seattle, Washington, USA (2013). https://aclanthology.org/D13-1141

Multimodal Learned Sparse Retrieval with Probabilistic Expansion Control

Thong Nguyen[1]([⊠])(iD), Mariya Hendriksen[2](iD), Andrew Yates[1](iD),
and Maarten de Rijke[1](iD)

[1] University of Amsterdam, Amsterdam, The Netherlands
{t.nguyen2,a.c.yates,m.derijke}@uva.nl
[2] AIRLab, University of Amsterdam, Amsterdam, The Netherlands
m.hendriksen@uva.nl

Abstract. Learned sparse retrieval (LSR) is a family of neural methods that encode queries and documents into sparse lexical vectors that can be indexed and retrieved efficiently with an inverted index. We explore the application of LSR to the multi-modal domain, with a focus on text-image retrieval. While LSR has seen success in text retrieval, its application in multimodal retrieval remains underexplored. Current approaches like LexLIP and STAIR require complex multi-step training on massive datasets. Our proposed approach efficiently transforms dense vectors from a frozen dense model into sparse lexical vectors. We address issues of high dimension co-activation and semantic deviation through a new training algorithm, using Bernoulli random variables to control query expansion. Experiments with two dense models (BLIP, ALBEF) and two datasets (MSCOCO, Flickr30k) show that our proposed algorithm effectively reduces co-activation and semantic deviation. Our best-performing sparsified model outperforms state-of-the-art text-image LSR models with a shorter training time and lower GPU memory requirements. Our approach offers an effective solution for training LSR retrieval models in multimodal settings. Our code and model checkpoints are available at github.com/thongnt99/lsr-multimodal.

1 Introduction

Learned sparse retrieval (LSR) [6,7,40] typically employs transformer-based encoders to encode queries and documents into sparse lexical vectors (i.e., bags of weighted terms) that are compatible with traditional inverted index. LSR has several nice properties. It provides an approach for effective and efficient neural retrieval, like dense retrieval, but with different advantages and trade-offs. For example, sparse representations have the potential to be interpretable because they are aligned with a vocabulary, and they leverage inverted index software rather than approximate nearest neighbor search [40]. Empirically, LSR also shows advantages over single-vector dense models on retrieval generalization benchmarks [6,16].

While LSR and dense retrieval are common in text retrieval, dense retrieval has taken the lead in multi-modal search. This is evident in state-of-the-art text-image pre-training methods like BLIP [22] and ALBEF [23], which rely on dense architectures.

T. Nguyen and M. Hendriksen—These authors contributed equally.

© The Author(s), under exclusive license to Springer Nature Switzerland AG 2024
N. Goharian et al. (Eds.): ECIR 2024, LNCS 14609, pp. 448–464, 2024.
https://doi.org/10.1007/978-3-031-56060-6_29

The preference for dense models arises because images, unlike text, consist of continuous pixel values, presenting a challenge when they are mapped to discrete lexical terms. For multi-modal LSR, LexLIP [51] and STAIR [2] are the only two recent methods that exhibit competitive results on standard benchmarks. However, both require complex multi-step training on extensive text-image pairs: LexLIP with up to 14.3 million pairs and STAIR with a massive 1 billion pairs, encompassing public and private data.

We approach the multi-modal LSR (MMLSR) problem by using a pre-trained dense model and training a small sparse projection head on top of dense vectors, using image-text dense scores as a supervision signal. Naively learning the projection layer leads to issues of (i) high dimension co-activation and (ii) semantic deviation. Issue (i) happens when text and image sparse vectors excessively activate the same output dimensions, forming a sub-dense space inside the vocabulary space. Issue (ii) means that produced output terms do not reflect the content of captions/images, making them not human-interpretable. To counter (i) and (ii), we propose a single-step training method with probabilistic term expansion control. By disabling term expansions, we force the projection to produce meaningful terms first, then gradually allow more term expansions to improve the effectiveness while also randomly reminding the model not to fully rely on expansion terms. This process is handled using Bernoulli random variables with a parameter scheduler to model the expansion likelihood at both caption and word levels.

Opting for dense to sparse projection, instead of training an MMLSR model from scratch, provides several advantages. First, it is aligned with the broader community effort to reduce the carbon footprint of training deep learning models [30]. By keeping the dense encoders frozen and learning a lightweight projection layer, we can avoid the double GPU training/inference cost of two models (dense & sparse) while having more flexibility. Our approach enables the pre-computation of dense text and image vectors, allowing easy integration or removal of the projection layer based on available (dense or sparse) software and infrastructure. Moreover, this transformation may shed light on the interpretability of dense vectors, possibly contributing to a deeper understanding of the fundamental distinctions between these two multi-modal retrieval paradigms.

To understand the effectiveness and efficiency of the proposed training method, we conduct extensive experiments on two dense multi-modal models (BLIP, ALBEF) and two scene-centric [14] datasets (MSCOCO [27], Flickr30k [48]). We analyze the problems of dimension co-activation and semantic deviation under different settings.

Our Contributions. The main contributions of our paper are: (i) We propose a line of research for efficiently converting a multi-modal dense retrieval model to a multi-modal LSR model.(ii) We train a lightweight projection head to convert dense to sparse vectors and show that our sparsified models are faithful to dense models while outperforming previous multi-modal LSR models. The training is efficient and does not require ground-truth labels.(iii) We identify the issues of high dimension co-activation and semantic deviation and propose a training method to address them.

2   Related Work

Learned Sparse Retrieval (LSR). Learned sparse retrieval is a family of neural retrieval methods that encode queries and documents into sparse lexical vectors that can be indexed and searched efficiently with an inverted index. There are many LSR approaches in the literature on text retrieval [7,39,49]; they are mainly built up from two types of encoder: MLP and MLM [40]. The MLP encoder uses a linear feedforward layer placed on top of the transformers's last contextualized embeddings to predict the importance of input terms (similar to term-frequency in traditional lexical retrieval). The MLP encoder has no term expansion capability. On the other hand, the MLM encoder utilizes the logits of the masked language model (MLM) for weighting terms and selecting expansion terms. Splade [6,7] is a recent state-of-the-art text-oriented LSR approach that employs the MLM encoder in both query and document side, while other methods [3,24,33] use MLP encoders on both sides or only on the query side. Although it seems to be more beneficial to have expansion on both queries and documents, a recent study [40] found that query and document expansion have a cancellation effect on text retrieval (i.e., having expansion on the document side reduces the usefulness of query expansion) and one could obtain near state-of-the-art results without query expansion.

Unlike prior work focused on converting sparse to dense representations for hybrid ad-hoc text retrieval [25,26], our work explores the reverse task of dense to sparse conversion in the multi-modal domain. This direction presents new challenges due to dimension co-activation and semantic deviation issues. Ram et al. [42] interpreted text dense retrieval by zero-shot projection from dense to vocabulary space using a frozen MLM head. Nguyen et al. [38] propose a simple sparse vision-language (VL) bi-encoder without query expansion and evaluate the performance on the image suggestion task. We aim for an efficient, effective, and semantically faithful drop-in sparse replacement of multi-modal dense retrieval, necessitating training of the projection layer.

Cross-Modal Retrieval. Cross-modal retrieval (CMR) methods construct a multi-modal representation space, where the similarity of concepts from different modalities can be measured using a distance metric such as a cosine or Euclidean distance. Some of the earliest approaches in CMR utilized canonical correlation analysis [11,18]. They were followed by a dual encoder architecture equipped with a recurrent and a convolutional component, the most prominent approaches in that area featured a hinge loss [8,46]. Later approaches further improved the effectiveness using hard-negative mining [5].

Later, the integration of attention mechanisms improved performance. This family of attention mechanisms includes dual attention [37], stacked cross-attention [20], bidirectional focal attention [28]. Another line of work proposes to use transformer encoders [44] for this task [36], and adapts the BERT model [4] as a backbone [9,52].

A related line of work focuses on improving the performance on CMR via modality-specific graphs [45], or image and text generation modules [12]. There is also more domain-specific work that focuses on CMR in fashion [10,19], e-commerce [13], cultural heritage [43], and cooking [45].

3   Background

Task Definition. We use the same notation as in previous work [1,50]. We work with a cross-modal dataset $\mathcal{D}$ that includes N image-caption tuples:

$\mathcal{D} = \left\{ \left(\mathbf{x}_{\mathcal{I}}^{i}, \{\mathbf{x}_{C_j}^{i}\}_{i=1}^{k} \right) \right\}_{i=1}^{N}$. Each tuple comprises an image $\mathbf{x}_{\mathcal{I}}$ and k associated captions $\{\mathbf{x}_{C_j}\}_{j=1}^{k}$.

The *cross-modal retrieval* (CMR) task is defined analogously to the standard information retrieval task: given a query and a set of candidates, we rank all candidates w.r.t. their relevance to the query. The query can be either a caption or an image. Similarly, the set of candidate items can contain either images or captions. CMR is performed across modalities, therefore, if the query is a caption then the set of candidates are images, and vice versa. Hence, the task comprises two subtasks:(i)*caption-to-image retrieval*: retrieving images relevant to a caption query, and (ii)*image-to-caption retrieval*: retrieving relevant captions that describe an image query.We focus on *caption-to-image retrieval* only as it is more challenging, as reported by previous research [22,23,51].

Sparsification-Induced Phenomena. In this work, we investigate two phenomena arising during the sparsification process: dimension co-activation and semantic deviation.

Definition 1 (Dimension co-activation). We define *dimension co-activation* as sparse image and caption representations activating the same output dimensions, creating a sub-dense space within the vocabulary. While co-activation is essential for matching captions with images and can be measured by FLOPs, *high co-activation* results in unnecessarily long posting lists, harming the efficiency of LSR. Establishing a clear threshold for *high co-activation* is challenging, but we observe that beyond a certain point, increased FLOPs yield minimal improvements in effectiveness. To quantify this effect, we use effectiveness metrics (e.g., R@k) in combination with the FLOPs metric:

$$\text{FLOPs} = \frac{1}{|C||\mathcal{I}|} \sum_{\mathbf{x}_C \in C} \sum_{\mathbf{x}_{\mathcal{I}} \in \mathcal{I}} \mathbf{s}_C^0 \cdot \mathbf{s}_{\mathcal{I}}^0 \tag{1}$$

where C and $\mathcal{I}$ are caption and image collections, $\mathbf{s}_C$, $\mathbf{s}_{\mathcal{I}}$ are sparse vectors of a caption $\mathbf{x}_C$ and an image $\mathbf{x}_{\mathcal{I}}$.

Definition 2 (Semantic deviation). We define *semantic deviation* as the disparity between the semantic information in the visual or textual query and that in the sparse output terms. High co-activation suggests (but does not guarantee) semantic deviation.

Measuring semantic deviation directly is challenging, so we use two rough proxies, *Exact@k* and *Semantic@k*, defined as follows:

$$Exact@k = \frac{1}{k} |\{t \mid t \in \mathbf{x}_C, t \in top_k(\mathbf{s}_C)\}| \tag{2}$$

$$Semantic@k = \frac{1}{k} \sum_{\mathbf{x}_t^i \in top_k(\mathbf{s}_C)} \max_{\mathbf{x}_t^j \in \mathbf{x}_C} \frac{f_{enc}(\mathbf{x}_t^i) \cdot f_{enc}(\mathbf{x}_t^j)}{\|f_{enc}(\mathbf{x}_t^i)\| \|f_{enc}(\mathbf{x}_t^j)\|}. \tag{3}$$

Exact@k measures the ratio of overlapping terms between the input caption and the top-k highest weighted output terms, providing a partial picture of semantic deviation without considering synonyms. *Semantic@k* complements *Exact@k* by calculating the averaged cosine similarity between static embeddings obtained using model $f_{enc}(\cdot)$ of top-k output terms and input caption terms. Higher values for both metrics suggest less semantic deviation, implying better alignment of output terms with input captions.

4 Methodology

4.1 Model Architecture

The architecture of our Dense2Sparse model is visualized in Fig. 1. Dense2Sparse takes an image and a caption as input, projecting them into a $|V|$-dimensional space, where each dimension represents the weight of a corresponding vocabulary entry. The key components include two dense encoders, an image encoder $f_\theta^{\mathcal{I}}(\cdot)$ and a caption encoder $f_\phi^{\mathcal{C}}(\cdot)$, as well as a multimodal sparse projection head $g_\psi(\cdot)$.

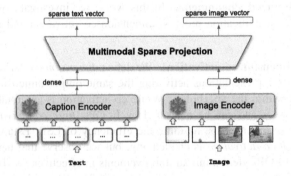

Fig. 1. The architecture of Dense2Sparse (D2S). The caption and image encoders are frozen, and the sparse projection is trained to project dense vectors to sparse vectors.

Dense Image and Text Encoders. The *dense image encoder* $f_\theta^{\mathcal{I}} : \mathcal{X} \rightarrow \mathcal{Z}$ takes an input image $\mathbf{x}_{\mathcal{I}}$ and maps it into a latent space $\mathcal{Z} = \mathcal{R}^d$: $\mathbf{z}_{\mathcal{I}} = f_\theta^{\mathcal{I}}(\mathbf{x}_{\mathcal{I}})$, where $\mathbf{z}_{\mathcal{I}} \in \mathcal{R}^d$. Similarly, the *dense text encoder* $f_\phi^{\mathcal{C}} : \mathcal{X} \rightarrow \mathcal{Z}$ takes an input text (caption) $\mathbf{x}_{\mathcal{C}}$, and maps it into a latent space $\mathcal{Z} = \mathcal{R}^d$: $\mathbf{z}_{\mathcal{C}} = f_\phi^{\mathcal{C}}(\mathbf{x}_{\mathcal{C}})$, where $\mathbf{z}_{\mathcal{C}} \in \mathcal{R}^d$. We obtain dense representations using BLIP and ALBEF as a backbone. Both encoders are frozen.

Multimodal Sparse Projection Head. The *multimodal sparse projection head* $g_\psi : \mathcal{Z} \rightarrow \mathcal{S}$ maps dense latent image and text representations into the sparse image and text vector space $\mathcal{S} = \mathcal{R}_{>0}^{|V|}$:

$$\mathbf{s}_{\mathcal{C}} = g_\psi(\mathbf{z}_{\mathcal{C}}) \quad \text{and} \quad \mathbf{s}_{\mathcal{I}} = g_\psi(\mathbf{z}_{\mathcal{I}}). \tag{4}$$

The multimodal sparse projection head comprises four steps. First, we project the d-dimensional dense vector $\mathbf{z}$ to an ω-dimensional dense vector: $\mathbf{z}_1 = \mathbf{W}_1\mathbf{z}$, where $\mathbf{W}_1 \in \mathcal{R}^{\omega \times d}$, $\mathbf{z} \in \mathcal{R}^d$, and $\mathbf{z}_1 \in \mathcal{R}^\omega$. Second, we apply layer normalization:

$$\mathbf{z}_2 = \frac{\mathbf{z}_1 - \mathbb{E}[\mathbf{z}_1]}{\sqrt{Var[\mathbf{z}_1] + \epsilon}} \cdot \gamma + \beta, \tag{5}$$

Algorithm 1. Multimodal LSR training with probabilistic expansion control

---

Input: image-caption pair $(\mathbf{x}_\mathcal{I}, \mathbf{x}_\mathcal{C})$, caption encoder $f_\phi^\mathcal{C}$, image encoder $f_\theta^\mathcal{I}$, sparse projection head g_ψ, loss function $\mathcal{L}$, and expansion rate function f_{incr}.

$p_i^v \leftarrow 1 - df_i^v$
$p_c \leftarrow 0$

for epoch **do**
  for batch **do**
    $\mathbf{z}_\mathcal{C} \leftarrow f_\phi^\mathcal{C}(\mathbf{x}_\mathcal{C}), \quad \mathbf{z}_\mathcal{I} \leftarrow f_\theta^\mathcal{I}(\mathbf{x}_\mathcal{I})$
    $\mathbf{s}_\mathcal{C} \leftarrow g_\psi(\mathbf{z}_\mathcal{C}), \quad \mathbf{s}_\mathcal{I} \leftarrow g_\psi(\mathbf{z}_\mathcal{I})$
    $\mathcal{E}_\mathcal{C} \sim \mathrm{Ber}(p_c), \quad \mathcal{E}_i^v \sim \mathrm{Ber}(p_i^v)$
    $\bar{\mathbf{s}}_\mathcal{C} \leftarrow \mathrm{EXPAND}(\mathbf{x}_\mathcal{C}, \mathbf{s}_\mathcal{C}, \mathcal{E}_\mathcal{C}, \mathcal{E}_i^v)$
    $\mathcal{L} \leftarrow \mathcal{L}(\bar{\mathbf{s}}_\mathcal{C}, \mathbf{s}_\mathcal{I}, \mathbf{z}_\mathcal{I}, \mathbf{z}_\mathcal{C})$
  end for
  $p_c \leftarrow f_{incr}(p_c), \quad p_i^v \leftarrow f_{incr}(p_i^v)$
end for

function EXPAND$(\mathbf{x}_\mathcal{C}, \mathbf{s}_\mathcal{C}, \mathcal{E}_\mathcal{C}, \mathcal{E}_i^v)$
  for $0 \leq i <$ batch_size **do**
    for $0 \leq k < |V|$ **do**
      if $v_k \notin \mathbf{x}_\mathcal{C}$ **then**
        $\mathbf{s}_{\mathcal{C}i,k} \leftarrow \mathbf{s}_{\mathcal{C}i,k} \cdot \mathcal{E}_\mathcal{C} \cdot e_k^v$
      else
        $\mathbf{s}_{\mathcal{C}i,k} \leftarrow \mathbf{s}_{\mathcal{C}i,k} \cdot \mathcal{E}_k^v$
      end if
    end for
  end for
  return $\mathbf{s}_\mathcal{C}$
end function

---

where $\mathbb{E}[\mathbf{z}_1]$ and $Var[\mathbf{z}_1]$ are the expectation and variance of $\mathbf{z}_1$, γ and β are learnable affine transformation parameters, and $\mathbf{z}_2 \in \mathcal{R}^\omega$. Third, we project z_2 to the vocabulary space $\mathcal{S} = \mathcal{R}_{>0}^{|V|}$: $\mathbf{s} = \mathbf{W}_2\mathbf{z}_2$, where $\mathbf{W}_2 \in \mathcal{R}^{|V|\times\omega}$, $\mathbf{z}_2 \in \mathcal{R}^\omega$, and $\mathbf{s} \in \mathcal{R}^{|V|}$. $\mathbf{W}_2$ is initialized with vocabulary embeddings similar to the transformer-masked language model. Fourth, we remove negative weights and apply a logarithmic transformation to the positive weights: $\mathbf{s} = \log_e(1 + \max(0, \mathbf{s}))$, where $\mathbf{s} \in \mathcal{R}_{>0}^{|V|}$. The resulting $|V|$-dimensional sparse vector is aligned with the vocabulary, and each dimension represents the weight of the corresponding vocabulary entry. This projection head is similar to the MLM head employed in previous work [6,33].

Probabilistic Expansion Control. Without any intervention, training the projection module with a standard contrastive loss could lead to high-dimension co-activation and semantic deviation as defined previously. This phenomenon affects the efficiency of an inverted index and the interpretability of the outputs. To mitigate this problem, we propose a single-step training algorithm with probabilistic lexical expansion control. It is described in Algorithm 1.

We define a Bernoulli random variable $\mathcal{E} \sim Ber(p)$, $p \in [0, 1]$ and use it to control textual query expansion. We consider a caption-level and a word-level expansion. The *caption-level expansion* is controlled by the random variable $\mathcal{E}_\mathcal{C} \sim Ber(p_c)$. If $\mathcal{E}_\mathcal{C} = 1$ the expansion is allowed, while $\mathcal{E}_\mathcal{C} = 0$ means the expansion is not allowed. Analogously, the *word-level expansion*, or the expansion to the i-th word in the vocabulary, is regulated by the random variable $\mathcal{E}_i^v \sim Ber(p_i^v)$.

The parameters p_c and p_i^v define the likelihood of caption-level and word-level expansion within a given training epoch. During training, we initially set the caption-level expansion probability, p_c, to zero. This initial value prevents the expansion of textual queries, forcing the model to project images onto relevant tokens belonging to the captions they were paired with. This approach facilitates the meaningful projection of dense vectors onto relevant words in the vocabulary. However, it adversely impacts retrieval effectiveness, as the model cannot expand queries. As a consequence, the model's ability to handle semantic matching is limited. To gradually relax this constraint, we use a scheduler that incrementally increases the value of p after each epoch until it reaches a maximum value of one in the final epoch. In each epoch, we sample the values of $\mathcal{E}$ per batch and enforce expansion terms to be zero when $\mathcal{E}_\mathcal{C}$ equals zero. Similarly, for word-level expansion, we initialize the expansion probability of the i-th word p_i^v to $1 - df_i^v$ where df_i^v is the normalized document frequency of vocabulary element v_i in the caption collection $\mathcal{C}$. This setting discourages the expansion of more frequent terms because they are less meaningful and can hinder the efficiency of query processing algorithms. We relax each p_i^v after every epoch, ensuring that it reaches a maximum value of one at the conclusion of the training process. The expansion rate increase after each epoch is defined as follows:

$$f_{\mathrm{incr}}(p) = \begin{cases} p + \frac{1}{\#\,epochs}, & \text{for caption-level expansion} \\ p + \frac{df_i^v}{\#\,epochs}, & \text{for word-level expansion.} \end{cases} \tag{6}$$

4.2   Training Loss

We train our Dense2Sparse using a loss that represents a weighted sum of a bidirectional loss and a sparse regularization parameter. The bidirectional loss is based on the following one-directional loss:

$$\ell^{(\mathcal{A} \to \mathcal{B})} = -\left(\frac{\exp(\mathbf{z}_\mathcal{A}^\mathsf{T} \mathbf{z}_\mathcal{B} / \tau)}{\sum_{\mathcal{I}*} \exp(\mathbf{z}_\mathcal{A}^\mathsf{T} \mathbf{z}_{\mathcal{I}*} / \tau)} \right) \log_2 \left(\mathrm{SoftMax}[\mathbf{s}_\mathcal{A}^\mathsf{T} \mathbf{s}_\mathcal{B}] \right),$$

where $\mathbf{s}_\mathcal{A} \in \mathcal{R}_{>0}^{|V|}$ and $\mathbf{s}_\mathcal{B} \in \mathcal{R}_{>0}^{|V|}$ are sparse vectors, $\mathbf{z}_\mathcal{A} \in \mathcal{R}^d$ and $\mathbf{z}_\mathcal{B} \in \mathcal{R}^d$ are dense vectors, and $\tau \in \mathcal{R}_{>0}$ is a temperature parameter.

The resulting loss is formalized to capture both bidirectional losses and sparse regularization. The overall loss $\mathcal{L}$ is defined as:

$$\mathcal{L} = (1 - \lambda) \underbrace{[\ell^{(\mathcal{I} \to \mathcal{C})} + \ell^{(\mathcal{C} \to \mathcal{I})}]}_{\text{bidirectional loss}} + \lambda \underbrace{\eta[L_1(\mathbf{s}_\mathcal{I}) + L_1(\mathbf{s}_\mathcal{C})]}_{\text{sparse regularization parameter}}, \tag{7}$$

where $\ell^{(\mathcal{I} \to \mathcal{C})}$ is an image-to-caption loss, $\ell^{(\mathcal{C} \to \mathcal{I})}$ is a caption-to-image loss; $\lambda = [0, 1]$ is a scalar weight, $\eta = [0, 1]$ is a sparsity regularization parameter, and $L_1(\mathbf{x}) = \|\mathbf{x}\|_1$ is L_1 regularization. It is worth noting that the loss utilizes dense scores for supervision, a strategy found to be more effective than using ground truth labels.

5   Experiments and Results

5.1   Experimental Setup

Datasets. We trained and evaluated our models on two widely used datasets for text-image retrieval: MSCOCO [27] and Flickr30k [41]. Each image in the two datasets is paired with five short captions (with some exceptions). We re-used the splits from [17] for training, evaluating, and testing. The splits on MSCOCO have 113.2k pairs for training, and 5k pairs for each validation/test set. Flickr30 is smaller with 29.8k/1k/1k for train, validation, test splits respectively. The best model is selected based on the validation set and evaluated on the test set.

Evaluation Metrics. To evaluate model performance and effectiveness, we report R@k where $k = \{1, 5\}$, and MRR@10 using the *ir_measures* [32] library.

Implementation and Training Details. The caption and image dense vectors of BLIP [22] and ALBEF [23] models are pre-computed with checkpoints from the larvis library [21]. We train our models to convert from dense vectors to sparse vectors on a single A100 GPU with a batch size of 512 for 200 epochs. The training takes around 2 h and only uses up to around 10 GB of GPU memory. We set the temperature τ to 0.001 and experiment with sparse regularization weights $\eta \in [1e - 5, 1e - 2]$.

Table 1. The effectiveness of sparsified models (D2S) and baselines. ($^\dagger p < 0.05$ *with paired two-tailed t-test comparing D2S to the dense model with Bonferroni correction*)

Model	MSCOCO (5k)				Flickr30k (1k)			
	R@1↑	R@5↑	MRR@10↑	FLOPs↓	R@1↑	R@5↑	MRR@10↑	FLOPs↓
T2I Dense Retrieval								
COOKIE [47]	46.6	75.2	–	–	68.3	91.1	–	–
COTS (5.3M) [29]	50.5	77.6	–	–	75.2	93.6	–	–
ALBEF [23]	53.1	79.3	64.3	–	79.1	94.9	86.6	–
BLIP [22]	**57.3**	**81.8**	**67.8**	–	**83.2**	**96.7**	**89.3**	–
T2I Sparse Retrieval								
VisualSparta	45.1	73.0	–	–	57.1	82.6	–	–
STAIR (zero-shot)	41.1	56.4	–	–	66.6	88.7	–	–
LexLIP (4.3M)	51.9	78.3	–	–	76.7	93.7	–	–
LexLIP (14.3M)	53.2	79.1	–	–	78.4	94.6	–	–
D2S (ALBEF, $\eta = 1e - 3$)	49.6†	77.7†	61.4†	18.7	74.2†	93.8†	82.6†	21.7
D2S (ALBEF, $\eta = 1e - 5$)	50.7†	78.2†	62.4†	74.2	75.4†	94.3†	83.6†	64.3
D2S (BLIP, $\eta = 1e - 3$)	51.8†	79.3†	63.4†	**11.5**	77.1†	94.6†	84.6†	**9.9**
D2S (BLIP, $\eta = 1e - 5$)	54.5†	80.6†	65.6†	78.4	79.8†	95.9†	86.7†	39.5

5.2 Results and Discussion

RQ1: How Effective and Efficient is the Proposed Method for Converting Dense to Sparse? We trained various Dense2Sparse models (D2S) using our proposed training method with different sparse regularization weights ranging from $1e - 5$ to $1e - 2$. Figure 2a illustrates the effectiveness and efficiency of these variations, with detailed results presented in Table 1. Firstly, we observe that increasing the sparse regularization weight enhances model efficiency (reduced FLOPs) but reduces its effectiveness (lower Recall and MRR). On the MSCOCO dataset, our most efficient sparse BLIP model

($\eta = 1e - 2$) achieves a R@1 of 47.2 and MRR@10 of 58.5 with the lowest FLOPs value of 1.6. Relaxing the regularization weight to $1e - 3$ results in an approximately 10% increase in R@1 to 51.8 and a similar rise in MRR@10 to 63.4, albeit at the expense of around 7 times higher FLOPs (less efficient).

Further relaxing the sparse regularization gradually brings the sparsified model's effectiveness closer to the original dense model, while reducing the efficiency. The most effective sparsified BLIP model with $\eta = 1e - 5$ performs competitively with the original dense version (54.5 vs. 57.3) and outperforms other dense baselines.

Additionally, we observe a diminishing gap between dense and sparsified models as we assess recalls at higher cutoff positions, such as $R@5$ and $R@10$. Similar trends are observed across different datasets, including Flickr30k and MSCOCO, as well as among different dense models, including BLIP and ALBEF. This indicates the broad applicability of our proposed approach to diverse datasets and models.

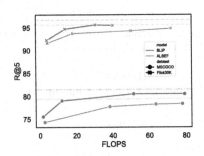

(a) Efficiency vs. effectiveness of sparsified models

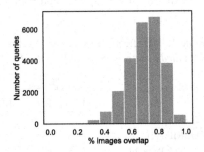

(b) Fraction of overlapping images in the top-10 by sparsified and dense BLIP model.

Fig. 2. Sparisified models compared to original dense models.

RQ2: How Does Our Sparsified Model Compare to State-of-the-Art Multi-modal LSR Models? In this research question, we compare our sparsified models with existing LSR baselines, namely Visual Sparta, STAIR, and LexLIP. Currently, neither the code nor the checkpoints for these baselines are publicly available. Therefore, we rely on the numbers reported in their respective papers for comparison, excluding the FLOPs.

STAIR and LexLIP are two of the most recent multimodal LSR approaches, both trained on large datasets, with STAIR utilizing 1 billion internal text-image pairs. In contrast, our proposed method leverages pretrained dense retrieval models to efficiently learn a lightweight sparse projection for converting dense vectors to sparse vectors.

The effectiveness of our methods and the baselines on MSCOCO and Flickr30k is presented in Table 1. Notably, our efficient model, D2S(BLIP, $\eta = 1e - 3$), performs competitively with LexLIP trained on 4.3 million text-image pairs at R@1. Its R@5 is slightly better than LexLIP (4.3M) and comparable to the LexLIP model trained on 14.3 million pairs. With a lower sparse regularization, our D2S(BLIP, $\eta=1e-5$) model significantly outperforms all baselines on both MSCOCO and Flickr30k. On MSCOCO, its R@1 is 21%, 5%, and 2.8% higher than the R@1 of Visual Sparta, LexLIP (4.3M), and LexLIP (14.3M), respectively. All our models outperform Visual Sparta and STAIR, although this comparison with STAIR uses a zero-shot setting, because we lack access to their code and checkpoints for fine-tuning STAIR further with in-domain data.

We kept the dense encoders frozen, so the effectiveness of our sparsified models is inherently bounded by the dense results. Our sparsified ALBEF models, for example, exhibit slightly lower overall effectiveness since their corresponding dense performance is lower than that of BLIP's dense scores. Nonetheless, our sparsified ALBEF models are also comparable with LexLIP variants.

RQ3: Does the Proposed Training Method Help Address the Dimension Co-activation and Semantic Deviation Issues? As discussed in Sect. 3, high co-activation increases posting list length, impacting inverted index efficiency. We examine this impact by analyzing FLOPs alongside model effectiveness metrics. Table 1 presents results for models trained with our method and three baseline variants, with fixed expansion rates of 0 and 1 in the first two baselines. The third baseline ($exp = c$) explores the influence of word-level expansion control, excluding it from our training method.

At an expansion rate of zero, models project the caption's dense vector only onto terms from the caption, with all other projections forced to zero. The image projector must then learn to align the image vector with terms in the paired captions. Conversely, setting exp to 1 gives the model the freedom to project onto any output vectors, making it more inclined toward dimension co-activation.

Table 2. The dimension co-activation effect of Dense2Sparse (D2S) variations.

Model (D2S variations)	MSCOCO (5k)				Flickr30k (1k)			
	R@1↑	R@5↑	MRR@10↑	FLOPs↓	R@1↑	R@5↑	MRR@10↑	FLOPs↓
(BLIP, $\eta = 1e - 3, exp = 0$)	45.5	73.0	57.3	2.8	68.9	89.5	77.8	3.0
(BLIP, $\eta = 1e - 3, exp = 1$)	53.4	80.0	64.6	49.1	79.5	95.5	86.4	50.3
(BLIP, $\eta = 1e - 3, exp = c$)	51.9	79.0	63.4	11.8	77.3	94.7	84.8	13.6
(BLIP, $\eta = 1e - 3, exp = c + w$)	51.8	79.3	63.4	11.5	77.1	94.6	84.6	9.9
(BLIP, $\eta = 1e - 5, exp = 0$)	47.2	74.4	58.8	3.2	72.3	91.8	80.7	3.5
(BLIP, $\eta = 1e - 5, exp = 1$)	55.9	81.3	66.8	343	81.4	96.0	87.7	213
(BLIP, $\eta = 1e - 5, exp = c$)	54.7	80.5	65.8	79.1	79.9	95.5	86.7	40.1
(BLIP, $\eta = 1e - 5, exp = c + w$)	54.5	80.6	65.6	78.4	79.8	95.9	86.7	39.5
(ALBEF, $\eta = 1e - 3, exp = 0$)	43.8	71.8	55.7	2.5	65.8	88.3	75.4	3.0
(ALBEF, $\eta = 1e - 3, exp = 1$)	50.9	78.4	62.5	68.2	75.7	94.2	83.8	61.9
(ALBEF, $\eta = 1e - 3, exp = c$)	49.7	77.7	61.5	38.3	74.6	93.7	82.8	17.9
(ALBEF, $\eta = 1e - 3, exp = c + w$)	49.6	77.7	61.4	18.7	74.2	93.8	82.6	21.7
(ALBEF, $\eta = 1e - 5, exp = 0$)	45.9	73.9	83.0	3.4	68.1	90.0	77.6	3.2
(ALBEF, $\eta = 1e - 5, exp = 1$)	52.4	78.7	63.7	283	77.2	94.6	84.8	210
(ALBEF, $\eta = 1e - 5, exp = c$)	51.2	78.3	62.8	77.9	76.4	94.8	84.0	71.7
(ALBEF, $\eta = 1e - 5, exp = c + w$)	50.7	78.2	62.4	74.2	75.4	94.3	83.6	64.3

In Table 2, rows with ($exp = 0$) show models with no expansion, resulting in remarkably low FLOPs, with each query averaging 2 to 3 overlapping terms with each document. However, disabling expansion reduces the model's ability for semantic matching, leading to modest effectiveness (45–47R@1 on MSCOCO and 68–72R@1 on Flickr30k with varying sparsity). Enabling non-regulated expansion ($exp = 1$) significantly improves model effectiveness (50–55 R@1 on MSCOCO and 75–79R@1 on Flickr30k with various regularization weights). However, this improvement comes at the cost of substantially increased FLOP scores, sometimes by up to 100 times, making sparsified vectors very computationally expensive. Ultimately, the resulting models behave like dense models, which is an undesired effect.

Our training method, which incorporates expansion control at the caption and word levels, is designed to gradually transition from one extreme ($exp = 0$) to the other ($exp = 1$). During training, we allow a likelihood of expansion, which increases progressively to over time. However, we also introduce random elements, represented by a random variable, to remind the model to remain faithful to the original captions/images.

The results, displayed in rows labeled with $exp = c + w$, demonstrate that our approach strikes a better balance between efficiency and effectiveness. It achieves competitive levels of effectiveness compared to models with $exp = 1$ while requiring only half or a third of the computational operations (FLOPs). For example, on MSCOCO with the BLIP model, Dense2Sparse ($\eta = 1e - 3$) achieves a performance of 51.8 R@1 (compared to 53.4 when $exp = 1$) with just 11.8 FLOPs, making it four times more efficient than the $exp = 1$ baseline. With the same setting, our method achieves 14% higher R@1 and 11% higher MRR@10 than the baseline with no expansion ($exp = 0$). Compared to the baseline without word-level expansion control, no significant differences are observed in terms of efficiency and effectiveness. Thus, caption-level expansion control alone seems sufficient for achieving reasonable efficiency and effectiveness. Similar results are noted across various settings, datasets, and dense models.

Sparse representations contain interpretable output dimensions aligned with a vocabulary. However, training a D2S model without our expansion regulation leads to semantic deviation, turning vocabulary terms into non-interpretable latent dimensions. We assess this effect using Exact@k and Semantic@k metrics (defined in Sect. 3), reporting results in Table 3 and providing qualitative examples in Table 4.

Table 3. Semantic deviation on different Dense2Sparse (D2S) variations. ($^{\dagger}p < 0.01$ *with paired two-tailed t-test comparing exp=c to exp=1*)

Model (D2S variations)	MSCOCO (5k)		Flickr30k (1k)	
	Exact@20	Semantic@20	Exact@20	Semantic@20
(BLIP, $\eta = 1e - 5, exp = c$)	$20.0^{\dagger}$	$60.1^{\dagger}$	$18.3^{\dagger}$	$58.0^{\dagger}$
(BLIP, $\eta = 1e - 5, exp = 1$)	6.9	48.5	3.2	40.7
(BLIP, $\eta = 1e - 3, exp = c$)	$25.0^{\dagger}$	$63.2^{\dagger}$	$23.1^{\dagger}$	$60.6^{\dagger}$
(BLIP, $\eta = 1e - 3, exp = 1$)	2.5	42.0	2.2	41.1
(ALBEF, $\eta = 1e - 5, exp = c$)	$20.5^{\dagger}$	$61.0^{\dagger}$	$19.2^{\dagger}$	$59.8^{\dagger}$
(ALBEF, $\eta = 1e - 5, exp = 1$)	5.6	43.5	1.2	40.5
(ALBEF, $\eta = 1e - 3, exp = c$)	$15.1^{\dagger}$	$51.3^{\dagger}$	$19.6^{\dagger}$	$56.4^{\dagger}$
(ALBEF, $\eta = 1e - 3, exp = 1$)	1.6	40.6	1.3	41.5

Table 4. Examples of semantic deviation. We show the top-10 terms per model.

Caption, Image	D2S ($\eta = 1e-3$, exp=c)	D2S ($\eta = 1e-3$, exp=1.0)
A man with a red helmet on a small moped on a dirt road	dirt, mo, motor, motorcycle, bike, red, riding, features, soldier, ##oot	, accent " yourself natural may while officer english ac
	mountain mountains bike bee dirt mo red path ##oot person man riding bicycle	accent ship natural de crown yourself " ra now wild
A women smiling really big while holding a Wii remote.	lady woman smile women remote laughing wii smiling video controller	, kai called forces rush lee war oil like ##h
	smile after green woman smiling sweater remote lady wii her	tall kai forces oil rush met war college thus there
A couple of dogs sitting in the front seats of a car.	dogs dog car backseat seat couple vehicle sitting two puppy	, electric stood forest national master help arts fc -
	dog car dogs puppy out vehicle pup inside early open	stood forest national electric master twice grant men para yet

Uncontrolled models (with $exp = 1$) exhibit lower Exact@20 and Semantic@20 than our expansion-controlled models ($exp = c$). In the top 20 terms of uncontrolled models, only one or none are in the original captions, while controlled models generate 3 to 5 caption terms. The low Semantic@20 of the uncontrolled models also suggests low relatedness of output terms to the caption terms. This implication could be further supported by the examples demonstrated in Table 4. Uncontrolled models generate random terms, while our method produces terms that more faithfully reflect captions and images. Most top-10 terms from our method are relevant to the input, including a mix of original terms and synonyms (e.g., "dog" vs. "puppy", "car" vs. "vehicle").

RQ4: Is the Sparsified Model Faithful to the Dense Model? This research question aims to analyze the faithfulness of sparsified models to their original dense models. We report in Table 5 the Pearson correlation calculated for various effectiveness metrics of dense and sparsified queries. The results show that the correlation between sparsified and dense models is consistently positive and tends to increase as we relax the sparse regularization. Furthermore, as we consider higher cutoff values (R@1, R@5, MRR@10), the correlation tends to increase as the performance gap between the two systems narrows. Manually comparing the top-10 ranked images of the most differing queries, we find that while the two models rank top-10 images differently, there are a lot of common images (including the golden image) that look equally relevant to the query. Figure 2b shows that a high ratio (average: 70%) of the top-10 images appear in both dense and sparse ranking lists. This analysis shows that the sparsified model is

Table 5. Correlation between dense and different variations of Dense2Sparse (D2S).

Model (D2S variations)	MSCOCO (5k)			Flickr30k (1k)		
	ρ-R@1↑	ρ-R@5↑	ρ-MRR@10↑	ρ-R@1↑	ρ-R@5↑	ρ-MRR@10↑
(BLIP, $\eta = 1e - 2$)	61.0	65.7	72.3	54.7	55.0	63.9
(BLIP, $\eta = 1e - 3$)	74.0	76.9	83.8	66.2	65.5	73.6
(BLIP, $\eta = 1e - 4$)	79.7	82.1	88.2	71.6	72.8	79.3
(BLIP, $\eta = 1e - 5$)	81.2	83.8	89.2	74.3	74.0	81.1
(ALBEF, $\eta = 1e - 2$)	64.4	68.7	75.5	57.7	57.0	67.5
(ALBEF, $\eta = 1e - 3$)	73.1	76.7	83.5	68.8	69.0	77.2
(ALBEF, $\eta = 1e - 4$)	78.1	80.7	87.2	73.2	74.6	81.3
(ALBEF, $\eta = 1e - 5$)	78.2	81.3	87.3	74.2	72.5	82.0

reasonably faithful to the dense model, suggesting that the sparse output terms could potentially be used for studying the semantics of dense vectors.

5.3 Retrieval Latency of Dense and Sparsified Models

We discussed the average FLOPs of sparsified models for retrieval efficiency. We now present query throughput and retrieval latency results in Table 6. Using Faiss [15] and PISA [31,35] on a single-threaded AMD Genoa 9654 CPU, the dense BLIP model with Faiss HNSW is exceptionally fast, outperforming D2S models with PISA. D2S models with query expansion (*exp=c*) are slower due to high FLOPs and possibly LSR known limitations [34]. Removing expansion terms (*exp=0*) improves latency (FLOPs similar to DistilSPLADE [6,7]) but is still approximately 30× slower than dense retrieval. To balance efficiency and effectiveness of D2S, we propose using the inverted index with original query terms for retrieval, followed by re-scoring with expansion terms. With our simple iterative implementation, this approach proves effective, especially for retrieving fewer images per query. Surprisingly, indexing D2S models with Faiss HNSW competes well with PISA, particularly at higher cut-off values (100, 1000).

Table 6. Retrieval latency (CPU - 1 thread) of D2S models on 123k MSCOCO images.

Model		Throughput (q/s)			Latency (ms)		
	FLOPS	@10	@100	@1000	@10	@100	@1000
Dense (BLIP, HNSW, Faiss)	–	13277	9739	7447	0.08	0.10	0.14
D2S (BLIP, $\eta = 1e - 3$, exp=c, PISA)	11.5	6	5	5	156.60	183.42	193.46
D2S (BLIP, $\eta = 1e - 3$, exp=0, PISA)	2.8	449	284	160	2.23	3.52	6.25
No Expansion >> Expansion	–	369	120	18	2.70	8.31	54.05
D2S (BLIP, $\eta = 1e - 5$, exp=c, PISA)	78.4	<1	<1	<1	>300	>600	>700
D2S (BLIP, $\eta = 1e - 5$, exp=0, PISA)	3.2	230	146	90	4.34	6.85	11.04
No Expansion >> Expansion	–	189	70	11	5.30	14.37	86.66
D2S (BLIP, HNSW, Faiss)	–	262	262	256	3.82	3.82	3.90

6   Conclusion

We have focused on the problem of efficiently transforming a pretrained dense retrieval model into a sparse model. We show that training a projection layer on top of dense vectors with the standard contrastive learning technique leads to the problems of dimension co-activation and semantic deviation. To mitigate these issues, we propose a training algorithm that uses a Bernoulli random variable to control the term expansion. Our experiments show that our Dense2Sparse sparsified model trained with the proposed algorithm suffers less from those issues. In addition, our sparsified models perform competitively to the state-of-the-art multi-modal LSR, while being faithful to the original dense models.

Acknowledgments. We thank our anonymous reviewers for their valuable feedback. This research was supported by Ahold Delhaize, the Hybrid Intelligence Center, a 10-year program funded by the Dutch Ministry of Education, Culture and Science through the Netherlands Organisation for Scientific Research, project LESSEN with project number NWA.1389.20.183 of the research program NWA ORC 2020/21, which is (partly) financed by the Dutch Research Council (NWO), project IDEAS with project number VI.Vidi.223.166 of the NWO Talent Programme, which is (partly) financed by the Dutch Research Council (NWO), and the FINDHR (Fairness and Intersectional Non-Discrimination in Human Recommendation) project that received funding from the European Union's Horizon Europe research and innovation program under grant agreement No 101070212. All content represents the opinion of the authors, which is not necessarily shared or endorsed by their respective employers and/or sponsors.

References

1. Brown, A., Xie, W., Kalogeiton, V., Zisserman, A.: Smooth-AP: smoothing the path towards large-scale image retrieval. In: Vedaldi, A., Bischof, H., Brox, T., Frahm, J.-M. (eds.) ECCV 2020. LNCS, vol. 12354, pp. 677–694. Springer, Cham (2020). https://doi.org/10.1007/978-3-030-58545-7_39
2. Chen, C., et al.: STAIR: learning sparse text and image representation in grounded tokens. arXiv preprint arXiv:2301.13081 (2023)
3. Dai, Z., Callan, J.: Context-aware sentence/passage term importance estimation for first stage retrieval. arXiv preprint arXiv:1910.10687 (2019)
4. Devlin, J., Chang, M.W., Lee, K., Toutanova, K.: BERT: pre-training of deep bidirectional transformers for language understanding. arXiv preprint arXiv:1810.04805 (2018)
5. Faghri, F., Fleet, D.J., Kiros, J.R., Fidler, S.: VSE++: improving visual-semantic embeddings with hard negatives. arXiv preprint arXiv:1707.05612 (2017)
6. Formal, T., Lassance, C., Piwowarski, B., Clinchant, S.: From distillation to hard negative sampling: Making sparse neural IR models more effective. In: Proceedings of the 45th International ACM SIGIR Conference on Research and Development in Information Retrieval, SIGIR 2022, pp. 2353–2359. Association for Computing Machinery, New York (2022)
7. Formal, T., Piwowarski, B., Clinchant, S.: SPLADE: sparse lexical and expansion model for first stage ranking. In: Proceedings of the 44th International ACM SIGIR Conference on Research and Development in Information Retrieval, pp. 2288–2292 (2021)
8. Frome, A., et al.: DeViSE: a deep visual-semantic embedding model. In: Proceedings of the 26th International Conference on Neural Information Processing Systems, vol. 2, pp. 2121–2129 (2013)

9. Gao, D., et al.: Fashionbert: text and image matching with adaptive loss for cross-modal retrieval. In: Proceedings of the 43rd International ACM SIGIR Conference on Research and Development in Information Retrieval, pp. 2251–2260 (2020)

10. Goei, K., Hendriksen, M., de Rijke, M.: Tackling attribute fine-grainedness in cross-modal fashion search with multi-level features. In: SIGIR 2021 Workshop on eCommerce. ACM (2021)

11. Gong, Y., Wang, L., Hodosh, M., Hockenmaier, J., Lazebnik, S.: Improving image-sentence embeddings using large weakly annotated photo collections. In: Fleet, D., Pajdla, T., Schiele, B., Tuytelaars, T. (eds.) ECCV 2014. LNCS, vol. 8692, pp. 529–545. Springer, Cham (2014). https://doi.org/10.1007/978-3-319-10593-2_35

12. Gu, J., Cai, J., Joty, S.R., Niu, L., Wang, G.: Look, imagine and match: improving textual-visual cross-modal retrieval with generative models. In: Proceedings of the IEEE Conference on Computer Vision and Pattern Recognition, pp. 7181–7189 (2018)

13. Hendriksen, M., Bleeker, M., Vakulenko, S., van Noord, N., Kuiper, E., de Rijke, M.: Extending CLIP for category-to-image retrieval in E-commerce. In: Hagen, M., et al. (eds.) ECIR 2022. LNCS, vol. 13185, pp. 289–303. Springer, Cham (2022). https://doi.org/10.1007/978-3-030-99736-6_20

14. Hendriksen, M., Vakulenko, S., Kuiper, E., de Rijke, M.: Scene-centric vs. object-centric image-text cross-modal retrieval: a reproducibility study. In: European Conference on Information Retrieval, pp. 68–85. Springer, Heidelberg (2023). https://doi.org/10.1007/978-3-031-28241-6_5

15. Johnson, J., Douze, M., Jégou, H.: Billion-scale similarity search with gpus. IEEE Trans. Big Data 7(3), 535–547 (2019)

16. Kamalloo, E., Thakur, N., Lassance, C., Ma, X., Yang, J.H., Lin, J.: Resources for brewing BEIR: reproducible reference models and an official leaderboard. arXiv preprint arXiv:2306.07471 (2023)

17. Karpathy, A., Fei-Fei, L.: Deep visual-semantic alignments for generating image descriptions. In: Proceedings of the IEEE Conference on Computer Vision and Pattern Recognition, pp. 3128–3137 (2015)

18. Klein, B., Lev, G., Sadeh, G., Wolf, L.: Fisher vectors derived from hybrid gaussian-laplacian mixture models for image annotation. arXiv preprint arXiv:1411.7399 (2014)

19. Laenen, K.: Cross-modal Representation Learning for Fashion Search and Recommendation. Ph.D. thesis, KU Leuven (2022)

20. Lee, K.H., Chen, X., Hua, G., Hu, H., He, X.: Stacked cross attention for image-text matching. In: Proceedings of the European Conference on Computer Vision (ECCV), pp. 201–216 (2018)

21. Li, D., Li, J., Le, H., Wang, G., Savarese, S., Hoi, S.C.: Lavis: a library for language-vision intelligence. arXiv preprint arXiv:2209.09019 (2022)

22. Li, J., Li, D., Xiong, C., Hoi, S.: Blip: bootstrapping language-image pre-training for unified vision-language understanding and generation. In: International Conference on Machine Learning, pp. 12888–12900. PMLR (2022)

23. Li, J., Selvaraju, R., Gotmare, A., Joty, S., Xiong, C., Hoi, S.C.H.: Align before fuse: vision and language representation learning with momentum distillation. Adv. Neural. Inf. Process. Syst. 34, 9694–9705 (2021)

24. Lin, J., Ma, X.: A few brief notes on deepimpact, coil, and a conceptual framework for information retrieval techniques. arXiv preprint arXiv:2106.14807 (2021)

25. Lin, S.C., Lin, J.: Densifying sparse representations for passage retrieval by representational slicing. arXiv preprint arXiv:2112.04666 (2021)

26. Lin, S.C., Lin, J.: A dense representation framework for lexical and semantic matching. ACM Trans. Inf. Syst. 41(4), 1–29 (2023)

27. Lin, T.-Y., et al.: Microsoft COCO: common objects in context. In: Fleet, D., Pajdla, T., Schiele, B., Tuytelaars, T. (eds.) ECCV 2014. LNCS, vol. 8693, pp. 740–755. Springer, Cham (2014). https://doi.org/10.1007/978-3-319-10602-1_48

28. Liu, C., Mao, Z., Liu, A.A., Zhang, T., Wang, B., Zhang, Y.: Focus your attention: a bidirectional focal attention network for image-text matching. In: Proceedings of the 27th ACM International Conference on Multimedia, pp. 3–11 (2019)

29. Lu, H., Fei, N., Huo, Y., Gao, Y., Lu, Z., Wen, J.: COTS: collaborative two-stream vision-language pre-training model for cross-modal retrieval. In: IEEE/CVF Conference on Computer Vision and Pattern Recognition, CVPR 2022, New Orleans, LA, USA, 18–24 June 2022, pp. 15671–15680. IEEE (2022)

30. Luccioni, A.S., Hernandez-Garcia, A.: Counting carbon: a survey of factors influencing the emissions of machine learning. arXiv preprint arXiv:2302.08476 (2023)

31. MacAvaney, S., Macdonald, C.: A python interface to pisa! In: Proceedings of the 45th International ACM SIGIR Conference on Research and Development in Information Retrieval (2022). https://doi.org/10.1145/3477495.3531656

32. MacAvaney, S., Macdonald, C., Ounis, I.: Streamlining evaluation with ir-measures. In: Hagen, M., et al. (eds.) ECIR 2022. LNCS, vol. 13186, pp. 305–310. Springer, Cham (2022). https://doi.org/10.1007/978-3-030-99739-7_38

33. MacAvaney, S., Nardini, F.M., Perego, R., Tonellotto, N., Goharian, N., Frieder, O.: Expansion via prediction of importance with contextualization. In: Proceedings of the 43rd International ACM SIGIR Conference on Research and Development in Information Retrieval, pp. 1573–1576 (2020)

34. Mackenzie, J., Trotman, A., Lin, J.: Wacky weights in learned sparse representations and the revenge of score-at-a-time query evaluation. arXiv preprint arXiv:2110.11540 (2021)

35. Mallia, A., Siedlaczek, M., Mackenzie, J., Suel, T.: PISA: performant indexes and search for academia. In: Proceedings of the Open-Source IR Replicability Challenge co-located with 42nd International ACM SIGIR Conference on Research and Development in Information Retrieval, OSIRRC@SIGIR 2019, Paris, France, 25 July 2019, pp. 50–56 (2019). http://ceur-ws.org/Vol-2409/docker08.pdf

36. Messina, N., Amato, G., Esuli, A., Falchi, F., Gennaro, C., Marchand-Maillet, S.: Fine-grained visual textual alignment for cross-modal retrieval using transformer encoders. ACM Trans. Multimedia Comput. Commun. Appl. (TOMM) 17(4), 1–23 (2021)

37. Nam, H., Ha, J.W., Kim, J.: Dual attention networks for multimodal reasoning and matching. In: Proceedings of the IEEE Conference on Computer Vision and Pattern Recognition, pp. 299–307 (2017)

38. Nguyen, T., Hendriksen, M., Yates, A.: Multimodal learned sparse retrieval for image suggestion. In: TREC (2023)

39. Nguyen, T., MacAvaney, S., Yates, A.: Adapting learned sparse retrieval for long documents. arXiv preprint arXiv:2305.18494 (2023)

40. Nguyen, T., MacAvaney, S., Yates, A.: A unified framework for learned sparse retrieval. In: European Conference on Information Retrieval, pp. 101–116. Springer, Heidelberg (2023). https://doi.org/10.1007/978-3-031-28241-6_7

41. Plummer, B.A., Wang, L., Cervantes, C.M., Caicedo, J.C., Hockenmaier, J., Lazebnik, S.: Flickr30k entities: collecting region-to-phrase correspondences for richer image-to-sentence models. In: Proceedings of the IEEE International Conference on Computer Vision, pp. 2641–2649 (2015)

42. Ram, O., Bezalel, L., Zicher, A., Belinkov, Y., Berant, J., Globerson, A.: What are you token about? dense retrieval as distributions over the vocabulary. In: Proceedings of the 61st Annual Meeting of the Association for Computational Linguistic, vol. 1: Long Papers, pp. 2481–2498. Association for Computational Linguistics, Toronto (2023)

43. Sheng, S., Laenen, K., Van Gool, L., Moens, M.F.: Fine-grained cross-modal retrieval for cultural items with focal attention and hierarchical encodings. Computers **10**(9), 105 (2021)

44. Vaswani, A., et al.: Attention is all you need. Adv. Neural Inf. Process. Syst. **30** (2017)

45. Wang, H., et al.: Cross-modal food retrieval: learning a joint embedding of food images and recipes with semantic consistency and attention mechanism. IEEE Trans. Multimedia **24**, 2515–2525 (2021)

46. Wang, L., Li, Y., Lazebnik, S.: Learning deep structure-preserving image-text embeddings. In: Proceedings of the IEEE Conference on Computer Vision and Pattern Recognition, pp. 5005–5013 (2016)

47. Wen, K., Xia, J., Huang, Y., Li, L., Xu, J., Shao, J.: COOKIE: contrastive cross-modal knowledge sharing pre-training for vision-language representation. In: 2021 IEEE/CVF International Conference on Computer Vision, ICCV 2021, Montreal, QC, Canada, 10–17 October 2021, pp. 2188–2197. IEEE (2021)

48. Young, P., Lai, A., Hodosh, M., Hockenmaier, J.: From image descriptions to visual denotations: new similarity metrics for semantic inference over event descriptions. Trans. Assoc. Comput. Linguist. **2**, 67–78 (2014)

49. Zamani, H., Dehghani, M., Croft, W.B., Learned-Miller, E., Kamps, J.: From neural reranking to neural ranking: Learning a sparse representation for inverted indexing. In: Proceedings of the 27th ACM International Conference on Information and Knowledge Management, pp. 497–506 (2018)

50. Zhang, Y., Jiang, H., Miura, Y., Manning, C.D., Langlotz, C.P.: Contrastive learning of medical visual representations from paired images and text. In: Proceedings of the Machine Learning for Healthcare Conference, MLHC 2022, Durham, NC, USA, 5–6 August 2022, Proceedings of Machine Learning Research, vol. 182, pp. 2–25. PMLR (2022)

51. Zhao, P., et al.: LexLIP: lexicon-bottlenecked language-image pre-training for large-scale image-text retrieval. arXiv preprint arXiv:2302.02908 (2023)

52. Zhuge, M., et al.: Kaleido-BERT: vision-language pre-training on fashion domain. In: Proceedings of the IEEE/CVF Conference on Computer Vision and Pattern Recognition, pp. 12647–12657 (2021)

Alleviating Confounding Effects with Contrastive Learning in Recommendation

Di You[✉] and Kyumin Lee[✉]

Worcester Polytechnic Institute, Worcester, MA, USA
{dyou,kmlee}@wpi.edu

Abstract. Recently, there has been a growing interest in mitigating the bias effects in recommendations using causal inference. However, Rubin's potential outcome framework may produce inaccurate estimates in real-world scenarios due to the presence of hidden confounders. In addition, existing works adopting the Pearl causal graph framework tend to focus on specific types of bias (e.g., *selection bias*, *popularity bias*, *exposure bias*) instead of directly mitigating the impact of hidden confounders. Motivated by the aforementioned limitations, in this paper, we formulate the recommendation task as a causal graph with unobserved/unmeasurable confounders. We present a novel causality-based architecture called Multi-behavior Debiased Contrastive Collaborative Filtering (MDCCL) and apply the front-door adjustment for intervention. We leverage a pre-like behavior such as *clicking an item* (i.e., a behavior occurred before the target behavior such as *purchasing*) to mitigate the bias effects. Additionally, we design a contrastive loss that also provides a debiasing effect benefiting the recommendation. An empirical study on three real-world datasets validates that our proposed method successfully outperforms nine state-of-the-art baselines. Code and the datasets will be available at https://github.com/queenjocey/MDCCL.

Keywords: Recommender system · causal inference · debiased recommendation · contrastive learning

1 Introduction

While we have witnessed the success of recommender systems in various domains (e.g., social platform [41], music sites [29], e-commerce platforms [30]), most recommendation models focus on fitting observed user-item historical interactions [10,11,15,24]. However, user-item interaction data, which forms the basis for training recommendation models, are observational rather than experimental [3]. A common practice in training conventional recommender systems is to consider the unobserved interactions as negative feedback, assuming the observed data are missing-at-random. However, user interaction data are always missing-not-at-random [19,27,40] in reality. While the matching-based method models

© The Author(s), under exclusive license to Springer Nature Switzerland AG 2024
N. Goharian et al. (Eds.): ECIR 2024, LNCS 14609, pp. 465–480, 2024.
https://doi.org/10.1007/978-3-031-56060-6_30

Fig. 1. A toy example where confounders mislead the recommendation result.

the correlation between a user and candidate items, it does not inherently reflect the true causal relationship between user-item interaction. The presence of confounders, such as item quality, can lead to misleading recommendation results. To demonstrate the impact of these confounders, we present a toy example in Fig. 1. The user's historical interactions on the left suggest a preference for comedy based on genre features. However, in this specific case, the user's choice of next movie is influenced by the director rather than the genre. Although there exists a strong correlation between the genre feature and user preference, genre serves as a confounder here, and the next consumed movie of the target user is actually driven by causation. Meanwhile, confounders induce bias, which leads to the bias effects in the correlations estimated from the observations.

There has been a surge of study exploring bias elimination in recent years [32,33,37,39,43,45]. For example, the inverse propensity score (IPS)-based approach has long been popular in the recommendation community [27,42], however, this line of work heavily relies on the estimation of the IPS score and suffers high variance issues. Another line of work formulates their research with a causal graph to describe causal relationships and conducts reasoning over the graph to estimate causal effect [33,37]. Existing works [33,37,43,49] adopting the Pearl causal graph framework mostly follow the following pattern: (1) identify a specific confounder first, and (2) propose a confounder-aware model to address the specific confounder. However, in real-world scenarios, it is unrealistic to identify specific confounders. Moreover, not all confounders are observable and measurable, which limits the efficacy of the existing tools. Fortunately, there is another tool named front-door adjustment [23], which allows us to deal with any types of confounders, including unobservable/unmeasurable confounders.

Motivated by the aforementioned analysis, we design a new causal graph as shown in Fig. 3 where unobserved/unmeasurable confounders exist with a mediator node. Then, we propose a novel Multi-behavior Debiased Contrastive Collaborative Filtering (**MDCCL**) framework that leverages the front-door adjustment to eliminate the bias effect induced by confounders. As for the choice of mediator, we utilize prior user feedback about items (i.e., click an item) to facil-

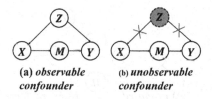

(a) *observable confounder*   **(b)** *unobservable confounder*

Fig. 2. X, Y, M and Z represent treatment variable, outcome variable, mediator variable and confounder variable, respectively. Figure (a) shows Z as observed and measurable, while in Figure (b), Z is unobserved or unmeasurable.

itate unbiased recommendation, which formulates our task as multi-behavior recommendation. Further, we design a debiased contrastive loss component to mitigate bias effect and improve recommendation accuracy. In the experiments, we compare our model with nine state-of-the-art baselines. Empirical experiments and in-depth analysis validate the effectiveness of MDCCL algorithm on both accurate recommendation and deconfounding.

2 Preliminaries

2.1 Task Formulation

$U = \{u_1, u_2, ..., u_k\}$ as a set of all users where $k = |U|$ is the total number of users, and $P = \{p_1, p_2, ..., p_n\}$ as a set of all items where $n = vertP|$ is the total number of items. Without loss of generality, we assume that the number of behaviors is T, and we use $Y_1, Y_2, ..., Y_T$ denotes behavior matrices if the user interacted with an item under behavior t, where $Y_1, Y_2, ..., Y_{T-1}$ are auxiliary behaviors, and Y_T is the target behavior. We consider interaction matrices in the binary form, in which each entry has value 1 if user u and item p interacted under behavior t, otherwise value 0. To simplify our analysis, in our work, we mainly discuss the target behavior, denoting as y and click behavior, serves as mediator, denoting as m. Bold versions of those variables, which we will introduce in the following sections, indicate their respective latent representations/embeddings.

2.2 Preliminaries on Causal Inference

In this section, we will briefly introduce some basic concepts and theorems in causal inference.

Definition 1: Causal Graph. Causal graph is a directed acyclic graph(DAG), where $G = (\mathcal{N}, \mathcal{E})$, describing the causal relationship. $\mathcal{N}$ represents a set of nodes, containing variables in U and P in recommendation; and $\mathcal{E}$ represents a set of edges, also known as the causal relations.

Definition 2: Backdoor adjustment [23]. Given an ordered pair of variables (X, Y) in a causal graph G, a set of variables Z satisfies the back-door criterion with respect to (X, Y) if Z satisfies the following conditions:

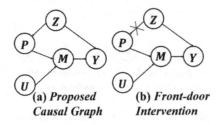

Fig. 3. Causal Graph of (a) the proposed model and (b) front-door intervention.

- No node in Z is a descendant of X;
- Z blocks every path between X and Y that contains an arrow into X.

With the help of a set of variables that satisfy the back-door criterion, we can adjust the effect of measured confounders. We take the causal graph in Fig. 2(a) as an example. Considering the treatment variable X and the outcome variable Y, we want to estimate the effect of X on Y, denoted as $P(Y = y|do(X = x))$. Due to the existence of confounder Z (i.e., Z is a parent node of X), we cannot conclude that $P(Y = y|do(X = x)) = P(Y = y|X = x)$. However, since variable Z satisfies the back-door criterion, we use it to adjust the effect, in other words, we are accounting for and measuring all confounders [23]. Therefore, we compute

$$P(Y = y|do(X = x)) = \sum_z P(Y = y|X = x, Z = z)P(Z = z) \qquad (1)$$

However, a serious limitation is that the above equation assumes that the confounder variables are all measurable and satisfy the backdoor criterion. However, unobservable and hidden confounders always exist in recommender systems [4,47]. Given this setting, we introduce the front-door criterion.

Definition 3: Frontdoor criterion and adjustment. Given an ordered pair of variables (X, Y) in a causal graph G, a set of variables M satisfies the front-door criterion with respect to (X, Y) if (X, Y) satisfies the following conditions:

- M intercepts all directed paths from X to Y;
- There is no unblocked path from X to M;
- X blocks all back-door paths from M to Y

If a set of variables M satisfies the front-door criterion related to an ordered pair of variables (X, Y), and if $P(x, z) > 0$, then the causal effect of X on Y is identifiable and is given by

$$P(y|do(x)) = \sum_m P(m|x) \sum_{x'} P(y|x', m)P(x') \qquad (2)$$

We take Fig. 2(b) as an example. In this case, the variable Z here is not measurable so that back-door adjustment cannot be directly applied. However, it satisfies the front-door criterion, allowing us to utilize the front-door adjustment to handle the unmeasurable confounder Z. Intuitively, the desired effect can be expressed as follows

$$P(y|do(x)) = \sum_m P(m|do(x))P(y|do(m)) \tag{3}$$

3 Methodology

3.1 Causal View of Deconfounding Recommendation

Figure 3 shows our proposed causal graph for interaction generation when the confounding feature exists. Next, we explain the semantics of the causal graph.

- Node U and P denote the user and item ID embeddings, respectively. In this work we only use ID feature.
- Node Z denotes the hidden confounder, which can be generated for various reasons (e.g., the producer's motivation, and the item's quality). Notice that the hidden confounder in our discussion is either unobservable or unmeasurable, thus the backdoor adjustment is not applicable.
- Node M denotes the mediator, which is pre-like/pre-purchase behavior (i.e., *click an item*) in this work. Node Y denotes the target interaction label (i.e., *purchase* or *like*);
- Edge $P \leftarrow Z \rightarrow Y$ denotes that hidden confounder Z affect both item features and happening of interaction, while it does not necessarily reflect users' real preference;
- Edge $\{P, U\} \rightarrow M \rightarrow Y$ denotes that user preference and item features jointly determine the level of user-item matching. And based on the front-door criterion, we can also observe that the prior feedback functions as a mediator in our task formulation (e.g., click $\rightarrow$ like on micro-video platforms.)

Most of conventional recommendation algorithms directly estimate the correlation $P(Y|U, P)$ using historical interaction data, which leads to biased estimation for recommended results. While previous causal models estimate the causal effect $P(Y|U, do(P))$, they overlook the effect of hidden confounder Z, thus bias issue still exists. In our work, we propose to simultaneously cut off the direct effect of $Z \rightarrow P$ and backdoor path $P \leftarrow Z \rightarrow Y$ to eliminate the confounding effect in our estimation.

3.2 Multi-behavior Debiased Contrastive CF

In this section, we discuss how to mitigate the confounding effect without measuring the confounder Z. We introduce prior feedback (i.e., *click*) as mediator as shown in Fig. 4.

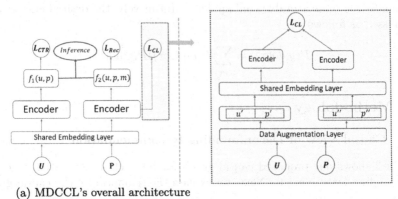

(a) MDCCL's overall architecture

(b) Contrastive Learning Structure

Fig. 4. Our proposed architecture where in Fig. 4a blue arrows indicate the training stage and the red arrows represent the inference stage. (Color figure online)

Intervention with Do-Calculus. Considering that the hidden confounder Z is unobservable or unmeasurable, we apply the front-door adjustment tool for user's preference estimation towards items. Specifically, we estimate the distribution as follows:

$$
\begin{aligned}
P(y|u, do(p)) &= \sum_{m} P(m|u, do(p)) \sum_{z} P(z)P(y|u, z, m) \\
&= \sum_{m} P(m|u, do(p))P(y|u, do(m))
\end{aligned}
\tag{4}
$$

In Eq. 4, the first term denotes the probability of mediator M being a set as m given certain item features, which reflects the causal effect of P on M. On the other hand, the second term denotes the probability of y when m happens, which is the causal effect of M on Y. The equation holds because of the backdoor criterion. Fortunately, both terms are measurable in our formulation.

Estimating $P(m|u, do(p))$. $P(m|u, do(p)) = P(m|u, p)$ since the backdoor path $P \leftarrow Z \rightarrow Y \leftarrow M$ is $d-$separated by collider Y, which means given Y, Z is independent of another set M. *Estimating $P(y|u, do(m))$* According to [23], blocking Z is equivalent to blocking P in the backdoor path $M \leftarrow P \leftarrow Z \rightarrow Y$. Also, according to our causal graph, M is independent of Z given P, and P is independent of Y given Z and M. Therefore, following prior causal recommendation work [43, 47], we can apply backdoor adjustment as follows:

$$P(y|u, do(m)) = \sum_z P(z)P(y|z, m, u)$$

$$\overset{(a)}{=} \sum_z \sum_p P(z|p)P(p)P(y|z, m, u)$$

$$\overset{(b)}{=} \sum_p \sum_z P(z|p)P(p)P(y|z, m, u, p) \tag{5}$$

$$\overset{(c)}{=} \sum_p (\sum_z P(y|m, z, u, p)P(z|p, m))P(p)$$

$$\overset{(d)}{=} \sum_p P(y|u, p, m)P(p)$$

We illustrate the derivation steps as follows:

- (a) holds due to $P(z) = P(z|p)P(p)$
- (b) holds since P is independent of Y given Z and M, thus we have $P(y|z, m, u) = P(y|z, m, u, p)$
- (c) is induced by $P(z|p) = P(z|p, m)$, since M is independent of Z given P
- (d) holds because of marginal distribution properties

Therefore, we derived the following equation to replace Eq. 4:

$$P(y|u, do(p)) = \sum_m P(m|u, p) \sum_{p'} P(y|u, p', m)P(p') \tag{6}$$

where we can get rid of $P(p')$ safely. Based on the above analysis, we present our deconfounding architecture in two stages: training and inference.

Deconfounded Training. In the training stage, as it showed in Fig. 4a, we shall estimate probability $P(m|u, p)$ and $P(y|u, p, m)$.

Since modeling $P(m|u, p)$ is equivalent to the well-known CTR prediction task, we parameterized it as $f_1(u, p)$, where $f_1(\cdot)$ can be any backbone model. On the other hand, $P(y|u, p, m)$ estimation is our main recommendation task, and can be decomposed into a late-fusion manner [31,33] without loss of generality,

$$f_2(u, p, m) = f_2'(u, m) * \sigma(f_2''(u, p)), \tag{7}$$

where f_2' and f_2'' are both backbone encoders, $\sigma(\cdot)$ is sigmoid function that introduces non-linearity for sufficient representation capacity of the fusion strategy. As it showed in Fig. 4a, we can use any existing model as backbone encoders to model in our framework. For simplicity, we adopt LightGCN for all components and take only the ID features of users and items as inputs.

The backbone model has different target values so that we estimate $P(m|u, p)$ and $P(y|u, p, m)$ as following:

$$\mathcal{L}_{CTR}(\mathcal{D}|\Phi) = - \sum_{(i, j^+, j^-)} log\sigma(s_{ij+} - s_{ij-}) + \lambda_\Phi \|\Phi\|_2 \tag{8}$$

$$\mathcal{L}_{Rec}(\mathcal{D}|\Theta) = - \sum_{(i,j^+,j^-)} log\sigma(o_{ij+} - o_{ij-}) + \lambda_\Theta \|\Theta\|_2 \qquad (9)$$

where (i, j^+, j^-) is a triplet of a target user, a positive item, and a negative item that is randomly sampled from the items set P. $\mathcal{D}$ denotes all the training instances. s_{ij+} and s_{ij-} are the respective positive and negative preference scores in CTR task. o_{ij+} and o_{ij-} are the respective positive and negative preference scores in the recommendation task. Φ and Θ are trainable parameters, and λ_Φ and λ_Θ are hyperparameters for regularization terms.

Contrastive Learning. We enhance robustness in learned representations by employing contrastive learning and data augmentation. A challenging procedure to apply contrastive learning in recommendation is to compose positive and negative pairs. We create a user-item bipartite graph and generate correlated views for each node, whether a user or an item, with its neighbor nodes under different behavior types by incorporating adaptive data augmentation techniques such as node drop. This approach intentionally reduces the impact of popular nodes while preserving isolated node information, mitigating popularity bias. These generated views are then input into the backbone model, treating views from the same node as positive pairs and views from different nodes as negative pairs. Furthermore, building upon recent work [36] on adaptive edge and node dropping, we adopt the idea of principle. This principle encourages representations to capture only the necessary information for the downstream task, minimizing mutual information between the original graph and generated views while maintaining recommendation performance.

To estimate mutual information between augmentation views, which encompasses both user and item perspectives, we utilize negative InfoNCE, as suggested by [8,21], which is equivalent to maximizing the lower bound of mutual information. We formally define our contrastive loss for representations as:

$$\mathcal{L}_{CL_u} = - \log \frac{\exp\left(s(u_i', u_i'')/\tau\right)}{\sum_{i'}^{k} \exp\left(s(u_i', u_{i'}'')/\tau\right)}, \mathcal{L}_{CL_p} = - \log \frac{\exp\left(s(p_j', p_j'')/\tau\right)}{\sum_{j'}^{n} \exp\left(s(p_j', p_{j'}'')/\tau\right)} \qquad (10)$$

where $\mathcal{L}_{CL_u}$ and $\mathcal{L}_{CL_p}$ are contrastive loss for user and item, respectively. $s(\cdot)$ denotes the cosine similarity function, and τ is the tunable temperature hyperparameter to adjust the scale for softmax. (u_i', u_i'') and $(u_i', u_{i'}'')$ are positive and negative user pairs, respectively. Similarly, (p_j', p_j'') and $(p_j', p_{j'}'')$ are positive and negative item pairs, respectively.

Formally, we incorporate the contrastive loss into our training schema as:

$$\mathcal{L}_{Total} = \mathcal{L}_{Rec} + \alpha * \mathcal{L}_{CTR} + \beta * (\mathcal{L}_{CL_u} + \mathcal{L}_{CL_p}), \qquad (11)$$

where α and β are hyperparameters controlling the effect of auxiliary tasks.

Inference. At the inference stage, we estimate the causal effect of the user-item pair with Eq. 6 and adopt a fusion strategy:

$$P(y|u, do(p)) = \sum_m P(m|u, p) \sum_p^{\prime} P(p^{\prime}) P(y|u, p^{\prime}, m)$$

$$= \sum_m f_1(u, p) * \sum_{p^{\prime}} f_2(u, p^{\prime}, m) P(p^{\prime})$$

$$= \sum_m f_1(u, p) f_2^{\prime}(u, m) \sum_{p^{\prime}} f_2^{\prime\prime}(u, p^{\prime}) P(p^{\prime}) \tag{12}$$

$$\propto \sum_m f_1(u, p) f_2^{\prime}(u, m) \sum_{p^{\prime}} f_2^{\prime\prime}(u, p^{\prime})$$

Notice that $\sum_{p^{\prime}} f_2^{\prime\prime}(u, p^{\prime})$ is a constant value given u that can be omitted, thus Eq. 12 reduces to $P(y|u, do(p)) = \sum_m f_1(u, p) f_2^{\prime}(u, m)$

3.3   Two Assumptions in Our Framework

In our proposed framework, we make two assumptions: Firstly, we focus on confounders positioned between users' click and like/purchase behaviors, such as item quality in the KuaiRec dataset(one of three datasets that we used), which encompasses various aspects, including resolution, content, and interestingness. Essentially, our introduced mediator remains independent of the influence of these confounders. Second, we acknowledge that confounders may affect auxiliary behaviors to varying degrees, but we simplify this assumption in our initial attempt to address confounder effects in multi-behavior recommendation, as articulated above. In future research, we will explore the impact of different confounders on various behaviors, which is beyond our current scope.

Table 1. Statistics of datasets.

Dataset	|Users|	|Items|	|Clicks|	|Likes/Purchases|	Overall Density (%)	Duration
Fliggy	2,730,201	104,342	32,444,647	1,160,723	0.01%	6 month
KuaiRec	7,176	10,729	12,530,806	1,124,378	17%	2 month
Adressa	31,123	4,895	1,437,540	998,612	1.59%	1 week

4   Empirical Study

4.1   Experimental Setup

Datasets. We evaluate all models on **three** public benchmark datasets collected from three real-world systems:

- *Fliggy Dataset* [28] is extracted from users' behavior logs at Fliggy in 2021, a prominent Chinese online travel portal Among various user behaviors, we only use *click* and *buy* to keep consistency with the other datasets.

- *KuaiRec Dataset* [5] is a dataset collected from the logs of the Kuaishou video-sharing mobile app. It features a "fully observed" user-item interaction matrix, minimizing missing values as each user has interacted with every video and provided feedback. Given the absence of explicit "like" behavior, we adopt the approach outlined in [5], considering a click with a **watch-ratio = play-duration/video-duration > 2** as "like" behavior.
- *Adressa Dataset* [6] is a news dataset. Following the prior study [13], we treat a click with dwell time > 30 s as "like" behavior.

The detailed information about the datasets is presented in Table 1.

Evaluation Protocol and Metrics. For data preprocessing, we adopted a popular *k-core* preprocessing step [9] (with $k = 5$), filtering out users and items with less than 5 interactions.

Following the prior works [17,30], each dataset is sorted by timestamp, and split to train/valid/test sets with corresponding 70%/10%/20% proportions. We used the same split for our model and baselines for a fair comparison. For evaluation, we followed [34,38] to sample 1,000 unobserved items, with which a user did not interact before a specific target behavior, considering them as negative items. Finally, we used them along with all positive items in the test set. We adopted Recall and NDCG as evaluation metrics.

Compared Baselines. We compared our proposed model with nine state-of-the-art recommendation models as follows:

- **MF-BPR** [24]: MF-BPR is a widely-used collaborative filtering baseline optimized by Bayesian personalized ranking (BPR) loss.
- **LightGCN** [10]: It is a graph neural network model that simplifies the original design of GCN so that it can fit better to recommendation applications.
- **IPW** [27]: It adds the standard Inverse Propensity Weight to reweight samples to alleviate item popularity bias.
- **Multi-DR** [42]: It uses Multi-task Inverse Propensity Weighting (Multi-IPW) estimator and Multi-task Doubly Robust (Multi-DR) estimator to mitigate selection bias and data sparsity in multi-behavior recommendation.
- **MACR** [37]: It is model-agnostic using a counterfactual reasoning method for eliminating popularity bias.
- **CR** [33]: It is a counterfactual inference-based method that addresses the clickbait issue. CR aims to capture unbiased user preferences without using like feedback. We use code released by the authors for implementation, using MMGCN as backbone. We use code released by the authors to reimplement experiments, where CR is also implemented based on MMGCN and takes exposure features as input.
- **PDA** [43]: It is a state-of-the-art method that performs de-confounded training while intervening the popularity bias during model inference. The authors provide two versions, where PD directly uses matching score for recommendation and PDA leverages predicted item popularity score in recommendation. We adopt the popularity-adjusted version in our work.

- **RD-DR** [4] accounts for the effect of unmeasured confounders on propensities, under the mild assumption that the effect is bounded.
- **HCR** [47] propose to leverages front-door adjustment to decompose the causal effect into two partial effects, which are independent from the hidden confounder and identifiable.

4.2 Implementation Details

We thoroughly tuned the baselines' hyperparameters to achieve optimal performance on the validation set. Early-stop training strategy was applied based on Recall@20 and NDCG@20 on the validation set with a patience of 20 epochs. We performed a grid search of the latent dimension size in the range of $\{16, 32, 64, 128\}$, the regularization weights (λ_Φ and λ_Θ) in the range of $\{0.00001, 0.0001, 0.001, 0.01\}$, and the learning rate in the range of $\{0.01, 0.005, 0.001, 0.0005, 0.0001\}$. We train all models with Adam optimizer [14]. After performing hyperparameter search, the learning rate was set to 0.0005, the batch size was set to 1024, and the size of the latent factor was set to 128. We tune the hyperparameters α and β in the range of $\{0.2, 0.4, 0.6, 0.8\}$ and ensures that $\alpha + \beta = 1$. The optimal combination of loss weight achieved with the combination of (CTR loss, CL loss) = (0.6, 0.4), indicating the importance of both auxiliary tasks.

4.3 Overall Performance and Ablation Study

The recommendation performance of baselines and our model is shown in Table 2. Among all baselines, bias-aware baselines perform better than conventional baselines in general. Remarkably, our MDCCL consistently outperformed all baselines in all three datasets. We underlined the best baseline. On average, our proposed model improved 9.72% at Recall@10 and 6.72% at NDCG@10 compared with the best baseline. A similar trend also appears at Top-20.

Table 2. Overall performance at top-10 on three real-world datasets. The best performance is in bold, the best baseline result is underlined. The last column shows relative improvement of our MDCCL over the best baseline.

		MF	LightGCN	IPW	Multi-DR	MACR	CR	PDA	RD-DR	HCR	MDCCL	Imp. %
KuaiRec	Recall@10	0.0054	0.0142	0.0108	0.0145	0.0239	0.0298	0.0264	0.0279	0.0312	**0.0332**	6.41%
	NDCG@10	0.0041	0.0113	0.0077	0.0127	0.0142	0.0279	0.0231	0.0261	0.0301	**0.0320**	6.31%
Adressa	Recall@10	0.0642	0.1034	0.0804	0.0919	0.124	0.1452	0.1021	0.1399	0.1573	**0.1771**	12.58%
	NDCG@10	0.0453	0.0794	0.0663	0.0817	0.1074	0.1078	0.0997	0.1126	0.1176	**0.1246**	5.95%
Fliggy	Recall@10	0.3712	0.3951	0.3862	0.4011	0.4295	0.4077	0.4023	0.4277	0.4261	**0.4732**	10.17%
	NDCG@10	0.1807	0.2077	0.1974	0.2115	0.2459	0.243	0.2015	0.2391	0.2442	**0.2653**	7.89%

Table 3. Ablation analysis.

	KuaiRec		Adressa		Fliggy	
	Recall@10	NDCG@10	Recall@10	NDCG@10	Recall@10	NDCG@10
MDCCL	*0.0332*	*0.032*	*0.1771*	*0.1246*	*0.4732*	*0.2653*
MDCCL w/o MB	0.0242	0.0217	0.1154	0.0832	0.4153	0.2273
MDCCL w/o CL	0.0295	0.0274	0.1477	0.0944	0.4425	0.2346
MDCCL w/o DA	0.0301	0.0293	0.1582	0.1021	0.4593	0.2395
LightGCN(Backbone model)	0.0142	0.0113	0.1034	0.0794	0.3951	0.2077

We observe that both IPW-based methods (*IPW, Multi-DR*) perform worse than other causal recommender models (*MACR, CR, PDA, RD-DR, HCR*). We postulate that this line of methods heavily relies on estimating a proper propensity score, which is non-trivial and typically suffers from high variance. *Multi-DR*, which augmented its architecture with an additional imputation model for robustness achieves better results. The counterfactual world constructed by *MACR* directly removes all natural direct effect from both the user-side and item-side regardless of whether it is harmful or not, which leads to its compromised accuracy. *PDA* models users' interest drift across time for bias adjustment, and turns to be effective for debiasing. *CR* achieves a competitive performance in most cases indicating that the clickbait issue has been a serious obstacle to produce accurate recommendation. *RD-DR* and *HCR* achieve impressive results indicating that hidden confounders impact on the recommendation quality.

To verify the effectiveness of each design in our framework, we developed three variants and summarize results in Table 3:

- **MDCCL w/o contrastive learning (CL):** We remove the contrastive learning while keeping deconfounded recommendation framework intact.
- **MDCCL w/o data augmentation (DA):** We use original subgraphs in contrastive learning without applying any data augmentation.
- **MDCCL w/o multi-behavior modeling (MB):** We set the α to 0, which disables auxiliary behavior modeling $f_1(u, p)$ – CTR – branch while keeping contrastive learning as an auxiliary task.

We can observe that the aforementioned designs all contribute positively to our proposed MDCCL, however, the importance varies greatly. Data augmentation (DA) contributes the least and degrades performance in one case. It makes sense because DA neither removes confounding features nor changes the modeling process, so inappropriate data augmentation may introduce noise and thus negatively affects the model prediction. Contrastive learning helps to further differentiate between similar items and learn more robust representations. Multi-behavior modeling makes the most contribution as it is the key component of the debiasing strategy in our proposed framework. The performance in Table 3 further proves the effectiveness of our proposed MDCCL.

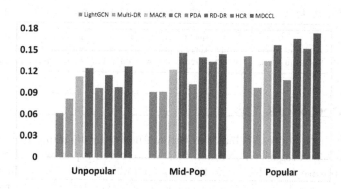

Fig. 5. Recall@10 at Adressa dataset grouped by popularity.

4.4 Debiasing Effect

We analyze the effectiveness of our model against biases via mitigating the impact of hidden confounders. In particular, as exemplars/case studies, we mainly show effectiveness of our model against popularity bias and exposure bias. However, the effectiveness is not limited to these two biases because our work focused on the hidden confounders.

Popularity Bias. We gauge popularity of each item using D_i/D_{total}, where D_i denotes the number of interactions an item involved, and D_{total} is the total interactions in each training set. Sorting items by popularity, we split the dataset into three subsets: unpopular, mid-pop, and popular, ensuring equal total popularity across subsets (Fig. 5). We evaluate Recall@10 by item popularity, expecting similar performance since the sum of popularity in each subset is equal. Remarkably, *MDCCL* consistently outperforms or competes well across all datasets, mitigating popularity bias effectively. Due to space constraints, we visualize the Adressa dataset only. In summary, all models excel on the popular subset, but conventional recommender *LightGCN* performs much worse on the unpopular subset compared to our model, which excels consistently in all subsets, indicating effective popularity bias mitigation.

Exposure Bias. In this experiment, we assess pretrained models under a test set without exposure bias. Fortunately, [5] provides a fully observed KuaiRec small matrix for testing, ensuring over a 99% exposure rate. It's important to note that there is no interaction overlap between the KuaiRec training and validation sets, and this test set. The experiment results in Table 4 demonstrate our model's continued superiority over baselines. This indicates the model's capacity to mitigate exposure bias influence and bolster recommendation robustness, although our model was training on data with various biases beyond exposure bias and did not exclusively focus on exposure bias.

Table 4. Performance of models on a test set without exposure bias.

		LightGCN	IPW	Multi-DR	MACR	CR	PDA	RD-DR	HCR	MDCCL	Imp. %
KuaiRec	Recall@10	0.0119	0.0105	0.0145	0.0272	0.031	0.0262	0.0316	0.0311	**0.0335**	**6.01%**
(no exposure bias)	NDCG@10	0.0092	0.0072	0.0121	0.0154	0.0291	0.0225	0.029	0.0294	**0.0312**	**6.12%**

5 Related Work

Researchers have addressed bias through methods like debiasing techniques. To name a few, [3] categorized common biases into seven types, with selection bias [12,27], exposure bias [18,20,22,48] , and popularity bias [37,43,44,46] being the most discussed. Existing approaches have either relied on heuristic rules [1,16,26] or have been sensitive to pseudo-labels for data imputation [25,35]. To mitigate bias in recommendation, IPS-based approaches [7,27,35], often combined with data imputation, were proposed and became popular. However, improper propensity scores can lead to inaccuracies and high variance. More recently, causality-based methods have been introduced to provide more accurate and explainable solutions. [39,43] involved causal intervention to remove bias factors from inference via $do - calculus$. Knowledge distillation methods [2] train a teacher model on the uniform dataset and then use it to guide the base model trained on biased dataset. Counterfactual-based methods [33,37] estimated causal effects by comparing the factual world with the counterfactual world, targeting different confounding features. Unlike the prior works, we have proposed leveraging a pre-like behavior such as click an item as the mediator to mitigate the bias effects caused by confounders in recommendation systems.

6 Conclusion

In this paper, we analyze confounding effect in recommender systems from the perspective of causal graph. Considering confounders in real-world scenarios are much more complex than assumptions in existing work. Therefore, we propose to utilize users' prior feedback as mediator and apply the front-door adjustment to free the influence of unobserved confounders from inference. Further, we develop an auxiliary contrastive learning task to ensure the robustness of learned representation. Extensive experiments prove the effectiveness of our model in both accuracy and estimating confounding effects.

References

1. Abdollahpouri, H., Burke, R., Mobasher, B.: Controlling popularity bias in learning-to-rank recommendation. In: RecSys, pp. 42–46 (2017)
2. Chen, J., et al.: AutoDebias: learning to debias for recommendation. In: SIGIR, pp. 21–30 (2021)
3. Chen, J., Dong, H., Wang, X., Feng, F., Wang, M., He, X.: Bias and debias in recommender system: a survey and future directions. ACM Trans. Inf. Syst. **41**, 1–39 (2020)

4. Ding, S., et al.: Addressing unmeasured confounder for recommendation with sensitivity analysis. In: KDD, pp. 305–315 (2022)
5. Gao, C., et al.: KuaiRec: a fully-observed dataset and insights for evaluating recommender systems. In: CIKM (2022)
6. Gulla, J.A., Zhang, L., Liu, P., Özgöbek, O., Su, X.: The Adressa dataset for news recommendation. In: WI (2017)
7. Guo, S., et al.: Enhanced doubly robust learning for debiasing post-click conversion rate estimation. In: SIGIR (2021)
8. Gutmann, M., Hyärinen, A.: Noise-contrastive estimation: a new estimation principle for unnormalized statistical models. In: AISTATS, pp. 297–304. JMLR Workshop and Conference Proceedings (2010)
9. He, R., McAuley, J.: Ups and downs: modeling the visual evolution of fashion trends with one-class collaborative filtering. In: WWW, pp. 507–517 (2016)
10. He, X., Deng, K., Wang, X., Li, Y., Zhang, Y., Wang, M.: LightGCN: simplifying and powering graph convolution network for recommendation. In: SIGIR, pp. 639–648 (2020)
11. He, X., Liao, L., Zhang, H., Nie, L., Hu, X., Chua, T.S.: Neural collaborative filtering. In: WWW, pp. 173–182 (2017)
12. Hernández-Lobato, J.M., Houlsby, N., Ghahramani, Z.: Probabilistic matrix factorization with non-random missing data. In: ICML (2014)
13. Kim, Y., Hassan, A., White, R.W., Zitouni, I.: Modeling dwell time to predict click-level satisfaction. In: WSDM, pp. 193–202 (2014)
14. Kingma, D.P., Ba, J.: Adam: a method for stochastic optimization. In: Bengio, Y., LeCun, Y. (eds.) ICLR (2015)
15. Koren, Y., Bell, R., Volinsky, C.: Matrix factorization techniques for recommender systems. Computer **42**(8), 30–37 (2009)
16. Li, Y., Hu, J., Zhai, C., Chen, Y.: Improving one-class collaborative filtering by incorporating rich user information. In: CIKM (2010)
17. Liang, D., Krishnan, R.G., Hoffman, M.D., Jebara, T.: Variational autoencoders for collaborative filtering. In: WWW, pp. 689–698 (2018)
18. Liu, D., et al.: Mitigating confounding bias in recommendation via information bottleneck. In: RecSys (2021)
19. Liu, D., Lin, C., Zhang, Z., Xiao, Y., Tong, H.: Spiral of silence in recommender systems. In: WSDM, pp. 222–230 (2019)
20. Ma, X., et al.: Entire space multi-task model: an effective approach for estimating post-click conversion rate. In: SIGIR (2018)
21. van den Oord, A., Li, Y., Vinyals, O.: Representation learning with contrastive predictive coding. arXiv preprint (2018)
22. Ovaisi, Z., Ahsan, R., Zhang, Y., Vasilaky, K., Zheleva, E.: Correcting for selection bias in learning-to-rank systems. In: WWW (2020)
23. Pearl, J., Glymour, M.M., Jewell, N.P.: Causal inference in statistics: a primer (2016)
24. Rendle, S., Freudenthaler, C., Gantner, Z., Schmidt-Thieme, L.: BPR: Bayesian personalized ranking from implicit feedback. In: UAI, pp. 452–461 (2009)
25. Saito, Y.: Asymmetric tri-training for debiasing missing-not-at-random explicit feedback. In: SIGIR (2020)
26. Saito, Y.: Unbiased pairwise learning from biased implicit feedback. In: SIGIR, pp. 5–12 (2020)
27. Schnabel, T., Swaminathan, A., Singh, A., Chandak, N., Joachims, T.: Recommendations as treatments: debiasing learning and evaluation, pp. 1670–1679. PMLR (2016)

28. Tao, W., et al. : SMINet: state-aware multi-aspect interests representation network for cold-start users recommendation. In: AAAI (2022)
29. Tran, T., Sweeney, R., Lee, K.: Adversarial Mahalanobis distance-based attentive song recommender for automatic playlist continuation. In: SIGIR (2019)
30. Tran, T., You, D., Lee, K.: Quaternion-based self-attentive long short-term user preference encoding for recommendation. In: CIKM, pp. 1455–1464 (2020)
31. VanderWeele, T.J.: A three-way decomposition of a total effect into direct, indirect, and interactive effects. Epidemiology **24**, 224–232 (2013)
32. Wang, W., Feng, F., He, X., Wang, X., Chua, T.S.: Deconfounded recommendation for alleviating bias amplification. In: KDD, pp. 1717–1725 (2021)
33. Wang, W., Feng, F., He, X., Zhang, H., Chua, T.S.: Clicks can be cheating: counterfactual recommendation for mitigating clickbait issue. In: SIGIR (2021)
34. Wang, X., He, X., Wang, M., Feng, F., Chua, T.S.: Neural graph collaborative filtering. In: SIGIR, pp. 165–174 (2019)
35. Wang, X., Zhang, R., Sun, Y., Qi, J.: Doubly robust joint learning for recommendation on data missing not at random. In: ICML (2019)
36. Wei, C., Liang, J., Liu, D., Wang, F.: Contrastive graph structure learning via information bottleneck for recommendation. In: NeurIPS, vol. 35, pp. 20407–20420 (2022)
37. Wei, T., Feng, F., Chen, J., Wu, Z., Yi, J., He, X.: Model-agnostic counterfactual reasoning for eliminating popularity bias in recommender system. In: KDD (2021)
38. Xin, X., He, X., Zhang, Y., Zhang, Y., Jose, J.: Relational collaborative filtering: modeling multiple item relations for recommendation. In: SIGIR, pp. 125–134 (2019)
39. Xu, S., Tan, J., Heinecke, S., Li, J., Zhang, Y.: Deconfounded causal collaborative filtering. arXiv arXiv:2110.07122 (2021)
40. Yang, L., Cui, Y., Xuan, Y., Wang, C., Belongie, S., Estrin, D.: Unbiased offline recommender evaluation for missing-not-at-random implicit feedback. In: RecSys, pp. 279–287 (2018)
41. You, D., Vo, N., Lee, K., Liu, Q.: Attributed multi-relational attention network for fact-checking URL recommendation. In: CIKM (2019)
42. Zhang, W., et al.: Large-scale causal approaches to debiasing post-click conversion rate estimation with multi-task learning. In: WWW (2020)
43. Zhang, Y., et al.: Causal intervention for leveraging popularity bias in recommendation. In: SIGIR (2021)
44. Zhao, Z., et al.: Popularity bias is not always evil: disentangling benign and harmful bias for recommendation. IEEE Trans. Knowl. Data Eng. **35**, 9920–9931 (2022)
45. Zheng, Y., Gao, C., Li, X., He, X., Li, Y., Jin, D.: Disentangling user interest and conformity for recommendation with causal embedding. In: Proceedings of the Web Conference 2021, pp. 2980–2991 (2021)
46. Zheng, Y., Gao, C., Li, X., He, X., Li, Y., Jin, D.: Disentangling user interest and conformity for recommendation with causal embedding. In: WWW (2021)
47. Zhu, X., Zhang, Y., Feng, F., Yang, X., Wang, D., He, X.: Mitigating hidden confounding effects for causal recommendation. arXiv arXiv:2205.07499 (2022)
48. Zhu, Z., He, Y., Zhang, Y., Caverlee, J.: Unbiased implicit recommendation and propensity estimation via combinational joint learning. In: RecSys (2020)
49. Zhu, Z., He, Y., Zhao, X., Caverlee, J.: Popularity bias in dynamic recommendation. In: KDD (2021)

Author Index

Printed in the United States
by Baker & Taylor Publisher Services